REAL GOODS

Alternative Energy Sourcebook

1991

Credits

Editor & Publisher: John Schaeffer
Authors: John Schaeffer, Robert Sardinsky, Randy Wimer, & Jim Cullen
Article Contributors: Mike Bergey, Alfred Canada, Clare & Gail Cochran, Paul Cunningham, Windy Dankoff, John Davey, Chris Greacen, Michael Hackleman, Creighton Hart Jr., Mike Mooney, Eric Nashlund, George Patterson, Joe Stevenson, Steve Stollman, & Jon Vara.
Design Director: Bob Steiner
Computer Wizardry: Michael Potts
Index: Stephen Bach
Technical Assistance: Jeff Oldham, Doug Pratt, & Douglas Bath
Proofreading: Andrew Alden, Eileen Enzler, Anne Mayea, & the Real Goods Staff
Photography: Sean Sprague & Tom Liden
Illustrations: Tom Jarvis & Kathy Shearn, Malcolm Wells, Steven Johnson
Typesetting: Debbie Robertson, Jill Slaven, Penny Tigner
Cover Photograph: Courtesy Arco Solar
Color Separations: ColorBrite
Printing: Consolidated Printing, Berkeley, CA

Special thanks & acknowledgments to: Jeff Oldham, Anne Mayea, Wayne Robertson, Windy Dankoff, Jim Allen, the entire Real Goods Staff, Nancy, Jesse, Sara & Ashley Schaeffer

ISBN 0-916571-01-7 - $14
Printed in U.S.A.

Real Goods Trading Corporation
966 Mazzoni Street
Ukiah, CA 95482
Toll Free Orders: 1-800-762-7325
Technical Assistance: 707-468-9214
Business Office: 707-468-9292
Fax Number: 707-468-0301

Printed on 50# Resolve Recycled Smooth Offset

About the Cover: Our cover depicts the Rancho Seco Nuclear Power Plant behind Arco Solar's 1-megawatt PV generating station, consisting of 3,584 PV panels. In early 1989, by a popular vote, Rancho Seco was *shut down forever*. The photograph gives vivid testament to the enduring and non-polluting power of photovoltaics. Nuclear power is decaying and PV is blossoming everywhere!

Foreword

Amory B. Lovins
Rocky Mountain Institute
Old Snowmass, Colorado

From 1979 through 1986, the United States got more than seven times as much new energy from savings as from all net increases in supply. Even more astoundingly, of those increases in supply, more came from sun, wind, water, and wood than from oil, gas, coal and uranium. Even as glossy magazine ads were dismissing renewable energy as unripe to contribute much of anything in this century, renewables came to provide some 11-12% of the nation's total primary energy (about twice as much as nuclear power), and the fastest-growing part. The only energy source growing faster was efficiency. Just the *increase* in renewable energy supplies during those years came to provide each year more energy than all the oil we bought from the Arabs. And efficiency, during 1973-86, came to represent an annual energy source two-fifths bigger than the entire domestic oil industry, which had taken a century to build; yet oil had rising costs, falling output, and dwindling reserves, while efficiency had falling costs, rising output, and expanding reserves.

To be sure, during about 1986-88 - the later years of what, in the telling phrase used of their own country by Soviet commentators, may be fairly called "the period of stagnation" - this momentum declined and in some respects stalled. The impressive successes of efficiency and renewables often fell victim to official hostility and collapsing oil prices (which were largely driven by the very success of energy efficiency). The Reagan Administration's rollback of efficiency standards for light vehicles immediately doubled oil imports from the Persian Gulf, effectively wasting exactly as much oil as the government hoped could be extracted each year from beneath the Arctic National Wildlife Refuge. While some electric utilities pressed ahead with good efficiency programs, additional electrical usage spurred by deliberate power-marketing efforts was officially projected, by the year 2000, to wipe out about two-thirds of the resulting baseload savings. The same Administration that touted the virtues of the free market

pressed home its strenuous efforts to deny citizens the information they needed to make intelligent choices. And Federal tax credits meant to help offset the generally much larger subsidies - totaling at least $50 billion per year - given to their competitors were generally abolished, while most of the subsidies to depletable and harmful energy technologies were maintained or increased, tilting the unlevel playing field even further. These and other distortions of fair competition gravely harmed many sectors of the renewable energy industries, often drying up distribution channels so that even sound, cost-effective options could no longer reach their customers. Leadership in some key R&D areas passed from America to Japan and Germany. Cynics began writing premature obituaries for the latest solar flash-in-the-pan.

Today, the best technologies on the market can save about three-quarters of all electricity now used in the United States.

Yet throughout that decade's rise, leveling, and sometimes stumbling of the keys to a safe, sane, and least-cost energy future - high energy productivity and appropriate renewable sources - a band of pioneers in Northern California sustained their vision of a way to give everyone fair access to a diverse tool kit for energy self-sufficiency. Through their dedication, thousands of people have had the privilege of discovering that the energy problem, far from being too complex and technical for ordinary people to understand, is perhaps on the contrary too simple and political for some technical experts to understand. The Real Goods team gave, and gives today, an equal opportunity to solve your piece of the energy problem from the bottom up.

Many did exactly that. They discovered that solar showers feel better, because you're not stealing anything from your kids. They found how to get greater security and high-quality energy services from a judicious blend of efficiency and renewables. Having a high do-to-talk ratio, they worked through trial and error, celebrated their inevitable mistakes, and found in Real Goods an effective way to share their experience with a wide audience. (There is, after all, no point repeating someone else's same old dumb mistakes when you can make interesting

Amory Lovins is currently director of research for the Rocky Mountain Institute. He is a consulting experimental physicist educated at Harvard and Oxford. Mr. Lovins has been active in energy policy in more than 15 countries and has briefed five heads of state. He has published a dozen books and over 100 papers. He is probably the most articulate writer on energy policy and strategy today. Newsweek called him "one of the western world's most influential thinkers." He has our vote for the next U.S. Energy Secretary!

new mistakes instead.) Piece by piece, with that uniquely American blend of idealism and intense pragmatism, they quietly built a grassroots energy revolution.

The tools of elegant frugality, the technical options that let you demystify energy and live lightly within your energy income, have long been available to anyone in principle. (There's an old Russian joke about the guy who asks whether he can buy various hard-to-find goods. Tired of being always told, "In principle, yes," he exclaims, "Sure, but where's this (Principle) shop that you keep talking about?") Real Goods turns the principle into practice: a useful selection of the things you were looking for, with essential background information, at fair prices, guaranteed, anywhere you want them delivered.

It's especially gratifying to see this catalog's nice blend of renewable energy supply options with increasingly efficient ways to *use* the energy. In most cases, the best buy is to get the most efficient end-use device you can, *then* just enough renewable energy supply to meet that greatly reduced demand. Our own

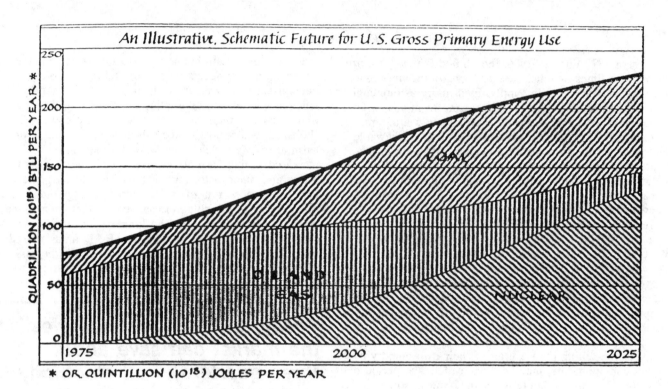

house/indoor farm/research center of nearly 4,000 square feet, for example, by harnessing some of the efficiency options in this catalog, uses so little electricity for the household lights and appliances - about $5 worth per month - that only about 400-500 peak watts of photovoltaics, a couple of thousand bucks' worth, could entirely meet those needs. Indeed, superinsulation and superwindows, by providing more than 99% of our space-heating passively in a climate that can get cold as -47°, raised our construction costs less than they *saved* us up front by eliminating the furnace and ductwork. All together, saving 99+ percent of our space and water heating load (the latter is virtually all passive and active solar, again made much cheaper by strong efficiency improvements), 90-odd percent of our electricity, and half our water together raised total construction costs only one percent, and paid back in the first ten months. It *looks* nice, too: we tramp in and out of a blizzard to be greeted by a jungleful of jasmine, bougainvillea, and a big iguana offering advanced lizarding lessons under the banana tree - then we remember, as we wipe the steam off our glasses, that there's no heating system; nothing's being used up. And that's all with six-year-old technology. Now you can do even better.

Today, in fact, the best technologies on the market can save about three-quarters of all electricity now used in the United States, while providing unchanged or improved services. The cost of that quadrupled efficiency: about 0.6 cents/kWh, far cheaper than just *running* a coal or nuclear plant (let alone building it). Thus the global warming, acid rain, and other side-effects of that power plant can be abated not at an extra cost but at a *profit*: it's cheaper to save fuel than to burn it. Similarly, saving four-fifths of U.S. oil by using the best technologies now demonstrated (about half of which are already on the market) costs only about $3 per barrel, less than just drilling to look for more. Such technologies obviously beat any kind of energy supply hands down. But what's the *next* best buy - the supply choices you need to make in partnership with efficiency in order to live happily ever after? Usually, if you buy the efficiency first, the next best buy will be the appropriate renewable sources. And if you count - as our children surely will - the environmental damage and insecurity that fossil and nuclear fuels cause, then well-designed renewables look even better.

That's not to say that every renewable technology makes sense anywhere. Wind machines and small hydro only work well in good sites. Photovoltaics can look competitive with grid electricity (which, by the way, receives tens of billions of dollars' annual direct subsidies and doesn't bear many of its costs to the earth) only if you're upwards of a certain distance (where we are, about a quarter-mile) from a power line - or closer if you count the benefits of greater reliability and higher power quality. (The Federal Aviation Administration is currently switching hundreds of ground avionics stations to

photovoltaics for exactly those reasons.) Every renewable application is site-specific and user-specific. No *single* renewable source can solve every energy problem. That's part of their strength and their charm.

A *single* compact fluorescent lamp will, over its life, keep out of the air a *ton* of carbon dioxide, twenty pounds of sulfur oxides, and various other nasty things.

But wherever you are, if you use energy in a way that saves you money, it's quite likely that *some* kind of renewable energy can further cut your bills, work as well or better, give you a good feeling, and set an example. What more can you ask?

Well, maybe something for the earth - and you get that too. A *single* compact-fluorescent lamp (see lighting section) replacing, say, 75 watts of incandescent lighting with 14-18 watts (but yielding the same light and lasting about 13 times as long) will, over its life, keep out of the air a *ton* of carbon dioxide (a major cause of global warming), twenty pounds of sulfur oxides (which cause acid rain), and various other nasty things. And far from costing extra, that lamp will make you tens of dollars *richer*, by saving more utility fuel, replacement lamps, and installation labor than it costs. Now put that saving into changing over to renewable energy supply and you're doing even better. Then take the time you used to put into working to pay your electric bill and put it instead into your garden, your compost pile, a walk, a fishing trip. Take the time you used to work to pay your doctor's bills and build a greenhouse. Invite your neighbors over to help munch your fresh tomatoes in February and tell them how you did it all. Then...

You get the idea. Here's a treasure-house of things *you* can use to improve both your own life and everyone else's. Read, enjoy, use, learn, and tell: implementation is left as an exercise for the reader.

Make Fuel Efficiency Our Gulf Strategy

The following piece by Amory and Hunter Lovins is reprinted with permission from the New York Times, December 3, 1990.

Are we putting our kids in tanks because we didn't put them in efficient cars? Yes: We wouldn't have needed any oil from the Persian Gulf after 1985 if we'd simply kept on saving oil at the rate we did from 1977 through 1985.

Even now we could still roll back the oil dependence that perpetually holds our foreign hostage and distorts other U.S. priorities in the Middle East. Just by aiming at greater efficiency, we could eliminate all gulf imports by using only an eighth less oil.

A greater place to start would be personal vehicles. Improving America's 19-mile-per-gallon household vehicle fleet by three miles per gallon could replace U.S. imports of oil from Iraq and Kuwait. Another nine miles per gallon would end the need for any oil from the Persian Gulf and, according to the Department of Energy, would cut the cost of driving to well below pre-crisis levels without sacrificing performance.

The Reagan Administration doubled 1985 oil imports from the gulf when it rolled back efficiency standards. Today's new cars average 29 miles per gallon; the fleet only 20. Yet 10 manufacturers have built and tested attractive, low-pollution prototype cars that get 67 to 138 miles per gallon. Better design and stronger materials make some of these safer than today's cars, as well as more nimble and peppy.

And efficiency needn't mean smallness: only 4 percent of past car-efficiency gains came from downsizing. Some of the prototype cars comfortably hold four or five passengers, and two of them are said to cost nothing extra to build.

Many other oil savings can help. Boeing's new 777 jet will use about half the fuel per seat of a 727. Technical refinements can save most of the fuel used by heavy trucks, buses, ships and industry. Insulation, draft-proofing and simple hot-water savings can displace most of the oil used in buildings. Superwindows that retain heat in the winter and reject it in summer could save each year up to twice as much fuel as we get from Alaska.

In all, we know how to run the present U.S. economy on one-fifth the oil we are now using, and the cost of saving each barrel would be less than $5. Even achieving just 15 percent of that potential oil savings would displace all the oil we've been importing from the gulf. Doing that requires only a small additional step. Since 1973, we've reduced our oil use per dollar of gross national product four and a half times as much as we'd need to reduce it today in order to eliminate all gulf imports.

How can we promote fuel efficiency? Higher gasoline taxes are a weak incentive to buy an efficient car, because gasoline costs four times less than the non-fuel costs of owning and running a car. And since the often higher purchase price of an efficient car about cancels out the lower gasoline bills, the total cost per mile for 20 - and 60-mile-per-gallon cars is about the same.

But the 40-mile-per-gallon difference, for cars and light trucks, represents more than twice America's imports from the gulf. If the security and environmental costs of inefficient cars had to be paid up front, buyers would choose more wisely. The best way is "feebates": When you register a new car, you pay a fee or get a rebate, depending on its efficiency. The fees would pay for the rebates.

Rebates for efficient cars should be based on the difference in efficiency between your new car and the old one - which you'd scrap, thus getting the most inefficient, dirtiest cars off the road first. That's good for Detroit, for the poor, for the environment, and for displacing gulf oil sooner.

The California legislature recently approved car feebates by a margin of 7 to 1; Connecticut, Iowa, Massachusetts are weighing them. Feebates are also being considered for new buildings in California, Massachusetts and the four northwestern states, and could be applied to trucks, aircraft, appliances and other energy-consuming goods. Unlike miles-per-gallon standards, feebates reward maximum performance and encourage businesses to bring superefficient models to market.

Energy efficiency is also the key to the decades-long transition to nondepleting, uninterruptible energy sources. Government studies confirm that sun, wind, water, geothermal heat, and farm and forestry wastes can cost-effectively provide, within 40 years, half as much energy as America uses today. Efficiency would raise that share and buy the time need for graceful conversion.

The military alternative to energy efficiency isn't cheap. Gulf jitters have added more than $40 billion a year to U.S. oil imports. Counting military costs, gulf oil now costs in excess of $100 a barrel.

The more than $20 billion net cost of U.S. forces in the gulf just from August through December 1990, if spent instead on efficient use of oil, could displace all the oil now imported from the gulf. It could also create jobs and wealth, improve America's trade balance, stretch domestic reserves, clean urban air, cut acid rain and global warming, and help the poor at home and abroad.

In 1989, the Pentagon used about 38 percent as much oil as the U.S. imported from Saudi Arabia, and estimated that its consumption could readily double or triple in a war. An M-1 tank get 0.56 miles per gallon. An oil-fired aircraft carrier gets 17 feet per gallon. And no good outcome - in dollars, oil, or blood - is in sight.

But from inside an efficient car, the gulf looks very different. From inside enough of them, its oil becomes irrelevant. National security, peacetime jobs in a competitive economy, and the environment demand immediate mobilization - not of tanks but of efficient cars, not of B-52's but of 777's, and not of naval guns but of caulk guns. - **Amory B. & Hunter Lovins**

Introduction

Welcome to Our 1991 AE Sourcebook!

As we continue into the final decade of the twentieth century, we contemplate the painful lessons of the last decade - Three Mile Island, Chernobyl, the Valdez oil spill and now Iraq. We're gratified to discover that alternative energy is finally becoming fashionable. As it becomes increasingly obvious that environmentalism must be the preoccupation of the nineties (if we are to survive into the next century) we are proud that we have been working on the right stuff.

It has been said that social activism runs in 30-year cycles and that the exuberance of the sixties will resurface in the nineties. We look forward to seeing environmental awareness blossom as we all work together to rescue our fragile planet. We are cheered to see that what we envisioned as an "environmentalist's fantasy" in earlier Sourcebooks is now happening routinely. In fact, this entire book is being produced with photovoltaic (PV) power from our solar-powered computer and laser printer. Nearly half a century after solar electricity was first developed for the space program of the 1950's, we are becoming ever more conscious that we each must do more to reduce resource consumption to sustainable levels for the benefit of the generations to come. PV is a prime example of a good technology: permanent, sustainable, benign, modular, maintenance-free, quiet, non-polluting, and clean. It is *in tune* with our planet, whereas other energies and technologies devour the planet for temporary cheap thrills. The typical alternative energy home (there are now over 40,000) uses between a third and a tenth the energy consumed by a conventional home. *Only 13 years ago we sold our first PV panel!* We can only hope that our society will learn the simple lessons of nature less painfully in the twenty-first century.

We've endeavored to demonstrate by example that appropriate technology is not a mystical mirage of the future, but a technology whose time is here. PV is everywhere - from the solar panel that runs our calculators and watches to the PV system pumping water from wells 1,000 feet deep with small efficient pumps so that flowers *do* bloom in the desert. Alternative energy opens a world of new possibilities: radio repeaters, livestock watering, electric fencing, ocean signal buoys, highway billboard lighting, and solar irrigation controllers, to name just a few. Calculators offer us a vivid example of PV's potential: it was only in 1978 that the first portable calculator hit the streets, but now they are everywhere. So it will be with PV.

Stock market watchers claim that the 1990's will be to photovoltaics what the 1970's and 1980's were to computers and electronics. The cat is out of the bag. PV is fashionable now - you can see it on BMW sunroofs running the cars' instrumentation and you can see it on RV's rolling down the highway in search of warmer climes. Even the commercial heavyweights (P.G.& E., Bechtel Corp., etc.) have finally figured out that it's often cheaper to get power from an array of PV's than to drop a lengthy transmission line or run a noisy generator.

We cannot expect PV prices to behave the same way

calculator prices have over the past eleven years. The wait for cheap is over, and we have lost. Prices have already begun to climb. Nevertheless, the 1990's will be to PV's what the 1980's were to electronics. Many PV manufacturers are now manufacturing photovoltaic modules around the clock. Customers' questions are no longer "will they work?" but "how many do I need?" and "when can I get them?" Ironic, isn't it, that the Three Mile Island, Chernobyl, the Valdez disaster, and the Iraqi war have consistently presaged an upswing in alternative energy sales?

Conservation is contagious. It has led us to discover hundreds of new ways to save energy. PV is just the beginning. In this 1991 AE Sourcebook we preview lots of new ideas and novel products, many of which you will find only here. They will save money, energy, and resources up front, and will make you feel better all the while that you use them! For example, did you know that the typical house refrigerator uses 3,000 watts and isn't even as efficient as refrigerators designed in the 1940's? Did you know that those refrigerators hyped with efficiency stickers on the appliance showroom floor are designed with the compressor on the bottom so that heat rises into the very box that it is supposed to cool? Contrast this with our Sunfrost solar refrigerators (compressor on top) which use only 300 watts - a tenth the power! Tell that to your local appliance dealer. If every American home had a Sunfrost, we could immediately shut down nineteen large nuclear power plants!

In line with energy and economic savings is the crux of our mission - to provide alternative energy products for a cleaner and saner world. The environmental snowball that began (got re-ignited for many of us) on Earth Day 1990 is still going strong. Not even the most starry-eyed among us could help but get a *little* cynical about what transpired in the final days leading up to Earth Day 1990. The bandwagon was packed with environmentalists all basking in a new found glow of green with no one left to protest against. Our favorite image that summarizes the situation was relayed by our friend at the EPA, whose booth in Washington was lodged between the booths of Exxon and Dow Chemical, both somehow believing they could atone for their environmental sins by showing up.

However, we prefer not to focus on the hypocrisy and negativity of the event, concentrating rather on what consciousness raising really did take place. We've all read the polls that show that 85% of Americans say they are now concerned about the environment and that 80% plan to recycle. We were greatly heartened by the response we received at all the Earth Day events in which we took part. A truly green economy requires that all products be audited for their "cradle to grave" effects. While it's certainly true that the current deluge of "green products" on the marketplace is riddled with hype, we have refined our mission to the task of separating out the "hip" from the "hype." We firmly believe we've culled a selection of environmentally sound, energy-efficient products that make the "green grade."

To challenge the efficacy of our product mix and to do a little self-analysis, we undertook a study of our sales for 1990 with an objective eye toward evaluating their overall environmental impact. The results should make our entire customer network proud!

In this Sourcebook you will find new and revolutionary technologies in lighting - DC and AC. New PL compact fluorescent technology puts out 4 times the light of the old incandescent lighting and lasts 20 times as long. It's been estimated that if the U.S. government outfitted every home in the nation with these revolutionary lights, we would be an energy exporting nation within 5 years!

Japan and Europe have used instantaneous tankless water heating technology for decades while Mr. and Mrs. Average American hoard (at great expense to themselves and the environment) 40 gallons of hot water through dark of night, over vacations and long weekends, just in case someone might want to wash a glass. Isn't it more reasonable to heat the water as it's needed? *When will our country come to its senses?*

Is it rational to speed the pathogens and bacteria in our toilet wastes into the biosphere on wings of water? Why do we spend thousands of dollars on myriads of septic tanks and leach fields, thousands of gallons of precious potable water, ultimately to pollute our ground water systems with our own bacteria? Over 300 Real Goods customers have just said No! and now recycle their own waste using the new technologies of composting toilets, many approved by the National Sanitary Foundation.

The cover we've chosen for this Sourcebook really says it all. In many ways it's an update of our last Sourcebook cover (which proved to be controversial beyond our wildest imaginings). The old Sourcebook cover showed an ecological utopian fantasy of humankind rising above the decay of a fossil-fuel based economy. Our new cover depicts the two cooling towers of the Rancho Seco Nuclear Power Plant behind Arco Solar's 1-megawatt PV generating station. As of this year *Rancho Seco has been shut down for good* and PV is blossoming everywhere. Taking you from fantasy to reality - that epitomizes our business.

Our strength at Real Goods lies in the diversity of our interests and contacts. We intend to remain the most thorough clearinghouse for all forms of energy sensible products on the planet. We relentlessly pursue new technologies, and endeavor to keep you informed of our finds. We know we can always do better, and we welcome your feedback. We hope to make your lives simpler and happier, to provide dramatic savings to our fragile environment, and to provide a thorough and valuable educational source along the path. Thanks for making it all work!

John Schaeffer

"Radiation leaks are caused by fools like me, but only God can build a nuclear reactor 93 million miles from the nearest elementary school." - Stewart Brand

Global Warming - Environmental Enemy #1

As I write this, America is involved in a reckless war with Iraq, thousands of lives are at risk, and thirty billion dollars have already been squandered. Saddam Hussein is really no more than a small symptom of our hopeless addiction to oil. Our energy problems will persist long after the last soundbite from Iraq airs on the evening news, until we forswear belligerent over-consumption and move off our collision course with environmental disaster.

Global warming is the number one problem facing our world today. *In fifty years, from 1937 to 1988, the world's annual energy consumption quintupled from 60 to 321 quads (quadrillion BTUs)*, and global emission rates for carbon mushroomed from one gigaton (one billion metric tons) to six gigatons per year. Atmospheric carbon dioxide levels have increased from 280 ppm (parts per million) to 350 ppm. Studies on people working in greenhouses containing elevated levels of carbon dioxide have shown that at 500 ppm, long term cardio-vascular effects may occur. As the concentration of greenhouse gases in the atmosphere increases, the overall warming effects continue to foul the environment, alter the climate (average temperatures will rise as much as two degrees Fahrenheit in the next 30 years), and threaten the habitability of our planet. At the current six gigaton carbon emission rate, the potentially devastating level of 500 ppm of environmental carbon dioxide will be reached by the year 2100 - our grandchildren will not thank us. If our current gluttony for energy doubles (if the second and third world match America's irresponsibility) these changes may happen soon enough for us to experience, within thirty years. The endangered ozone layer is a change we already live with: sunbathing at the beach has become an extinct pleasure we recollect for our children.

In 1988 the United Nations convened a panel of scientists called the Intergovernmental Panel on Climate Change (IPCC) to study global warming. These scientists concluded that "rates [of warming] are likely to be faster than [the rates at which] ecosystems can respond, possibly leading to substantial reductions in biological diversity." Today's global carbon emission rate must drop by at least two-thirds to stabilize the atmospheric concentration of carbon as carbon dioxide. This can only be accomplished through a drastic reduction in the use of fossil fuels for energy production.

Armed with the IPCC's discoveries, Sweden plans to phase out nuclear power and substantially reduce carbon dioxide emissions. Seven European nations are organizing a regional global warming treaty. Sadly, the U.S. government (led by John Sununu) persists in stonewalling the report, whining that "some species would 'thrive' in a warming world!" (*mabye cockroaches, slime mold, and politicians!*) The world's heavyweight champion producers of carbon dioxide, the Soviet Union (18%) and the United States (20%), are the *only two governments advocating no action*.

How do global warming and oil wars tie together and why are we discussing them in our alternative energy catalog? There is really only one answer to the global warming riddle, and it also answers the question of how to avoid reckless oil wars - *energy efficiency!* Full practical use of the electricity-saving technologies now available would save about 75% of all power used. According to Amory Lovins, of the *Rocky Mountain Institute*, energy efficiency at European and Japanese levels would save America about $200 billion per year - enough to pay off the national debt by the year 2000. As a bonus, lowering our energy consumption would stabilize global warming and give the atmosphere and biosphere a chance to heal. *If we used oil as efficiently as Japan we would save seven million barrels of oil per day. Yet our government spends five times as much on our military adventure in the Persian Gulf each week as it budgets for renewable energy for all of 1990!*

If the government won't do it, we must. We can begin to reverse these negative trends by making energy conscious personal choices in the products we buy. In this catalog you will find energy-efficient light capsules, water saving showerheads, Sunfrost refrigerators, solar-electric modules, and many other products which help to curb our energy appetite and can go a long way toward making a dent in reversing global warming and our dependence on fossil fuel based energy systems. In 1990 alone our customers, through their purchases, kept over 102.7 million pounds of carbon dioxide out of the air. It's a small dent (it would take 19,000 purchases like that to reduce one gigaton) but a very positive step in the right direction and sets an example to the rest of the world. (As energy costs explode, we get a dividend: energy efficiency saves *us* money.)

Yet our government spends five times as much on our military adventure in the Persian Gulf each week as it budgets for renewable energy for all of 1990!

As individuals we do as much as we can. The major culprits in global warming are automobiles, which produce from a third to half the damaging emissions. Real Goods supports development of a marketable and affordable electric vehicle conversion program, attacking the problem at its source. (See our chapter on electric vehicles.)

The global warming disaster won't end until you, our readers, take action and spread the word about the short-sighted and self-destructive path our government is taking. Beyond that, we all need to focus on education, helping everyone we come in contact with to understand the gravity of this problem. Our children and grandchildren depend on us. - **John Schaeffer**

Real Goods Products Environmental Impact Report

In order to truly evaluate the impact of your purchases, we asked for the help of the Natural Resources Defense Council (NRDC) in San Francisco and the Rocky Mountain Institute in Snowmass, Colorado to help us quantify the energy impact of what you bought. Our thanks to Chris Calwell (NRDC) and Rick Heede (RMI) for helping us with the formulas and number crunching. Here's a list of the highlights of what you purchased in 1990:

- 180,508 watts of solar-electricity (4,325 solar panels)
- 13,800 watts of wind-electricity (43 wind generators)
- 9,000 watts of hydro-electricity (9 hydro plants)
- 46 Sunfrost refrigerators
- 251,849 watts of compact fluorescent lighting (16,000 lights)
- 528 tankless water heaters
- 36 solar hot water collectors
- 4,254 low-flow water-saving shower heads
- 4,245 water-saving faucet aerators
- 182 low flush toilets
- 193 composting (no-flush) toilets
- 2,486 sets of toilet dams
- 147,744 rolls of toilet paper from recycled paper
- 9,090 rolls of paper towels from recycled paper
- 23,880 boxes of facial tissue from recycled paper

Our goal was to analyze the impact these purchases had on the environment for the life of the appliance or the light bulb. Here are some of the energy conversion data that we used to make these calculations:

It takes 11,000 Btu's to generate 1 kilowatt hour (kWh)
Carbon dioxide emissions per kWh = 1.5 lb./kWh
Coal required to produce 1kWh = 1 lb./kWh
57% of U.S. electricity comes from coal-fired plants
6.25 million Btus = 1 barrel of oil
Average U.S. cost of 1 kWh = $0.08
1 car = 500 gallons gasoline per year
1 low-flow shower head = 466 kWh/year in heat savings
1 low-flow shower head = 14,000 gal/year water savings
1 faucet aerator = 4,000 gal/year water savings
1 set toilet dams = 5,475 gal/year water savings
135 trees = 1 ton of pulp

The following are the results of actual energy and environmental savings directly brought about by you, our customer network, by your purchases in 1990:

- **You have saved 68,491,160 kWh of energy!**
  This represents:
    - 68,491,160 pounds of coal or
    - 127,000 barrels of oil or
    - 5,334,000 gallons of gasoline (*enough to keep 10,688 cars off the road!*)
- **You have kept 102.7 million pounds of carbon dioxide out of the air (a major cause of global warming).**
- **You have kept 14.7 million pounds of sulfur oxides (which cause acid rain) out of the air.**
- **You have saved 1 billion, 857 million gallons of water.**
- **You have saved 180,000 trees from destruction.**
- **You have saved $6,849,116 on energy expenses.**

CONGRATULATIONS TO YOU ALL!

Our Goal For the Nineties - Eliminate 1 Billion Pounds of Carbon Dioxide

We're taking on our most ambitious goal yet. With your help we plan to eliminate one one billion pounds of carbon dioxide from our atmosphere by the turn of the century. This will have a major impact on global warming, the preservation of natural habitats - and indeed on life on earth itself.   Global warming is the number one environmental problem facing our planet today. (See article above) The only way to solve the problem is for the world to drastically slow down its consumption of fossil fuel. Americans have an insatiable appetite for fossil fuel - coal, oil, and natural gas power our cars, our homes, and produce most of our electricity. Natural occurring carbon dioxide (CO_2) and other gasses trap heat and keep the earth warm. The burning of fossil fuels is releasing trillions of pounds of extra carbon dioxide into the atmosphere each year, which will likeley raise temperatures, causing extensive flooding, more frequent droughts, hightened sea levels, and disrupting farming. Habitats will shrink as plants and animals struggle

1 billion lbs.
900 million
800 million
700 million
600 million
500 million
400 million
300 million
200 million
100 million

to adapt to the rising temperatures.

We can stop wasting energy. The way to do this is to drive cars that use less gas, recycle, insulate our homes, install compact fluorescent light bulbs, and purchase energy-efficient appliances. In the first year of the 1990s we're already 10% toward our goal. You have already saved over 100 million pounds of carbon dioxide from being spewed into our atmosphere. We're working on our goal in conjunction with the Union of Concerned Scientists (26 Church Street, Cambridge, MA 02238; 617/547-5552) who have the goal of Americans collectively reducing one billion pounds of CO_2 in 1990. Printed below is a "Personal Diet Tally Sheet" with which you can record your savings. For our part, we will focus on energy saving products. We hope you'll use the tally sheet to help the U.C.S. and us achieve our goals!

In order to monitor our progress we will print our billion pound CO_2ometer in each successive **AE Sourcebook** and *Real Goods Catalog*.

The Billion Pound Diet
Personal Diet Tally Sheet

By committing to a few of the energy-saving actions below, you can help reduce carbon dioxide (CO_2) emissions by one billion pounds a year. Your actions will not only improve the environment but will also reduce our nation's dependence on foreign oil. The numbers below are approximate yearly national averages; to learn how we arrived at them, see the back of this sheet.

1. Drive and Ride Less. I will:

☐ eliminate _____ miles of car travel over the next year.

Total = _____ pounds of CO_2 (= the number of miles of car travel eliminated)

2. Tune Up. I will:

☐ tune my car regularly and maintain proper tire pressure.

Total = 1,000 pounds of CO_2

3. Recycle. I will:

☐ recycle _____ aluminum cans (typical CO_2 reduction = 34 pounds per 100 cans).

☐ recycle _____ glass bottles (typical CO_2 reduction = 30 pounds per 100 bottles).

☐ recycle _____ pounds of paper (typical CO_2 reduction = 20 pounds per 100 pounds).

Total = _____ pounds of CO_2

4. Insulate. I will:

☐ wrap a home hot-water heater (typical CO_2 reduction = 1,200 pounds for electric heaters; 400 pounds for gas heaters).

☐ insulate the attic of a house (for a 6-room house, typical CO_2 reduction = 1,800 pounds if oil-heated; 1,390 pounds if gas-heated; 4,430 pounds if electrically heated).

Total = _____ pounds of CO_2

5. Replace Light Bulbs. I will:

☐ replace _____ incandescent lights with compact fluorescents (typical CO_2 reduction = 110 pounds per light).

☐ replace _____ high-watt incandescents with lower-watt incandescents (each 10-watt reduction typically eliminates 22 pounds of CO_2 emissions).

Total = _____ pounds of CO_2

6. Reduce Hot-Water Use. I will:

☐ wash my clothes in cold water (typical CO_2 reduction = 250 pounds per person with an electric hot-water heater; 110 pounds with gas).

☐ install a low-flow showerhead (typical CO_2 reduction = 225 pounds per person with an electric hot-water heater; 99 pounds with gas).

☐ turn my hot-water heater down 10 degrees (typical CO_2 reduction = 240 pounds with an electric heater; 106 pounds with gas).

Total = _____ pounds of CO_2

7. Other Actions. To calculate the CO_2 savings of other actions, see the back of this sheet.

Total = _____ pounds of CO_2

- -

Total Personal CO$_2$ Reduction

Name: _____

Address: _____

_____

_____

Impact of my planned action:

_____ pounds (from #1)
_____ pounds (from #2)
_____ pounds (from #3)
_____ pounds (from #4)
_____ pounds (from #5)
_____ pounds (from #6)
_____ pounds (from #7)

Total = _____ pounds of CO_2

Further Information on the Billion Pound Diet

The Billion Pound Diet is sponsored by the Union of Concerned Scientists and Student Pugwash USA. For further information, contact the national coordinating office: Union of Concerned Scientists, 26 Church Street, Cambridge, MA 02238 (tel. 617-547-5552).

How We Arrived at the Numbers on the Diet Tally Sheet

The numbers on the diet tally sheet are approximate yearly national averages and are rounded for convenience. Because tens of thousands of people across the nation will participate in The Billion Pound Diet, the use of average numbers will produce an accurate total. However, your personal CO_2 reduction will vary from the national average depending upon your region, the size of your home or apartment, the efficiency of your car and appliances, and your personal habits.

Here is how we computed our CO_2 reductions:

❖ **Automobiles**. Cars produce approximately 20 pounds of CO_2 for each gallon of gas they use. Since the average car on the road gets 20 miles per gallon, it produces one pound of CO_2 each mile. A well-tuned car with properly inflated tires can use up to 20% less gas than a poorly tuned car with underinflated tires.

❖ **Electricity**. A variety of energy sources—coal, water power, oil, natural gas, nuclear power, wind, and solar power—are used to generate electricity. Since only some of these produce CO_2, different electric companies emit different amounts of CO_2. A recent report from the National Audubon Society concluded that, on average, one kilowatt-hour of electricity produces 1.5 pounds of CO_2.

❖ **Recycling**. Industry associations provided data on the energy savings for recycling activities. We used the weight of average-sized cans and bottles.

❖ **Attic insulation**. It takes 600 gallons of oil, 840 therms of gas, or 19,700 kilowatt-hours of electricity to heat the average 6-room house for a year. Attic insulation should save 15% of this.

❖ **Light bulbs**. Our numbers assume that each light bulb is on 4 hours a day and that the average compact fluorescent is 50 watts less than the incandescent it replaces.

❖ **Hot water**. We assumed the average hot-water heater uses 4,000 kilowatt-hours of electricity annually or 240 therms of gas.

How to Figure the Impact of Other Actions You Take

You can use either of two approaches. You can start with the numbers on the tally sheet and adapt them to your proposed action, or you can determine your actual energy savings in gallons of oil, therms of natural gas, or kilowatt-hours of electricity. Then compute the resulting CO_2 reduction. Use the following:

> **heating oil and gasoline** = 20 pounds of CO_2 per gallon
> **electricity** = 1.5 pounds of CO_2 per kilowatt-hour
> **natural gas** = 11 pounds of CO_2 per therm.

If You Plan to Make Major Purchases

By purchasing energy-efficient products, you can make a significant difference in your environmental impact. You will also save money in the long run. For example, a car that gets at least 35 miles per gallon of gas produces 1,560 pounds less CO_2 annually than the average new car, assuming it is driven 10,000 miles per year. A new energy-efficient refrigerator that uses 900 kilowatt-hours of electricity a year produces 900 pounds less CO_2 than a typical 10-year-old refrigerator, which uses 1,500 kilowatt-hours. If you plan to buy a car or refrigerator during the next year, pledge to purchase an energy-efficient model. Then figure your potential yearly CO_2 savings and include this number on your Billion Pound Diet tally sheet.

As an addendum to our Real Goods Products Environmental Impact Report, we thought we'd present a very interesting "Energy Index" that is quite enlightening.

Thanks to Chris Calwell and Judy Christrup for compiling this information.

Home Energy Index

Extra annual energy cost of taking a daily bath instead of a low-flow shower (electric heat): $103

Extra cost of the daily bath with gas-heated water: $35

Months between bulb changes for a typical 60 watt incandescent: 4

Months between changes for an equivalent, 15 watt screw-in fluorescent: 56

Annual cost (bulbs + electricity) to light your hallway with the incandescent: $12.76

Annual cost with the fluorescent: $5.83

Annual energy cost of cooking 4 frozen dinners per week separately in a gas oven: $8.32

Cost to cook the four dinners simultaneously each week in the oven: $2.08

Percent of the U.S. Population that cares: (less than 1%)

Average annual energy bill for America's hot tubs: $200,000,000

Annual energy cost of commuting 10 miles by car, 5 days per week, in congested traffic) (45 minutes each way) $468.00

Typical annual energy cost of picking up, watching, and returning a video every weekend: $25.15

Tons of paper clips sold in the U.S. in one year that were actually used to clip paper: 2000

Tons of paper clips broken or twisted by people while talking on the phone: 1400

Number of aluminum cans that can be recycled with the energy required to make a new one from virgin bauxite: 20

Number of 1000 MW baseload power plants needed to run America's 99 million refrigerators in 1986: 22

Power plants needed to run 120 million refrigerators in the year 2000 with NAECA-87 appliance standards: 20

Number of solar calculators sold in the U.S. in 1983: 60,000,000

Number of continuously lit watt lightbulbs their annual electrical output could power: 1.4

Annual energy bill of the United States: $420 billion

Annual savings attributable to energy efficiency improvements made since 1973: $130+ billion

Percent growth in the U.S. economy, 1973-1986: 33+ percent

Percent growth in U.S. energy consumption, 1973-1986: 0

Barrels of oil imported to the U.S.: 6.8 million per day

Barrels of oil saved by energy efficiency improvements made since 1973: 13 million per day

Market price of Middle East Oil, January 1988: $18 per barrel

Price of Middle East Oil, if U.S. Military costs incurred there are included: $170 per barrel

Average generating cost of nuclear power, per kilowatt hour: 10-13 cents

For electricity in general: six cents

For energy conservation programs: one to four cents

Price of oil, at the rate paid for nuclear power: $240 per barrel

Amount the U.S. Treasury spent in 1987 of energy conservation research and development: $200 million

Amount the U.S. Treasury spends each year on subsidies to the nuclear power industry: $15 billion

Value of U.S. nuclear power plants that have been abandoned: $15 billion

Contribution of new coal and nuclear power to U.S. capacity, 1985-1986: 0.7 quadrillion BTU's

Contribution of energy efficiency to total U.S. capacity, 1985-1986: 5.1 quadrillion BTU'S

Countries ranking above the United States in percentage of GNP devoted to energy conversation R&D: Italy, Japan, Canada, West Germany, United Kingdom and Sweden.

Year in which the joint DOE industry research program on compact fluorescent high efficiency lamps was halted: 1981

Year in which General Electric began distributing Japanese-made compact fluorescent high efficiency lamps: 1985

Amount saved in one year if the U.S. converted to best available lighting technology: $30 billion

Cost of upgrading energy efficiency of World Bank headquarters lighting: $100,000

Amount of World Bank headquarters saved in electricity costs in 1984: $500,000

Prices of photovoltaic solar cell in 1954 $600/peak watt; 1976 $44/peak watt; 1986: $5.25/peak watt

Amount of oil the U.S. would have to import, to meet present demand, if the average MPG of all cars in the U.S. was 42 MPG: none

Average MPG of all cars in use in the U.S. in 1973: 13, in 1985: 25

MPG of the prototype Toyota AXV: 98

MPG of 1987 Mustang V-8: 14

Barrels of oil saved between 1975-1985 because of efficiency improvements in cars and light trucks: 2,400,000/day

Barrels of oil produced by the entire state of Alaska: 1,500,000 per day

Amount of time it takes for the average American car to pollute the atmosphere with its own weight in carbon: 1 year

Our Policies

Subscribe to Real Goods

We have discovered that there are two kinds of people in Real Goods' world. First, we have you old time mainstays who have supported us from the beginning, generally live off the grid, and subscribe to our philosophy (or, at least, humor us.) You loyal believers tend to order every six months or so, and are the backbone of our business. Many of you have already become subscribers. Your purchases have sustained Real Goods through the years.

When we only had three or four thousand people on our mailing list we could afford to mail to all our customers every time. Many postal increases later (which, we believe, discriminate against us catalogers,) we pay 65¢ to mail a *Real Goods Catalog* to each one of you. Multiply that by well over 100,000 customers, figure at least three major mailings per year plus at least three subscriber mailings announcing sale merchandise and newsworthy information, and you get over $500,000 - a hefty chunk of change! Upwards of 3,000 customers live in other countries where it costs as much as $8 per issue to mail the *Real Goods Catalog.*

A $25 annual subscription ($40 foreign) to Real Goods will get you the three issues of the *Real Goods Catalog,* the annual update of the AE Sourcebook (cover price - $14), plus a free gift valued at $10 or more. You'll also receive our regular subscriber newsletter mailings and all other communications we may send. We're putting a lot of extra effort this year into broadening our subscriber base and developing a brand new newsletter called "*The Real Stuff,*" that will be available to subscribers only. We often send special offerings to subscribers with savings of up to $100 per item that are advertised nowhere else.

We no longer mail to non-subscribing foreign customers. The new subscriber system is our way of identifying our preferred customers and ensuring that we continue to stay in contact. Our subscriber program has been wildly successful and we intend to continue *and intensify* our special mailings and offerings for subscribers only.

Real Goods Catalogs

We publish the *Real Goods Catalogs* three or four times every year. These catalogs include sections on new products, updates to products currently in the AE Sourcebook, and lots of feedback from customers on products and the way they work, as well as our ever-popular "**Readers' Forum**" with interesting and meaty responses on politics and lots of other topics. You'll also find new programs, special promotions, closeouts, and our popular "*Unclassified Ads*" section featuring lots of our customers' AE bargains. The *Real Goods Catalogs* are bulk mailed to all recent U.S. customers. We encourage out-of-country customers and those of you who haven't bought from us in the past year to **subscribe** to *Real Goods.* (See subscription terms above)

Pricing Policy

We've struggled with the pricing issue for years. It's difficult to stay on top of the marketplace, researching the competition at every turn. We bend over backwards to develop a fair pricing structure on all of our products and have never been known to "gouge" anyone. Hundreds of customers' letters rave and praise in testament that our product knowledge and service are second to none.

This time we find there are three kinds of people in our world: Eighty percent of you who have dealt with us before trust that our prices are a fair value for products received and services rendered. We appreciate you! Fifteen percent of you check us against one or two competitors just to make sure we're in line and you generally end up buying from us. We appreciate you, too, perhaps even more, because your purchases are a vote of confidence that our knowledge and service is worth the small difference. Then there are the last 5% of you whom we call "professional shoppers" - who religiously collect every alternative energy catalog ever published, including J.C. Whitney and the Amish/Mennonite sales flyers. (Don't feel embarrassed, many of us are slimy shoppers too!) It is for these five percent that our pricing policy was established! Quite simply stated, **We promise to beat any price in the USA on any alternative energy item that we sell.** It's our way of assuring you that you never need to shop anywhere else!

Since the Sourcebook is updated only annually, there are inevitable price changes that will occur from one printing until the next. The periodic *Real Goods Catalogs* are designed to be price updates to the Sourcebook. Naturally, *all prices are subject to change without notice.*

Quantity Discount & Dealer Program

We've put lots of energy into developing and refining our quantity discount program over the past several years. In 1978, when we were the first retailer of solar panels in California, there was no need to sell to other dealers. We had a virtual *corner* on the market and could provide PV all around the country with no competition. In the last several years, and in particular the last year (following EarthDay 1990), new alternative energy and environmental products dealers have proliferated across the country. We believe this to be a very healthy trend, as it gives us all visibility, credibility, and greatly promotes renewable energy resources.

We encourage and support all of you new solar dealers in a spirit of cooperation rather than competition, in the firm belief that more business fosters even more business. With this in mind, we've set up a quantity discount structure for dealers and non-dealers alike that allows you to purchase many of our items in quantity at reduced prices. Our telephone staff will be happy to quote you quantity pricing.

Engineering Services

Word is out that our technical services are world class, and we find ourselves doing more and more design work. Some of our customers are new to PV, getting ready to move away from the power lines, and want to be sure they put in the right system in the right way. Other customers are construction or solar sophisticates, but know that we have the latest data and the best access to solar information, and use us to augment their design capacity. We welcome the opportunity to engineer whole systems from planning to installation. A brochure outlining how this service works is in production, but here's how it works:

The first step in planning your system is assessing your needs, capabilities, limitations, and working budget. To arrive at this, you may respond to a questionnaire we've designed to help us understand your circumstances, or you may send us an explanation of what you think we should know. Be sure to include a complete list of what energy needs you expect, estimating high to anticipate expansion. This should include an inventory of all lighting, water-pumping, refrigeration, tools, and appliances you intend to use. We also need site information, including geographical location, project siting, access to a *"solar window"* away from trees and mountains, and potential for hydro-electric and wind development. Don't forget to include a phone number, so one of our Applications Engineers can call you (with no obligation, of course.) The ensuing design cycle will be between you and your Applications Engineer - in effect, he will be working for you.

Once he determines your needs, he designs a system which corresponds to your needs, be it a stand-alone solar system, or a hybrid system utilizing a back-up fossil-fueled generator, wind and hydro. Next, we produce a schematic drawing for you, outlining how all the components fit together. Your engineer will work with you until your system has been designed and refined. He will (when you say the word) order the parts and supervise assembly of your system at our end, assuring you that you get precisely what you need. If you're working with a licensed contractor or are knowledgeable and capable of your own installation, our services may end when we ship your equipment off to you. More often, you and your contractor will need some more hand holding, and our full-time engineering department can provide full working drawings and installation instructions, as well as providing phone assistance. If needed, we can also provide a professional and experienced installation team to complete the installation.

The first hour is free and part of our service in making the sale. Additional engineering time is billed at $40 per hour. We generally require a 10% retainer fee or credit card number to get started engineering a system after the first hour. Our satisfied customers assure us that our technical support saved them money, time, and hassle, and was the best investment they made. Please feel free to call us for clarification and costs for our services.

An Invitation to Our Customers

New Product Suggestions

As a way of staying in closer touch with your needs, we instituted the *New Product Suggestion Program* several years ago, and this program has been an unqualified success. The program works wonders for us all: think of it as 300,000 investigative professional buyers working to broaden our product mix, seeking new products to enhance the quality of off-the-grid life. You buyers get the opportunity to pick up some free credit and cheap products by suggesting useful products.

The program works like this: SEND US IDEAS FOR NEW PRODUCTS that you think we should carry in our catalog - include manufacturer's name, address, and phone number, along with any information you may have on the product. We are particularly interested in your experience with the product and its manufacturer. If we include your product in our next catalog, we'll issue you a $25.00 credit, and you can buy the product at our cost. (Sorry, but this offer applies only to the *first* suggestion we get for any particular product!) It's a great way to turn all our creative minds to the benefit of the entire off-the-grid network. **Thanks!** Send all your new product ideas to our regular address **Attn: New Products.**

Wanted: A Few Good Writers

We try to make the *Real Goods Catalog* readable from cover to cover, more than just a sales rag. We know there are thousands of talented writers out there with vast storehouses of knowledge relating to all aspects of alternative energy, and we're eager to inspire you to pull those keyboards and pens out of the closet. Putting our money where our mouth is, we're offering gift certificates valued from $25 to $100 to any customer who sends in **an article** that we use in a future edition of the *Real Goods News* or in the *AE Sourcebook*. Send your articles to our regular address **Attn: Article Submissions.**

Catalog Covers & Photo Invitation

We sometimes run out of cover ideas for our catalogs, and we need help. We're sure that **lots of you have photogenic homesteads and power systems** which deserve international attention. (*Real Goods Catalogs* go to addresses in over 50 countries on 5 continents.) Help us keep our graphics exciting. If we choose your **artwork for a cover** of the *Real Goods Catalog* we'll send you a $100 gift certificate. Inclusion in the inside or for future editions of the *AE Sourcebook* earns a gift certificate valued at $25. Please submit either black and white glossy photographs (8" x 10" for the cover) or screened artwork. If you'd like the artwork returned, please include an SASE. Send to **Attn: Photo Submissions.**

[When we can't find a satisfying graphic, we often resort to cover antics we may later regret. Save us from ourselves! Send in your photographs and graphical ideas.]

Come Visit Our Showroom

We invite you all to stop in and visit our newly renovated showroom in Ukiah next time you're passing through Northern California. We're only two hours north of the Golden Gate Bridge, at the north edge of the wine country, and smack dab in the middle of the solar ghetto (aka the Solar Capitol of the World!) Follow the map; take the North State Street offramp off of U.S. Highway 101.

In our showroom, you'll find almost everything from inside this Sourcebook, many of them installed so you can see how they work. We try to make our showroom a veritable Exploratorium of Alternative Energy so you can understand and compare. Our showroom hours are 9am - 5pm Monday through Friday and 10am - 4pm on Saturday. Our staff will be happy to help you design an alternative energy system or answer any questions you may have about our products. If you require engineering or intensive technical assistance, please call beforehand to make an appointment.

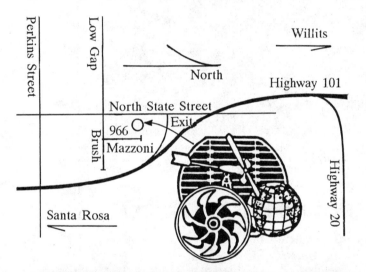

Come visit our showroom and see the Sourcebook come to life. We look forward to meeting you!

AE Sourcebook Updates

The AE Sourcebook is updated annually with new products, policies, and pricing. Because prices change frequently, the *Real Goods Catalog* serves as the pricing update for our products, and prices in the AE Sourcebook go out of date.

Quantity & Educational Discounts on Sourcebooks

The AE Sourcebook is available to schools, religious organizations, energy centers, government agencies, and retailers at the quantity prices below. These are significant discounts off of the cover price of $14.

1-5 copies	$10.00
6-24 copies	$ 7.50
25+ copies	$ 6.50

Home Power Magazine

We've incorporated eleven articles from Home Power Magazine into this Sourcebook. Home Power has become the favored periodical of the off-the-grid network (besides our catalogs of course!). We have culled out what we consider to be their finest articles. Because all these articles are copyrighted ©, you must contact them for reprinting rights. We highly recommend that you subscribe. Information is available from Home Power Magazine; POB 130-RG, Hornbrook, CA 96044; 916/475-3179.

Sourcebook Refund

The price that you've paid for your 1991 AE Sourcebook is fully refundable with your first $100 order. Simply include your original paid receipt with your order and deduct it from the total.

REAL GOODS GUARANTEE:

1. Our merchandise represents the best value available anywhere. We promise to beat any price in the USA on any alternative energy product that we sell as long as you provide adequate proof of a lower price before the sale.

2. We promise to treat you fairly. You can return any stocked item for full cash if returned in original condition & packaging within 90 days of purchase for any reason whatsoever. No questions asked except for: "How can we do better next time?"

Something New Under the Sun. It's the Bell Solar Battery, made of thin discs of specially treated silicon, an ingredient of common sand. It converts the sun's rays directly into usable amounts of electricity. Simple and trouble-free. (The storage batteries beside the solar battery store up its electricity for night use.)

Bell System Solar Battery Converts Sun's Rays into Electricity!

Bell Telephone Laboratories invention has great possibilities for telephone service and for all mankind

Ever since Archimedes, men have been searching for the secret of the sun.

For it is known that the same kindly rays that help the flowers and the grains and the fruits to grow also send us almost limitless power. It is nearly as much every three days as in all known reserves of coal, oil and uranium.

If this energy could be put to use — there would be enough to turn every wheel and light every lamp that mankind would ever need.

The dream of ages has been brought closer by the Bell System Solar Battery. It was invented at the Bell Telephone Laboratories after long research and first announced in 1954. Since then its efficiency has been doubled and its usefulness extended.

There's still much to be done before the battery's possibilities in telephony and for other uses are fully developed. But a good and pioneering start has been made.

The progress so far is like the opening of a door through which we can glimpse exciting new things for the future. Great benefits for telephone users and for all mankind may come from this forward step in putting the energy of the sun to practical use.

BELL TELEPHONE SYSTEM

Winston Smith, our resident mad artistic genius, came across this gem while searching through some 1950s National Geographic magazines. We're still waiting (nearly 40 years later) for Ma Bell to fulfill her promises.

Chapter 1
Systems Planning - An Overview

One of the most important keys to independence as we enter into the 1990's will be energy self-sufficiency. Utility rates most likely will continue to rise as various shortages develop, making the independent home power system more and more attractive economically. The cost of interfacing with existing utility infrastructure is prohibitive for many modern homesteaders in rural areas. Even for the urban dweller, low-voltage power systems will offer freedom from increased utility bills and utility system failures.

An alternative energy or stand-by power system for your home or business requires a careful investigation of the generating sources available. You will want to design a system that best utilizes your natural resources, whether they are solar, wind, or water. You will also want to make sure that you choose the best equipment for your power needs.

This AE Sourcebook is designed to help you make the best choices, and will provide you with a great deal of information on types of systems and equipment.

Also talk to your friends and neighbors who have been building their own power systems; they will have a lot of practical do's-and-don'ts to offer from their own experience.

We feel that the equipment featured in this catalog is the very finest available. Most of it has been tested by our staff or our customers under actual living conditions over the last fifteen years. If we haven't had much experience with something that we offer, we've noted it in the text.

The most successful low-voltage, direct-current, power-generating systems are of a hybrid design. That is, they incorporate more than one power-generating source.

Photovoltaics

Each photovoltaic panel will produce power from sunlight and will average a certain electrical output each day over the course of a year. Multiply the output average in your area times 30 days times each solar panel to determine monthly kilowatthour capability. The photovoltaic section of the catalog contains a map showing average daily peak solar radiation you can expect in your area. It also contains complete instructions for site evaluation and technical specifications on solar panel output under various conditions.

Hydro-electric

Small hydro-electric systems require at least 10 gpm (gallons per minute) of water at 50 ft of "head" (fall) minimum to deliver any usable power. The power generated is usually DC for battery storage and ranges from 20 watthours to 500 watthours. Larger 120V-240VAC, 50 & 60 Hz systems are available in the 2.0 kWh and up power delivery range. They require from 16 to 100 ft of head and from more than 800 gpm to more than 2,000 gpm. If you have the water, hydro can be the most cost-effective system available. See the hydro-electric section of the catalog for site evaluation techniques and equipment specifications.

Traditional Generators

Used in a new and cost-effective way, the traditional generator can cut the costs of a stand-alone PV system by as much as two-thirds. Diesel fuel and propane gas conversion can cut costs and maintenance. The unique Photo-Gen Set hybrid application combines a sophisticated micro-processor with the generator, a special high-powered battery charger, inverter, an automatic transfer switch, and storage batteries. The result is a system that automatically monitors itself. When battery voltage falls below a specified point, the micro-processor automatically signals the generator/battery charger to take over from PV's, wind, etc. and rapidly recharge the batteries. 120V or 240VAC is available always through the inverter or generator. The automatic transfer switch keeps one or the other always on line to your AC main. See the Generator section of the catalog, in Chapter 2, for complete specifications.

Beginning to Size Your System

The heart of the independent power systems described in this Sourcebook is the battery system where a small quantity of power is delivered over time and is stored. The stored power can be drawn out in large quantities as needed over short periods of time. The generators used usually produce DC (direct current) electricity because AC (alternating current) cannot be stored.

Knowing your own consumer demand is necessary before you can choose the appropriate charging system, batteries, or compatible equipment and accessories. You must be knowledgeable about the appliances and accessories you will want to use in your environment and how much power each accessory will consume per day. The charts on the following pages are designed to help you arrive at just that knowledge. When completed, you can compare your probable electrical use to how much electricity can be stored and generated. These considerations are the first step as you begin to use this book and design your own independent power system. They are all your choices.

Some appliances may have their power rating in terms of kW or kilowatts. If this is the case, convert kilowatts to watts by multiplying the rating in kilowatts times 1,000. For example, 1.2 kW is equal to 1,200 watts. The watt, which is the product of voltage and current and is the unit of measure for electrical power, is

An island PV system from one of our customers in Rangiroa, French Polynesia.

most commonly used to describe the rate of power consumed by an AC electrical appliance. Don't confuse the watt, which is a rate of power consumption, with the watthour, which is a quantity of power consumed. For example, if power is consumed at a rate of 250 watts for 2.5 hours, then the amount of power that has been used is (250 x 2.5) 625 watthours. In order to keep the numbers on a more reasonable scale, the quantity of power consumed is usually expressed as kilowatthours (kWh) and is obtained by dividing watthours by 1,000.

DC appliances are usually rated in terms of the amount of current they draw when operating. This can easily be misleading because an appliance that draws 10 amps (the amp, an abbreviation for ampere, is the unit of measure for electrical current) at an applied voltage (the unit of measure of electrical pressure) of 12 volts, uses only half as much power as another appliance that draws 10 amps at 24 volts. Comparisons of appliances that operate at different voltages can only be made after multiplying the current drawn times the voltage applied to get the rate of power consumption in watts for each appliance. However, since all of the DC appliances in any given environment (your home or car, for example) are usually supplied by the same voltage, another convenient term often used to describe the quantity of power consumed is the amphour.

One reason that the amphour is so convenient is that in the storage battery systems, even though more power is actually consumed in recharging a battery than is delivered to the appliance as the battery is being discharged, if 10 amphours is used, 10 amphours will be required to recharge the battery replacing the power that was removed. Losses are made up because the battery is charged at a higher voltage than it produces while being discharged.

Another reason that the amphour is useful when dealing with a storage battery system is that most often battery capacity is stated in amphours.

In calculating your daily power requirement for the purpose of sizing a storage battery for a DC system, the current in amps drawn by each appliance is multiplied by total time in hours the appliance is used each day to get the number of amphours required per day for that appliance.

As a general rule of thumb, an environment that requires 60 kWh per month (2.0 kWh or less each day) may be more efficient and economical using a 12V or 24V system. More than 60 kWh per month usually indicates the need for powerful AC appliances which are more efficiently operated on a 24V or 48V system. Most likely, you will end up with an AC/DC system with AC current operating appliances and equipment not practical to convert to DC and a DC circuit for some lighting and communications. Then if the AC side should fail, the DC circuitry will provide light and sound.

You can start small and reach your independent power goal over a few years.

Many energy saving devices are available to help cut down on electrical needs. Solar/photovoltaic, solar thermal hot water systems, wood stoves equipped with hot water jackets, gas refrigeration, catalytic radiant gas heating, the new light capsule technology for AC and DC lighting, and more are all listed in this Sourcebook. Use them! They are the secret to planning an affordable alternative energy system.

Our expert staff is available to answer your questions by phone or in person. Complete systems design and consultation is available and charged for on an hourly basis. Refer to "Our Policies" at the beginning of this Sourcebook.

Use the charts to determine how much electrical power your choice of appliances and equipment will consume per day and multiply that by the frequency of use to arrive at an average monthly kWh figure. If an appliance is used every day of the month, we use a multiplier of 30.42 days to determine the average monthly power demand. This is because photovoltaic (PV) sizing depends on the average daily peak solar radiation available annually.

Match the power consumption figure you arrive at to the various power-generating systems available. It will be up to you to choose one or more generators to generate the power you need. Factors in that choice will be the location of your site in relation to wind, hydro, or solar power feasibility and your own economics. All our systems are modular in concept. You can start small and reach your independent power goal over a few years. Or you can do it all at once. The output and specifications of each generator are listed in this book.

If an appliance uses some other form of electricity such as 115 volts AC, through an inverter which ultimately is to be supplied by the battery, then you must consider it as a DC device. Use the current drawn by the inverter (or other device) while the appliance is operating

when calculating its daily amphour requirement. An example would be a computer that runs on 115VAC plugged into an inverter that is being powered by the battery. The computer may require 150 watts to operate which is supplied by the inverter, and the inverter may draw 16 amps while the computer is connected and running. If you use the computer for 3.5 hours each day, then 56 amphours will be used by the computer every day and an average of 56 amphours will have to be returned to the battery every day just to make up for what the computer uses.

Thus, in figuring the size of the battery and the type and size of the charging system, the computer-inverter combination is considered as a DC appliance along with all the other DC devices. Those appliances that will be run exclusively on AC (not derived from the battery system) should be segregated and their power requirements figured separately.

Inverters

An inverter provides 117V to 240V, 60 or 50 Hz (hertz = cycles per second) AC electricity from a primary 12V, 24V, 36V, 48V, or 110VDC battery system. The size inverter needed will be determined by the surge load, running watts, and type of AC home or business equipment you chose to operate, be it 120V or 240V. This will also dictate the primary DC voltage of your system as well as the battery configuration and storage capacity necessary. To select the proper inverter, compare your power and equipment use figures from the worksheet to the specifications detail for each inverter. See the Inverters and Battery section in Chapter 2 of the catalog for complete technical details and specifications.

Surge Load

Be conscious when sizing an inverter of the surge load. Surge load normally refers to the large amount of power required by inefficient AC motors. This large power demand lasts for a second or less to achieve running speed. The amount of "surge" power required by a particular unit can be seen listed opposite AC capacitor and split-phase motors. Both the inverter and any traditional generator will require sizing according to surge load requirements. Example: if a refrigerator starts up at the same time you turn on a clothes washer, blender, and water pump, there will be a surge load equal to the sum of the surge current for each of them.

Your inverter or generator must be sized to deliver this surge load demand. Check the specification plate on your AC appliance motors to find wattage, amps, or hp. This will be the running power. If you cannot determine which type motor your appliance has, ask your appliance dealer. Also check the rpm's of your AC motor. If it is 3,450, increase the surge load wattage by 15%. For hard starting or high inertia motors, i.e., compressors and blowers, add another 25% (40% altogether). The AC motors listed in the worksheet assume 1,750 rpm, the DC motor, 3,600 rpm.

Calculating System Loads

Now that you've gotten a good general idea of how the nuts and bolts of an alternative energy system work from the overview, it's time to sit down with pencil and paper and actually calculate what you will need in your home power system. The calculation is easy, but first there are some basics you need to know.

Conservation

Amory Lovins' concept of "Negawatts" is nowhere as important as it is with an alternative energy system. Energy conservation gives new meaning to "A watt saved is many pennies earned." The use of energy efficient appliances and lighting as well as non-electric alternatives wherever possible can make solar electricity a cost-competitive alternative to gasoline generators and, in some cases, utility power.

It makes no sense to spend thousands of dollars on PV panels only to dissipate that power away with a wasteful old styled compressor-on-the-bottom refrigerator! Take the time to study the quantum strides that have been made in lighting technology and refrigeration technology. You need to understand that conventional electric cooking, space heating, and water heating equipment use a prohibitive amount of electricity.

The answer of course is to use wood for space heating, gas (propane or natural) for instantaneous water heating, gas or the highly efficient SunFrost refrigerator for refrigeration, and a microwave oven or a gas stove for cooking. While it is indeed possible to use conventional electric refrigerators with an alternative energy system, you will need at least 20 solar panels compared to 2-3 for a Sunfrost!

Wattage Requirements of Common Appliances

Description	Wattage
Refrigeration	
22 cu ft auto defrost	490
(normally used 14.4 hr per day)	
12 cu ft Sunfrost Refrigerator	58
(normally used approx. 6-9 hr per day)	
Water Pumping:	
AC Jet Pump, 165 gal per day,	500
20 ft well depth	
DC Flojet pump for house pressure system	60
(typical use is 1-2 hr per day)	
Econosub submersible pump	50
(typical use is 6 hr per day)	
Entertainment/Telephones	
TV (25-inch color)	130
TV (19-inch color)	60
TV (12-inch black & white)	15
Satellite system, 12 ft dish	
with auto orientation/remote cntrl.	45
VCR	30
Laser disk	30
Stereo (Avg. volume)	15
CB (Avg. hrly use)	10
Cellular telephone	24
Radio telephone (avg. hrly use)	10
Electric player piano	30
General Household	
Typical fluorescent light (60W equivalent)	15
Incandescent light of 60W brightness	60
Intellevision	27
Electric clock(s)(3) and	12
Clock radio(s)(2)	10

	Wattage
Iron (electric)	1500
Clothes washer	450
Dryer (gas)	250
Central vacuum	750
Furnace fan (yearly avg.)	240
Alarm/security system	3
Kitchen Appliances	
Dishwasher	1500
Trash compactor	1500
Can opener (electric)	100
Microwave	750
Exhaust fans (3)	144
Coffee pot (electric)	1200
Food processor	1200
Toaster	1200
Coffee grinder	100
Office/Den	
Computer/Printer/Moniter/Modem	80
Electric eraser	100
Phone dialer	4
Typewriter	200
Adding machine	8
Electric pencil sharpener	100
Hygiene	
Hair dryer	1500
Waterpik	90
Whirlpool bath	750
Hair curler	750
Miscellaneous	
Electric blanket (yearly avg.)	240
Garage door opener	300

Tools: (annual avgs.)		Space:	
AC table saw, 2.0 hp	2250	Gas or wood stove with water heat	24
AC grinder, 1/2 hp	600	exchanger, chill chaser & low-voltage	
AC lath, 3/4 hp	732	circulating pump (avg. yearly use)	
AC drill, 1/8 hp	120	Air conditioning: (avg. annual use)	105
		(1500 sq ft using DC evaporative	
		cooler)	

Heating:

Water:	
Solar thermal panel/wood (coal)	24
stove heat exchanger w/low voltage	
circulating pump	

Calculating Your System Loads

In the following format systematically list all the appliances that you intend to use (refer to chart above), the wattages of each one, and the hours per day you intend to use them. Remember that some appliances like microwaves, trash compacters, and blenders will be used far less than one hour per day, so use the fraction of the hour. List the appliance regardless if it is to run DC direct or AC from the inverter.

Appliance	Watts	Hrs/Day	Watthours/Day

An additional full sized worksheet can be found in the appendix at the back of this book.

TOTAL WATTHOURS PER DAY _____

Multiply by 1.2 to account for inverter & battery efficiency losses

Total Watthours per Day
(with efficiency loss)   _____

As a general rule of thumb, if your total required watthours per day is less than 2,000 watthours you probably will be using a 12 volt system. If it is higher than 2,000 watthours you may want to go with a 24 Volt system because of the increased amperage and potential line losses in your copper wire with a 12V system.

Whatever system voltage you choose, now it is time to divide your daily watthour requirement (with efficiency losses) by the system voltage to arrive at your required amphours per day for your system.

$$\frac{\text{Watthours/Day}}{\text{System Voltage}} = \text{Amphours/Day}$$

Now that you know how many **amphours per day** your system requires, all that remains is to determine where you are going to get that power. In a PV-only charging system the calculation is easy.

- First look at the Solar Insolation Chart to determine how many average hours per day of sun you will get.
- Next determine the PV module that you will use. For ease of calculation it's best to pick a 48W, 16V panel which yields an even 3 amps.
- Next multiply 3 amps times the hours per day in your area to arrive at the total number of amphours per day provided by one solar panel.
- Next divide your required load by this figure to arrive at the number of solar panels you'll need.

That was simple enough. The calculation is basically the same for wind, hydro, and gasoline generators. Simply take the amphours provided by each power device and gang them together so that you come up with the amphours per day of load that you require. We recommend you carefully examine our Remote Home Kits 1-4 in the next Chapter to see how some examples in real life relate to the number of solar panels required.

The only other consideration left is the amount of battery storage required, and that is addressed fully in Chapter 3 of this Sourcebook.

Real Goods helped design and provided all the materials for this unique project in North Carolina. The main structure is a "Biodome" for which we supplied 156 Sovonics 20 watt PV modules. Two 5-HP Stirling Generators provide the back up power. Six Trace 48 volt, 2200 watt inverters with stacking interfaces were provided for the AC power. Also pictured is the control room with the battery bank consisting of 24 each 2-volt Absolyte cells wired in series to produce 1,000 amphours at 48 volts.

Chapter 2
Power Generating Devices and Whole Systems

Photovoltaics

Solar Electric Generation

Sunlight is energy that is permanent, free, non-polluting, totally quiet and universally available. The sun is a giant power plant that generates radiant energy at an enormously high kilowatt (kW) rate estimated to be a staggering 110-trillion kW. Even though less than one billionth of this energy is intercepted by the earth, every 10 sq. feet of the earth's surface facing the sun is estimated to receive about 1000 watts at mid-day.

A photovoltaic device or silicon solar cell converts light into DC (direct current) electricity. It does not use heat from the sun as does thermal solar hot water. In fact, the higher the ambient temperature, the less efficient a solar electric cell becomes. The most common commercially available solar cell is a small wafer or ribbon of semiconductor material, usually silicon. One side of the semiconductor material is electrically positive (+) and the other side is negative (-). When light strikes the positive side of the solar cell, the negative electrons are activated and produce a tiny unit of electrical current.

Photovoltaic Technology

When a group of solar cells are connected or the semiconductor ribbon material is applied to a predetermined surface area, a solar module is created, and the cells are encapsulated under tempered glass or some other transparent material. Quantitative electrical output is determined by the number of cells or ribbon material connected together within the module and then further determined by the number of modules connected together. More than one module connected together is called a solar array.

There are two primary photovoltaic technologies in use today: crystalline and amorphous silicon. Crystalline silicon was developed for the space program in the early 1950's and has actually seen very little improvement in over 30 years. Silicon is grown in large blocks and the thin wafers that become the solar cells are sliced off. The conversion efficiency of crystalline silicon is a maximum of 13%. The great advantage of crystalline silicon is that there has been virtually no evidence of voltage degradation over time since the 1950's! Most reputable PV manufacturers warrant their crystalline modules for a minimum of 10 years, yet expect a life in excess of 40 years! Crystalline modules can be found in two varieties: single crystal (like Hoxan and Siemens) and polycrystalline (like Kyocera and Solarex). There is very little if any difference between them in performance.

Amorphous silicon appears to be the wave of the future. It is created by a diffusion process whereby the silicon material is vaporized and deposited onto a glass or stainless steel substrate. Because of this process, there are unlimited possibilities for solar generation using amorphous silicon. Skylights, bay windows, office windows on skyscrapers, roofing material, and sunroofs on vehicles are all potential electrical generators with this new technology. Amorphous silicon, unfortunately, at this point has a maximum of 6% conversion efficiency, requiring nearly twice the surface area as crystalline silicon to produce the same amount of power. Also thus far with research and development still in its infancy, amorphous silicon has degraded significantly over time. Arco Solar introduced its first amorphous G-4000 power module in late 1986 only to recall it a year later. Nevertheless, the future is very bright for amorphous PV technology. Because of the manufacturing process, amorphous technology is far cheaper to produce than crystalline technology, particularly when quantities increase. The flexibility of amorphous PV also brings significant advantages, allowing the modules to be custom fitted to countless applications. One further advantage of amorphous is its property of greater output in low light conditions.

Gallium-Arsenide is one more technology of photovoltaics that is currently under intense research. It has the advantage of conversion efficiencies in the 30% range, but thus far is far too expensive to be practical.

A PV Power System For Your Home

A properly planned photovoltaic system consists of six basic components. They are: (1) a solar array (solar panels) based on the number of kWh (or amphours) you will use each day, with back-up or secondary generator (wind, hydro, etc.), (2) the choice of a roof, ground, pole, or passive tracker mounting, (3) a charge controller and safety disconnect system, (4) a metering panel, (5) battery storage, and (6) an inverter (if desired).

A battery storage system is needed to act as a buffer between the solar array and your home on nights and sunless days. Although a solar array will generate some electricity on cloudy days and even under a full moon, the output will vary greatly on both a daily and seasonal basis. A battery system smooths out some of the variation.

There are exceptions to the need for a battery storage system for solar electric power generation. Some appliances and equipment run directly from the power produced from one or more solar modules. These are called "sun-synchronous" or "array-direct" devices. You will find some of them listed under "Solar Cooling" in Chapter 4 featuring solar-powered fans and evaporative coolers, others in "Water Pumping & Storage" in Chapter 4 featuring solar-powered water pumps.

On the following pages you will find a selection of photovoltaic modules by different manufacturers, most of whom we have worked with for many years. These modules are highly reliable and warranted by their respective manufacturers for up to 12 years; however you can expect a 30-year life span or more with a possible power degradation of 10% over that time. There are no residential systems that have been in service for 30 years, so we really can't say for sure just how long a module will last, but it looks like the life expectancy is substantial. *We often come up with used or "recycled" modules that are very attractively priced and will last for many years. If you are interested in these PV bargains ask our phone staff or keep your eyes on our thrice-per-year catalog updates.*

How Many PV Modules Do I Need?

Below is a map of the United States that shows the average daily peak solar radiation in all major zones. If you look at your own zone and see 4.5, you can expect to receive 4.5 hours per day of bright, direct sunlight on the average each day over the course of a year. At winter solstice, however, you may receive as little as 2 to 3 hours per day average peak solar radiation until sometime in February. Conversely, at summer solstice, June 21, you may receive 6 hours or more peak solar radiation on the average each day all the way through September or October. Of course averages are not infallible, and we have included this map for guidance only. You know your area best. For instance, some people live high in the mountains where there is more sunshine than in the valley below them. Therefore, you may find more or less sunshine than the map lists, so plan accordingly. The Real Goods engineering staff has sophisticated software on board to accurately pinpoint your particular ecological niche and provide you with exact figures for solar generation potential.

In the winter, a back-up generating source should be considered, or if this is to be a stand-alone PV system, sufficient battery storage must be employed for the shorter and cloudier days of winter.

Look now at the specification for each PV module listed. Note that the output of the various modules is listed in amps and watts.

For a stand-alone PV system simply choose the size module you want. Multiply its output in amps times the number of peak solar daily radiation hours for your location on the map. If you choose a passive tracker, in most cases you can add 30% or more to the output. Divide that number into the total number of amphours that you need to arrive at the total number of PV's required. Example: if you need 180 amphours per day for your power needs; and you've chosen a Hoxan 4810 @ 3 amps and you live in Hawaii (6 hr peak sun per day): 6 times 3 = 18 amphours per day for each solar panel; 180 divided by 18 = 10 Hoxan 4810s required, or 8 with a tracker mount.

If you want to simplify the process, order one of our starter kits, which are all fully modular and can be added onto at any time.

Mean Daily Solar Radiation

This map shows the average daily peak solar radiation in all major zones in the USA. If your zone shows 4.5, you can expect to receive 4.5 hours per day of bright, direct sunlight on the average each day over the course of a year. Multiply the number in your area by the number of amps output of your solar panel (or projected solar panel) to determine the total number of amphours per day you'll get. Example: 4 Arco M-75 solar panels at 3.12 amps each times 5.6 hours per day (assuming you live in Hawaii) will give you about 70 amphours per day. Remember, you'll get lots more hours per day in the summer than in the winter!

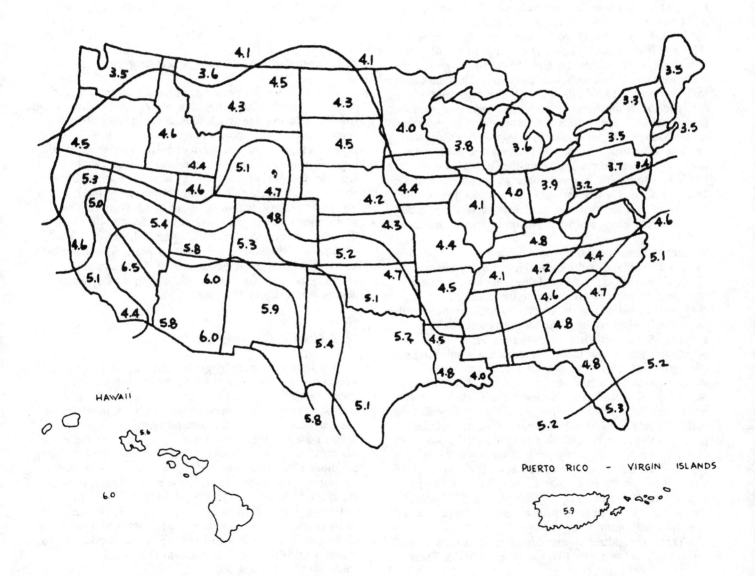

How To Determine Which PV Voltage Is Best

All PV manufacturers make modules in several different voltages for different applications. This needn't be confusing so long as these applications are understood. Three "standard" voltages have emerged over the years: **14.5V** (Siemens M-65, Kyocera K-45), **16V** (Siemens M-75, Kyocera K-51, & Hoxan 4810), and **17.1V** (Siemens M-55 and Solarex MSX-60) with slight variations in voltage for different manufacturers. In the electricity vs. water analogy, voltage represents pressure or potential energy. Think of the battery as a pressure tank and the PV panel as the voltage pump.

To fill the 12V battery, higher pressure (voltage) must be pumped into the tank (battery) to stabilize it at 12 volts. A typical 12V battery reaches its optimum charge around 14.5V. Thus if we use a 14.5V PV panel to charge the battery system there is virtually no room for voltage loss or voltage "drop." 14.5V PV modules are called "self-regulating" because theoretically they taper the charge down to a minimum when the battery reaches full charge. These self-regulating modules (Siemens M-65 & Kyocera K-45) have been heavily promoted over the years by the RV (recreational vehicle) industry as the panacea for full-time RV'ers. There is one serious flaw in their thinking. RV's typically hover in warm climates like the Southwest desert where the high temperatures invariably decrease the PV module's voltage output. This can easily cause the lower 14.5V modules to begin regulating too early, seriously diminishing the output when the battery is in its final charging cycle. Other potential voltage loss can occur through long wire runs with improperly sized wire. In these situations even the slightest voltage drop can cause significant power loss when the "pump" is only 14.5V to begin with. The only time we recommend using a 14.5V module is where the user has no intention of ever increasing to more than two modules, where the wire is sized appropriately, and where temperatures do not exceed 100° F.

For 95% of the systems we design we recommend using a **16V** solar panel like the Hoxan 4810, Siemens M-75, or Kyocera K-51. 16V modules must have a charge controller to prevent overcharging the battery from charging beyond 14.5 volts. Operating without a controller, particularly when you're using all the power you're generating, will not harm the batteries, but if you go away for several weeks at a time thus removing a balancing load, your batteries will get fried. On a one- or two-module system, a charge controller is not absolutely necessary. Generally you can leave a one-module system for a week or so without damaging a battery provided the battery is rated at 100 A/hr or more. We recommend placing a switch in the line to one module of a two-module system to disconnect one of them when leaving a system unattended for up to one week. If a system of any size will be left for more than a week, install a charge controller. The other obvious advantage to 16V panels is that you are forgiven for a little bit of voltage drop when your wire isn't quite large enough.

There are several applications where we recommend using 17.5V solar panels like the Siemens M-55. These higher voltage panels begin to put out power at 5% of noonday sun instead of 10% of noonday sun like the lower voltage modules. Therefore, many of our customers in far northern climates like British Columbia and Alaska prefer these modules. A second application for 17.5V modules is where voltage drop is unavoidable - houses already wired with long wire runs, for example, where it's more economical to buy a 17.5V module and lose 1.5 or so volts and still come out with plenty of power to charge your batteries. The third and probably most widely used application for high-voltage modules is in PV-direct (sun-synchronous) systems. Using deep-well pumps, ceiling fans, or motors, often a maximum speed can only be obtained using the highest voltage possible.

Can I Mix & Match Different Solar Panels?

As a general rule it is fine to mix different solar panels together in the same array, but try to keep the voltages within 5-10% of each other. If you put a 17.5V panel in parallel with a 14.5V panel the overall voltage will decrease to 14.5V. You won't hurt anything - you'll only lose a little power.

Tilt Angle for PV Module Mounting

Your PV's will operate at optimum when they are placed perpendicular to the path of the sun. Always mount your PV's facing due south (due north in the southern hemisphere). You should change the tilt angle at least two times every year. At the summer solstice set them at your latitude minus 15° and at the winter solstice at your latitude plus 15° at least. Many new PV users lately have chosen to eliminate this step by simply sizing the PV array for the worst-case winter scenario and leaving it. This would mean setting your array to 60-65° in a 40° north latitude and location never changing it. More power will be delivered if changed four times a year. Modules placed within 5 degrees of perpendicular to the sun operate within .9962% maximum allowable output. Beyond a 5 degree error, dramatic losses occur.

Do I Need a Blocking Diode?

Customers frequently get differing opinions on whether or not to use a blocking diode on their solar systems. A blocking diode is an electronic device that allows current to flow in only one direction, much like a one way "check valve" in a water system. Without this device power will flow backwards at night from the batteries to the PV's where it will dissipate. In performing the diode's function some voltage loss occurs from the resistance. Generally you will lose more voltage through the diode than you will lose in night-time losses for systems employing up to four modules. For systems of five or more modules we always recommend using a blocking diode (preferably a low-resistance Schottky diode), if your charge controller does not perform this function automatically as most do.

A List of PV Absolutes

Do not use concentrators or reflectors to enhance module output unless you are located in a very cold climate. Modules operate best from 63° F. to 120° F. Concentrators will decrease the efficiency of the module due to heating.

Do not place plexiglas, glass, etc. as a shield in front of the panels. This will act as a filter, reducing the light, increasing the temperature, decreasing the output.

Do not mount the panels flat on the roof. They need air circulation under the panels.

Do not place any part of the module in the shade. No light, no electricity.

Do site your location to determine where the best solar window is at your home site.

Do wash off the modules from time to time. *This is the only regular maintenance requirement.*

Environmental Benefits of PV

What is the overall benefit of PV to the environment in terms of energy savings? We did some calculations to determine the savings in electricity (kWh), barrels of oil, pounds of coal, pounds of carbon dioxide, and sulfur dioxide. We calculated these savings figures for one 50-watt module and for the total amount of PV modules that you bought from us last year. You can see from the following table that the effect is significant!

Analysis of the Environmental Savings of Photovoltaic Modules (Solar Electric Panels)

	Savings of One 50 Watt Solar Panel	Savings of Real Goods' Customers' Purchases in 1990
kWh Saved per year	90 kWh	657,000 kWh
kWh Saved per life of solar panel	2,700 kWh	20 million kWh
Barrels of oil saved over lifetime of solar panel	4.8 barrels of oil	35,000 barrels of oil
Pounds of Coal saved over life of solar panel	2,700 lbs. of coal	20 million lbs. of coal
Pounds of Carbon Dioxide kept out of the air over life of solar panel	4,000 lbs of Carbon Dioxide (a major cause of global warming)	30 million lbs. of Carbon Dioxide
Pounds of Sulfur Dioxide kept out of the air over life of solar panel	23.3 lbs. of Sulfur Dioxide	170,000 lbs. of Sulfur Dioxide

Formulas upon which these calculations are based:

1 PV Module = 50 Watts
1 PV Module will last 30 years
It takes 11,000 Btus to generate 1 kWh of Electricity
6.25 million Btus = 1 barrel of oil
Coal required to produce 1 kWh = 1 lb.
Carbon dioxide emissions per kWh = 1.5 lb.

Having It Both Ways

The following article by our computer wizard, Michael Potts, explains the ease, joy, and economic savings of living both on and off of the grid.

I live in a small town on the North Coast of California. We joke about being on the edge of the continent, but power lines run down Main Street and I get most of my power the easy way: from the grid. We get some weather up here, and it takes the power out regularly a dozen times a year, sometimes for a day at a time, and so I put quite a bit of thought into surviving the times we are on our own - gas cooktop, passive solar and wood heat, gravity fed water, solar hot water. Kerosene lighting is romantic, but it is inconvenient to scurry around in the dark to find the fragile lamp and the elusive match. There must be a better way.

I designed my house when 12-volt lighting meant automotive lighting, and my power supply was a car battery and a trickle charger plugged into the wall. Fortunately, I had plenty of wire, and ran a lot of duplicate circuits, thinking I might like to use alternative energy more extensively down the road; here I am, down the road, and I am glad I buried all that copper in the walls! Because the times have changed, a kilowatt from the grid costs five times what it did and promises to go higher, and I couldn't live without the reliability of the 12-volt system.

The first low voltage light bulb to go on was a light above the bed. When the power failed, it gave off enough light for me to find flashlight or lamp and match. But I soon discovered that the light was perfect for reading - why not use it all the time? Why not add a 12-volt digital clock, so the time would always be correct, even after a power failure? There's an uncanny correspondence between high winds and power outages here on the edge, and the anemometer - the device that tells how fast the wind is blowing - always went off about the time the winds got really interesting: why not put it on the low voltage system? Easy - and I got rid of a little transformer that was converting 110v AC into 12v DC with that little extra inefficiency we've grown to love.

The uses of low-voltage power, and the justifications for using it, are many and multiplying. Energy self-sufficiency is, perhaps, the best. At the most trivial, it just feels good to have a hand on generating my own power. In my work as a writer and computer consultant, it also saves me money: time when the lines are down but I can work anyway, because my computers run on the 12-volt system, and work saved that I used to lose when the grid went down. I enjoy the reliability of a small, centralized system. I confess to a small twinge of superiority when the lights all around me go dark, but my house remains workable.

The Nuts and Bolts

Designing low-voltage circuits into a house still on the drawing board costs very little, and will add only slightly to the electrician's bill. You must try to think of the places where alternative energy will be of use, and provide the branch circuits. The 'All Electric Home' of the fifties uses electricity in profligate ways, where a 'remote home' makes the most of what is available, and so the alternative system should provide just enough. A well integrated system will allow a degree of swapping back and forth between AC and DC circuits just by changing fixtures and connections at source and destination. When the wiring is all done, it should look to you, your electrician, and the building inspector, like an over-wired house. You should accept from the beginning that, no matter how carefully you plan, your needs or the technologies will change.

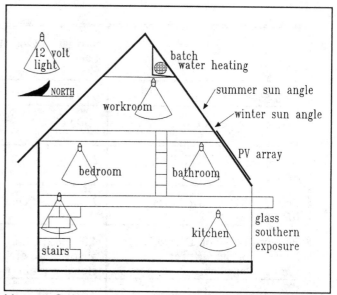

House Schematic

Retrofitting an existing house - adding a 12-volt system to a house already wired for conventional power - is as complicated as rewiring a house, and could involve ripping out walls and all manner of unpleasantness ...it simply may not be worth it. Plan a limited application, or incorporate it into any renovation plans.

The power required to perform a function is comparable, from 110v AC to 12v DC: if it takes a 12 gauge wire to provide conventional power, it will take the same wire to power a low-voltage lamp, and the hardware at both ends is nearly the same. (The authorities, in their never-ending struggle to stand in the way of progress, want you to use 8 gauge wire - build a copper mine in your walls - and hoarding strategic metals that way can't hurt, but 12 gauge is quite sufficient for a modest lighting circuit.) Buying more of the same wire, boxes, and connectors offers economies of scale, and you (or your electrician) already know how to deal with the running of it. Part of the planning phase must go to researching the availability and best source of alternative devices. Low voltage lighting is well developed, but there are fewer appliances. If power outages are a major problem, you should plan your AC system to allow branches to be cut over from the grid to an inverter, so essential services (like microwaves and blenders) still operate.

The single best source of information is Real Goods in Ukiah - their Alternative Energy Sourcebook and Remote Home Power Kits Manual provide an encyclopedic listing of low-voltage devices and the whys and hows of installing alternative energy systems. Whatever you do, you must remember to work safely with low voltage systems; although are inherently safer (you could not electrocute yourself as conveniently) there is plenty of power to manage.

The Elements of a System

An alternative energy system consists of a power source (or several sources), storage device, transmission paths, and the tools and fixtures that turn power into something useful. To put together a useful system, you must consider the whole system, always keeping the goal in mind. For me, the goal was enough light to get around with and an uninterruptible power supply for my computers. I made a schematic of my house, identified the fixtures and their power requirements, and used (naturally) a spreadsheet program to make revising the calculations simpler. I specified a generous system to allow myself room to add to the system, and this is what I came up with:

Power uses	wattage	hrs/day	WHrs/da	excluding computers
Lighting	50	3	150	150
Instrumentation	25	24	600	600
Computers	300	6	1800	
Tools	50	2	100	100
	total wattage requirement		2650	850
12	volt system amperage		221	71
5	hours average sun	panels:	15	5

System Requirements

Fifteen panels would be too many for the space allotted, and so I elected to run the computers from a system attached to the grid through a charger, and the lights, tools, and instruments from a second, 'honest' system. (I had not planned for the computers, and so this simplified the retrofit wiring.) Storage capacity has to bridge the gap between the time the source is lost and comes back on line. On the computer system, the source is the grid,

and I wanted at least two hours to complete my work in an orderly manner. Storage required: 2 100-amp-hour gel-cell batteries. For the lights and instrumentation, where the source is the sun, I needed to be able to go about a week with negligible charging - the longest a serious stormy patch lasts in these parts. Again, about 200 amp-hours of storage should tide me over.

If it takes this much trouble to gather my own power, I should be frugal in using it, and that guided my selection of wiring and fixtures. The larger the wire, the lower the transmission loss, and so I wired with 12/2WG wire - 12 gauge, 2 conductor plus ground, the electrician's standard household wire. I like the intensity and color temperature of halogen lighting, and so that was my choice for task lighting. For wider area lighting, the high efficiency PL fluorescent technology is the only rational choice - it uses a quarter the energy of its incandescent equivalent, and, lasts ten 10 times as long. (While I'm being frugal, I might apply the same logic to my 110v AC lighting and save a bundle - see box.)

Wiring the 12 volt fixtures is very straightforward: take the same precautions you would with 110 volt wiring with respect to overcurrent protection with a fuse box or distribution center, use 12v DC switches, conventional wire-nuts to make splices, make all splices in boxes, and use clamps to protect wires from sharp metal edges if you are using metal splice boxes. 'Cigarette lighter' type plugs, are ungrounded, and are frowned upon by the authorities. Since 12 volt DC is so benign, you may be tempted (knowing you have fused the circuit conservatively and can be careful to avoid shorting the wires) to work the system 'hot' and get immediate confirmation when you've got something wired in: it lights up! I am told this is a bad and reckless habit; some humans experience burns and worse even with low voltage power, so work it cold - pull the fuses. Observe the polarity with more care than with conventional AC wiring, because you run the risk of frying delicate instrumentation if you get it backwards. If in doubt, use a multimeter to establish polarity, and use red electricians tape liberally to mark the positive side.

Is It Worth It?

There is no doubt that it costs more to run an alternative system - in the short run, and with nothing else considered. After all, you're building the generation capacity that the utility company provides as well as the consumer end of things. The utilities hasten to tell us about economies of scale, but there is reason to suspect that subsidies play a big part in the real equation, and there are many hidden costs to fossil-fuel and nuclear power generation as well. But I was curious to know just how much this thing would cost. Back to the spreadsheet.

If I assume that power costs will continue to escalate at about the current rate, I predict that I'll see the following rates on future electric bills:

In other words, using the realistic model as a basis for analysis, costs will double about every four years from the present rate of just over 13¢ a kilowatt-hour, crossing the 30¢/KwH line as early as 1997.

To build a system to completely satisfy my alternative needs - so I would be able to survive if the utility company folded its tent and stole away into darkness - would cost about $9000, and would perform as follows:

Projected North Coast Electric Rates
(per Kilowatt hour)

	conservative		realistic	
		% up		% up
1986	.0881*	.02%	.0881	.02
1988	.0994*	12.82%	.0994	12.82
1990	.1188*	8.87%	.1188	8.87
1992	.1470	10.99%	.1514	13.00
1994	.1768	10.99%	.1933	13.00
1996	.2158	10.99%	.2468	13.00
1998	.2634	10.99%	.3152	13.00
2000	.3215	10.99%	.4025	13.00
2002	.3924	10.99%	.5139	13.00
2004	.4789	10.99%	.6562	13.00
2006	.5845	10.99%	.8379	13.00
2008	.7134	10.99%	1.0699	13.00
2010	.8707	10.99%	1.3662	13.00
2012	1.0628	10.99%	1.7445	13.00
2014	1.2971	10.99%	2.2275	13.00
2016	1.5831	10.99%	2.8443	13.00
2018	1.9322	10.99%	3.6319	13.00
2020	2.3583	10.99%	4.6376	13.00

* actual

What is the Point?

Enough sunlight falls on the exposed southern face of my house to provide for modest electric and hot water needs; it would take only a minor realignment of my priorities for me to live within my own capacity. The same is true for my neighbors almost everywhere in rural and suburban America; elsewhere around the globe, what I would consider a sufficiency would be thought a surfeit. The energy equation has had some its key terms shifted - the real cost of a barrel of crude used to generate the bulk of America's energy may be $30 (today's market price for West Texas) or $80 (Carl Sagan's guestimate) or $200 - $500 (GreenPeace's pessimistic assessment) - and the trend is not likely to reverse. Energy costs will rise, and the only argument is about whether it will be a linear, a geometric, or an exponential curve. (My assumptions have taken the middle ground.) My favorite columns in the table above are the second and fourth, which show that I have locked in a reasonable rate for my power, and that my capital expenditures are negligible after the initial outlay: a healthy economic profile, particularly when compared with the uncertainties of the public energy picture. There are undoubtedly hidden horrors in the photovoltaic closet - what chemicals despoil what streams near the factories where silicon wafers are fabricated? how much power does it take to make a silicon wafer? - and I hope I will find the answers to that concern, along with solutions, in this magazine.

The point, simply, is that we need not uproot carbon compounds it took nature millennia to get buried just to enjoy ample power. A grass roots grid, community PV arrays and distributions channels, and a sharing of technology, can back us out of the ugly corner we seem painted into. Since the myth of cheap power has evaporated, those of us who are mainstreaming our power from the grid must reassume responsibility for our energy needs.

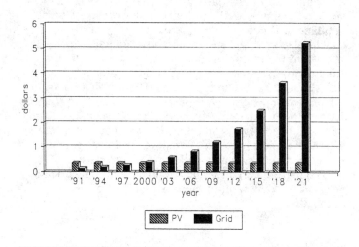

Projected Power Cost
per Kilowatt-Hour

A generous PV System

AmpHrs/Day	225
AmpHrs/Year	82181
KWHrs/Year	986
System cost	$10,500
Cost/KWHr	0.3549

			Energy Cost		Cost
	PV KwHr	Grid KW	PV	Grid	(Profit)
1991	0.3549	0.1340	9000	132	8868
1992	0.3549	0.1514	0	149	8719
1993	0.3549	0.1711	0	169	8550
1994	0.3549	0.1933	0	191	8359
1995	0.3549	0.2184	0	215	8144
1996	0.3549	0.2468	0	243	7901
1997	0.3549	0.2789	0	275	7626
1998	0.3549	0.3152	0	311	7315
1999	0.3549	0.3562	0	351	6964
2000	0.3549	0.4025	0	397	6567
2001	0.3549	0.4548	0	448	6118
2002	0.3549	0.5139	0	507	5612
2003	0.3549	0.5807	0	573	5039
2004	0.3549	0.6562	0	647	4392
2005	0.3549	0.7415	0	731	3661
2006	0.3549	0.8379	1500	826	4335
2007	0.3549	0.9468	0	934	3401
2008	0.3549	1.0699	0	1055	2346
2009	0.3549	1.2090	0	1192	1154
2010	0.3549	1.3662	0	1347	-193
2011	0.3549	1.5438	0	1522	-1715
2012	0.3549	1.7445	0	1720	-3435
2013	0.3549	1.9713	0	1944	-5379
2014	0.3549	2.2275	0	2196	-7575
2015	0.3549	2.5171	0	2482	-10057
2016	0.3549	2.8443	0	2805	-12862
2017	0.3549	3.2141	0	3169	-16031
2018	0.3549	3.6319	0	3581	-19612
2019	0.3549	4.1041	0	4047	-23658
2020	0.3549	4.6376	0	4573	-28231
2021	0.3549	5.2405	0	5167	-33398
total cost over 30 years			10500	43898	

Here is a system we designed for Carol Salisbury in Ahualoa on the big island of Hawaii. It consists of 40 Arco M-75 48 watt PV modules installed on four Zomeworks 24-module trackers. We set up the system in a 24-volt configuration with 1050 amphours of battery storage at 24 volts dc. One Trace 2024 was employed for the main system loads which draw 227 amphours per day. An additional Trace 612 was installed and dedicated to the computer system. Two 24-volt Enermaxers were employed for the charge controller.

Photovoltaic Modules

Unlike many distributors and retailers of solar panels in the market place, Real Goods does not have a vested interest in any one manufacturer. We try to keep ourselves completely objective and offer you as wide a selection of PV products as there are responsible manufacturers capable of delivering them. The PV market has changed continually since we sold our first module in 1978 and many manufacturers have come and gone. Photowatt, Exxon, Mobil, and Solavolt are no more. Arco Solar, which has accounted for over 80% of our module sales in the last 13 years, has sold out to Siemens of Germany. Sovonics of Michigan is going into a joint venture with a Japanese company (Canon). That leaves Solarex (Maryland) and Solec of Los Angeles as the only remaining U.S. PV manufacturers. Kyocera of Japan has been coming on strong with a fine line of polycrystalline modules in the past several years with an industry-leading 12-year warranty. Hoxan of Japan has recently begun to strongly affect the PV marketplace with a renewed commitment and extremely fine quality modules. Solar panels are basically generic - we're not talking Yugo vs. Mercedes here. For the most part a watt is a watt and the PV module you buy, regardless of manufacturer, is going to put out its rated power for many years to come. Real Goods guarantees the lowest price in the USA on PV modules, and will beat any other price you find.

Hoxan (Heliopower) PV Modules

Hoxan operates the world's largest and most highly automated PV manufacturing plant. Hoxan PV modules are very close in looks and performance to Siemens modules. All Hoxan modules come with a full 10-year power output warranty. These have been our standby modules in our remote home packages for most of 1990 and we're very pleased with their performance as well as the companies commitment to quality.

H-4810 Hoxan

The Hoxan H-4810 is nearly identical in output to the Siemens M-75. It uses 36 high efficiency square single crystal cells on a thick anodized aluminum frame with highly transparent tempered glass. The 4810 produces 48 watts of power at 16.2 volts to generate 3.0 amps of current. The 4810 measures 16.5" wide by 37.25" long by 1.25" thick.

HOXAN 4810		
rated watts	48.0	@ 25°C
rated power	volts: 16.2	amps: 3.0
open circuit volts	21.3	@ 25°C
short circuit amp	3.2	@ 25°C
size (LxWxD)	37.2 X 16.5 X 1.3	
construction	single crystal, tempered glass	
warranty	10 years	

11-401   Hoxan 4810 (Qty 1-3) $349
           Hoxan 4810 (Qty 4-19) $339
           Hoxan 4810 (Qty 20+) $329

Siemens (Arco Solar) Photovoltaic Modules

Recycled Arco Modules

We've sold over 6,000 recycled Arco modules in the last year. These modules are 6-7 years old and were used on a power plant in Southern California. They must be purchased in sets of three in order to charge 12V battery systems. Each module is rated at 5.33V and puts out approximately 32 watts. In sets of three wired in series, they put out 16 volts at approximately 96 watts of power to produce about six amps. All modules are warranted to put out within 12% of 96 watts for a period of five years. The original modules are 12" wide x 48" long, but framed they measure 36-5/8" wide x 48-5/8" long.

The modules are typically brown colored at the solar cells themselves. The browning is actually the EVA (the bonding material that holds the cell to the glazing). Once the cells have browned there will usually be no further discoloration nor loss of power. Tests have shown that the worst case is a power reduction of 10% from a brand new to a browned module. Therefore what you buy now as a recycled module will have very minute if any future power loss. *A mounting structure is not included.*

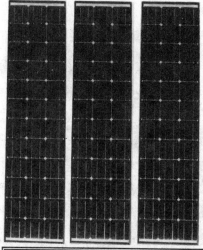

RECYCLED ARCO TRI-LAMS		
rated watts	96	@ 25°C
rated power	volts: 16.2	amps: 5.88
open circuit volts	20.4	@ 25°C
short circuit amps	6.3	@ 25°C
size (LxWxD)	approx. 48" x 12" x 1.4"	
construction	single crystal, tempered glass	
warranty	12% of 96 watts 5 years	

11-130   Recycled Tri-Lams (96 watts) (1-3) $479
           Recycled Tri-Lams (96 watts) (4-19) $459
           Recycled Tri-Lams (96 watts) (20+) $429

Siemens M-75

The Siemens M-75 produces 48 watts at 15.9 volts. The M-75 is efficient, attractive, easy to install, and comes with a wired-in bypass diode in each junction cover. The M-75 consists of 33 cells in series. Each module comes with an easy to understand instruction manual and a 10-year warranty. With its high amperage (3 amps at load) it is an excellent module for stand-alone systems.

SIEMENS (ARCO) M-75		
rated watts	48	@ 25°C
rated power	volts: 15.9	amps: 3.02
open circuit volts	19.8	@ 25°C
short circuit amps	3.4	@ 25°C
size (LxWxD)	48 X 13 X 1.4	
construction	single crystal, tempered glass	
warranty	10 years	

11-101   Siemens M-75 (Qty 1-3) $449
         Siemens M-75 (Qty 4-19) $439
         Siemens M-75 (Qty 20+) $419

WE'LL BEAT ANY PRICE IN THE U.S.A. - DON'T HESITATE TO ASK!

Siemens M-40

The Siemens M-40 is an underrated version of the M-75. It is exactly the same size as the M-75, but when Siemens tests their modules, those that come out less than 48 watts get classified as M-40s, guaranteed to produce at least 40 watts at 15.7 volts for a current of 2.55 amps.

11-102   Siemens M-40 (1-3 qty) $389
         Siemens M-40 (4-19 qty) $379
         Siemens M-40 (20+ qty) $359

Siemens M-55

The M-55 is Siemens' most powerful standard module, producing 3.05 amps with 53 watts at 17.4 volts, and consisting of 36 cells in series. It is ideal for water pumping applications where higher voltage is required, or for providing extra voltage for long wire runs. It is the best module to use in extremely hot climates as high-temperature voltage drop is kept at a minimum. It is slightly more efficient in low light conditions as it begins producing power at 5% of noonday sun rather than the standard 10%. We sell lots of M-55's to Alaska and Washington.

SIEMENS (ARCO) M-55		
rated watts	53	@ 25°C
rated power	volts: 17.4	amps: 3.05
open circuit volts	21.7	@ 25°C
short circuit amps	3.4	@ 25°C
size (LxWxD)	50.9 X 13 X 1.4	
construction	single crystal, tempered glass	
warranty	10 years	

11-105   Siemens M-55 (Qty 1-3) $499
         Siemens M-55 (Qty 4-19) $489
         Siemens M-55 (Qty 20+) $479

SIEMENS (ARCO) M-40		
rated watts	40 +	@ 25 C
rated power	volts: 15.7	amps: 2.55
open circuit volts	19.5	@ 25 C
short circuit amps	3.0	@ 25 C
size (LxWxD)	48 X 13 X 1.4	
construction	single crystal, tempered glass	
warranty	10 year	

Siemens M-65

The M-65 is designed primarily for RV, marine, and remote home usage; only one module is employed. The M-65 consists of 30 cells wired in series. It is a "self-regulating" module that decreases its current output from 3 amps to less than 1/2 amp when the battery approaches full charge, eliminating the need for a charge controller, but often seriously limiting the module's output. Not recommended for very hot climates where temperatures often exceed 100° F. The Siemens M-65 produces 43 watts at 14.6 volts for a current of 2.95 amps.

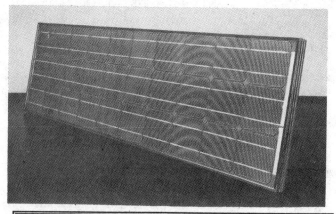

SIEMENS (ARCO) M-65		
rated watts	43	@ 25°C
rated power	volts: 14.6	amps: 2.95
open circuit volts	18.0	@ 25°C
short circuit amps	3.3	@ 25°C
size (LxWxD)	42.6 x 13 x 1.4	
construction	Single crystal, tempered glass	
warranty	10 years	

11-103   Siemens M-65 (Qty 1-3)   **$439**
        Siemens M-65 (Qty 4-19)   **$429**
        Siemens M-65 (Qty 20+)   **$409**

Siemens M-20

The M-20 is a compact, 20-watt self-regulating module ideal for RV's, boats, and remote homes where needs are minimal, use is intermittent, or space is limited. As the battery approaches full charge, the M-20's current output decreases from a rate of 1.37 amps to less than 1/4 amp, eliminating the need for a charge controller. Siemens recommends at least 70 amphours of battery storage for each M-20 module. Five-year warranty. 14.6 volts, 1.5 amps.

SIEMENS (ARCO) M-20		
rated watts	22	@ 25°C
rated power	volts: 14.6	amps: 1.5
open circuit volts	18.2	@ 25°C
short circuit amps	1.65	@ 25°C
size (LxWxD)	22.4 X 13 X 1.4	
construction	single crystal, tempered glass	
warranty	5 years	

11-107   **Siemens M-20**   **$249**

Siemens M-35

The M-35 is an underrated version of the Siemens M-65 and it produces a guaranteed 37 watts at 14.5 volts making 2.56 amps of current. It is identical in size and configuration of the M-65.

SIEMENS (ARCO) M-35		
rated watts	37	@ 25 C
rated power	volts: 14.5	amps: 2.56
open circuit volts	18.1	@ 25 C
short circuit amps	3.0	@ 25 C
size (LxWxD)	42.6 X 13 X 1.4	
construction	single crystal,tempered glass	
warranty	10 year	

11-104	Siemens M-35 (Qty 1-3)	$369
	Siemens M-35 (Qty 4-19)	$349
	Siemens M-35 (Qty 20+)	$339

Siemens M-50

The M-50 is an underrated version of the Siemens M-55, producing at least 48 watts at 17.3V for a current at load of 2.78 amps. It is identical in size and configuration to the M-55.

SIEMENS (ARCO) M-50		
rated watts	48	@ 25˚C
rated power	volts: 17.3	amps: 2.78
open circuit volts	21.6	@ 25˚C
short circuit amps	3.2	@ 25˚C
size (LxWxD)	50.9 X 13 X 1.4	
construction	single crystal,tempered glass	
warranty	10 year	

11-106	Siemens M-50 (Qty 1-3)	$459
	Siemens M-50 (Qty 4-19)	$439
	Siemens M-50 (Qty 20+)	$429

Siemens T-5

The T-5 produces 5 peak watts at 14.5 volts for 350 mA of current. It is useful for small applications like our solar gate opener.

11-124	Siemens T-5	$89

Siemens G-50

The G-50 produces 2.5 watts at 14.5 volts for 170 mA of current. It's ideal for very small applications.

11-123	Siemens G-50	$59

Kyocera Photovoltaic Modules

Kyocera is a Japanese manufacturer of PV modules. They make an excellent polycrystalline panel with a conversion efficiency over 13% (about the same as Hoxan & Siemens) and a 12-year warranty.

Kyocera K51

The K51 is Kyocera's standard "building block" module similar to the Hoxan 4810 and the Siemens M-75. It produces 51 watts at load using 16.9 volts for an amperage of 3.02 amps.

KYOCERA K-51		
rated watts	51.0	@ 25˚C
rated power	volts: 16.9	amps: 3.02
open circuit volts	21.2	@ 25˚C
short circuit amps	3.25	@ 25˚C
size (LxWxD)	38.8 X 17.5 X 1.4	
construction	multicrystal, tempered glass	
warranty	12 years	

11-201   Kyocera K51 (Qty 1-3) $419
　　　　 Kyocera K51 (Qty 4-19) $409
　　　　 Kyocera K51 (Qty 20+) $399

Kyocera K63

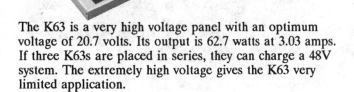

The K63 is a very high voltage panel with an optimum voltage of 20.7 volts. Its output is 62.7 watts at 3.03 amps. If three K63s are placed in series, they can charge a 48V system. The extremely high voltage gives the K63 very limited application.

Kyocera K45

The K45 is very similar to the Siemens M-65. It is a self-regulating module with an optimum voltage of 15.0 volts. Its output is 45.3 watts for a maximum current of 3.02 amps. It is ideal when only one or two modules are used, since a charge controller need not be employed. The same cautions apply as do to M-65's: don't use in extremely hot climates or when wire isn't adequately sized.

KYOCERA K-45		
rated watts	45.3	@ 25˚C
rated power	volts: 15.0	amps: 3.02
open circuit volts	18.9	@ 25˚C
short circuit amps	3.25	@ 25˚C
size (LxWxD)	34.6 X 17.5 X 1.4	
construction	multicrystal, tempered glass	
warranty	12 years	

11-202   Kyocera K45 (Qty 1-3) $389
　　　　 Kyocera K45 (Qty 4-19) $369
　　　　 Kyocera K45 (Qty 20+) $359

KYOCERA K-63		
rated watts	62.7	@ 25˚C
rated power	volts: 20.7	amps: 3.03
open circuit volts	26.0	@ 25˚C
short circuit amps	3.25	@ 25˚C
size (LxWxD)	47 X 17.5 X 1.4	
construction	multicrystal, tempered glass	
warranty	12 years	

11-215   Kyocera K63 (Qty 1-3) $559
　　　　 Kyocera K63 (Qty 4-19) $549
　　　　 Kyocera K63 (Qty 12+) $539

MSX30 Solarex Unbreakable Module

The "Light" series of Solarex modules are made without glass and are very thin, lightweight, and highly portable. Typically they're used for boating, camping, RV's, scientific expeditions, railroad signaling, and mobile communications. The MSX30 - Light produces 30 watts of peak power at 17.8 volts for a current of 2.0 amps, weighing in at less than 5 lbs! Ten-year warranty. 24.38"L x 19.63"W x 0.38"D. 4.5 lbs.

11-511   MSX30 Solarex $399

MSX18 Solarex Unbreakable Module

The MSX18 - Light produces 18.6 watts of power at 17.8 volts, for a current of 1.06 amps. 19.63"L x 17.57"W x 0.38"D. 3.5 lbs.

11-512   MSX18 Solarex $239

MSX10 Solarex Unbreakable Module

The MSX10 - Light produces 10 watts of peak power at 17.5 volts for a current of 0.57 amps. 17.63"L x 10.63"W x 0.38"D. 1.8 lbs.

11-513   MSX10 Solarex $139

SOLAREX MSX-30		
rated watts	30	@ 25°C
rated power	volts: 17.8	amps: 1.68
open circuit volts	??	@ ??°C
short circuit amps	??	@ ??°C
size (LxWxD)	24.25 X 19.5 X .38	
construction	multicrystal, no glazing	
warranty	10 year	

SOLAREX MSX-18		
rated watts	18.6	@ 25°C
rated power	volts: 17.8	amps: 1.06
open circuit volts	??	@ ??°C
short circuit amps	??	@ ??°C
size (LxWxD)	19.63 X 17.63 X .38	
construction	multicrystal, no glazing	
warranty	10 year	

SOLAREX MSX-10		
rated watts	10	@ 25°C
rated power	volts: 17.5	amps: .57
open circuit volts	??	@ ??°C
short circuit amps	??	@ ??°C
size (LxWxD)	17.63 x 10.63 x .38	
construction	multicrystal, no glazing	
warranty	10 year	

MSX60 Solarex Module

The MSX60 is Solarex's production solar panel. It has the highest power output, 60 watts, of any standard sized commercially available power module. Solarex uses polycrystalline cells coated with a titanium dioxide anti-reflective material. The MSX60 is built with a strong, rugged frame of corrosion-resistant, bronze-anodized aluminum. It will produce 60 watts of power at 17.1 volts to produce 3.5 amps of current. Ten-year warranty. 43.5"L x 19.75"W x 2"D.

11-501	Solarex MSX60 (Qty 1-3)	$439
	Solarex MSX60 (Qty 4-19)	$429
	Solarex MSX60 (Qty 20+)	$419

MSX56 Solarex Module

The MSX56 is the next largest Solarex module. It will produce 56 watts of peak power at 16.8 volts to produce 3.35 amps of current. It comes with standard Solarex features including the 10-year warranty. It measures 43.65" x 19.76" x 2.13" and weighs 16 lbs.

11-506	Solarex MSX56 (Qty 1-3)	$419
	Solarex MSX56 (Qty 4-19)	$409
	Solarex MSX56 (Qty 20+)	$399

MSX53 Solarex Module

The MSX53 is the next largest Solarex module. It will produce 53 watts of peak power at 16.7 volts to produce 3.2 amps of current. It comes with standard Solarex features including the 10-year warranty. 43.65"L x 19.76"W x 2.13"D. 16 lbs.

11-502	Solarex MSX53 (Qty 1-3)	$389
	Solarex MSX53 (Qty 4-19)	$369
	Solarex MSX53 (Qty 20+)	$359

SOLAREX MSX-60		
rated watts	60	@ 25°C
rated power	volts: 17.1	amps: 3.5
open circuit volts	20.8	@ 25°C
short circuit amps	3.8	@ 25°C
size (LxWxD)	43.65 X 19.76 X 2.13	
construction	multicrystal, tempered glass	
warranty	10 year	

SOLAREX MSX-56		
rated watts	56	@ 25°C
rated power	volts: 16.8	amps: 3.35
open circuit volts	20.8	@ 25°C
short circuit amps	3.6	@ 25°C
size (LxWxD)	43.65 X 19.76 X 2.13	
construction	multicrystal, tempered glass	
warranty	10 year	

SOLAREX MSX-53		
rated watts	53	@ 25°C
rated power	volts: 16.7	amps: 3.22
open circuit volts	20.6	@ 25°C
short circuit amps	3.4	@ 25°C
size (LxWxD)	43.65 X 19.76 X 2.13	
construction	multicrystal, tempered glass	
warranty	10 year	

MSX40 Solarex Module

The MSX40 is similar in output to the old Arco M-73. It produces 40 watts at 17.2 volts to deliver 2.34 amps of current. It comes with the standard 10-year Solarex warranty. 30.07"L x 19.76"W x 1.97"D. 10 lbs.

11-503	Solarex MSX40 (Qty 1-3)	$339
	Solarex MSX40 (Qty 4-19)	$329
	Solarex MSX40 (Qty 20+)	$319

SOLAREX MSX-40		
rated watts	40	@ 25˚C
rated power	volts: 17.2	amps: 2.34
open circuit volts	21.1	@ 25˚C
short circuit amps	2.53	@ 25˚C
size (LxWxD)	30.07 X 19.76 X 1.97	
construction	multicrystal, tempered glass	
warranty	10 year	

Car Charger

Our car charger is a self-contained, self-regulated, portable battery charger. When exposed to the sun, it provides a continuous trickle charge to your battery in order to maintain optimal power and performance. Batteries typically leak or discharge power at a rate of up to 15-40% per month. Parking a car for a long period always carries the risk of finding a dead battery when you return. The charger is very conservatively rated at one watt (although we consider them to be two watts) and all you need to do to use it is place it on your dashboard and plug it into your cigarette lighter. An 8' cord is included. Size: 13.5" L x 7" W x 1/2" D. No more dead batteries when you leave your car at the airport. It works great for tractors and farm vehicles too.

| 11-123 | Car Charger | $39 |

10 Watt Amorphous Chronar PVs

We have just made a large purchase of brand new Chronar 10 Watt PV modules and are offering them at a very attractive price. Chronar has recently discontinued manufacturing in the U.S. These modules are amorphous and typically will degrade in output by as much as 20% in the first year. They were originally rated at 12 watts, but allowing for degradation we are very conservatively calling them 10 watt modules. They measure 12-1/8" x 36-1/8" and they have a polycarbonate frame. Output is 14.5 volts at 690 mA at load. At these prices they won't last long!

11-555 Chronar 10W $69

Upscale Solar Electric

Solar-powered, environmentally-conscious residence in Carlisle, MA., designed and built by Solar Design Associates of Harvard, MA.

Features include active solar water heating, passive solar heating and cooling, solar generated electricity by photovoltaics, high-efficiency, dual-compressor heat pump system, earth sheltering, internal thermal mass and super-insulation.

Solar-powered, environmentally-conscious residence in eastern Massachusetts designed and built by Solar Design Associates of Harvard, MA.

Features include active solar water and space heating, passive solar heating and cooling, solar generated electricity by photovoltaics, radiant heat distribution, a solar greenhouse, waste recycling, earth sheltering, internal thermal mass and super-insulation.

Steve Strong of Solar Design Associates was kind enough to provide us with these wonderful photographs of solar homes. While serving as an engineering consultant on the Alaskan pipeline project in the early 1970's, Steve Strong quickly became convinced that there were easier, less-costly, more environmentally-desirable ways to provide comfort and convenience to the consumer than "going to the ends of the earth to extract the last drop of fossil fuel." He founded Solar Design in 1974.

Solar Design Associates of Harvard, MA is a group of Architects and Engineers dedicated to the design of sustainable, environmentally-responsive residences and other buildings, and the engineering and integration of renewable energy systems.

Over the past fifteen years, Solar Design has designed dozens of homes and buildings across the country which provide delightful living/working environments while requiring little or no purchased energy. The homes share a common reliance on photovoltaics (PV) to produce their electricity.

Solar Design designs the buildings and engineers their energy support systems together as an **integrated effort**. This integrated approach is essential to creating a high-performance building, especially if it's to be energy-independent.

Solar Design recently completed a 100-kilowatt PV demonstration program sponsored by New England Electric which includes a neighborhood of 30 PV-powered homes in the small Massachusetts town of Gardner tied to the local power grid. "While we were at it, we also did the Town Hall, the library, the community college, a furniture store and even the local Burger King," Steve

The south roof of this solar house in Brookline, MA is made of glass panels, most of which generate electricity directly from the sun using photovoltaics (PV). Other panels produce thermal energy which is used to heat the home or the domestic hot water. There is no fossil fuel burned, no pollution from combustion by-products, etc.

The house, designed by Solar Design Associates of Harvard, MA., also features earth sheltering, passive solar gain, super-insulation, low-emissivity, inert-gas-filled glazing, an earth-coupled heat pump system, internal thermal mass and radiant barriers to limit colling requirements.

reports. The project is a first and has received world-wide attention.

Steve Strong is the author of *The Solar Electric House* which is available in the book section. You can reach Steve at Solar Design Associates, Harvard, Massachusetts, 01451-0242.

A Proposal for Large Scale Solar Electric Generation

The following is a proposal recently submitted to the United States Congress by Alfred H. Canada, who has long been a member of the I.E.E.E. power engineering society, and a consultant to the Department of Energy.

Background

Strategic electrical energy decency appears possible through installation of large-scale contemporary photovoltaic generator modules, e.g., fixed-tilt-panel arrays at installed nameplate capacity densities in excess of 150 Megawatts per section of land. In cogeneration with agriculture on tilled land, reducing the crop yield to about 85%, SVG avoids conventional generation's encroachment on virgin lands and streams, while allowing robotics-agronomy with the array powered access-tractors running on the rows of tracked structure needed to mount and service the SVG Array generator modules.

Four pivotal criteria for least cost planning obtain in cost effectiveness evaluations of any electric generation options for the 21st Century. The Solar Voltaic Generation option, considered against these criteria for Energy Decency: - Abundance, on-site inexhaustible "solar fuel"; e.g., at Southwest Desert locations, using current 14% solar-to-electric conversion and a gross annual capacity factor of 0.25 suggests that if in a single Solar Voltaic Generation Plant 93 miles by 93 miles array, such an SVG array would generate enough electrical energy to supply the entire National electricity consumption (2,854,168 gigawatt-hours, 1989). Furthermore, a mere 40% increase in that area of SVG Plant Array could charge the batteries in economically feasible fuel for the Nation's total automobile-miles if in the emerging 1.5 kwh/mile electric car.

Affordable, zero-cost-solar-fuel, delivered at the SVG Plant site, converted to electric power in solar voltaic cell panels mounted on a simple track and stand structure, should in large-scale manufacture (estimated worldwide 1990 Module production 52-55 megawatts) provide a plant cost under $1000 per kilowatt of installed nameplate capacity. This assuming that manufacturing costs and production quantities of the photovoltaic cells can soon be in range of a comparable electrical device, the "99 cent sale", 1/2 sq. ft. area, 4-foot fluorescent lamp, for cell modules at $40 to $75 per square meter.

SVG can go on line incrementally, revenue producing as fast as installed or replacement in an SVG plant approaching a century service life.
Amicable in the Environment --- benign, silent, self-contained process, solar photon energy is converted to electricity --- actually removing atmospheric carbon dioxide when in co-generation with crops. Nor does SVG entail the conventional generation environmental impacts of mining, refining, transporting the fuel, plant decommissioning and waste disposal. Attainable --- now well demonstrated, for a high expectancy of success; no research and development is needed beyond implementing vast SVG module production and the subsequent market driven improvements in the solar voltaic cell panels.

Design and large-scale construction presently hinge on an industry and government willingness to undertake: - 1) a National Energy Decency Strategy lease-cost-planning use of domestic solar fuel abundance of the solar-fuel applied in least-cost exceeding that of conventional fuels, toward the SVG Plants to displace most of the needed conventional generation also providing power for both highway and rail transportation, 2) a significant Federal Power Agency demonstration, such as to engineer a "Hydro-Solar Integration", such as doubling the present Bonneville Power administration capacity by 2001 with Solar Voltaic Generation. This is approximately 65 "ranches" of 2500 Acres (40,000 megawatts) with crops "cogeneration"; distributed along existing transmission corridors, constituting a renewal and enlargement of the Bonneville Power costs to expand the "yardstick" against the extortion of oil pegged electricity prices and for electrical energy decency for the national and the global environment.

Reasonable Capacity Costs

As large-scale markets, mass-production methods, and competition are defined, there is little doubt that the cost of a square meter of photovoltaic cells can in this decade reach the **PV Five Year Plan Goal** of $40 to $75; $80 to $150 for the proposed SVG Module --$266 to $500 per kilowatt of nameplate capacity. A comparable processing complexity and materials electrical device, the mass produced four foot fluorescent lamp (1/2 ft2 area), often retails for 99 cents; that being only $21 for a square meter area of tight packed fluorescents.

To the generator Modules add the modest costs of the land, an energy management system, O&M for an automated plant, and the charged-line structure to deliver the Joules to the power net and the plant costs will be near $500/kW-inc and 4 cents/kWh (at a utility fixed charges rate of 20%) and less for public utility assessment or pay-as-you-go financed plants.

With $30/BBL Oil becoming the world coin-of-the-realms --- and wars, consider the economics of SVG Generation in "barrels-equivalent". At 17 BBLs investment in a kilowatt of SVG capacity to generate 2200 kilowatt-hours annually, the yield is 3.6 barrels (600 kWh/BBL a normal oil-fired plant efficiency) for a 22% net annual return on investment. An Advanced-NUP Plant kilowatt of capacity at an expected 120 BBLs investment, even at a 60% capacity factor, yields only 8.7 BBLS (5256 kWh) for a 7% gross return, from which deduct four or five BBLs for fuel, decommissioning and millennia waste storage toward an unacceptable net return in any national energy strategy to displace oil. Oil- and coal-fired plants are simply no contest in coin-of-the-realms, providing a negative rate of return after deducting fuel BBL for BBL from the gross yield. - **Alfred H. Canada**

Pre-Packaged PV Power Systems

The beauty of solar electric power systems lies in their modularity. Solar panel installations are like building-block constructions that can always be added to and enhanced in the future. Unlike most consumer items that have obsolescence built in, you can use your solar panels virtually forever and add on as power needs and budget capabilities expand.

We'll be happy to sit down with you and plan out your system no matter how small or large, either in person or by phone. To make getting started easier, we've assembled some pre-packaged kits that are properly sized for various applications and that include the necessary hardware to bring you solar electricity. We've been selling these pre-engineered systems for two years now and have received rave reviews from purchasers.

Photograph courtesy Arco Solar

RV Kits

The following RV kits are packaged by Siemens Solar and are designed for minimal charging capacities (up to 14.6 amphours per day).

RV Kit 1

Usage: Trickle charger for starter battery
Components: Siemens G-50 solar panel, 25 ft wire, installation accessories & instructions.

Power produced: 12 watthours per day

12-101   RV Kit 1 $109

10" x 10" x ⅜"

RV Kit 2

Usage: Trickle charger for starter battery and some small appliances.
Components: Siemens T-5 solar panel, 25 ft wire, installation accessories & instructions.

Power produced: 25 watthours per day

12-102   RV Kit 2 $129

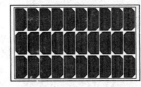

14" x 13" x ⅜"

RV Kit 3

Usage: Battery charger plus more small appliances
Components: Siemens M-20 solar panel, 25 ft wire, installation accessories & instructions.

Power produced: 110 watthours per day

22" x 13" x 1½"

12-103   RV Kit 3 $399

RV Kit 4

Usage: Battery charger plus power for small and large appliances.
Components: Siemens M-65 solar panel, 25 ft wire, installation accessories and instructions.

Power produced: 175 watthours per day

43" x 13" x 1½"

12-104   RV Kit 4 $539

Remote Home Kits

We've spent considerable time and energy fine-tuning our Remote Home (RH) Kits over the last four years. We've produced a comprehensive and easy to understand owner's manual and installation guide that includes complete instructions and safety guidelines for all the Remote Home Kits. We've refined the kits themselves to a state we're very happy with, including our much larger and deluxe Remote Home Kit 4 for larger power consumers who are looking for state-of-the-art equipment. All power figures are calculated at 5 peak sun hours per day. If you'd like to receive a copy of the owner's manual and preview our systems, please send in $5 (see book section).

Remote Home Kit 1 (RH-1)

Usage: Part-time summer or weekend cabin, larger RV, or boat. Will operate 12V sound system, 12V TV, 12V cassette player, 12V lighting, and some 12V water pumping.

Components:
- 2 ea. 48 watt PV modules
- 1 ea. M8 Bobier charge controller
- 1 ea. Two-panel mounting structure
- 2 ea. 6V x 220 amphour deep cycle battery
- 1 ea. #4 Battery interconnect
- 1 ea. 10-16V voltmeter
- 1 ea. 0-10A ammeter
- 1 ea. 30 amp, 2-pole disconnect (safety switch)
- 1 ea. Newark six position fuse box
- 1 ea. 12 position buss bar
- 1 ea. Remote Home Kit Owner's Manual

Power Produced: 360 watthours per day

12-201-KT   Remote Home Kit 1 (RH-1) $1095
Batteries are shipped freight collect

Remote Home Kit 2 (RH-2)

Usage: Full-time home using conservation measures like propane refrigerator/freezer, propane cook stove, propane water heater, and energy efficient DC lighting. System will operate lighting, 110V or 12V TV/VCR, satellite dish receiver, computer system, limited water pumping, sewing machine, drill, small vacuum, small microwave, juicer, blender, and any 110V appliance up to 600 watts for 30 minutes.

Components:
- 4 ea. 48-watt PV modules
- 1 ea. latitude adjustable four-panel mount
- 1 ea. Trace C-30A charge controller
- 1 ea. 30A 2-pole fused disconnect w/fuses
- 1 ea. UL-listed DC load center w/2 15A breakers
- 2 ea. 15 amp circuit breakers
- 1 ea. Two Channel Digital Volt-Amp Monitor
- 4 ea. 6 volt, 220 amphour deep cycle batteries
- 4 ea. #4 battery interconnects
- 1 ea. Trace 612 12V to 110V power inverter
- 1 ea. Remote Home Kit Owner's Manual

Power Produced: 720 watthours per day

12-202   Remote Home Kit 2 (RH-2) $2,795
12-212   RH-2 Kit with no Batteries $2,475
Batteries are shipped freight collect

Trace 2012 Upgrade - Many customers have found our RH-2 Kit to be properly sized for their needs but want to be able to run larger appliances than the Trace 612 inverter will allow. We offer the 2012 upgrade for an additional **$460** for those in that situation and we include two #4/0 five foot inverter cables.

12-222   RH-2 Kit w/Trace 2012 $3,275
Batteries shipped freight collect

Remote Home Kit 3 (RH-3)

Usage: Full-time larger home with more occupants and/or power requirements. Will power all household needs including vacuum, washer, microwave, large color TV, computer, and most power tools. Also water pumping including some deep well pumping.

Components:

    8 ea. 48 watt PV modules
    1 ea. eight-module mounting structure
    1 ea. Trace C-30A charge controller
    1 ea. UL-listed DC load center w/2 15A breakers
    2 ea. 20 amp circuit breakers
    1 ea. 30A, 2-pole fused disconnect w/30A fuses
    1 ea. Two Channel Digital Volt-Amp monitor
    ■ 8 ea. 6-volt, 220 Amphour deep cycle batteries
   10 ea. #4 gauge battery interconnects
    1 ea. Trace 2012 2,000-watt power inverter
    1 ea. Trace 2012 cable
    1 ea. 300A DC Fuse assembly
    1 ea. Remote Home Kit Owner's Manual

Power Produced: 1440 watthours per day

| 12-203 | Remote Home Kit 3 (RH-3) | $5190 |
| 12-213 | RH-3 Kit with no Batteries | $4595 |

All Batteries are shipped freight collect

ON SUBSTITUTIONS: We realize that everyone's needs are different and that our kits may not fit perfectly for all systems. With this in mind, we welcome substitution, wherever you think it's necessary. Many of you will want to add one of our generators (Makita or Onan) as a backup system. Please call us for a modified price quote and feel free to let us help engineer your system - we've got lots of experience! If you want to deduct an item from a Remote Home Kit, deduct its price less 10% from the total price of the kit. This is because the package price is discounted.

Remote Home Kit 4 (RH-4)

This is our state-of-the-art system using only the finest components - we didn't cut any corners for price!

Usage: Will power a full-scale home with the comfort level of a city dwelling hooked up to the power line. It will power vacuum cleaner, washing machine, microwave, large color TV, VCR, satellite dish receiver, computer and printer, most all power tools, and water pumping.

Components:

   15 ea. 48 watt PV modules
    2 ea. eight-module mounting structures
    1 ea. SCI Mark III Volt/Amp Meter
    1 ea. SCI Mark III Charge Controller
    1 ea. Double enclosure for above
    ■ 1 ea. Chloride 675N33 1,476 amphour battery
    1 ea. Trace 2012 inverter
    1 ea. Trace 2012 Cable
    1 ea. 300A DC fuse assembly
    1 ea. UL-listed DC load center w/12 circuit breakers
    2 ea. 20 amp circuit breakers
    1 ea. 60A fused disconnect w/fuses
    1 ea. Remote Home Kit Owner's Manual

Power Produced: 2700 watthours per day

12-204	Remote Home Kit 4 (RH-4)	$ 9995
12-234	RH-4 with Trace 2012SB	$10,195
12-237	RH-4 with no batteries	$ 7,495
12-238	RH-4 w/2012SB w/o batteries	$ 7,695

Batteries are shipped freight collect

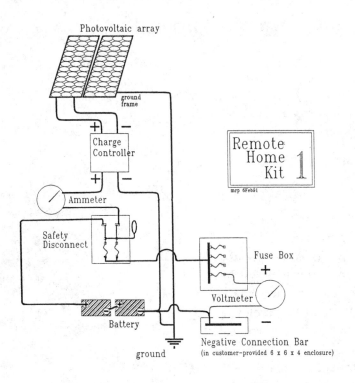

Photovoltaic array

Charge Controller

Remote Home Kit 1

mrp 6Feb91

ground frame

Ammeter

Safety Disconnect

Battery

Fuse Box

Voltmeter

Negative Connection Bar
(in customer-provided 6 x 6 x 4 enclosure)

ground

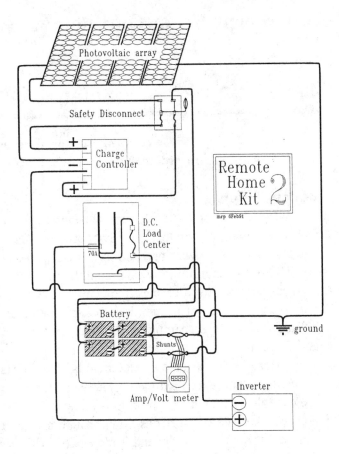

Photovoltaic array

Safety Disconnect

Charge Controller

Remote Home Kit 2

mrp 6Feb91

D.C. Load Center

70A

Battery

Shunts

Amp/Volt meter

ground

Inverter

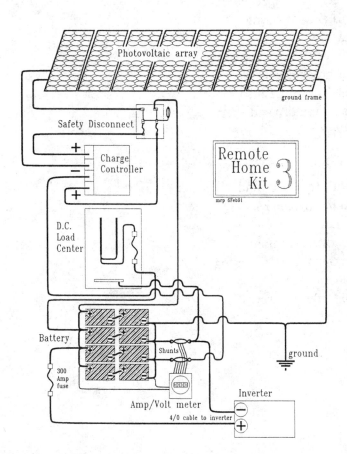

Photovoltaic array

ground frame

Safety Disconnect

Charge Controller

Remote Home Kit 3

mrp 6Feb91

D.C. Load Center

Battery

Shunts

300 Amp fuse

Amp/Volt meter

ground

Inverter

4/0 cable to inverter

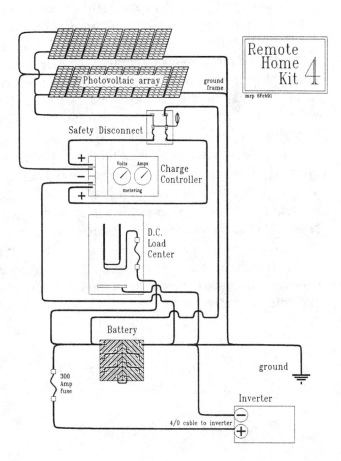

Photovoltaic array

ground frame

Remote Home Kit 4

mrp 6Feb91

Safety Disconnect

Volts Amps

metering

Charge Controller

D.C. Load Center

Battery

300 Amp fuse

ground

Inverter

4/0 cable to inverter

An Excerpt From Our Remote Home Owner's Manual

Safety

**Please read this section thoroughly
before proceeding to the installation sections.**

Battery Safety

Batteries are powerful devices. Properly installed and maintained, they will provide many years of safe power storage. But you must respect your batteries and take every precaution to assure their safe use. Although 12-volt DC power is inherently safer than its jittery cousin, 110-volt AC, it has its own dangers. Used carelessly or improperly, the technology can cause burns, explosions, not to mention back problems associated with carelessly hefting the batteries. Simple precautions, outlined below, will keep you safe and healthy.

Battery hazards - acid or electrical burns, fire hazards from poor connections and explosive battery gas - are the primary concern. Batteries are capable of moving an enormous amount of amperage through metal objects. A surge of electrons flow through a piece of metal in a short time will heat the metal rapidly and it may melt. This is especially dangerous on the unfused portion of the system (the battery terminals and all wiring between the battery and the circuit breaker or fuse box).

Burns and arcs are also possible in the fused section of the system, although this rarely happens because the fuse or circuit breaker will interrupt the circuit before much damage can be done. It is unusual to receive a shock from a 12-volt system. Keep hands clean and dry and there will not be much chance of shock, because the low voltage will not overcome the resistance of human skin.

Battery acid (the fluid inside the battery cells) is a highly corrosive blend of sulfuric acid and pure water. Avoid contact with the fluid, and keep it away from children. *Do not allow battery acid to touch skin, clothes, or eyes. See the precautions on the battery warning label, attached to the battery.* Wash the outside of the battery case every two or three months with a solution of baking soda and water. This will remove the acid coating left there by the acid mist venting from the battery cell caps during charging. **Do not get any of this cleaning solution inside the battery, it can ruin the battery cells.**

Install batteries in a well ventilated enclosure. A bottom air vent lets fresh air in and a top vent (at the highest point in the battery box) lets the explosive hydrogen gas out. [There is a sample plan for a battery enclosure in the appendix.]

Do not expose batteries to direct sunshine for long periods. The ultraviolet rays in sunshine will weaken the plastic of the battery case, possibly causing a cracked case and an acid spill.

Battery installation is only a small part of the total kit installation but it is a critical part of the system, and worth taking the time to do properly. Once the batteries are properly installed the only maintenance required is to check the battery fluid level every couple of months and add distilled water if necessary.

Electrical Safety

Follow these safe electrical practices regardless of the voltage of your system:

1. **Always turn off all power sources and loads.** Remove all fuses, turn off all circuit breakers, etcetera, before working on the battery or any other circuit.

2. **Cover the non-working end of all tools used around batteries with plastic electrical tape.** Wrap all exposed metal on wrenches except the part which contacts the nut or bolt. On screwdrivers leave only the tip exposed. Doing this will prevent the tool from short circuiting the battery. If a wrench or screwdriver were to accidentally drop across the battery terminals, the tool and the terminals could melt. The heat generated by the resulting short circuit could cause the battery to explode.

3. **Battery Acid Spills:** Keep several large boxes of baking soda and a large container of water near the battery. If battery acid spills sprinkle baking soda onto the spill. When foaming stops acid is neutralized. If battery acid contacts your clothes or body, rinse it off with plenty of water. Time is of the essence, **- the faster you flood the area with water, the less damage will be done.**

4. **Eye protection is mandatory** for anyone working on or around a battery. Wear a clear full-face shield. If you get battery acid in your eyes, immediately flush them with water for at least 15 minutes. Then seek medical attention.

5. **Wear rubber gloves** when working on or around batteries. Check gloves for holes each time before putting them on.

6. The temperature of a lead-acid battery should never be allowed to exceed 110°F. Temperatures above 110° can cause the battery plates to warp, possibly short out, and cause an explosion.

Why Fusing is Necessary

The primary purpose of the fuse is to prevent fire. It does this by protecting wiring from overheating and melting its insulation.

All battery circuits should be properly fused in order to protect wires from melting, causing an electrical fire during an overload or short circuit.

The size of wire used determines the amperage rating of the fuse needed to protect that wire.

Fuse Tables

Wire Size*	Fuse amperage
AWG # 14	15 AMP
AWG # 12	20 AMP
AWG # 10	30 AMP

All wires listed are copper.
AWG: American Wire Gauge

* Wire larger than chart listing may be used. Wire gauge shown represents minimum safe wire size.

NEVER USE A FUSE RATED FOR MORE AMPERAGE THAN THE WIRE CONNECTED TO IT.

Note: Two-Screw Box Connectors must *ALWAYS* be used to protect wires that pass through knock-out holes. This should be done not only to comply with the National Electrical Code, but for your own safety.

All buildings vibrate and wires passing across sharp metal edges will eventually short out when their insulation is cut into by the metal. This can cause a fire. Non-compliance with National Electrical Code standards may void your fire insurance.

(see drawing on page 32 for more information)

more Remote Home Kit excerpts on page 396

PV Mounting Structures and Trackers

*Here's an article by **Jon Vara** in Vermont on mounting and tracking photovoltaic arrays.*

Mounting A Photovoltaic Array

How do you plan to mount your solar array? If you're like most people, you'll spend less time thinking about that then about batteries, charge controllers, panels, and other flashy items. Choosing a suitable array mount, however, is just as important. It requires juggling a number of variables, including the latitude of the site, the prevailing weather, topography, and orientation of any buildings on the site. Consider the following questions before selecting a mount, and you'll be in a better position to make a wise choice.

To Track Or Not To Track?

Tracking array mounts are designed to keep the array aimed directly at the sun throughout the solar day, either automatically--as in the freon-charged Zomeworks trackers, and some DC motor-driven trackers--or manually, as in the passive Mantracks array mounts. By maximizing the amount of sunlight that strikes the panels throughout the day, a tracking mount can increase overall power production by as much as 50% during the summer months.

During the winter months, however, when the sun follows a shorter path across the sky, the increase will be much less. In the northern U.S., in fact, where winter days are short, the increased power production will be negligible for a good part of the year. If you live north of 40 degree latitude or so, and if you need more power in the winter--when the demand for lighting is greatest--a tracking mount may provide a power surplus in the summer without any significant gain in the winter. Under those circumstances, it is more cost-effective to spend the money that would otherwise go into a tracker on a larger fixed array.

Tracking is most productive where there is plenty of sun all year, especially when the greatest power demand occurs in summer. That is often the case where refrigeration or water pumping for irrigation are major loads. If that describes your situation, a tracking mount will more than pay its own way.

Fixed Latitude Or Adjustable?

In the same way that the ideal side-to-side orientation of solar array changes throughout the day, the optimum incline--or vertical angle--varies seasonally. At the spring and fall equinoxes--on about March 21st and September 21st--a solar array receives maximum exposure to the sun if its tilt corresponds to the local latitude. At the summer solstice, on June 21st, the ideal inclination is that of the local latitude plus approximately 15 degrees; at the winter solstice, on December 21st, it is the local latitude minus 15 degrees. (A solar panel lying flat on its back has an inclination of 0 degrees, while one standing upright on its edge has an inclination of 90 degrees.)

As a result, many array mounts are designed to allow easy tilt adjustment, enabling the user to increase the array output by correcting the angle of the array several times a year. That is simple and worthwhile in a ground-mounted array. In a less readily accessible wall--or roof-mounted array, however, hauling out the ladder four times a year may be more risk and aggravation than it is worth.

In such cases, it is best to fix the angle of the array so that solar gain is greatest when it is most needed. In the north, that will usually mean tilting the array steeply to make the most of the low winter sun; while an array dedicated primarily to summertime water pumping would best be placed at a low sun angle all year. You may choose to fix the array at the optimum angle for spring and fall production or the best year-round average.

Free-Standing Or Building-Mounted?

The most important advantage of a free-standing, ground-mounted array has already been mentioned---its accessibility. An array that is located close to the ground is easy to install--particularly for those made anxious by heights--simple to clean and inspect, and can readily be adjusted from season to season.

A ground-mounted array is not always a practical option, however. The required mounting structure may take up an unacceptable amount of space in your yard. The available space may receive too much shade from trees or nearby buildings, or may be to far away to allow efficient transmission of the power produced, particularly in 12-volt systems. Consequently, solar arrays are often on rooftops, where they will stand clear of obstructions and make for short transmission lines between the panels and the indoor battery bank. A roof-mounted array is also less conspicuous visually, in most cases, and less vulnerable to damage from batted baseballs or stones kicked up by the lawn mower.

Inevitably, though, there are a few drawbacks as well. Attaching the panel mount to the roof will require drilling holes in the roofing, which may cause leakage unless the holes are properly sealed or flashed. The problem is compounded in the north, where snow and ice buildup at the edge of the roof may cause meltwater to back up and work its way through unsealed bolt holes.

Worse, snow and ice can exert tremendous force against the array in sliding from the roof. That can be minimized by locating the array close to the ridge, where little snow will accumulate above it, but that will make it difficult or impossible to brush snow from the surface of the panels without climbing onto the roof. Mounting the panels so that they stand the long way will encourage the snow to slide off on its own. Even so, if a heavy snow is followed by a period of clear, cold weather, the panels may remain blanketed by snow for days on end, just when you need the power most.

If a wall of your house faces due south, or nearly so, you can avoid these problems by mounting the panels on the wall instead. When contemplating a wall-mounted array, however, make sure the panels will not project beyond the overhang of the roof, where they will likely be in the path of sliding snow and falling icicles. At the same time, they must be far enough below the overhang to avoid summer shade, and spaced far enough apart so that the uppermost panels do not throw shade on those below. Unless the roof has a very broad overhang, it is usually necessary to align the long axis of the panels with the edge of the roof to prevent them from projecting too far. (Snow will not slide off readily in that direction, but it is usually possible to sweep the panels clear from below with a long-handled broom.) Long,narrow panels, such as those made by Arco and Hoxan, are more easily tucked beneath the eaves than wider, shorter designs, such as those manufactured by Kyocera or Solarex. It's a good idea to choose the style of mount you plan to use before choosing your panels.

Build Or Buy?

The longevity of your solar array--probably the most expensive part of your solar-electric system--depends largely on the mounting structure that supports it. A dependable, commercially-made array mount is cheap insurance. Still, after sifting through all the variables above, you may be unable to find a manufactured mount that precisely fits your needs, or you may wish to build your own mount--or have it built--to save money.

One strong, versatile, and extremely simple way to build array mounts is with slotted steel angle, which consists of sections of 12-or 14 gauge steel angle with regularly spaced slots and holes already in place. It's usually purchased in ten-or twelve-foot lengths, which can be cut with a hacksaw and bolted together, erector-set fashion, with supplied nuts and bolts. To find a local supplier, look in the yellow pages under "Shelving," "Racks," or "Material Handling Equipment." Since the material is designed to be taken apart and re-used again and again, many dealers offer used angle at a discount.

Shelving and materials-handling dealers are also a good source for pipe rail fittings. These resemble the galvanized steel pipe fittings used in plumbing, but instead of being threaded onto the pipe, they simply slip into place and are secured with a set screw tightened with an allen wrench. Pipe mounts are generally more expensive than those fashioned from slotted angle, but they are neater in appearance and easier to adjust, thanks to the wide variety of hinged and sliding fittings available. Sizes range from 1/2" to 2" or more; for a typical residential solar array, 1" or 1-1/4" fittings are about right. -Jon Vara

[Technician's Note: Look at the dollar per watt price of tracking when deciding if tracking makes more economic sense than a fixed mount with maybe an extra module or two - **Jeff Oldham**]

Zomeworks Track Racks

The passive solar tracker will increase the output of your PV panels by up to 50%. There are no drive motors, gears, or pistons to wear out. They are as dependable as gravity and the heat of the sun. Easy to install and seasonally adjustable, the Track Rack comes with a 10-year warranty.

The north-south axis is seasonally adjustable for top performance all year long. All joints are silver soldered and there are no plastic drive components. The Zomeworks track rack is constructed of painted mild steel. Stainless steel racks are available for extra cost. The racks are designed to withstand 30 lb/sq ft wind loading. The specially-made shock absorbers dampen motion in high winds.

Tests conducted over a 12-month period by New Mexico State University at 34° latitude showed an improvement in electrical output by tracking around an adjustable N-S axis of 29% over a fixed latitude mount. The improvement ranged from 19% in November to 42% in June and July. This means that using an eight-module track rack is like having 1.6 extra modules in the winter and 3.2 in the summer, or an average of 2.3 extra modules for the year.

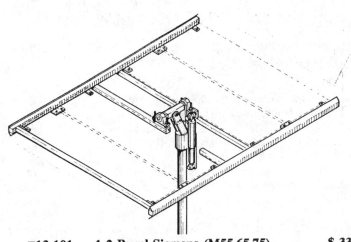

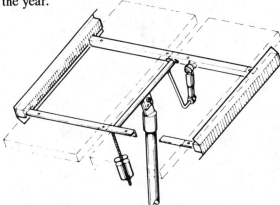

■13-101	1-2 Panel Siemens (M55,65,75)	$ 330
■13-111	1-2 Panel Hoxan	$ 350
■13-121	2 Panel Kyocera (K51/K63)	$ 350
■13-131	1-2 Panel Solarex (MSX60)	$ 350
■13-102	4 Panel Siemens (M55,M75)	$ 540
■13-112	4 Panel Hoxan	$ 545
■13-122	4 Panel Kyocera K51	$ 545
■13-123	4 Panel Kyocera K63	$ 690
■13-132	4 Panel Solarex (MSX60)	$ 690
■13-103	6 Panel Siemens (M55,M75)	$ 670
■13-113	6 Panel Hoxan	$ 625
■13-126	6 Panel Kyocera K51	$ 625
■13-127	6 Panel Kyocera K63	$ 820
■13-133	6 Panel Solarex (MSX60)	$ 805
■13-104	8 Panel Siemens (M55,65,75)	$ 805
■13-114	8 Panel Hoxan	$ 825
■13-128	8 Panel Kyocera K51	$ 825
■13-124	8 Panel Kyocera K63	$1,020
■13-134	8 Panel Solarex (MSX60)	$ 990
■13-105	12 Panel Siemens (M55,65,75)	$ 990
■13-115	12 Panel Hoxan	$1,050
■13-125	12 Panel Kyocera K51	$1,050
■13-129	12 Panel Kyocera K63	$1,250
■13-135	12 Panel Solarex (MSX60)	$1,240
■13-106	14 Panel Siemens (M55,65,75)	$1,175

(all Track Racks shipped freight collect from New Mexico)
Stainless Steel Track Racks are available. Call for quote

■13-171	Marine Bearings	$20

This graph and table show the results of a year-long test which started in summer 1983. The test was conducted by the New Mexico Solar Energy Institute at the Southwest Residential Experimental Station.

AVERAGE DAILY ENERGY OUTPUT (WATT-HOURS)

	TRACKING Adj. Elev∗ Angle	FIXED Adj. Elev∗ Angle	FIXED Fixed Elev Angle	TRACKING Fixed Elev Angle
JUL 1983	270	207	184	242
AUG 1983	263	204	195	252
SEP 1983	219	179	178	219
OCT 1983	209	172	172	204
NOV 1983	226	196	185	215
DEC 1983	183	160	149	168
JAN 1984	193	167	155	177
FEB 1984	255	207	197	244
MAR 1984	271	211	209	265
APR 1984	304	224	219	296
MAY 1984	286	217	200	267
JUN 1984	250	194	172	225
YEAR LONG DAILY AVERAGE	**244**	**195**	**185**	**231**

∗ 6 Adjustments in one-year test period

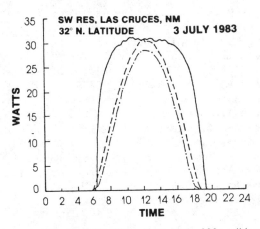

SW RES, LAS CRUCES, NM
32° N. LATITUDE   3 JULY 1983

———— Tracking, tilt = 5° South, Output = 338 watt-hours
– – – – Fixed, tilt = 5° South, Output = 241 watt-hours
–·–·– Fixed, tilt = 32° South, Output = 212 watt-hours

Rapid Response West Canister Shadow Plate

Zomeworks has developed this shadow plate which speeds morning tracker wake-up. It's available for track racks in the field and as an option for new track racks. It uses a special 3M reflective film adhered to a wider shade to reflect more sunlight onto the canister. We have found that this special shadow plate will cut wake-up time by approximately 40%.

■ **13-161 Shadow Plate 8-12-14 module racks $115**
■ **13-162 Shadow Plate 4-6 module racks $ 95**
■ **13-163 Shadow Plate for 2 module racks $ 75**
Shipped freight collect from New Mexico
You must specify the mount and type of module you have!

Roof-Mounted Zomeworks Track Racks

The new Zomeworks Track Rack for roof-mounted photovoltaic arrays are made with "driver" and "slave" axles which hold four modules each. Each driver axle can operate itself plus one slave axle. Pairs of driver and slave axles are linked together on a roof. A minimum of four feet is required between racks on the roof. Roof-mount trackers are made to mount at the angle of the existing roof slope. If a different elevation angle is desired, custom end-stands will be required.

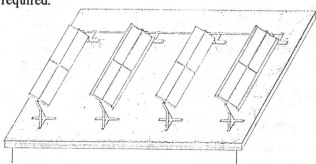

Example: To track 4 modules order one driver axle; to track 8 modules, order one driver axle and one slave axle; to track 12 modules, order two driver axles and one slave axle.

■ **13-201 Driver Axle for Siemens $605**
■ **13-202 Slave Axle for Siemens 250**
■ **13-203 Driver Axle for Hoxan 645**
■ **13-204 Slave Axle for Hoxan $365**
■ **13-205 Driver Axle for Kyo K-51 $665**
■ **13-206 Slave Axle for Kyo K-51 $295**
■ **13-211 Driver Axle for Solarex 665**
■ **13-212 Slave Axle for Solarex 295**
All Shipped freight collect from New Mexico

Top-of-Pole Fixed Zomeworks Racks

Zomeworks makes a standard pole-mounted fixed rack of painted hot rolled steel and heavy-duty pipe gimbals with horizontal and vertical axes of rotation which are locked in place by hex-head set-bolts. These racks are easier to install than lean-to style racks because careful alignment is not necessary. You can spin the rack on top of the pole to face it south. Stainless steel module mounting hardware is provided with each rack. Poles are not included. *Item numbers below show how many and which module the rack will hold.*

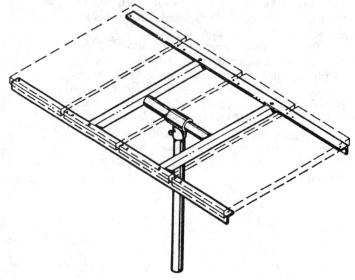

■ 15-400	1 Siemens	$ 60
■ 15-401	2 Siemens	65
■ 15-406	2 Hoxan or K-51	72
■ 15-407	2 MSX-60 or K-63	82
■ 15-408	3 Siemens	145
■ 15-402	4 Siemens	155
■ 15-409	3 Hoxan or K-51	165
■ 15-410	3 MSX-60 or K-63	175
■ 15-411	4 Hoxan or K-51	170
■ 15-412	4 MSX-60 or K-63	190
■ 15-403	6 Siemens	185
■ 15-413	6 Hoxan or K-51	200
■ 15-414	6 MSX-60 or K-63	275
■ 15-404	8 Siemens	270
■ 15-415	8 Hoxan or K-51	295
■ 15-416	8 MSX-60 or K-63	325
■ 15-405	12 Siemens	460
■ 15-417	12 Hoxan or K-51	490
■ 15-418	12 MSX-60	525
■ 15-419	14 Siemens	515

All Shipped freight collect from New Mexico

Specifications For Pole Mounted Track Racks (TRPM) & Top of Pole Racks (FRPT)

	model	steel pole size (O.D.)	minimum pole height above ground	hole depth	hole diameter (fill w/concrete)
TRPM01,02/ AR, SX, HX, KY	FRPT01/AR	2" SCH 40 (2.38")	48"	30"	10"
	FRPT02/AR 01/SX 01/KY 01/HX	2" SCH 40 (2.38")	48"	30"	10"
	FRPT02/SX 02/KY 02/HX	2" SCH 40 (2.38")	48"	30"	10"
TRPM04/ AR, HX, KY51	FRPT04/AR 03/SX 03/K51 03/HX	2.5" SCH 40 (2.88")	48"	32"	12"
	FRPT04/K51 04/HX	2.5" SCH 40 (2.88")	48"	32"	12"
TRPM06/ AR, HX, KY51	FRPT04/SX 06/AR 04/K63	3" SCH 40 (3.5")	48"	35"	12"
TRPM04/ SX, KY63	FRPT06/K51 06/HX	3" SCH 40 (3.5")	52"	35"	12"
TRPM08/ AR, HX, KY51	FRPT08/AR 06/SX 06/K51	4" SCH 40 (4.5")	56"	44"	16"
TRPM06/ SX, KY63	FRPT08/K51 08/HX	4" SCH 40 (4.5"	58"	44"	16"
TRPM12/ AR, HX, KY51	FRPT12/AR 08/SX 08/KY63	5" SCH 40 (5.56")	66"	50"	18"
TRPM08/ SX,KY63	FRPT12/K51 12/HX	5" SCH 40 (5.56")	66"	50"	18"
TRPM14/AR TRPM12/SX	FRPT14/AR 12/SX	6" SCH 40 (6.63")	66"	52"	20"

AR = Arco (Siemens)
SX = Solarex
HX = Hoxan
KY = Kyocera (or "K")

Ground or Roof Mount Fixed Zomeworks Racks

Zomeworks has added to its line of mounting structures these sturdy, economical racks made of heavy gauge mild steel, painted with a satin black urethane paint. Their telescoping struts offer quick and easy seasonal adjustment from 15 to 65 degrees. They're designed to withstand 30 lb/sq ft wind loading. Stainless steel module mounting hardware is provided with each rack. The customer or installer provides the anchor bolts. They are designed for ground or roof mounting.

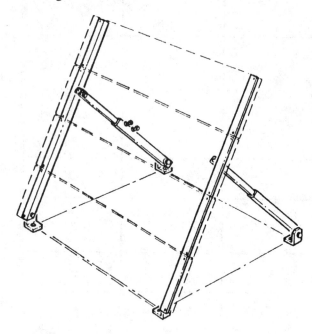

In the first series of fixed mounts the modules are mounted with their length horizontal and they are stacked vertically. Listed below are the number and type of modules each rack will handle. The K series refers to Kyocera K-51s or K-63s.

■13-501	2 Siemens	$ 88
■13-502	2 Hoxans, 2 K51, or 2 K63	105
■13-503	2 Solarex	108
■13-504	3 Siemens	105
■13-505	3 Solarex	117
■13-506	4 Siemens	115
■13-507	4 Hoxans, 4 K51, or 4 K63	120
■13-508	4 Solarex	126
■13-509	6 Siemens	126
■13-510	5 Hoxan or 5 K-51	126
■13-511	6 Hoxan or 6 K-51	220
■13-512	5 Solarex or 5 K-63	220
■13-513	6 Solarex or 6 K-63	260

The next series of mounts is the "low-profile" series. Module length is vertical and they are stacked side-by-side.

■13-521	8 Siemens (Low Profile)	210
■13-522	6 Hoxans or 6 K-51 (Low Profile)	225
■13-523	6 Solarex or 6 K-63 (Low Profile)	240
■13-524	9 Siemens (Low Profile)	250

The last series of mounts stack the modules vertically. They are stacked half on top and half on the bottom.

■13-531	8 Siemens	$275
■13-532	6 Solarex or 6 K-63	275
■13-533	8 Hoxan or 8 K-51	290
■13-534	10 Siemens	380
■13-535	12 Siemens	420
■13-536	12 Hoxans or 12 K-51	440
■13-537	12 Solarex or 12 K-63	470
■13-538	14 Siemens	470

All Shipped freight collect from New Mexico

Side-of-Pole Zomeworks Racks

These racks are adjustable from 0° to 90°. Two hose clamps securely hold the rack to the side of a 1-1/2" to 3" OD pipe. The item codes below with the prices specify which modules each rack will hold.

■13-301	1 Siemens M-20	$ 52
■13-311	1 Siemens M65-M75-M55	65
■13-315	1 Hoxan or K-51	70
■13-316	1 MSX-60 or K-63	75
■13-321	2 Siemens M65-M75-M55	110
■13-322	2 Hoxan or K-51	130
■13-323	2 MSX-60 or K-63	135
■13-331	3 Siemens M65-M75-M55	145
■13-341	4 Siemens M65-M75-M55	175

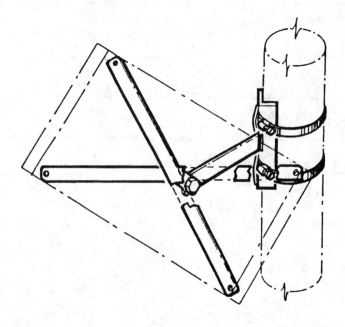

Photograph courtesy Arco Solar

Solarpivot

Solarpivot is a simple, sturdy pole mount for your PV panel(s), which allows you to manually track the sun. Whether you wish to adjust your panels throughout the day, or merely seasonally, this mount is effective and inexpensive. It's constructed completely out of structural aluminum up to 3/8" thick with all stainless steel bolts that securely mount to any post (steel or wood). It is engineered to withstand 70 mph winds. It pivots without touching pins, bolts, or wingnuts, moving easily to aim it anywhere you want. The unit is shipped fully assembled, except for the aluminum arms that the panels bolt to. Mounted by a frequently used path or door, it takes only seconds to track the sun for maximum energy production. No tools are needed. Solarpivots are currently in use from Hawaii to New York and from Texas to Alaska. The Solarpivot currently fits 1-4 Siemens (Arco) M-series modules, 1-3 Solarex, or 1-3 Kyocera K-series modules. **You must specify which modules when ordering.**

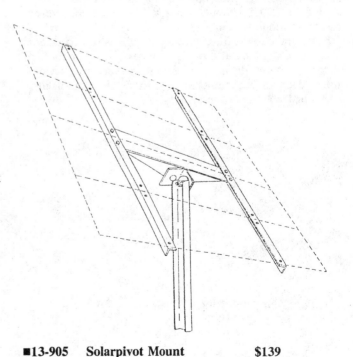

■13-905   **Solarpivot Mount** $139

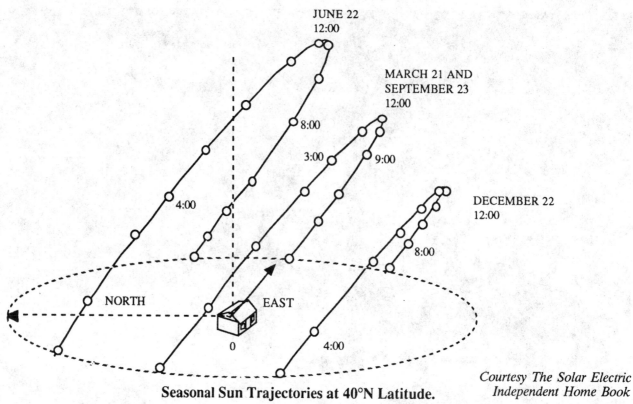

JUNE 22
12:00

MARCH 21 AND
SEPTEMBER 23
12:00

8:00

DECEMBER 22
12:00

3:00   9:00

4:00

8:00

NORTH   EAST

0   4:00

Seasonal Sun Trajectories at 40°N Latitude.

*Courtesy The Solar Electric
Independent Home Book*

Aluminum Solar Mounts

For a simple installation, it's hard to beat the economy of
our aluminum PV mounting structures. Each mount will
hold up to four Siemens (Arco) M-Series modules, three
Kyocera modules, 3 Solarex modules (MSX-53, 56, 60), or
3 Hoxan modules. Aluminum angle rails adjust to three
seasonal tilt angles. It mounts on a deck, wall, or roof. All
stainless steel bolts and hardware is included to attach
your solar panels. The installer provides his own anchor
bolts. Instructions are included.

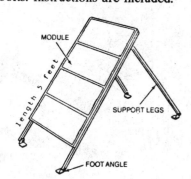

MODULE

length 5 feet

SUPPORT LEGS

FOOT ANGLE

13-701   Solar mount for 4 Modules $105

Dear Sirs,
      I love the idea behind your company. Here's one
of the many solar powered navigational markers on Rainy
Lake, Northern MN bordering Canada. Use it if you
want. All this summer, the 12 watt panel I purchased
from you, kept up both deep cycle batteries on my sail
boat. Thanks. This summer coming up I'll be developing
my land. I'll be going with all solar power.I'll be in touch
about that.
- J. Wydra, Keewatin, MN

EchoLite High Security Solar Module Mount

The EchoLite is new, a very rugged and well designed mounting structure that will accommodate any one of the Arco, Kyocera, or Solarex modules. Each kit consists of two pairs of extruded, corrosion-resistant aluminum with integral stainless steel fasteners. One pair attaches to the PV module and the other to any flat surface, such as a wall, roof, or deck. EchoLite makes a mount available for two modules as well.

The 1000 series is for flat non-adjustable mounting and the 2000 series is for adjustable tilt. EchoLite is available for a number of different modules. EchoLite is easy to assemble, adjust, secure or move. Each bracket is designed for stand-alone or array mounting. Additional base brackets are available for multi-use applications. The Locking Fastilt (LF) option provides both good security and instant tilt angle adjustment. The Tamperproof (T) option provides high security and comes with its own specialized tool. We find that 30% of our customers order the Locking Fastilt option with their units. **All Echolite mounting systems now come with a full 10 year** warranty.

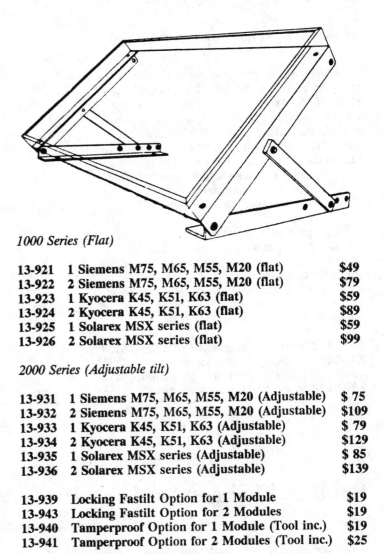

1000 Series (Flat)

13-921	1 Siemens M75, M65, M55, M20 (flat)	$49
13-922	2 Siemens M75, M65, M55, M20 (flat)	$79
13-923	1 Kyocera K45, K51, K63 (flat)	$59
13-924	2 Kyocera K45, K51, K63 (flat)	$89
13-925	1 Solarex MSX series (flat)	$59
13-926	2 Solarex MSX series (flat)	$99

2000 Series (Adjustable tilt)

13-931	1 Siemens M75, M65, M55, M20 (Adjustable)	$ 75
13-932	2 Siemens M75, M65, M55, M20 (Adjustable)	$109
13-933	1 Kyocera K45, K51, K63 (Adjustable)	$ 79
13-934	2 Kyocera K45, K51, K63 (Adjustable)	$129
13-935	1 Solarex MSX series (Adjustable)	$ 85
13-936	2 Solarex MSX series (Adjustable)	$139

13-939	Locking Fastilt Option for 1 Module	$19
13-943	Locking Fastilt Option for 2 Modules	$19
13-940	Tamperproof Option for 1 Module (Tool inc.)	$19
13-941	Tamperproof Option for 2 Modules (Tool inc.)	$25

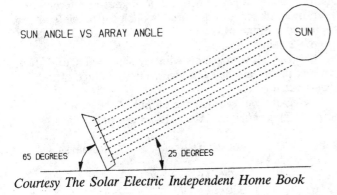

SUN ANGLE VS ARRAY ANGLE

SUN

65 DEGREES   25 DEGREES

Courtesy The Solar Electric Independent Home Book

RV Mounting Brackets

Here is a very simple mounting bracket for the top of a recreational vehicle. Two Z-brackets are made of 14 gauge anodized aluminum and will hold one Arco module (or any other module if you re-drill the holes).

13-901   RV mounting brackets (2) $17

Mantracks Manual PV Trackers

Mantracks is a very rugged manually operated sun tracking device. It will adjust to the most efficient angle of incidence at any time of day, all year, anywhere in the world, and provides a cost-effective alternative to the Zomeworks automatic freon tracker. The only tool required to maintain it is a crescent wrench, and it can be locked in place to withstand extremely high winds or to avoid hail damage. It is designed to slide on a 3" ID steel pipe cemented in the ground.

By mounting an ammeter on the Mantracks hub, you can get a graphic demonstration of the effects of tracking watching the ammeter needle vary with the angle of the sun. By manually moving the Mantracks three times per day you will get 90% of your optimum output. **Lifetime Guarantee.**

13-601 2 Module Mantracks (No holes) $230
13-602 2 Module Mantracks (W/holes) $250
13-603 4 Module Mantracks (No holes) $250
13-604 4 Module Mantracks (W/holes) $265

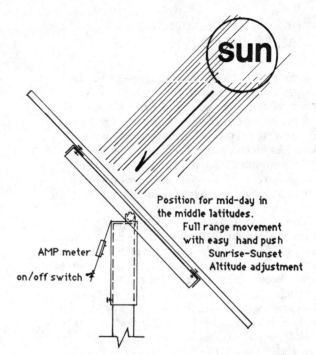

Please specify module type if you want the holes drilled. Inquire for larger mounts for up to 8 modules.

Random Thoughts on Solar Dependence

The following is a short article submitted by Eric Nashlund of Copper Center, Alaska. It deals with problems that Eric experienced in his alternative energy childhood. Now that he has entered AE adolescence, he thought he could help novices avoid his foolish mistakes. Eric points out that ignorance is not bliss in this dawning age of energy consciousness.

Being a habitual experimenter, it was inevitable that I would hook up to the sun. Three years have gone by and now I have more questions than answers. Perhaps, I can shed some light and prevent others from duplicating my mistakes.

Electricity is an addictive drug. We grow up with "Reddy Kilowatt" providing an enviable standard of living on the world scale. Perhaps, we are slightly twisted at birth. However, society is slowly starting to notice those of use who choose to live light on the land.

Still, we had better heed this addiction and design our PV systems to protect our investment and reduce the additional demands on our natural resources. You can pay now or pay again later. YOU WILL EXPAND YOUR SYSTEM.

Buy the largest wire you can afford. One, you will reduce voltage losses to a minimum. Two, you have prepared for future expansion of your system. From the panels to the controller, I recommend nothing smaller than #8 gauge; preferably #4 gauge. The same goes for the controller to battery bank. House wiring should be no smaller than #12 gauge with the #10 gauge preferable. Keep in mind, leave some circuits open for possible larger appliances. Some of these will draw 10-50 amps requiring up to #2 gauge wiring.

Buy all your panels with the same voltage. You can mix them, but they will give you the voltage of the lowest rated panel.

Don't believe that the output of the panel is gospel. Right now, my Arco M-75's are producing:

Volts	Location	Remarks
13.34	Safety Disconnect	with load
13.16	Batteries	
21.2	Safety Disconnect	no load

Please note, I'm getting 7.2 amps out of 9.06 typical. The weather is -33 degrees C, clear and sunny. For those unfamiliar with the M-75, its typical no load voltage 19.8 volts.

As you can see, PV systems are dynamic and perform different from the specifications. I'm' not condemning the specs; just use them as a guideline in designing your installation. If you use the typical amperage output times 80 percent, you will be close to the average output.

Buy the right batteries the first time. I bought a "Cat" battery first. It worked, but I couldn't let the voltage drop below 12.5 volts. I spent $150 for roughly 25 amp-hours. I finally bought a pair of Trojan L-16s for $370 and I get about 175 amp-hours usable. For the sake of your batteries, don't cycle them more than 50 percent. They will last a lot longer. [Ed. Note: Batteries will last longer if only cycled to 50% depth of discharge, but will actually end up costing more per usable amphour. The accepted belief is that cycling to 75-80% depth of discharge is the most cost-effective way to use lead acid batteries]

Invest in meters. These relatively inexpensive devices will allow you to keep tabs on the performance of your system. I recommend, at a minimum, a voltmeter across your batteries and an ammeter in series with your panels. For trouble shooting, get a good digital volt-ohm meter. These are available for under a hundred dollars.

When you buy your controller, buy big. My first controller was a 10 amp. Later, I had to buy a 30 amp to be able to expand. Fortunately, I was able to use the 10 amp controller at a cabin I'm building. Your other option is to parallel controllers to get the amp capacity you need.

An inverter is a large investment that is soon outgrown. Always plan for expansion. My inverter is a Heart 300X that is no longer made. It runs the XT compatible computer I'm typing this on. It will not run power tools larger than a small drill. You can get by on a small inverter to start with, but leave room to install a large one. You will find, for small loads, a small inverter is more efficient. Always keep in mind, an inverter's continuous output is usually 2/3's of its rated capacity. Some inverters are rated by continuous output. I have a friend that burned out his big-name brand inverter three times in three years before learning this. Fortunately, it was under warranty.

When you buy light, TANSTAAFL [there ain't no such thing as a free lunch, ed.] Go directly to fluorescent or PL lighting. Yes, I know they're expensive, but they make up for it in efficiency and lifespan. Quartz-halogen and Hi-intensity lights will get you by for task lighting, but at a high energy cost per light output. If you are lighting a large area, use large lights. These put out more light per watt then smaller one. As an example, my 8 watt fluorescents put out 400 lumens and my 13 watt PL's put out 860 lumens. I can now get by with 2 PL's instead of 4 fluorescents and I save 6 watts. Granted, this isn't a great savings, but energy conservation is just as important in an alternative energy home as the on-line home.

In the three years I've been PV, I've lost four "Thin-lite" and two "Sunalex" ballasts. The "Thin-lites" lasted 1-3 years and the "Sunalex" lasted two weeks. Being an experimenter, I tried to solve this problem. I found the problem to be voltage. Using a 6 amp car battery charger, an "Interstate" deep cycle gell cell battery and a digital volt-ohm meter, I recorded the following:

TIME		VOLTAGE
10:55 AM	12.6	BATTERY VOLTAGE
	12.9	CHARGER CONNECTED
1:30 PM	13.97	
2:10 PM	14.34	
2:30 PM	14.51	CHARGER DIS-CONNECTED
3:00 PM	13.00	
3:04 PM	14.21	CHARGER RECONNECTED
4:04 PM	14.89	CHARGER DISCONNECTED
4:37 PM	13.03	BATTERY FLOAT VOLTAGE

As you can see, the battery voltage can exceed 14 volts during charging. This is true for chargers and PV controllers. Upon checking the write-up on "Sunalex ballasts in the "Real Goods Sourcebook," I found that these ballasts would be destroyed by voltages in excess of 14 volts. The mystery was over. I usually charge at night when lighting is required. I had inadvertently burned out the "Sunalex" ballasts and probably the "Thinlites". So, do your charging in the daytime or pay for the ballasts twice.

I recently ordered an industrial battery charger to take the place of my automotive type. After careful examination, I fully believe industrial chargers are more cost and energy efficient in an alternative energy household. Your generator will run a substantially less time and pay OPEC a smaller price for a depleting resource.

I have named several manufacturers in this article. I can endorse these businesses if you use their products in accordance with their specifications. The user needs to exercise common sense and read the specifications to determine if a product will be suitable for your application. These manufacturers can't be expected to protect you from your willing ignorance. Caveat Emptor is very true for the ignorant.

- Eric Nashlund, Copper Center, AK

Wind Generators

We've found over the years at Real Goods that wind generation captivates intense interest, but there are relatively few situations where it is actually practical. The main reason for this is insufficient wind speed. With photovoltaic prices inching slowly downward, and with the ultimate noiseless, maintenance-free performance of PV's, we only sell about one wind generator per 100 solar panels. Nevertheless if you live in a high wind area, it makes good sense to utilize this valuable resource.

There are some important considerations in installing a wind generator: site evaluation, energy conservation, and proper choice of equipment.

Site Evaluation

First inquire about local building code requirements. Permits are required in many communities for structures higher than 20-25 ft. Wind plants make noise, so check with your neighbors also. If you are in close proximity to your neighbor, a little discussion may go a long way towards eliminating future problems.

For a wind generator to be useful, you will need an average monthly wind speed of 8 to 14 miles per hour (mph) at your site. Small wind plants like the Windseeker II will require a minimum of 11 mph average monthly wind speed.

Generally, your wind plant will be mounted on top of a sturdy tower 30 to 85 feet high in order to take maximum advantage of prevailing winds. Roof mounting of your system is not recommended. A wind generator

will transmit a great deal of vibration throughout a building and one can easily lose a roof in high winds.

An anemometer is the instrument used to measure wind speed. As the anemometer turns at the top of your temporary mast, it measures the distance a column of air moves over your site in a given period of time and registers the count on a digital meter. Daily readings are advisable and totaled each month for at least three months.

The government weather stations in your area can provide average monthly wind speed figures, comparable to your own, but for the whole year. Contact at least three weather stations in your area and arrange to compare your records to theirs. Starting with your first month, compare their averages to yours. Then you can find another figure that is the ratio between your monthly wind speed and the average of all their monthly wind speeds. As an example: if your winds during a three-month period are 1.30 times as great as the average of the three closest weather stations for the same month, then you know your average monthly wind speed is 1.30 times greater than their combined average wind speed. If those three weather station reports average 8 mph, the average monthly wind speed at your site will be 10.4 mph (8 mph x 1.30 = 10.4 mph), do this averaging for each of the 12 months in the year.

You will also want to know the highest wind speed gust likely to be experienced at your site because your wind generator and tower must be built to withstand violent winds. Your local weather station can also supply data on storm winds and gusts. Let's assume the fastest mile per hour wind in the last 25 years at these stations was 60. With the 1.30 adjustment ratio for your site, your fastest gust in the same time can be calculated using the following formula: 60 x 1.30 x 1.33 = 103.74 mph. (1.33 is the "gust constant" figure used by weather stations.) Your chance of experiencing a nearly 100 mph wind within the next 25 years is about 100%. Therefore, your wind generator and tower support must be designed for safety and insurance purposes.

If you have trouble assembling enough weather information in your area, surface weather observations from local airports, the Coast Guard, and forestry people can be obtained on microfiche for a moderate charge from the U.S. Department of Commerce, Environmental Data and Information Service, National Climatic Center, Asheville, NC 28801. A microfiche film report is much cheaper than its printed counterpart. Microfiche can be viewed on a special machine that most Ace Hardware and auto parts stores have. One of these local store owners should be amenable to letting you view the data on their machine.

Before selecting the equipment you need, here are some additional guidelines regarding the site that must be met:

1. Generally, the wind plant should be located within 100 feet of the house because of potential voltage drop in the transmission lines of DC power generators.

2. Nothing should be close that would interfere with maximum winds. Extend the tower high enough so that the wind plant reaches 20 feet above the nearest obstruction for at least 500 feet around. This will reduce or eliminate turbulence from trees and roof tops which can negatively affect the output of a wind generator. Turbulence not only negatively affects output, but greatly multiplies stresses on the machine. The constant shifting of direction causes gyroscopic stresses that will compromise both reliability and safety. A valley may appear to be a good place to trap downhill winds, but they usually come from only one direction. Depending on seasonal winds may be all right if you have photovoltaic, hydro, or some other alternative charging device interfaced with your system, but it won't do if you plan for wind as your only power source.

3. Towers must be strong and must be properly guyed with wire and grounded against lightning strikes. A wind generator's rated output is usually for sea level air density. Higher altitudes have lower air density and require higher wind speeds to achieve a given output. Temperatures also have some effect on output. This density ratio altitude (DRA) chart will help you to get some idea of the real output of our wind generators according to your site's altitude. We are using 60° F. to provide a typical norm.

Altitude, Sea Level	DRA (60° F.)
2,500 ft	1.000
5,000 ft	0.912
7,500 ft	0.756
10,000 ft	0.687

From the chart, take the DRA figure nearest your own site's altitude and multiply it by the rated output of a wind generator. As an example, a Winco rated at 450 watts at a given wind speed at sea level would deliver 340 watts at 7,500 feet altitude (450 W x 0.756 DRA = 340.2 W).

You should equip any wind generator with a blocking diode and proper voltage regulator to provide battery protection. The diode will prevent the generator from becoming a motor which would deplete the batteries when there is no wind. The SCI photovoltaic regulator cannot be used to regulate a wind generator.

Although a wind-plant's tower would appear to be a good grounding system, it is not. The concrete base and anchor points are poor conductors, so the tower needs to be grounded with an 8-ft copper grounding rod and connecting strap.

Water pumping windmill in Tanzania, Africa. Photo by Sean Sprague

Wind Generators

The following article was submitted by Mike Bergey of Bergey Windpower. It offers a very thorough introduction to wind energy.

A Little History

The wind has been an important source of energy in the U.S. for a long time. The mechanical windmill was one of the two "high-technology" inventions (the other was barbed wire) of the late 1800's that allowed us to develop much of our western frontier. Over 8 million mechanical windmills have been installed in the U.S. since the 1860's and some of these units have been in operation for more than a hundred years. Back in the 1920's and 1930's, before the REA began subsidizing rural electric coops and electric lines, farm families throughout the Midwest used 200-3,000 watt wind generators to power lights, radios, and kitchen appliances. The modest wind industry that had built up by the 1930's was literally driven out of business by government policies favoring the construction of utility lines and fossil fuel power plants (sound familiar?).

In the late 1970's and early 1980's intense interest was once again focused on wind energy as a possible solution to the energy crisis. As homeowners and farmers looked to various electricity producing renewable energy alternatives, small wind turbines emerged as the most cost effective technology capable of reducing their utility bills. Tax credits and favorably federal regulations (PURPA) made it possible for over 4,500 small, 1-25 kW, utility-intertied wind systems to be installed at individual homes between 1976 and 1985. Another 1,000 systems were installed in various remote applications during the same period. Small wind turbines were installed in all fifty States. None of the small wind turbine companies, however, were owned by oil companies or other industrial giants, so when the federal tax credits expired in late 1985, and oil prices dropped to $10 a barrel two months later, most of the small wind turbine industry once again disappeared. The companies that survived this "market adjustment" and are producing small wind turbines today are those whose machines were the most reliable and whose reputations were the best.

Windfarms

Starting in the early 1980's larger wind turbines were developed for "windfarms" that were being constructed in windy passes in California. In a windfarm a number of large wind turbines, typically rated between 100-400 kW each, are installed on the same piece of property. The output of these units is combined and sold under contract to the utility company. The windfarms are owned by private companies, not by the utilities. Although there were some problems with poorly designed wind turbines and overzealous salesman at first, windfarms have emerged as the most cost effective way to produce lots of electrical power from solar energy. There are now over 16,000 large wind turbines operating in the California

windfarms and they produce enough electricity to supply a city the size of San Francisco. Large wind turbine prices are coming down steadily and even conservative utility industry planners project massive growth in windfarm development in the coming decade, most of it occurring outside California. One recent study actually called North Dakota the "Saudi Arabia of wind energy." With the federal governments "hands-off" energy policy, however, a key question is whether the thousands of large wind turbines that will be installed in the years ahead will be built in the U.S. or imported.

The Cost Factor

Photovoltaic technology is attractive in many ways, but cost is not one of them. Small wind turbines can be an attractive alternative to those people needing more than 100-200 watts of power for their home, business, or remote facility. Unlike PV's, which stay at basically the same cost per watt independent of array size, wind turbines get less expensive with increasing system size. At the 50 watt size level, for example, a small wind turbine would cost about $9.00/watt compared to $6.60/watt for a PV module. This is why, all things being equal, PV is less expensive for very small loads. As the system size gets larger, however, this "rule-of-thumb" reverses itself. At 250 watts the wind turbine costs are down to $3.50/watt, while the PV costs are still at $6.60/watt. For a 1,500 watt wind system the cost is down to $1.93/watt and at 10,000 watts the cost of a wind generator (excluding electronics) is down to $1.15/watt. The cost of regulators and controls is essentially the same for PV and wind. Somewhat surprisingly, the cost of towers for the wind turbines is about the same as the cost of equivalent PV racks and trackers.

For homeowners connected to the utility grid, small wind turbines are usually the best "next step" after all the conservation and efficiency improvements have been made. A typical home consumes between 800-2,000 kWh of electricity per month and a 5-10 kW wind turbine or PV system is about the right size to meet this demand. At this size wind turbines are much less expensive.

Wind Energy

Wind energy is a form of solar energy produced by uneven heating of the Earth's surface. Wind resources are best along coastlines, on hills, and in the northern states, but usable wind resources can be found in most areas. As a power source wind energy is less predictable than solar energy, but it is also typically available for more hours in a given day. Wind resources are influenced by terrain and other factors that make it much more site specific than solar energy. In hilly terrain, for example, you and your neighbor may have the exact same solar energy. You could also have a much better wind resource than your neighbor because your property is on top of the hill or it has a better exposure to the prevailing wind direction. Conversely, if your property is in a gully or on the leeward side of a hill, your wind resource could be substantially lower. In this regard, wind energy must be considered more carefully than solar energy.

Wind energy follows seasonal patterns that provide the best performance in the winter months and the lowest performance in the summer months. This is just the opposite of solar energy. For this reason wind and solar systems work well together in what are called "hybrid systems". These hybrid systems provide a more consistent year-round output than either wind-only or PV-only systems.

Wind Turbines

Most wind turbines are horizontal-axis propeller type systems. Vertical-axis systems, such as the egg-beater like Darrieus and S-rotor type Savonius type systems, have proven to be more expensive. A horizontal-axis wind turbine consists of a rotor, a generator, a mainframe, and, usually, a tail. The rotor captures the kinetic energy of the wind and converts it into rotary motion to drive the generator. The rotor usually consists of two or three blades. A three blade unit can be a little more efficient and will run smoother than a two blade rotor, but they also cost more. The blades are usually made from either wood or fiberglass because these materials have the needed combination of strength and flexibility (and they don't interfere with television signals!).

The generator is usually specifically designed for the wind turbine. Permanent magnet alternators are popular because they eliminate the need for field windings. A low speed direct drive generator is an important feature because systems that use gearboxes or belts have generally not been reliable. The mainframe is the structural backbone of the wind turbine and it includes the "slip-rings" that connect the rotating (as it points itself into changing wind directions) wind turbine and the fixed tower wiring. The tail aligns the rotor into the wind and can be part of the overspeed protection.

A wind turbine is a deceptively difficult product to develop and many of the early units were not very reliable. A PV module is inherently reliable because it has no moving parts and, in general, one PV module is as good as the next. A wind turbine, on the other hand, must have moving parts and the reliability of a specific machine is determined by the level of the skill used in its engineering and design. In other words, there can be a big difference in reliability, ruggedness, and life expectancy from one brand to the next.

Towers

A wind turbine must have a clear shot at the wind to perform efficiently. Turbulence, which both reduces performance and "works" the turbine harder than smooth air, is highest close to the ground and diminishes with height. Also, wind speed increases with height above the ground. As a general rule of thumb, you should install the wind turbine on a tower such that it is at least 30 ft above any obstacles within 300 ft. Smaller turbines typically go on shorter towers than larger turbines. A 250 watt turbine is often, for example, installed on a 30-50 ft tower, while a 10kW turbine will usually need a tower of 80-100 ft.

That's Jesse and his friend Cheesy who climbed to the top (60") and put their little hands into the generator to put the blades on.

The least expensive tower type is the guyed-lattice tower, such as those commonly used for ham radio antennas. Smaller guyed towers are sometimes constructed with tubular sections or pipe. Self-supporting towers, either lattice or tubular in construction, take up less room and are more attractive but they are also more expensive. Telephone poles can be used for smaller wind turbines. Towers, particularly guyed towers, can be hinged at their base and suitably equipped to allow them to be tilted up or down using a winch or vehicle. This allows all work to be done at ground level. Some towers and turbines can be easily erected by the purchaser, while others are best left to trained professionals. Anti-fall devices, consisting of a wire with a latching runner, are available and are highly recommended for any tower that will be climbed. Aluminum towers should be avoided because they are prone to developing cracks.

Remote Systems Equipment

The balance-of-systems equipment used with a small wind turbine in a remote application is essentially the same as used with a PV system. Most wind turbines designed for battery charging come with a regulator to prevent overcharge. The regulator is specifically designed to work with that particular turbine. PV regulators are generally not suitable for use with a small wind turbine. The output from the regulator is typically tied into a DC source center, which also serves as the connection point for other DC sources, loads and the batteries. For a hybrid system the PV and wind systems are connected to the DC source center through separate regulators, but no special controls are generally required. For small wind turbines a general rule-of-thumb is that the AH capacity of the battery bank should be at least 6 times the maximum charging current.

Being Your Own Utility Company

The federal PURPA regulations passed in 1978 allow you to interconnect a suitable renewable energy powered generator to your house or business to reduce your consumption of utility supplied electricity. This same law requires utilities to purchase any excess electricity production at a price ("avoided cost") usually below the retail cost of electricity. In about a dozen states with "net energy billing options" small systems are actually allowed to run the meter backwards, so they get the full rate for excess production.

These systems do not use batteries. The output of the wind turbine is made compatible with utility power using either a special kind of inverter or an induction generator. The output is then connected to the household breaker panel on a dedicated breaker, just like a large appliance. When the wind turbine is not operating, or it is not putting out as much electricity as the house needs,

the additional electricity is supplied by the utility. Likewise, if the turbine puts out more needs, the additional electricity is supplied by the utility. Likewise, if the turbine puts out more than the house needs the excess is instantaneously sold to the utility. In effect, the utility acts as a very big battery bank.

Hundreds of homeowners around the country who installed 4-12 kW wind turbines during the go-go tax credit days in the early 1980's now have everything paid for and enjoy monthly electrical bills of $8-30, while their neighbors have bills in the range of $100-200 per month. The problem, of course, is that these credits are long gone and without them most homeowners will find the cost of a suitable wind generator prohibitively expensive. A 10 kW turbine (the most common size for homes), for example, will typically cost $19,000-24,000 installed. For those paying 10 cents/kilowatt-hour or more for electricity in an area with an average wind speed of 10 mph or more, and with an acre or more of property, a residential wind turbine may be worth considering. There is a certain thrill that comes from seeing your utility meter turn backwards.

Performance

The rated power for a wind turbine is not a good basis for comparing one product to the next. This is because manufacturers are free to pick the wind speed at which they rate their turbines. If the rated wind speeds are not the same then comparing the two products is very difficult. Fortunately, the American Wind Energy

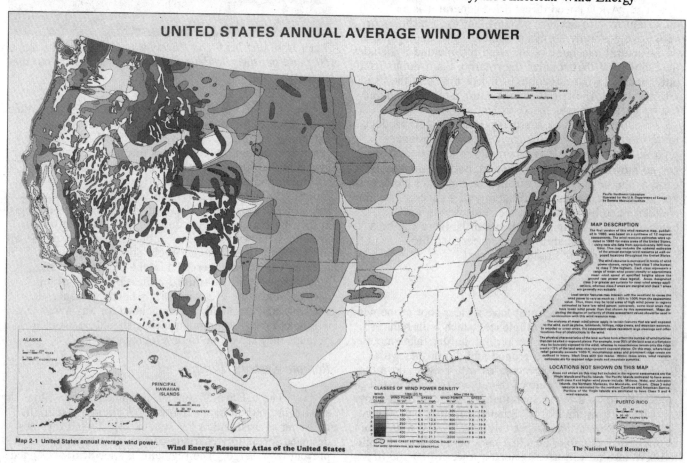

Map 2-1 United States annual average wind power.

Wind Energy Resource Atlas of the United States

The National Wind Resource

Association has adopted a standard method of rating energy production performance. Manufacturers who follow the AWEA standard will give information on the Annual Energy Output (AEO) at various annual average wind speeds. These AEO figures are like the EPA Estimated Gas Mileage for your car, they allow you to compare products fairly, but they don't tell you just what your actual performance will be ("Your Performance May Vary").

Wind resource maps for the U.S. have been compiled by the Department of Energy. These maps show the resource by "Power Classes" that mean the average wind speed will probably be within a certain brand. The higher the Power Class the better the resource. We say probably because of the terrain effects mentioned earlier. On open terrain the DOE maps are quite good, but in hilly or mountainous terrain they must be used with great caution. The wind resource is defined for a standard wind sensor height of 33 ft (10 m), so you must correct the average wind speed for wind tower heights above this height before using AEO information supplied by the manufacturer. Wind turbine performance is also usually derated for altitude, just like an airplane, and for turbulence.

As a rule of thumb, wind energy should be considered if your average wind speed is above 8 mph for a remote application and 10 mph for a utility-intertied application.

Keeping Current

The best way to keep current with the progress of wind energy development, both small and large scale, in the U.S. is to join the American Wind Energy Association (777 North Capitol St., NE, Suite 805, Washington, DC 20002; (202) 408-8988). A $35/year individual membership brings a newsletter and an opportunity to help push legislation to promote the increased use of wind energy. - **Mike Bergey.**

Bergey Wind Generators

For those with the need or the ability to generate more wind power than the Windseeker can offer, we're happy to offer the Bergey Windpower line with two basic models at 1500 and 10,000 watts. Bergey is one of few American windplant manufacturers that has survived the post-1986 tax credit crash and is still thriving and growing. Both of these units are wonderful in their simplicity: No brakes, pitch changing mechanism, gearbox, drive shaft or brushes. There is an automatic furling design that forces the generator and blades sideways at wind speeds over maximum production, but still maintains maximum rotor speed, so the generator keeps producing. Start up speed is 7.5 mph, cut-in wind speed is 7 mph, maximum wind speed is 120 mph. The 1500W unit weighs 275 lbs.

At the SERI (Solar Energy Research Institute) test site in Rocky Flats, Colorado, a Bergey 10kW unit survived wind gusts of 120 mph or possibly more as SERI's anemometer was carried away at that point. In an

independent test of four wind power manufacturers by Wisconsin Power & Light over a 3-year span, Bergey scored a top 99.4% availability. Down-time was only due to a direct lightning strike. Recommended maintenance is this" once a year walk out to the tower and look up. If the blades are turning, everything is OK! Tower not included in price.

■16-201	Bergey 1500W Turbine (24V)	$3,250
■16-202	Bergey 1500W Turbine (12V)	$3,495

Shipped freight collect from Oklahoma

■16-205	Bergey Control Unit (12,48,120V)	$995
■16-206	Bergey Charge Control Unit (24V)	$895

CALL FOR ADDITIONAL INFORMATION ON THESE AND LARGER UNITS - 60' TO 120' TOWERS ALSO AVAILABLE.

Windseeker II
250-Watt Wind Generator

The Windseeker II 250-watt wind generator fills the giant gap in the wind generator industry created when **Winco** discontinued its 200-watt Windcharger. This reliable and affordable small wind turbine is specifically designed for use in small independent home wind or hybrid power systems, especially in combination with photovoltaic panels. If you have an average annual windspeed of at least 8 mph, the Windseeker will work for you.

The Windseeker II uses a 12 pole, 3 phase alternator (based on the Ford alternator) modified for high output and efficiency at low rpm. The unit controls the torque applied to the specially developed rotor with an electronic circuit which adjusts field strength for maximum power output, without stalling the rotor. This allows the Windseeker II to begin generating power in lighter winds (7.5 mph cut-in speed) and to continue efficient power production to its maximum (300 watts at 30 mph).

The Windseeker is extremely lightweight (18 lb) and made of corrosion-resistant cast aluminum. The fail-safe tilt-up furling windforce regulator allows stability even in wind conditions in excess of 120 mph. A "smart" regulator is included with the unit, which compensates for wiring line loss and provides a conditioning tapered charge to batteries. If the voltage output reaches 20V the regulator will cut it off completely. It also senses for line loss and monitors the voltage at the regulator - if it is too low it will raise the voltage at the turbine.

The Windseeker II is available in 12 or 24 volt configuration. The Windseeker now comes standard in marine grade metals for longer durability and uses near the ocean.

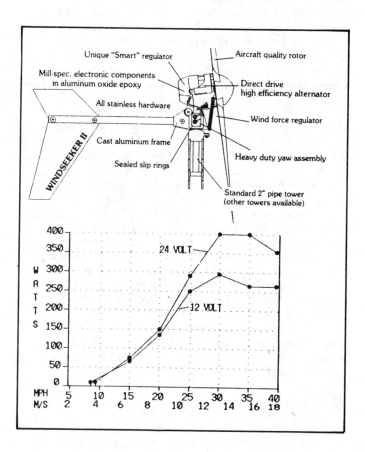

| 16-101 | Windseeker II - 12V | $875 |
| ■16-102 | Windseeker II - 24V | $975 |

Batteries for Wind Generators

The battery system size should be designed to allow you to operate under average loads for at least three days with no wind power. If wind is to be your only source of power, be careful not to plan too large a battery capacity, i.e., one that could not be recharged from the wind within 4-5 days. Otherwise your battery system will be in danger of never attaining a fully charged state. Use true deep cycle, high antimony, lead-acid batteries for a stand-alone wind system. See the *Battery chapter* for more details about battery types and selection.

Things That Work:
The Windseeker II

Here's an article that recently appeared in Home Power Magazine under the "Things That Work" section. It is written by Windy Dankoff who manufactures and supplies us with the Flowlight Slopump and Booster pump.

I'm spoiled. I "grew up" with the wind power revival of the 70's, and began my business rebuilding the Cadillac of wind generators, the "Jacobs Wind-Electric Plants" built in the 40's and 50's. They ran my shop for several years. Newer wind machine designs have rarely proven so reliable. The vast majority have been absolute disasters, so I'm a hardened skeptic when it comes to trusting anything new. As PV's got cheaper in the early 80's, I happily "retired" from wind power, phasing out all but maintenance for my old customers. Their rebuilt Jacobs machines were the only ones I could trust.

A year ago, word blew my way (from another old Jacobs rebuilder in Flagstaff) that "Wind Seeker" was cranking out reliable amps around the Northern Arizona mountains. I waited. I heard more good things. I talked to the factory, and ordered the instruction manual. Wind Seeker's manual is thorough, and easy to follow. It treats tower placement, siting, energy use, and other aspects of system design in a streamlined, efficient manner. Could it be that the machine works as gracefully as the manual? At the crusty old age of 37, it was going to take SOME CONVINCING to bring me out of wind power retirement! I bought one.

The Wind Seeker II bears genetic similarities to our 1942 Paris-Dunn 12-volt 200-watt wind machine. Made for running radios on the farm, the Paris-Dunn was the most durable of the very small machines made during the old days of wind power. Ours survived a second incarnation, blasted by a decade of wild New Mexico winds. The key to its durability is a simple, classic speed-governing system, the generator tips upward in winds over 30 mph to spill excess wind and relieve pressure on the structure. In winds over 40 mph, it resembles a helicopter, and sounds like a 200 lb. hummingbird. Wind Seeker uses the same basic system, but with much 1980's refinement. You can hardly hear its swishing sound above the sound of the wind itself.

Wind Seeker's pivoting speed governor is best for such small machines. It is FAR superior to the Air Brake governor used by the popular "Windcharger". Windcharger is another 1940's design that ceased production early in 1989.) Comparing the Wind Seeker to Windcharger is like comparing a Toyota Tercel to a rattling Model T with bad brakes. Wind Seeker II is one quarter the weight, and highly resistant to icing and lightning hazards. It is also easy to shut down by hand with a control cable.

Wind Seeker II is quality-constructed from aluminum castings, and uses a "smart" controller built onto the machine. The controller allows optimum use of light winds, and compensates for wiring loss as it provides the proper taper-charge to your batteries. (I was delighted to watch the current interrupting on my ammeter as the voltage reached the perfect 14.3V.) Output (12V model) is 21 watts at 10 mph, 75 watts at 15 mph, 175 watts at 20 mph, 325 watts at 25 mph, and 400 watts at 27-30 mph governing speed. Wind-charger watts simply add to whatever power may be coming from PV or other energy sources.

The propeller blade is an efficient taper, twist design that is extremely light in weight to reduce stresses and vibrations. The electrical heart is a modified Ford alternator. Parts and service for it are easily available from automotive sources. Wind Seeker's special field-control circuit is the key to adapting an automotive alternator to wind power (along with rewinding its coils for lower speed operation). Most attempts at using alternators for wind power have failed for lack of such a controller.

Wind Seeker can mount on a common antenna tower available from your local communications supplier, or on a simple mast made of 2" iron pipe with guy wires. (Roof mounting is not recommended). Your tower MUST clear all surrounding obstacles by at least 15 ft. to avoid turbulent wind, or you will not obtain satisfying results (this goes for ALL wind machines). No additional controls are required except for a fuse and (Optional) ammeter. It has a two-year warranty, and looks like it can last for decades.

Wind Seeker II is a very small machine, intended primarily for supplementary power in combination with photovoltaics or other energy sources. Wind power tends to fill in the gaps, during storms, and during many nights as well. So, if you live on the coast of Maine, or the dark winter wilds of Michigan, Montana, or the Yukon, or even in sunny New Mexico, consider the WIND SEEKER II to complete your energy system.

Hydro-Electric Systems

Water wheels working in locations between Maine and Georgia numbered in the thousands during the nineteenth century. Used mostly in mills, their construction required both money and hard work, even without being able to generate electricity.

Among the most important technological developments made in the last century is the water wheel designed by Lester Pelton in 1880. The Pelton wheel uses cup-like buckets at regular intervals. A jet nozzle protruding from the end of a pipe shoots water into the buckets with enough force to turn the wheel. The wheel then turns a generating device.

As with today's photovoltaic and wind generating systems, low-voltage hydro-electric technology can take advantage of water to generate electricity more economically than connecting to the electrical grid.

One such machine is the Burkhardt or Harris turbine, which operates on the Pelton principle. It is reliable, low-cost equipment that can return your investment faster than any other alternative energy device sold today, if you have adequate water.

The small hydro-electric system interfaces well with photovoltaics too, providing much needed power in the winter months, when water is usually abundant and sunshine scarce.

Many user-installed Burkhardt and Harris Peltons in the United States and Canada have been in operation over the past few years with good success and customer satisfaction.

Overview

Designed for light to medium household power requirements, hydro-electric systems are appropriate for small streams that produce from 4 to 400 gallons of water per minute (gpm) and where the head (vertical fall) is a *minimum* of 10 feet (at 100 gpm). At 400 feet of head only 4 GPM is needed to produce useable power.

The standard system is designed for charging 12-volt DC batteries from a modified Delco automotive alternator; 24, 36, 48 or 100- volt DC battery configurations may be charged from a modified Burkhardt or Harris system using a permanent magnet DC generator and can be special ordered.

The Burkhardt and Harris turbines are not designed to produce the 120/240-volt 60 cycle AC used by the typical home. One reason is that AC electricity cannot be stored. Another is the high cost of stabilizing the alternating current at precisely 60 Hz in the face of continuously changing loads. The DC power output of the standard system range from about 20 to 500 watts. (1.6 to 42 amps at 12VDC).

Site Evaluation

The operation of a water-driven generator requires water pressure, which builds under the force of gravity as a body of water moves. There are two aspects of water movement, head and flow. You need to know both to determine if there is enough force available in your stream to produce usable electricity.

Measuring the head of your water source means calculating the vertical distance water falls from a dam or other source to the spot where you will install the hydro-electric system. The most likely spot downhill from your water source will be as close to the house as possible, ideally no more than 100 feet away because your batteries will be located at the house. A greater distance than 100 feet will translate into extremely large, more costly wire to transport the electrical current with minimum power loss. (For every 100 feet of distance, you have 200 feet of wire.)

Begin your measurements by working back toward the water source using this primitive but effective process. You will need a carpenter's level and a long board or straight stick. Take the board or stick, stand it upright and mark it at eye level. Then measure the length of the stick from your mark to the ground. Let's say it is 5 feet. Now stand on your likely spot with the measuring stick. Using the carpenter's level look uphill across the mark toward your water source. Pick a landmark between yourself and the source on the same level as the mark. Walk over to it and repeat the process. When you have done it 15 times, using the 5-ft mark, you will have measured a water head of 75 feet. If you have a friend with a stick marked at the same height, you can speed up the sighting by rotating on each other's landmarks, exchanging places as you go up the hill.

Of course if you or a friend have access to aircraft equipment, borrowing or renting a sensitive altimeter will really speed things up.

The small hydro-electric system interfaces well with PV in the winter, when water is abundant and sunshine scarce.

Once you have checked for sufficient head, you need to check the water's flow, which should be the minimum necessary to produce usable power (see chart). To measure the flow, block the stream, then take a length of 1-1/2" pipe and stick it into the water right at the point where you plan to have the dam. With the other end of the pipe in a 5-gallon container, check your watch to see how long it will take to fill the container. If it takes no more than 30 seconds, you have 10 gpm.

If either the head or flow falls short of the minimum requirement, you can balance one against the other to a degree. Greater head permits slower flow, and faster flow reduces the head requirement.

If piping already exists from your water source, install a pressure gauge and a rainbird sprinkler nozzle (1/4", 5/16", 1/2") at the lowest end of the pipe. Turn on the water and the pressure gauge will indicate the dynamic p.s.i. you will have at the generator. Experiment with different size nozzle jets. Then you can determine the power output from the charts in this section.

The batteries should be located at the house for two good reasons; (1) to keep them warm and retain maximum storage capacity and (2) the charging voltage being higher than normal battery voltage, less voltage loss will occur during transmission through a given wire size. Such loss should not exceed 4% for best results. See "Wire & Connectors" in Chapter 3 for more details.

What Your Hydro System Will Cost

Total hydro system costs are determined by four factors:

1. **Size of Turbine:** Turbine costs vary from $695 to $1,050 depending on the number of nozzles you need. Higher output systems require greater battery capacity, about $150 per 100 watts of output.

2. **Length of Pipe:** This varies from under $25 for small, short, steep pipes in mountainous country to over $1,000 for large long pipes on flat land.

3. **Distance from turbine to point of use:** Wire cost is dependent on system output, transmission distance, and voltage. It can vary from under $10 to over $1,000. For very long distances a high-voltage transmission system is needed. This costs from $600 to $1,600 depending on conditions.

4. **Power Conditioning Equipment:** All but the smallest systems need voltage regulation to protect batteries from overcharging. Costs range from $100 to $400. Inverters are often needed to power AC loads in a house for a cost of $500 to $1,200.

Here is a hydro cost breakdown for a typical mountain cabin:

Site Conditions:		Costs:	
Head	100 ft	Pipe	$100
Flow	15 gpm	Turbine	$695
Pipe length	300 ft	Batteries	$150
Pipe size	2" PVC	Wire, etc.	$ 15
Distance	30 ft		
Output	100 W	**Total Costs**	**$960**

Note that this system will produce 100 watts times 24 hours or 2,400 watthours per day, or 200 amphours per day. To generate this much power using PV would require 11 48-watt PV panels generating for 5 hours per day at a cost in excess of $3,000 for the solar panels alone.

Hydro Site Evaluation

In order to accurately size your potential hydro-electric site we need specific information from you. We will take this information and feed it into our computer with a hydro sizing program. Send in the following information with an SASE and $10 and we'll be happy to size your potential site and recommend a system for you. Allow two weeks processing time for this service.

1. Your head, or drop in elevation from source to turbine site.
2. Flow in gallons per minute to be used.
3. Length, size, and condition of pipe to be used.
4. Distance from turbine to point of power use.

17-001 Hydro Site Evaluation $20

Sizing Your System's Potential

Short of having our computer size your system's potential, the chart below will give you a good idea of how much output in watts you may expect for different gpm's (gallons per minute) for varying feet of head. Multiply the number of watts by 24 to get the total number of watts generated in one day. Divide this by 12 for 12V systems to get your amphour per day output.

Feet of Head

G.P.M.	25	50	75	100	200	300	600
3	-	-	-	-	40	70	150
6	-	-	10	20	100	150	300
10	-	15	45	75	180	275	550
15	-	50	85	120	260	400	800
20	25	75	125	190	375	550	1100
30	50	125	200	285	580	800	1500
50	115	230	350	500	800	1200	-
100	200	425	625	850	1500	-	-
200	275	515	850	1300	-	-	-

For Standard Delco, expect 20% to 30% lower output.

Hydro Siting

*The following excellent article on hydro plant sizing by Paul Cunningham is reprinted from **Home Power Magazine** (#8, Dec. 88/Jan. 89, Page 17-19).*

Many people have access to some form of running water and are wondering just how much power, if any, can be produced from it. Almost any house site has solar electrical potential (photovoltaic). Many sites also have some wind power available. But water power depends on more than the presence of water alone. A lake or well has no power potential. The water must be FLOWING. It also must flow from a high point to a low one and go through an elevation change of at least three or four feet to produce useable power. This is called the head or pressure, usually measured in feet or pounds per square inch (PSI). The flow is measured in gallons per minute (GPM) or for those blessed with larger flows, cubic feet per second (CFS).

At most sites, what is called run of river is the best mode of operation. This means that power is produced at a constant rate according to the amount of water available. Usually the power is generated as electricity and stored in batteries and can be tied to an existing PV or other system. The power can take other forms: shaft power for a saw, pump, grinder, etc.

Both head and flow are necessary to produce power. Even a few gallons per minute can be useful if there is sufficient head. Since power = Head X Flow, the more you have of either, the more power is available. A simple rule of thumb to estimate your power is Head (in feet) X Flow (in gpm) /10 = Power (in Watts). This will give you a rough idea of the power available at the average site and reflects an overall efficiency of 53%. This is a typical output for a well designed system. For example: if your head is 100 feet and the flow is 10 gpm, then 100 X 10/10 = 100 watts. Keep in mind this is power that is produced 24 hours a day. It is equivalent to a PV system of 400-500 watts - if the sun shines every day. Of course, the water may not run year round either. So it is apparent how a combined system can supply your power needs on a continuous basis.

Determining Head & Flow

Let's start with the head since that is easier than the flow and will give you confidence to continue. The best method to determine the head is also the easiest and can be used at any site. It is also very accurate. It involves using a length of hose or pipe in the neighborhood of 1/2" diameter. You can start anywhere along the brook and proceed upstream or down. First submerge the upstream end in the water and weigh it down with a rock or something similar. With the top end fixed in place underwater you move the rest of the pipe downstream. When you have reached the end, it is now time to start the water flow through the pipe. This may require you to suck on the end. Once flow is established and all air bubbles are removed, slowly raise the pipe upward until the flow ceases. When this point has been reached, use a tape measure to measure the distance from

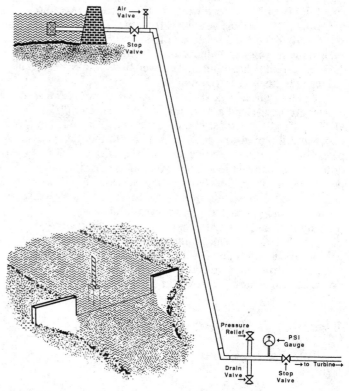

the end of the pipe to the surface of the water. This reading is the head for the stretch of brook. The pipe then becomes a convenient measure of horizontal run if you use a standard length like 100 feet. If you are working with a brook longer than your length of pipe, then simply carry the pipe to the next section to measure and repeat the procedure as required, starting where you ended before.

It is probably best to "map" more of the brook than you intend to use. This will give you a good overall idea of your site and may reveal some surprises.

Measuring flow is a little more difficult. This should probably be done in more than one place too. This is because most streams pick up water as they go. Therefore choosing the best spot for your system requires careful consideration of several things.

There are several ways to measure flow; here are two. In both cases, the brook water must all pass through either a pipe or a weir. The weir system uses an opening that the water flows through and measuring the dept of water gives the flow. The first involves a technique very similar to the head measuring technique. You must divert all of the water into a short length of pipe. This will usually require the use of a dam in order to pack dirt around the intake end. Pipe size may be from 1" to 6" depending on the flow rate. Once that is done the water is directed into a bucket or other container of known volume. The time required to fill it is then noted and this is converted into GPM.

The weir technique is more involved so if the pipe plan works--fine. This consists of setting a bulkhead in the stream with an opening cut in it. The water level is measured as it flows over and with the aid of charts the flow is determined.

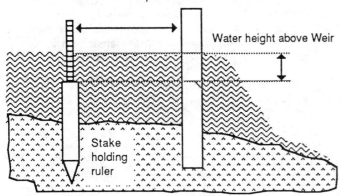

1 to 4 feet between depth ruler & weir

Water height above Weir

Stake holding ruler

Many materials can be used for the weir but sheet metal is the easiest to make since the thickness is slight. Wood requires a beveled edge for accuracy. A stake is driven into the stream bed a foot or so upstream of the weir and level with the bottom of the notch. This is the point the depth of water is measured since the level drops somewhat at the weir opening.

Water flow should be measured several times during the year. Once a month will give a good idea of how much power can be expected year round. The 50% efficiency rule applies to sites with heads greater than 30-40 feet or so. At lower heads everything becomes more difficult. Turbine and pipes become larger and speeds of rotation decrease.

The diameter and length of pipeline can now be determined once you have an idea of the potential power output of your site. It is assumed that you are planning on using a TURBINE and will generate ELECTRICITY. Other courses of action are possible but will not be discussed now.

Weir Measurement Table

Weir Depth inches	Table shows flow in cubic feet per minute. Width of Weir in inches					
	1	2	4	8	12	24
0	0.00	0.00	0.00	0.00	0.	0.
1	0.40	0.80	3.20	25.60	307.	7373.
2	1.13	2.26	9.04	72.32	868.	20828.
3	2.07	4.14	16.56	132.48	1590.	38154.
4	3.20	6.40	25.60	204.80	2458.	58982.
5	4.47	8.94	35.76	286.08	3433.	82391.
6	5.87	11.74	46.96	375.68	4508.	108196.
7	7.40	14.80	59.20	473.60	5683.	136397.
8	9.05	18.10	72.40	579.20	6950.	166810.
9	10.80	21.60	86.40	691.20	8294.	199066.
10	12.64	25.28	101.12	808.96	9708.	232980.
11	14.59	29.18	116.72	933.76	11205.	268923.
12	16.62	33.24	132.96	1063.68	12764.	306340.
13	18.74	37.48	149.92	1199.36	14392.	345416.
14	20.95	41.90	167.60	1340.80	16090.	386150.
15	23.23	46.46	185.84	1486.72	17841.	428175.
16	25.60	51.20	204.80	1638.40	19661.	471859.
17	28.03	56.06	224.24	1793.92	21527.	516649.
18	30.54	61.08	244.32	1954.56	23455.	562913.
19	33.12	66.24	264.96	2119.68	25436.	610468.
20	35.77	71.54	286.16	2289.28	27471.	659313.

A rough average of the stream flow can be made after you have made measurements at different times of the year. Most sites will have periods of very high flow that don't last long and times of very low or no flow at all. You need a pipeline capable of handling a reasonable flow average.

Let us use an example of a typical site and see what is involved. Assume your measurements show that 100 feet of head is available over a distance of 1,500 feet. The water will be taken from the high end of the pipe and discharged at the low end through the turbine at a point as close to the brook as is reasonable. This will give you the maximum head available. Exceptions to this will be where the discharge water is to be used for another purpose (aquaculture, irrigation).

Assume for the example that a flow of 30 gpm is available most of the year. Any pipeline will produce maximum power when the pressure drop due to friction is 1/3 of the pressure when no water is flowing. The pressure available under conditions of water flow is called the NET or DYNAMIC head. The pressure under conditions of no flow is the STATIC head. The difference between these two is the loss due to friction. Therefore the larger the pipe the better.

For the example you will require a pipeline that has no more than a head loss of 100/3 or 33.3 feet (over 1,500'). This is 33.3/15 or 2.22 feet of head loss per 100 feet of pipe. Since this flow rate will probably allow the use of fairly small pipe, let's use the chart for polyethylene. Two inch pipe gives a flow loss of .77 feet per 100 feet and 1 1/2 inch gives 2.59. From this information the 1 1/2 inch looks a little small and with the 2 inch we can use up to almost 55 gpm before the power drops off (50 gpm = 1.98' head loss and 55 gpm = 2.36 feet head loss/100').

So the choice of 2 inch pipe will cause a pressure drop of .77/100 X 1,500 = 11.55' head loss or a NET head of 100 - 11.55 = 88.45 feet at a flow of 30 gpm.

Water must be channeled into the intake end of the pipe. This may require a minimal dam sufficient to raise the water level a foot or so. It is useful to make a small pool off to one side of the main flow for this so that the trash (leaves, twigs, sand) will largely bypass the inlet. The inlet can be covered with window screen and need only be a simple wooden frame to support the screen and have a hole for the pipe to enter.

To facilitate draining the pipe, valves can be fitted as shown. A valve the size of the pipe can be installed just downstream of the intake. This is followed by a small air inlet valve to allow the water to exit and prevent pipe collapse. At the turbine end of the pipe a valve should be installed just before the turbine with a pressure gauge upstream of it. This will enable you to stop the flow and determine the pressure under both static and dynamic conditions. Another valve may be added on a tree to drain the pipe without running the turbine. A pressure relief valve can be added in higher pressure systems. Keep in mind that even if you are always careful to shut the stop valve slowly, the pressure can still rise suddenly for at least two reasons. A piece of trash may plug the nozzle or air pockets may discharge causing the water to speed up and then slow down abruptly when water hits the nozzle. Some respect for the forces involved will help protect your system.

Another area that may require protection is the aquatic environment your system intrudes upon. Remember that your water needs should not cause the stream level to become too low. Many areas also have legal guidelines for the use and diversion of stream water.

Friction Loss- Polyethylene (PE) SDR-Pressure Rated Pipe

Pressure loss from friction in psi per 100 feet of pipe.

Flow GPM	NOMINAL PIPE DIAMETER IN INCHES							
	0.5	0.75	1	1.25	1.5	2	2.5	3
1	0.49	0.12	0.04	0.01				
2	1.76	0.45	0.14	0.04	0.02			
3	3.73	0.95	0.29	0.08	0.04	0.01		
4	6.35	1.62	0.50	0.13	0.06	0.02		
5	9.60	2.44	0.76	0.20	0.09	0.03		
6	13.46	3.43	1.06	0.28	0.13	0.04	0.02	
7	17.91	4.56	1.41	0.37	0.18	0.05	0.02	
8	22.93	5.84	1.80	0.47	0.22	0.07	0.03	
9		7.26	2.24	0.59	0.28	0.08	0.03	
10		8.82	2.73	0.72	0.34	0.10	0.04	0.01
12		12.37	3.82	1.01	0.48	0.14	0.06	0.02
14		16.46	5.08	1.34	0.63	0.19	0.08	0.03
16			6.51	1.71	0.81	0.24	0.10	0.04
18			8.10	2.13	1.01	0.30	0.13	0.04
20			9.84	2.59	1.22	0.36	0.15	0.05
22			11.74	3.09	1.46	0.43	0.18	0.06
24			13.79	3.63	1.72	0.51	0.21	0.07
26			16.00	4.21	1.99	0.59	0.25	0.09
28				4.83	2.28	0.68	0.29	0.10
30				5.49	2.59	0.77	0.32	0.11
35				7.31	3.45	1.02	0.43	0.15
40				9.36	4.42	1.31	0.55	0.19
45				11.64	5.50	1.63	0.69	0.24
50				14.14	6.68	1.98	0.83	0.29
55					7.97	2.36	0.85	0.35
60					9.36	2.78	1.17	0.41
65					10.36	3.22	1.36	0.47
70					12.46	3.69	1.56	0.54
75					14.16	4.20	1.77	0.61
80						4.73	1.99	0.69
85						5.29	2.23	0.77
90						5.88	2.48	0.86
95						6.50	2.74	0.95
100						7.15	3.01	1.05
150						15.15	6.38	2.22
200							10.87	3.78
300								8.01

Numbers in Bold indicate 5 Feet/Second Velocity

Friction Loss- PVC Class 160 PSI Plastic Pipe

Pressure loss from friction in psi per 100 feet of pipe.

Flow GPM	NOMINAL PIPE DIAMETER IN INCHES										
	1	1.25	1.5	2	2.5	3	4	5	6	8	10
1	0.02	0.01									
2	0.06	0.02	0.01								
3	0.14	0.04	0.02								
4	0.23	0.07	0.04	0.01							
5	0.35	0.11	0.05	0.02							
6	0.49	0.15	0.08	0.03	0.01		Bold Numbers indicate				
7	0.66	0.20	0.10	0.03	0.01		5 Feet per Second Velocity				
8	0.84	0.25	0.13	0.04	0.02						
9	1.05	0.31	0.16	0.05	0.02						
10	1.27	0.38	0.20	0.07	0.03	0.01					
11	1.52	0.45	0.23	0.08	0.03	0.01					
12	1.78	0.53	0.28	0.09	0.04	0.01					
14	2.37	0.71	0.37	0.12	0.05	0.02					
16	3.04	0.91	0.47	0.16	0.06	0.02					
18	3.78	1.13	0.58	0.20	0.08	0.03					
20	4.59	1.37	0.71	0.24	0.09	0.04	0.01				
22	5.48	1.64	0.85	0.29	0.11	0.04	0.01				
24	6.44	1.92	1.00	0.34	0.13	0.05	0.02				
26	7.47	2.23	1.15	0.39	0.15	0.06	0.02				
28	8.57	2.56	1.32	0.45	0.18	0.07	0.02				
30	9.74	2.91	1.50	0.51	0.20	0.08	0.02				
35		3.87	2.00	0.68	0.27	0.10	0.03				
40		4.95	2.56	0.86	0.34	0.13	0.04	0.01			
45		6.16	3.19	1.08	0.42	0.16	0.05	0.02			
50		7.49	3.88	1.31	0.52	0.20	0.06	0.02			
55		8.93	4.62	1.56	0.62	0.24	0.07	0.02			
60		10.49	5.43	1.83	0.72	0.28	0.08	0.03	0.01		
65			6.30	2.12	0.84	0.32	0.09	0.03	0.01		
70			7.23	2.44	0.96	0.37	0.11	0.04	0.02		
75			8.21	2.77	1.09	0.42	0.12	0.04	0.02		
80			9.25	3.12	1.23	0.47	0.14	0.05	0.02		
85			10.35	3.49	1.38	0.53	0.16	0.06	0.02		
90				3.88	1.53	0.59	0.17	0.06	0.03		
95				4.29	1.69	0.65	0.19	0.07	0.03		
100				4.72	1.86	0.72	0.21	0.08	0.03	0.01	
150				10.00	3.94	1.52	0.45	0.16	0.07	0.02	
200					6.72	2.59	0.76	0.27	0.12	0.03	0.01
250					10.16	3.91	1.15	0.41	0.18	0.05	0.02
300						5.49	1.61	0.58	0.25	0.07	0.02
350						7.30	2.15	0.77	0.33	0.09	0.03
400						9.35	2.75	0.98	0.42	0.12	0.04
450							3.42	1.22	0.52	0.14	0.05
500							4.15	1.48	0.63	0.18	0.06
550							4.96	1.77	0.76	0.21	0.07
600							5.82	2.08	0.89	0.25	0.08
650							6.75	2.41	1.03	0.29	0.10
700							7.75	2.77	1.18	0.33	0.11
750							8.80	3.14	1.34	0.37	0.13
800								3.54	1.51	0.42	0.14
850								3.96	1.69	0.47	0.16
900								4.41	1.88	0.52	0.18
950								4.87	2.08	0.58	0.20
1000								5.36	2.29	0.63	0.22
1500									4.84	1.34	0.46
2000										2.29	0.78
2500										3.46	1.18
3000											1.66

Saltery Cove Hydro Electric Project

The following article was submitted by Clare & Gail Cochran in Ketchikan, AK. about their unique Hydro-Electric system. They'll be happy to answer detailed questions if you'll write them at Saltery Cove Cabins, PO Box 6981, Ketchikan, AK 99901.

Pelton wheel in generator house

The site of this hydro power plant is in Saltery Cove located on the east side of Prince of Wales Island in southeast Alaska. The nearest town is Ketchikan, 35 nautical miles to the east. This area is accessible by sea plane or boat only. The site is remote and self sufficient.

Annual rainfall is approximately 100 inches per year and snowfall varies from 0 to 50". The operation of the hydro system depends exclusively on precipitation as the stream is a runoff system. This hydro electric site was once used by the Straits Packing Co. in the 1930's. Evidence of this project still remains. A portion of the old dam still stands and sections of 16" ID wood stave pipe are found on the hillside.

A log crib rock fill type dam made of native cedar impounds 500,000 cubic feet of water. 740 feet of 8" PVC penstock moves the water from the dam to the generating site. Net head is 60 feet.

Equipment consists of:

1. **Lima alternator** - brushless, synchronous, self regulated. 1800 RPM 60 Hz single phase 120-140 volts 10 KW.

2. **Pelton wheel - 18"**

3. **Full load governor**

4. **Voltage meter**

5. **Heat sinks** - Two 240 volt 500-watt baseboard heaters, one switchable.

Water flow through the wheel is 225 - 450 gallons per minute. The housing is built so that 2 jets are used. Depending on the amount of water available 1-3/8", 1-1/4", or 1" jets are used. Actual power output is 600 watts with one jet or 1200 watts with two jets.

The system works as follows - The water directed by the jets drives the Pelton wheel, which in turn, drives the alternator at its rated 1800 RPM. The electrical output is then routed directly to the full load governor. The governor then distributes the electrical load to the heat sink (baseboard heater). Additional power demands are then regulated by the governor which distributes the load.

Example - If only the heater is used all the power is absorbed there. When a light is turned on, the governor subtracts the needed power from the heater to the light.

Possible over/under voltage conditions within the system are regulated by the over/under voltage regulator. This system is activated if a drop below 110V or over 145V is experienced. Care is taken not to overload the system by being aware of how much power is available and using appliances that stay within the limits. Sometimes it is necessary to turn off one appliance before turning on another. A voltage meter is mounted in the house for monitoring available power. If an over/under situation arises, the voltage regulator activates the water deflector solenoid, the deflectors drop in front of the jets and the wheel stops. At this time the shunt trip breaker cuts off the power at 110V or 145V protecting electrical equipment from high or low voltage.

Backup power is provided by a generator and 12V batteries when there is not enough water in the creek to run the hydro electric charging.

Old dam from 1930's
8" PVC penstock

Kennedy Creek Hydroelectric Systems

By Richard Perez, Home Power Magazine

In the six thousand foot tall Marble Mountains of Northern California, it rains. Wet air flows straight from the Pacific Ocean only forty miles distant. This moist ocean air collides with the tall mountains and produces over sixty inches of rainfall annually. Add this rainfall with the spectacular vertical terrain and you have the prefect setting for hydroelectric power. This is the story of just one creek in hydro country and of five different hydro systems sharing the same waters.

Kennedy Creek

Kennedy Creek is on the west drainage of 4,800 foot tall Ten Bear Mountain. The head waters of Kennedy Creek are located in a marsh at 2,500 feet of elevation. The headwaters are spread out over a 10 acre area and the power of Kennedy Creek doesn't become apparent until its waters leave the marsh. After a winding course over five miles in length, Kennedy Creek finally empties its water into the Klamath River at about 500 feet elevation. This gives Kennedy Creek a total head of 2,000 vertical feet over its five mile run.

The volume of water in Kennedy Creek is not very great. While we weren't able to get really hard data as to the amount of water, the residents guessed about 500 gallons per minute. Kennedy Creek is not large by any standards. It varies from two to eight feet wide and from several inches to about four feet deep. We were able to cross it everywhere and not get of feet wet. The point here is that you don't need all that much water if you have plenty of vertical fall.

The Kennedy Creek Hydro Systems

Kennedy Creek supports five small scale hydroelectric systems. Each system supplies electric power for a single household. Each system uses the water and returns it to the creek for use by the next family downstream.

These systems are not new comers to the neighborhood; they have been in operation for an average of 7.6 years. These systems produce from 2.3 to 52 kilowatt-hours of electric power daily. Average power production is 22 kWh daily at an average installed cost of $4,369. If all the hydroelectric power produced by all five Kennedy Creek systems is totaled since they were installed, then they have produced over 305 megawatt-hours of power. And if all the costs involved for all five systems are totaled, thèn the total cost for all five systems is $21,845. This amounts to an average of 7¢ per kilowatt-hour. And that's cheaper than the local utility. One system, Gene Strouss's, makes power for 3¢ a kilowatt-hour, less than half charged by the local utility.

All the power production data about the Kennedy Creek hydroelectric systems is summarized on the table below. All cost data is what the owners actually spent on their systems. Being country folks, they are adept at shopping around and using recycled materials. The cost figures do not include the hundreds of hours of labor that these hydromaniacs have put into their systems.

So let's take a tour of the Kennedy Creek Hydros starting at the top of the creek and follow its waters downward to the Klamath River.

KENNEDY CREEK HYDROS

Hydroelectric System's Operator	System's Age in Years	Average Power Output in Watts	Daily Power Output in kWh.	Total Power made in kWh.	System Cost	System Power cost to date $ per kWh.
Gary Strouss	6	2,040	49	107,222	$8,795	$0.08
Stan Strouss	8	180	4	12,614	$3,520	$0.28
Gene Strouss	9	2,166	52	170,767	$5,950	$0.03
Max&Nena Creasy	6	97	2	5,098	$1,295	$0.25
Jody&Liz Pullen	9	120	3	9,461	$2,285	$0.24
AVERAGES	7.6	921	22	61,033	$4,369	$0.18
TOTALS	38	4,603	110	305,163	$21,845	

Kennedy Creek as a Power Producer
Total All Systems Cost / Total Power All Systems Made to Date
in Dollars per kiloWatt-hour ($ / kWh.) $0.07

Gary Strouss

Gary Strouss wasn't home the day that Bob-O, Stan Strouss, and I visited Gary's hydroelectric site. Gary is a contractor and off about his business. So as a result, we got this info from his brother Stan and father, Gene (the next two systems down Kennedy Creek).

Gary's hydroelectric system uses 5,300 feet of four inch diameter PVC pipe to deliver Kennedy Creek's water to his turbines. The head in Gary's system is 280 feet. In hydro lingo, head is the number of VERTICAL feet of drop in the system. Static pressure is 125 psi at the turbines.

Gary uses two different hydroelectric generators. One makes 120 vac at 60 Hz. directly and the other produces 12 VDC. The 120 vac system is very similar to the one his father, Gene Strouss uses and is described in detail below. Gary's 120 vac system produces 3,00 watts about eight months of the year. During the summer dry periods, Gary switches to the smaller 12 Volt hydro.

The 12 VDC system uses a Harris turbine that makes about 10 Amperes of current. The Harris turbine is fed from the same pipe system as the larger 120 vac hydro.

Gary's home contains all the electrical conveniences, including a rarity in an AE powered home- an air conditioner! The 120 vac hydro produces about 48 kilowatt-hours daily, so Gary has enough power for electric hot water and space heating.

Stan Strouss

Stan's hydro is supplied by 1,200 feet of 2 inch diameter PVC pipe. His system has 180 feet of head. In Stan Strouss's system this head translates to 80 psi of static pressure, and into 74 psi of dynamic pressure into a 7/16 inch diameter nozzle.

Stan uses a 24 Volt DC Harris hydroelectric system producing three to ten Amperes. Stan's hydro produces an average of 180 Watts of power. This amounts to 5,400 Watt-hours daily. The system uses no voltage regulation.

The DC power produced by the hydro is stored in a 400 Ampere-hour (at 24 VDC) C&D lead-acid battery. These ancient cells were purchased as phone company pull-outs eight years ago. Stan plans to use an inverter to run his entire house on 120 vac. Currently. he uses 24 VDC for incandescent lighting. When I visited, there was a dead SCR type inverter mounted on the wall and Stan was awaiting delivery of his new Trace 2524.

Stan's system is now eight years old. The only maintenance he reports is replacing the brushes and bearing in his alternator every 18 months. That and bears wrecking his water intake filters.

Stan and his father, Gene, own and operate a sawmill and lumber business from their homesteads. This business, along with raising much of their own food, gives the Strouss families self-sufficiency.

Gene Strouss

Gene Strouss's hydroelectric system is sourced by 600 feet of six inch diameter steel pipe connected to 1,000 feet of four inch diameter PVC pipe. Gene got an incredible on the 20 foot lengths of steel pipe, only $5 a length.

A twelve inch diameter horizontal cast steel Pelton wheel translates the kinetic energy of moving water into mechanical energy. The Pelton wheel is belted up from one to three and drives an 1,800 rpm, 120 vac, 60 Hz. ac alternator. All power is produced as 60 cycle sinusoidal 120 vac. The Pelton's mainshaft runs at a rotational speed of between 600 and 800 rpm. The output of the alternator is between 1,500 to 2,500 watts out depending on nozzle diameter. At an annual average wattage of 2,000 watts, Gene's turbine produces 48,000 watt-hours daily.

The pipe delivers 60 psi dynamic pressure into a 9/16 inch in diameter nozzle, for summertime production of 1500 watts at 70 gallons per minute of water through the turbine. In wintertime with higher water levels in Kennedy Creek, Gene switches the turbine to a larger,13/16 inch diameter nozzle. Using the larger nozzle reduces the dynamic pressure of the system to 56 psi and produces 2,500 watts while consuming 90 gallons per minute.

Gene's system is nine years old. The only maintenance is bearing replacement in the alternator every two years. Gene's system uses no batteries, all power is consumed directly from the hydro. Gene keeps a spare alternator ready, so downtime is minimal when it is time to rebuild the alternator. Regulation is via a custom made 120 vac shunt type regulator using a single lightbulb and many parallel connected resistors. Major system appliances are a large deep freezer, a washing machine, 120 vac incandescent lighting, and a television set.

Gene's homestead is just about self-sufficient (which is why he needs his freezer). Hundreds of Pitt River Rainbow trout flourish in a large pond created by the Pelton wheel's tail water. The trout love the highly aerated tail water from the hydro turbine. Gene grew 100 pounds of red beans for this winter and maintains two large green houses for winter time vegetables. Gene Strouss also keeps a large apple orchard. Gene raises chickens and this, with the trout, make up the major protein portion of his diet. His major problem this year was bears raiding the apple orchard and destroying about half of the 250 trees. For a second course, the bears then ate up over sixty chickens, several turkeys, and a hive of honey bees. Gene called his homestead, "My food for wildlife project."

Max and Niña Creasy

Seven hundred feet of two inch diameter PVC pipe sources a Harris hydro turbine with two input nozzles. Static pressure at the turbine is about 80 psi from a vertical head of 175 feet. It produces five to eight Amperes depending on the availability of water. Max and Niña use 100 feet of #2 USE aluminium cable to feed the hydro power to the batteries.

Max and Niña's system uses two Trojan L-16 lead-acid batteries for 350 Ampere-hours of storage at 12 VDC. All usage is 12 Volts directly from the battery. Max and Niña don't use an inverter. The system uses no voltage regulation and overcharging the batteries has been a problem. Power production is 97 Watts or 2,328 Watt-hours daily.

The major appliances used in this system are halogen 12 VDC incandescent lighting, television, tape deck and amplifier. This system has been operation for the last six years. Niña reports two year intervals between bearing and brush replacement in their alternator.

Max works with the US Forest Service and Niña runs a cottage industry making and selling the finest chocolate truffles I have ever eaten.

Jody and Liz Pullen

Jody and Liz's hydro system uses 1,200 feet of 2 inch diameter PVC pipe to bring the water to the turbine. Jody wasn't sure of the exact head in the system and without a pressure gauge it was impossible to estimate. The system works, producing more power than Jody and Liz need, so they have never investigated the details.

The turbine is a Harris 12 volt unit that produces between six and ten Amperes. Jody normally sets the Harris current output at six to ten Amps so as not to overcharge his batteries. An average output figure for this system is about 120 Watts or 2,800 Watt-hours daily. The power is carried from the hydro turbine via 480 feet of 00 aluminium USE cable to the batteries.

The batteries are located in an insulated box on the back porch. The pack is made up of four Trojan T220 lead-acid, golf cart batteries. The pack is wired for 440 Ampere-hours at 12 VDC. This system uses no voltage regulation and Jody has to be careful not to overcharge the batteries. Jody uses all power from the system via his Heart 1000 inverter. He also uses a gasoline fueled generator 4,000 watt for power tools (skil saw & table saw) and the washing machine. These tools require 120 vac and more power than the 1000 watt inverter can deliver.

Jody and Liz have used this hydro system for their power for the last nine years. They report the same biannual alternator rebuild period. Jody runs a fishing and rafting guide business on the Klamath River called Klamath River Outfitters, 2033 Ti Bar Road, Somes Bar, CA 95568 • 916-469-3349. Liz is just about finished her schooling and will soon be a Registered Nurse.

What the Kennedy Creek Hydros have discovered

Hydroelectric systems are more efficient the larger they get. Consider that the smaller systems have the higher power costs. The largest system, Gene Strouss's, operates at an incredibly low cost of 3¢ per kilowatt-hour. And that's the cost computed to date. Gene fully expects his hydro system to continue producing electricity for years to come.

The level of user involvement in these systems is low after their initial installation. While digging the trenches and installing the pipe takes both time and money, after it's done it is truly done. Only regular maintenance reported was bearing and brush replacement and trash rack cleaning.

The battery based DC hydros all showed signs of battery overcharging. Voltage regulation is the key to battery longevity in low voltage hydro systems.

A parting shot

As Bob-O and I were driving down Ti Bar Road on our way home, we passed the Ti Bar Ranger Station run by the US Forest Service. They were running a noisy 15 kw. diesel generator to provide power for the ranger station. Which is strange because they are at the very bottom of the hill with over two thousand feet of running water above them. And they have five neighbors above them who all use the hydro power offered by the local creek.

The practical and effective use of renewable energy is not a matter of technology. It is not a matter of time. It is not a matter of money. Using renewable energy is just doing it. Just like the folks on Kennedy Creek do.

Burkhardt & Harris Turbines

We refer repeatedly to Burkhardt & Harris as if the companies were interchangeable. In reality Don Harris is Harris Hydroelectric and John Takes is Burkhardt Turbines. Both companies use virtually the same components, have the same prices, and the same warranty. Both are well-proven reliable systems.

When calculated on a cost-per-watt basis, the average hydro-electric generator costs as little as ONE-TENTH as much as a solar (photovoltaic) system of equivalent power. Solar only generates power when the sun is shining; hydro generates power 24 hours a day.

The generating component of the Burkhardt & Harris turbines is an automotive alternator (Delco or Autolite, depending on system requirements) equipped with custom wound coils appropriate for each installation. The rugged turbine wheel is a one-piece Harris casting made of tough silicon bronze. There are hundreds of these wheels in service with NO failures to date.

The aluminum wheel housing serves as a mounting for the alternator and up to four nozzle holders. It also acts as a spray shield, re-directing the expelled water into the collection box.

Burkhardt & Harris turbines vary in cost depending on site specifications. Typically the hydro generator unit costs under $1000. A whole system, including pipe, wire, and battery set is usually less than $2,000.

Burkhardt & Harris turbines are available in several nozzle configurations to maximize the output of the unit. The particular number of nozzles that you need is a function of the available flow (gpm) and the existing pipe diameter. Here is a chart with some general rules for sizing the number of nozzles on the system you will need, but bear in mind that we need to size your system exactly.

gpm	Number of Nozzles
5-40	1
40-75	2
75-150	4
150+	8

Figures in gallons/min.
Feet of Head

# of nozzles	25	50	75	100	200	300	600
1	17	25	30	35	50	60	85
2	35	50	60	70	100	120	170
3	52	75	90	105	150	-	-
4	70	100	120	140	200	-	-
8	140						

The above flow figures can be exceeded by up to 50% with a loss of efficiency of up to 20%.

Burkhardt & Harris Hydro Turbines

■17-101	1 nozzle Turbine	$ 695
■17-102	2 nozzle Turbine	$ 795
■17-103	4 nozzle Turbine	$ 950
■17-108	8 nozzle Turbine	$1,050

■17-131	High output alternator	add $160
■17-132	24V operation	add $ 50
■17-133	Low head (less than 60')	add $ 50
■17-134	Extra Nozzles	$ 5

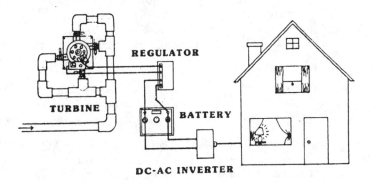

Pelton Wheels

For the small hydro-electric do-it-yourselfer, we offer reliable and economical Pelton wheels. Harris silicon bronze Pelton wheels resist abrasion and corrosion far longer than polyurethane or cast aluminum wheels. A nozzle jet of up to 1/2 inch can be used.

■17-202   Silicon Bronze Pelton Wheel $290

Gasoline Generators

For many years the traditional gasoline or diesel powered generator was the *only* economical "alternative" for generating electricity. Fuel was so cheap it offset the maintenance costs. However, in 1973, and almost each year since right up to the Iraqi oil crisis of 1990, fuel costs have soared and inflation has sent generator prices and maintenance costs up too.

Today the traditional generator still plays an important part in the scheme of producing affordable alternative energy, but in a new, more energy-sensible way - the machine is operated only occasionally to provide low-voltage power. The key to this variation on the system is a high-powered battery charging device that plugs into any traditional AC generator. When the generator operates, the charger can deliver up to 110 amps of DC power into a battery system.

With the proper size generator you can run the battery charger once a week or so, do other chores, such as water pumping, vacuuming, or clothes washing and charge your batteries at a very rapid rate simultaneously. Using a generator in this way greatly prolongs the life of the machine and drastically reduces fuel and maintenance costs, not to mention the alleviation of that hideous noise pollution! It provides abundant AC electricity for major appliances and tools and it permits full enjoyment of a low-voltage system too. When used in a "photo-gen" hybrid system the machine can reduce the cost of a photovoltaic system by as much as two-thirds.

Many people start out with a traditional generator, using it to begin construction on their home and outbuildings and for water pumping, the life blood of any homestead. They then add photovoltaic, wind, or hydro-electric as their budget permits over the years, finally relegating the generator to a complete backup status.

Before choosing your generator, you must consider the following:
- Local regulations
- Kilowatt capacity
- Fuel type
- Optional features

Since the generators we are referring to produce AC electricity, they are governed by local electrical codes. The first step is to find out what those regulations are. Then use a qualified licensed electrician to do the installation.

The kilowatthours and surge capability of the machine is another consideration. You should refer to "Inverters" in Chapter 3 to determine your maximum load for this purpose. The rated horsepower of a generator assumes operation at sea level. For each 1,000 feet of altitude,

derate the output by 3-1/2%. (A 3,000-watt generator operating at 2,000 ft altitude would deliver only 2,790 watts.) For LPG or propane fuel, derate the output another 10%.

Generators that burn diesel fuel are more economical than those that burn gasoline. Maintenance is lower because diesel machines are more efficient and they avoid conventional ignition systems. However, the initial cost is higher.

Conversion to LPG, propane, or natural gas can be economical too. Propane-fired generators burn clean thereby reducing maintenance and repair costs substantially, and since most homes will use propane for cooking and refrigeration, it works in nicely.

Some of the built-in features and options available for generators that you might want to consider looking for include:
- Manual, 12-volt electric, remote, or automatic load demand start. This is necessary if you choose a photo-gen system. Of course the convenience of electric or remote or automatic load demand in any situation is obvious.
- Continuous duty rating. This feature ensures efficient operation at the fully loaded rate on a constant basis. All our generators are rated as such.
- Full wattage output. Protect yourself from generators that distribute 120VAC on two circuits, which means you get only half the rated capacity on one. (Example: some inexpensive generators rated at 2,000 watts and equipped with two outlets are really capable of delivering only 1,000 usable watts because you can't combine the two outlets - say to run a 2,000-watt tool.) All our generators are rated for full wattage output.
- High motor starting. Special windings deliver extra surge power for induction and capacitor motors.
- Inherent voltage regulation. The voltage output adjusts automatically to match the required load.
- Idler control. Fuel consumption is reduced by automatic adjustment to match the load. One caution here: a low wattage motor alone (such as used in a can opener) may not cause the generator to react sufficiently. To correct, simply add a 25-watt bulb to the load.

Buying a generator can be a lot like buying an automobile. The body is there, but it's not only the whistles and bells that are an extra cost. The basic price of a generator rarely includes the charging battery (for remote start models), a gas container to hold the fuel, or gauges to help you to monitor the operation of the generator. All this should definitely be taken into consideration when determining what the final tab will be when the generator is in place and ready to go.

PV-Hybrid Block Diagram.

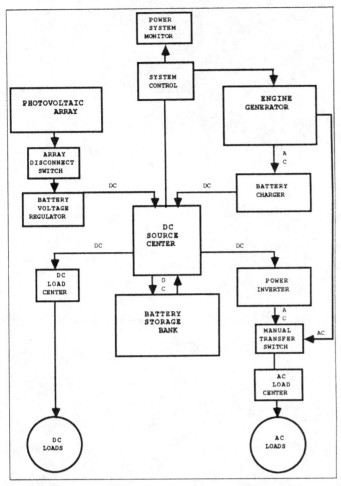

Courtesy The Solar Electric Independent Home Book

Onan Generators

Onan is one of the oldest generator manufacturers in the USA. They have an impeccable reputation for quality, longevity, and service. They make lots of different lines and lots of different models. We've chosen to feature the heavy duty Gen Sets that we recommend for backup and primary generators for remote homes. We of course have access to the entire Onan line which includes generators of 20 kW and higher capacity, diesel models, etc. Space prevents us from listing them here, but don't hesitate to contact us for further information.

Onan Gen Sets

These are vertically exhausted generators, with a cast iron block. All units are remote start, operating at 1,800 rpm on 120/240V, 60 hertz, one phase, four wire. Gasoline fueled and air-cooled, they can be converted for propane gas operation.

LPG / Gasoline Conversion Kit

All of the following Onans come standard with gasoline fuel capability. They can be ordered converted to dual propane (LPG) or gasoline capability for the following costs:

- ■18-301 LPG Conversion for 4.0 kW $139
- ■18-302 LPG Conversion for 5.0 kW $139
- ■18-304 LPG Conversion for 7.5 kW $179

Vacu-Flo Cooling For Onans

Vacu-Flo Cooling is a highly recommended option for all air-cooled Onan Generators. It converts the engine from standard pressure cooling to a suction cooling system. This system uses a specially designed direct-drive centrifugal blower housed in a sheet metal scroll with a built-in air outlet duct. The centrifugal blower draws air through the generator and over the engine cooling surfaces, then

discharges the heated air through the air outlet.

It is particularly essential that Vacu-Flo be used when installing an Onan in a small generator house or compartment. Heat, odors, and dangerous fumes are expelled from the compartment providing a safe installation.

■18-201 Vacu-Flo for CCK Models (4.0 & 5.0 kW) $ 74

■18-203 Vacu-Flo for Series JB (7.5 kW) $228
Shipped freight collect

4.0 CCK-3CR 4.0 kW Onan Generator

Standard equipment includes mounted control panel with start-stop rocker switch; fused battery charging circuit with two-step voltage regulator; remote control terminals; start disconnect relay; AC output box; muffler; flexible exhaust connection; fuel line; battery cables; vibration isolators and low oil pressure shut down. Delivers 4.0 kVA at 1.0 PF and 17.0 amps.

■18-101   4.0 CCK-3CR Onan $3,256
Shipped freight collect from manufacturer

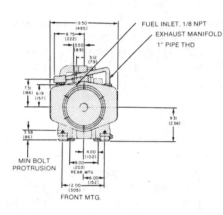

Portable Models: Overall dimensions—length 39.25 in. (996.9 mm); width 25.63 in. (651.2 mm); height 22.88 in. (581.2 mm); approximate net weight 322 lb (151 kg)
Remote Start Models: Approximate net weight 297 lb (135 kg)

Dimensions in parentheses are millimeters

Caution: These drawings provided for reference only.

Operating Data

4.0 CCK (60 Hz)
3.5 CCK (50 Hz)

Hertz	Engine Power Min. bhp (kW)	Piston Speed fpm (mm/s)	Air Cooled Air ft³/m (m³/min)	Combustion Requirement Air ft³/m (m³/min)	Generator Cooling Air ft³/m (m³/min)	Gasoline gph (L/h) at load- 0	2/4	4/4	Natural Gas at full load Btu/ft³ (MJ/m³)	ft³/h (m³/h)	Propane at full load Btu/ft³ (MJ/m³)	ft³/h (m³/h)
60 (1800 r/min)	10.2 (7.61)	900 (4572)	500 (14.2)	21 (0.60)	100 (2.83)	0.41 (1.55)	0.57 (2.16)	0.75 (2.84)	1000 (37.25)	90 (2.55)	2500 (93.13)	40 (1.13)
50 (1500 r/min)	9.0 (6.7)	750 (3810)	417 (11.8)	18 (0.51)	83 (2.35)	0.41 (1.55)	0.56 (2.12)	0.70 (2.65)	1000 (37.25)	81 (2.29)	2500 (93.13)	31 (0.88)

Average Fuel Consumption

Design: Onan revolving armature, 4-pole, self excited, inherently regulated. Drip-proof construction. Permanently aligned to engine.
Insulation System: Class F per NEMA MG1-1.65 definition. Insulating varnish conforms to MIL-I-24092
Temperature Rise: Per NEMA MG1-22.40 at rated load
Cooling: Direct drive axial flow blower
Armature: Laminated electrical steel stack, keyed and press fitted to shaft. Heavy insulated copper wire windings. Balanced.

Stator: Laminated electrical steel, one piece construction. Field coils wound directly onto pole shoes
Collector Rings: Brass alloy. Machine finished
Commutator: Hard drawn silver-copper alloy, machine finished
Brushes: AC, electrographitic; DC, metal-graphitic
Bearing: Double sealed, prelubricated ball-bearing

Model: Onan CCK Gasoline
Design: 4-cycle, L-head, two cylinders horizontally opposed
Displacement: 49.8 in³ (816 cm³)
Bore: 3.25 in. (82.55 mm)
Stroke: 3.0 in. (76.2 mm)
Compression Ratio: 5.5 to 1
Cooling: Pressure air-cooled
Fuel System: Downdraft carburetor with diaphragm fuel pump, fuel lift 4 ft (1.22 m) Hand choke on portable models, electric choke on remote start models
Governor: Enclosed, flyball, cam gear-driven, speed regulation 5%

Ignition Systems: Flywheel magneto ignition on portable models, 12-volt battery ignition on remote models
Starting Systems: Portable models—hand start Readi-Pull, automatic rope recoil. Stop button on blower housing. Remote start models—12-volt, 3-wire, negative ground
Lubrication: Positive displacement lube oil pump Oil capacity 4 qt (3.79 L)
Bearings: Two main
Cylinder Heads: Cast aluminum alloy
Valves: Intake, carbon steel Rotating exhaust, austenitic steel with hard chrome-cobalt alloy facing

5.0 CCK-3CR 5.0 kW Onan Generator

Standard equipment includes: mounted control panel with start-stop rocker switch; fused battery charging circuit with two-step voltage regulator; remote control terminals; start disconnect relay; AC output box; muffler; flexible exhaust connection; fuel line; battery cables; vibration isolators and low oil pressure shut down. Delivers 5.0 kVA at 1.0 PF and 21.0 amps.

■18-102   5.0 CCK-3CR Onan $3,816
Shipped freight collect from manufacturer

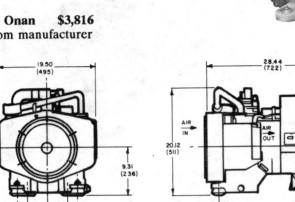

Portable Models: Overall dimensions—length 39.25 in. (996.7 mm), width 25.63 in. (651.0 mm) height 22.88 in. (581.2 mm), **net weight 357 lbs (161.9 kg)**.

Remote Start Models: **Approximate net weight 322 lb (146.1 kg)**

Caution: These drawings provided for reference only.

DIMENSIONS IN ( ) ARE MILLIMETRES

Operating Data   5.0 CCK (60 Hz)   4.0 CCK (50 Hz)

Hertz	Engine Power Max. bhp (kw)	Piston Speed fpm (mm/s)	Air Cooled ft³/m (m³/min)	Combustion Requirement Air-ft³/m (m³/min)	Generator Cooling ft³/m (m³/min)	Gasoline, gph (L/h) at load- 0	2/4	4/4	Natural Gas at full load Btu/ft³ (MJ/m)	ft³/h (m³/h)	Propane at full load Btu/ft³ (MJ/m³)	ft³/h (m³/h)
60 (1800 r/min)	10.2 (7.61)	900 (4572)	500 (14.2)	21 (0.60)	100 (2.83)	0.37 (1.40)	0.63 (2.38)	0.88 (3.33)	1000 (37.25)	115 (3.26)	2500 (93.13)	47 (1.33)
50 (1500 r/min)	9.0 (6.7)	750 (3810)	417 (11.8)	18 (0.51)	83 (2.35)	0.37 (1.40)	0.58 (2.20)	0.78 (2.95)	1000 (37.25)	90 (2.55)	2500 (93.13)	36 (1.02)

Average Fuel Consumption

Design: Onan AC revolving armature, 4-pole, self excited, inherently regulated. Drip-proof construction. Permanently aligned to engine.
Insulation System: Class F per NEMA MG1-1.65 definition. Insulating varnish conforms to MIL-I-24092
Temperature Rise: Per NEMA MG1-22.40 at rated load
Cooling: Direct drive axial flow blower
Armature: Laminated electrical steel stack, keyed and press fitted to shaft.
Heavy insulated copper wire windings. Balanced

Stator: Laminated electrical steel, one piece construction. Field coils wound directly onto pole shoes.
Collector Rings: Brass alloy. Machine finished
Commutator: Hard drawn silver-copper alloy, machine finished
Brushes: AC, electrographitic; DC, metal-graphitic
Bearing: Double sealed, prelubricated ball-bearing

Model: Onan CCK Gasoline
Design: 4-cycle, L-head, two cylinders horizontally opposed
Displacement: 49.8 in³ (816 cm³)
Bore: 3.25 in. (82.55 mm)
Stroke: 3.0 in. (76.2 mm)
Compression Ratio: 5.5 to 1
Cooling: Pressure air-cooled
Fuel System: Downdraft carburetor with diaphragm fuel pump, fuel lift 4 ft (1.22 m)
Hand choke on portable models, electric choke on remote start models
Governor: Enclosed, flyball, cam gear-driven. Speed regulation 5%

Ignition System: Flywheel magneto ignition on portable models, 12V battery ignition remote models
Starting System: Portable models—hand start Readi-Pull, automatic rope recoil
Stop button on blower housing
Remote start models - Exciter cranked, 12V, 3-wire negative ground
Lubrication: Positive displacement lube oil pump Oil capacity 4 qt (3.79 L)
Bearings: Two main
Cylinder Heads: Cast aluminum alloy
Valves: Intake, carbon steel
Exhaust, rotating, austenitic steel with hard chrome-cobalt alloy facing

7.5 JB-3CR 7.5 kW

Standard equipment includes: vibration isolators, muffler, flexible exhaust connection, air cleaner with reuseable polyurethane (oil-wetted) element, spin-on full flow lube oil filter, oil pressure gauge, removable lifting bracket; mounted control box with charge rate ammeter, start-stop switch, exciter field circuit breaker, manual reset, remote control terminals, mounted connection box, battery cables. Delivers 7.5 kVA at 1.0 PF and 31.3 amps.

■18-104　7.5 JB-3CR　$5,028
Shipped freight collect from Manufacturer

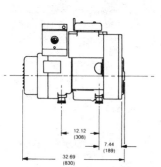

REMOVABLE LIFTING BKT
EXH OUTLET 1-1/4 PIPE TAP

26.62 (676)

12.12 (308)
7.44 (189)
32.69 (830)

11.50 (292)
17.88 (454)

Caution: This drawing supplied for reference only.

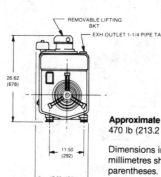

Approximate net weight:
470 lb (213.2 kg)

Dimensions in inches; millimetres shown in parentheses.

WHEN EQUIPPED WITH OVERSPEED SWITCH, ADD 1.50" (38) TO OVERALL LENGTH

Operating Data 7.5 JB (60 Hz)　6.0 JB (50 Hz)

Hertz	Engine Power Max. bhp (kw)	Piston Speed fpm (mm/s)	Air Cooled Air ft³/m (m³/min)	Combustion Requirement Air ft³/m (m³/min)	Generator Cooling Air ft³/m (m³/min)	Average Fuel Consumption								
						Gasoline gph (L/h) at load-					Natural Gas at full load		Propane at full load	
						0	1/4	2/4	3/4	4/4	Btu/ft³ (MJ/m³)	ft³/h (m³/h)	Btu/ft³ (MJ/m³)	ft³/h (m³/h)
60 (1800 r/min)	15.0 (11.2)	1087 (5522)	560 (15.9)	25 (0.71)	160 (4.53)	0.6 (2.27)	0.68 (2.6)	0.84 (3.2)	1.0 (3.8)	1.24 (4.7)	1000 (37.25)	134 (3.8)	2500 (93.13)	68 (1.92)
50 (1500 r/min)	12.5 (9.3)	9.06 (4602)	467 (13.3)	21 (0.60)	133 (3.77)	0.6 (2.27)	0.66 (2.50)	0.77 (2.91)	0.90 3.41	1.04 3.94	1000 (37.25)	110 (3.1)	2500 (93.13)	53 (1.50)

Design: Onan brushless revolving field 4-pole alternator rigidly coupled to engine, permanently aligned, drip proof construction.
Amortiseur windings and skewed stator minimizes field heating and voltage harmonics.
Dynamically balanced rotor.
Windings impregnated with 100% solid epoxy resin for improved cooling and complete environmental protection.
Twelve leads brought out on brushless, broad range, reconnectible 3-phase alternators; user reconnect to obtain required voltage.

Model: Onan JB Gasoline
Design: 4-cycle, 2-cylinder, vertical in line
Displacement: 60 in³ (983.4 cm³)
Bore: 3.25 in. (82.55 mm)
Stroke: 3.625 in. (97.07 mm)
Compression Ratio: 6.5 to 1
Cooling System: Pressure air cooled
Fuel System: Horizontal carburetor
Mechanical fuel pump with bowl and screen filter
Fuel lift 6 ft (1.83 m)
Automatic electric choke

Bearing: Double sealed prelubricated ball bearing
Cooling: Direct drive centrifugal blower
Exciter System: Brushless exciter with 8 pole stator mounted in endbell
Encapsulated rotating rectifier assemblies protect from adverse environments and easily accessible through endbell
Permanent magnet embedded in exciter stator field pole ensures alternator voltage buildup
Insulation: Class F per NEMA MG1-1.65,
Insulating varnish conforms to MIL-I-24092, Grade CB, Class 155°C

Ignition: 12V battery
Lubrication: Positive displacement, lube oil pump
Oil capacity 3.5 qt (3.31 L) including filter
Starting: Battery, remote, 12V 3-wire, negative ground
Solenoid-shift starter
Bearings: Two main
Cylinders and Crankcase: Single unit alloy iron casting
Removable cast iron oil base
Cylinder Head: Cast aluminum alloy
Valves: Overhead, free to rotate

Makita Generators

For years, Makita power tools have been the first choice of professionals, and a staple in the Real Goods product line. Now this high level of quality has been expanded to a new line of generators. All Makita generators come standard with a first-quality Wisconsin-Robin motor with a Super-quiet USDA-approved spark arrester muffler. They also include water-deflecting louvers - not just on top but all around. Auto decompression means ultra-easy pull starts. Circuit interrupters shut down the power the instant there's an electrical leak. All Makita generators have low-oil auto shut-down, easy access spark plug, heavy-steel gauge, and condenser voltage regulation. All Makita generators will charge at 8 amps into a 12V battery. Simply stated the Makita offers more regulation and conveniences than any 3,600 RPM generator on the market - including Honda! *The starting battery is not included on any model.*

Makita 2400 Watt Generator (G2400R)

The G2400R will deliver 2400 watts at 4.6 hp. It has 120VAC and 12VDC voltage output. The 2.6 gallon fuel tank capacity is good for 7.4 hours of continuous operation. Recoil start. A local Ukiah generator dealer will convert the G2400R unit to propane for around $300. If you can't have it done locally, ask us.

18-501 Makita 2400W Generator(G2400R) $1,150
Shipped freight collect

Makita 5500 Watt Generator w/Electric Start (G5501R)

The G5501R will deliver 5500 watts at 10 hp. It has dual voltage output (120/240VAC and 12VDC.) The 4.1 gallon fuel tank capacity will power the generator for 5 hours of continuous operation. Idle control automatically reduces engine RPM under no load for reduced fuel consumption and engine wear. Electric start is standard but the battery is not included. A local Ukiah generator dealer will convert the G5501R unit to propane for around $300. If you can't have it done locally, ask us.

Makita 3500 Watt Generator (G3500R)

The G3500R will deliver 3500 watts at 7.2 hp. It has convenient dual voltage output (120/240VAC) or 12VDC. The 3.4 gallon fuel tank will run the generator for 6.1 hours of continuous operation. Idle control automatically reduces engine RPM under no load for reduced fuel consumption and engine wear. Recoil start.

18-502 Makita 3500W Generator(G3500R) $1,575
Shipped freight collect

Makita 3500 Watt Electric Start (G3501R)

All the same features as the G3500R but with electric start added (no battery included).

18-503 Makita 3500W Elec. Start(G3501R) $1,845
Shipped freight collect

18-504 Makita 5500W Elec. Start(G5501R) $2,175
Shipped freight collect

Makita U.S.A., Inc.

GENERATOR SPECIFICATION CHART

SPECIFICATIONS

MODEL	G2400R	G3500R	G3501R	G5501R
GENERATOR				
AC Voltage	120V - 60HZ	120V/240V - 60HZ	120V/240V - 60HZ	120V/240V - 60HZ
Maximum Output Current	2400W - 20A	3500W - 29.2A/14.6A	3500W - 29.2A/14.6A	5500W - 45.8A/22.9A
Rated Output Current	2000W - 16.7A	3000W - 25A/12.5A	3000W - 25A/12.5A	4800W - 40A/20A
DC Voltage	12V - 8.3A	12V - 8.3A	12V - 8.3A	12V - 8.3A
Receptacles	A	A,B,C	A,B,C	A,B,C
ENGINE				
Maximum H.P. @ 3600 RPM	4.6	7.2	7.2	10.0
Displacement	183CC	273CC	273CC	388CC
Fuel Tank Capacity	2.6 Gal.	3.4 Gal.	3.4 Gal.	4.1 Gal.
Continuous Operation	7.4 Hrs.	6.1 Hrs.	6.1 Hrs.	5.0 Hrs.
Noise Level @ Rated Output, 7M, 60HZ	74dB	72dB	72dB	72dB
SIZE				
Dimensions L × W × H	21.7″ × 14.8″ × 19.0″	21.7″ × 16.3″ × 21.1″	21.7″ × 17.5″ × 21.1″	25″ × 17.6″ × 23.5″
Dry Weight	100 lbs.	124 lbs.	135 lbs.	179 lbs.

FEATURES

	G2400R	G3500R	G3501R	G5501R
GENERATOR				
Brushless Design	•	•	•	•
Condenser Type Voltage Regulator	•	•	•	•
AC & DC Circuit Breakers	•	•	•	•
Full Power Switch		•	•	•
ENGINE				
Low Oil Automatic Shut-off	•	•	•	•
Automatic Idle Control		•	•	
Electronic Ignition	•	•	•	•
Noise Suppression	•	•	•	•
Easy Start Recoil Starter	•	•	•	•
Auto Decompression	•	•		
Electric Start			•*	•*
Spark Arrestor Muffler	•	•	•	•
Fuel Gauge	•	•	•	•
Shock Mount Vibration Isolation	•	•	•	•
Roll Cage Frame	•	•	•	•

ACCESSORIES

	G2400R	G3500R	G3501R	G5501R
OPTIONS				
Portability Kit	•	•	•	•
Lifting Hooks	•	•	•	•

RECEPTACLES

A
One Duplex 120V - 20A GFCI
(Ground Fault Circuit Interrupter) Receptacle
NEMA 5 - 20R

B
One 120V - 30A Locking Recpetacle
NEMA L5 - 30R

C
One 240V - 20A Locking Recpetacle
NEMA L14 - 20R

*Battery not included.

GSC-190-75M   MA-111-90

Big Uns & Lil Uns

Richard Perez

One of the greatest advantages to using sunshine to make electricity is freedom— freedom to live where we want and how we want. All we have to do is lightly tap Mama Nature for a smidgen of her endless energy. Our system's size depends on us, on our needs and desires. If you don't need the power, then you don't have to either produce it or pay for it. Here's an article about a large system that meets large needs and a smaller one that meets smaller needs. Both work and are cost-effective. Both point out the freedom and flexibility built into photovoltaic systems.

Lil Uns

When Bill and Jean Andrews moved to their mountain home in June of this year, they were ready to leave many conveniences behind. Well, Bill was and Jean remained to be convinced... Bill, a retired logger, and Jean love the peaceful beauty of the high country. At 4,500 feet in the Siskiyou Mountains of southwestern Oregon, their home has a panoramic view of snow-covered Mt. Shasta some 60 miles away. Bill & Jean's home is on a south facing slope surrounded by tall douglas fir and ponderosa pine trees. There's a spring that flows into a small pond in their front yard, home to a least a million frogs and tadpoles.

Their 80 year old log cabin is located about 2.5 airline miles from commercial electricity. The nearest paved highway is over five long, rough, muddy, and deeply rutted miles away. Electrical alternatives, other than running in the commercial grid at $70,600, included an engine/generator and photovoltaics (PVs). They choose to use a stand alone PV system for essential electrical chores like communication and lighting.

Bill & Jean's Electrical Consumption. Every system starts with a thorough survey of the appliances, Bill & Jean's was no exception. They sought help from Electron Connection in specifying and installing their system. In Bill & Jean's case the list of appliances was very short. They only need electricity for two functions, communications and lighting. The chart below details the appliances and their consumption.

We decided right off to put all the electrical appliances on 12 VDC and not to use an inverter in this system. With such small scale consumption, PVs are easily capable of producing all the energy without the necessity of a back-up engine/generator. Twelve Volt lighting is readily available in either fluorescent or incandescent models. Just about all 2-way radios, either CB or radiotelephone, are available in 12 VDC powered models. Electrical power consumption averages about 500 Watt-hours daily.

Bill & Jean's PV System. This system has only two major components, PV panels and batteries. The PVs produce the power and the batteries store it. Very simple and very direct and very inexpensive. The two Trojan L-16Ws form a battery that will store about 6 sunless days of power for Bill and Jean. Each Kyocera 48 Watt PV panel will produce about 250 Watt-hours per sunny day in this location. There is NO generator in this system, PVs are the ONLY power input. The cost of this system is detailed in the spreadsheet and chart below. Please note that the low cost of this system is due to Bill & Jean's very small electrical consumption. Once again, if you don't consume the energy, then you don't have to generate it, store it or convert it. Stand alone PV system cost is directly proportional to the amount of power required from the system. Note also that Bill & Jean had us install a rack for four PV panels eventhough they now only use two panels. In stand-alone PV systems this is a very good idea. As the system's electrical consumption grows (and it always seems to), then adding more panels is simple and direct.

No.	Item Description	Cost	%
2	Kyocera 48W. Photovoltaic Panels	$712	52%
2	Trojan L-16W Batteries	$550	40%
1	PV Mounting Rack (4 panels)	$100	7%
1	Battery/Inverter Cable	$15	1%

Total System Cost $1,377

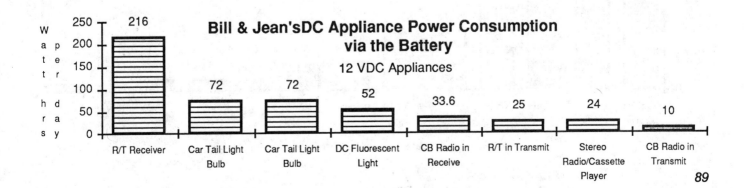

Bill & Jean's DC Appliance Power Consumption via the Battery

12 VDC Appliances

Watt hours/day

- R/T Receiver: 216
- Car Tail Light Bulb: 72
- Car Tail Light Bulb: 72
- DC Fluorescent Light: 52
- CB Radio in Receive: 33.6
- R/T in Transmit: 25
- Stereo Radio/Cassette Player: 24
- CB Radio in Transmit: 10

Big Uns

Jim and Laura Flett moved to the Siskiyou Mountains in 1980. This is the very same mountain range that Bill & Jean inhabit, but far enough south to be in California rather than Oregon. Jim is a farrier and Laura is a physical therapist, both run their own businesses out of their backwoods home. Their 80 acre homestead along Camp Creek is home to Jim, Laura, their two children Saylor and Dana, two horses- Shorty and José, and numerous other critters. Jim and Laura moved to the hills for the same reasons that most of us have- freedom, a clean unspoiled environment, and some basic peace & quiet.

Jim & Laura's home is located about 2 miles from the nearest commercial electricity. At today's rates, the power company wants about $60,000 to run in the lines. It's easy to see why Jim & Laura decided to make their own power.

Jim & Laura's Power Consumption

Jim and Laura's home is a large ranch house equipped with the conveniences needed for effective country living. They consume electricity both as 120 vac from their inverter and as 12 VDC directly from the batteries. The chart below details the major consumers of inverter produced 120 vac power.

The 12 VDC appliances are powered directly from the batteries and are detailed in the chart below.

Total power consumption adds up to about 2,000 Watt-hours per day, including inverter inefficiency and several small intermittently used appliances not listed in the charts.

When Jim and Laura first moved to the mountains, their electrical system was much smaller and sourced by a single engine generator. During the eight years before they invested in PVs, an inverter and a much larger battery pack, they learned well the lessons of conservation. Even now, they religiously perform the small tasks that make their system so efficient and effective. Tasks like, turning off lights that are not in use, using efficient lighting and placing it where illumination is needed. While Jim & Laura use their system like veteran energy misers, the visitors to their home are unaware that it's not plugged into the grid. Some visitors leave the house without ever knowing that the electricity they used there was solar produced and battery stored.

Jim & Laura' house is interesting from an electrical standpoint because the home is totally wired for both 12 VDC and 120 vac. Everywhere there is a 120 vac wall receptacle (and there are lots of them because the house is wired to NEC code), there is a 12 VDC outlet. Everywhere there is a permanently mounted 120 vac light, there is also a permanently mounted 12 VDC light. The 12 VDC wiring system was done as follows. Two 0 gauge copper cables feed DC energy from the battery to a buss in the attic. This buss is made from #2 copper wire and runs about 65 feet along the attic crawl space of this single story ranch type house. Each 12 VDC outlet or light is individually connected to the buss (no daisy-chaining like 120 vac circuits). **All connections to the buss are soldered!**

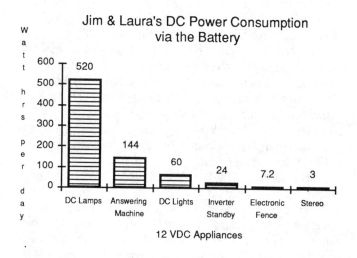

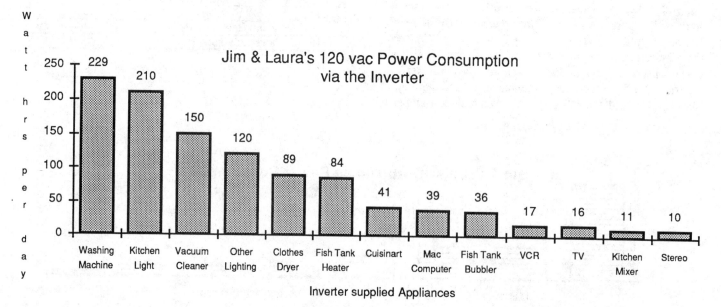

Jim & Laura's PV/Engine System

First off, this system is very different from Bill and Jean's because it didn't happen all at once or as an integrated unit. It grew gradually over the years with their needs and as the equipment became affordable. As such, Jim & Laura's system contains items like redundant generators that are only very occasionally used now.

The main power source is eight 48 Watt Kyocera PV modules. This photovoltaic array produces about 23 Amperes in full sun, or about 2,000 Watt-hours daily. The PV produced electricity is passed through a Heliotrope CC-60 PWM charge controller (see HP#8 for a review of the CC-60) which prevents overcharging the batteries. Power storage is in six Trojan L-16W batteries with a total capacity of 1,050 Ampere-hours at 12 VDC. This storage is sufficient to run the system for over 5 days with no energy input. The L-16Ws are equipped with Hydrocap catalytic converters to reduce gassing into the house and water consumption (see HP#11 for a report on Hydrocaps). This system uses a 1.2 kW. Heart inverter/battery charger to convert the battery stored power into 120 vac for household use.

Two engine/generators are used in this system. The first is a 120/240 vac Honda ES6500 generator. This unit produces 6,500 watts and can both directly source the system and recharge the batteries via the Heart inverter's battery charger. Jim had very good things to say about this two cylinder, overhead valve and cam, water cooled generator. He's found it to be very quiet, reliable and easy on fuel. The second generator in this system is a Mark VI type 100 Ampere, 12 VDC alternator setup. This powerplant uses a 5hp. Honda OHV single cylinder engine to turn a 100 Ampere Chrysler automotive alternator. The unit is controlled by Electron Connection's Mark VI alternator field controller. See HP#2 for a complete description of the Mark VI system and how to build one. The DC plant was used extensively to recharge the batteries before Jim & Laura installed their PV panels. Since the PVs arrived, both generators are getting a real vacation with only light use on the ES6500 and almost no use on the

Mark VI system. Jim uses the ES6500 for large tools and Laura uses it on "cleanup days" when the washing machine, clothes dryer, and vacuum cleaner are all running at once.

The spreadsheet and chart below show the costs of Jim & Laura's system. Please note that this cost breakdown includes **all** the power equipment used in the system. If the system were specified now, then at least one of the generators could be deleted. However, this system grew up before PVs were affordable and hence, extra generators.

Several other aspects of Jim & Laura's system are interesting although they have nothing to do with electricity. Jim uses a Dempster wind powered jack pump to move water from his deep well to a holding tank behind the house. Jim & Laura's extensive vegetable garden is drip water and micro sprinkled from a gravity flow system sourced by nearby Camp Creek.

Big Uns & Lil Uns

Both the systems described here use PVs for power input and both store power in batteries. The cost and amount of power produced & consumed is different. The Big Un cycles about 2,000 Watt-hours daily and the Lil Un cycles about 500 Watt-hours daily. The higher cost of the larger system is due to its increased flexibility and capabilities. The Big Un uses an inverter (and the increased battery capacity to power it) that allows Jim and Laura essential 120 vac appliances like a computer & printer for their businesses, and appliances for Jim's kitchen wizardry. And other small essentials like a heater and bubbler for the fish tank (while this may not seem essential to some, it certainly does to the fish and 4 year old Saylor Flett who loves them). The Big Un also supports large appliances like a washing machine, gas fired clothes dryer, and a vacuum cleaner. The engine/generators assure that that Jim & Laura won't run out of power during extended cloudy periods or even extended visits from switch-flipping city folks.

And there's good things to say about the Lil Uns too. Small stand alone PV systems are supremely reliable and very cost effective. The Lil Un delivers essential power for lighting and communications that otherwise have to be sourced by a noisy, expensive and unreliable engine/generator. The Lil Un is very reliable because it doesn't depend on complex, expensive electronics like an inverter, but uses power directly from the battery as 12 VDC. Since the only power input is solar, the Lil Un has virtually no maintenance other than occasionally watering the batteries.

So, ya pays yer money and makes yer choice. The size, complexity and cost of a power system depends on what you require of it. Those with simple low powered requirements can have what they need for very little cost. Those requiring more will have to use larger systems that cost more. But the essential feature here is that you don't have to have any more than you need. And that's up to you to decide. Freedom...

No.	Item Description	Item Total	%
1	Honda ac Generator	$2,860.00	27%
8	Kyocera PV Panels	$2,848.00	27%
6	Trojan L-16W Batteries	$1,650.00	16%
1	Heart Inverter/Charger	$1,335.00	13%
1	DC Engine/Alternator	$1,100.00	10%
2	PV Mounting Racks	$275.00	3%
1	Heliotrope Controller	$219.75	2%
9	Battery/Inverter Cables	$183.73	2%
18	Hydrocaps	$102.60	1%

Total System Cost $10,574.08

Chapter 3

Energy Storage & Management

Batteries

Battery storage is the foundation of an alternative energy system and often the least understood component. It serves as a buffer to provide electricity during the times no energy is being generated or a system is undergoing maintenance or repair.

Admittedly, the battery is today the weakest link compared to the other electronic marvels incorporated in the overall alternative energy package, but that seems to be slowly changing. In the meantime, we must turn to traditional battery technology now nearly 100 years old; however, it is tried and true and today requires surprisingly little maintenance.

The battery you choose to supply your energy demand will have a predetermined electrical storage capacity known as amphours, and one cannot store more energy than a battery's storage capacity. However, some flexibility can be gained and the net storage capacity increased by combining more than one battery. We will describe this procedure called paralleling and series wiring of batteries a little later.

When in use a battery discharges amphours, and the rate of this discharge over a certain period of time and at a certain temperature determines its amphour capacity. This measurement varies with different manufacturers and

is intended as a guide only and is not to be taken as an absolute guarantee. As an example, a manufacturer may discharge a battery over a 20-hour period at 10 amphr at an ambient temperature of 82° and rate the battery at 200 amphr. If you discharge the battery at home using 5 amphr at an ambient temperature of 68°, the capacity of the battery will be entirely different. All things being equal for a new battery, you will probably have slightly more amphours available. If you discharge the battery at 20 amphr in a 40° environment, the amphr capacity will be less. (The warmer a battery is kept, the closer it will be to storing the manufacturer's amphour rating.) How much more or less capacity one would have in a battery under the above circumstances cannot be measured with the tools available to most users, so that is why we caution that the amphour capacity listed by a manufacturer is given only to loosely help determine the storage capacity of one size compared to another. It is, however, close enough to size home power systems. In the alternative energy industry the standard for rating batteries is at the 20 hour rate, and all of our batteries listed in the Sourcebook will be at that rate.

A typical manufacturer will list a battery according to type, guarantee or service adjustment in months, hourly discharge rate, and, for lead-acid batteries, surge power. You will be concerned with the first three categories. The surge power can be ignored because it refers to the amps available to start a car or truck at 0° F.

Often you will want to determine how many watthours a battery will store from its amphour rating. To do so, multiply the voltage times amphour rate. Example: a 12V battery rated at 220 amphr stores approximately 2,640 watthours or 2.64 kWh (12V x 220 amphour = 2,640 watthours).

There are four types of batteries you will be considering for your alternative energy system: lead-acid, lead-calcium, lead-antimony and nickel-cadmium (*nicad's*).

Nicads

Nickel-cadmium batteries (nicads) are the least likely to be recommended or considered strictly because of their cost. New, they can run $1,700 for 65 amphours with a life expectancy of 25 to 30 years or more based on more than 3,000 deep cycles. We've recently found some reconditioned nicads which makes the price far more reasonable and within the realm of possibility for AE systems. From time to time we do come across other pre-owned nicads rated "aircraft worthy" and make a good buy which we pass on to our customers through announcements in our regular newsletters. Watch for these announcements or give us a call to see if we have them. Conversely, if you happen to come across a supply, let us know and we will work with you to test and distribute them.

Nicads are generally good for all types of alternative energy systems; however, their potassium-hydroxide electrolyte is not at all compatible with lead-acid, lead-antimony or lead-calcium battery electrolyte so the two dissimilar battery types cannot be used in the same system. Nicads are generally available as a 12V battery or 1.2VDC cell which can be connected in series to achieve the desired net voltage. Nicads can be discharged down to 100% depth of discharge, and can be brought back to a full charge without denigrating the batteries' condition for a 20-30 year period.

Lead-Antimony Motive Power Deep Cycle

Lead-antimony motive power deep cycle batteries are next in line for high life expectancy and they will last many years (usually 15 to 20) based on their 3,000+ cycles available to an 80% depth of discharge. They are more costly than lead-calcium and lead-acid batteries, but more economical than nicads and are generally good for all types of higher output and industrial alternative energy systems. They are excellent for PV-generator hybrid systems. Our Chloride batteries fit into this category.

Lead-antimony batteries are available as a 2VDC cell which can be connected in series to achieve the desired net voltage. A five-year limited warranty is typical with this class of battery.

Lead-Calcium Shallow Cycle

Lead-calcium shallow cycle batteries are next in line after the lead-antimony for high life expectancy and they will last many years also (usually 10 to 15). However, their rating of 1,500+ cycles depends on allowing no more than a 25% discharge to occur before recharging begins so they are really not "deep cycle" but are "shallow cycle" batteries. They are more costly than lead-acid batteries but more economical than nicads and lead-antimony. They are good for all types of alternative energy systems where automatic or manual charging is begun at the 25% discharge level.

Lead-calcium batteries are available in 12V and 2VDC cells which can be connected in series to achieve the desired net voltage.

A five-year limited warranty is available on some models, one year prorata on others.

Once in a while we are offered lead-calcium shallow cycle 2V batteries that have been retired from telephone company stand-by service, usually because the company's equipment has been updated and new battery configurations are required. The amphour rating varies from 400 to 1,680. When we have them, we are able to offer them at low prices, so again, watch for our newsletter announcements or give us a call or write about availability, price, and shipping.

Lead-Acid Batteries

Lead-acid batteries are the most commonly available and most economical to buy and use in an alternative energy system if the right selection is made and proper maintenance is performed by the user.

Basically there are three groups of lead-acid batteries: car, commercial heavy-duty (trucks, buses, etc.) and commercial extra-heavy-duty, deep cycle (golf carts and electric vehicles).

Car Battery Group

A regular lead-acid battery will not last long in a home power system because it can be subjected to no more than forty 100% discharges before failure occurs and it may fail earlier than that depending on the manufacturer. Car batteries are designed to be constantly recharged by an automobile alternator and thus not particularly well suited for home power systems. Some manufacturers are calling their car batteries "deep cycle" which is misleading. They are not true deep cycle batteries but really heavy-duty batteries with thicker plates and some antimony which will last about 50% longer than a regular car battery. They can be used successfully for a limited time, but we recommend that marine type deep cycle batteries such as our group 24 and 27 Interstate battery with a 6% antimony content be used in that system. The car type battery is available in a 12V configuration only but can always be put in series or paralleled to increase voltage or amperage. The limited prorata warranty is only for 12 months, with a three-month replacement warranty. Typically these batteries last between 3 and 5 years when taken care of.

Commercial Heavy-Duty Group

Commercial, heavy-duty lead-acid batteries can be used successfully in small photovoltaic and hydro-electric systems. They usually contain no more antimony than 2%. This is important as the antimony content of a lead-acid battery will resist low-current charging devices. The higher the antimony content, the more resistance there is to photovoltaic input.

A good rule of thumb is to use the commercial heavy-duty group of batteries where the continuous charging current from your alternative energy system is below 120 watts (10 amps in a 12V system, 5 amps in a 24V system). The commercial, heavy-duty type battery is available in a 12V configuration only. We currently offer no batteries from this group.

The limited prorata warranty is usually for 12 months, 90 days replacement due to faulty workmanship or material and the balance on a prorata basis.

Commercial, Extra-Heavy-Duty, Deep Cycle Group

These are the true "deep cycle" group of lead-acid batteries. They have an antimony content of at least 6%, thicker and larger plates, and are rated to handle 500 or more 100% deep discharge cycles and be recharged at high amperage rates (80-85% of their rated capacity for up to an hour). These batteries are recommended for photovoltaic systems producing a continuous 10 amps or more, wind-electric generators, hydro-electric, and photo-gen hybrid systems.

The commercial, heavy-duty type deep cycle lead-acid battery is available in a 12V and 6VDC configuration.

The limited prorata warranty is usually for 12 months, 90 days replacement due to faulty workmanship or material and the balance on a prorata basis.
The length of time a battery will last depends on the temperature and the number and depth of charge cycles it is subjected to, rather than years of service. Maintenance and battery care are important to battery life expectancy, too. Our L-16, 350 amphour 6V and our T-105, 220 amphour 6V batteries fall into this group.

Sizing Your Battery System

From the introduction to the catalog, you will have arrived at the kilowatthours or amphours your system will use each day. An inefficiency loss figure of 20% should be added to your kWh consumption. Much of that loss is due to battery inefficiency. You must put more electricity back into the battery than you took out. Look at it as a service charge. That service charge will increase as the battery gets older.

Choosing the correct battery will be determined by several factors.
- The rated capacity of each battery.
- The kilowatthours (or amphours) consumed each day.
- The number of days you plan to operate the dwelling without any input.
- The charge controller or voltage regulation your system employs.

Batteries are available from us rated from 65 amphr (lead-acid) to 1,476 amphr (lead-antimony and lead-calcium). Almost any amphour capacity is available on a special order basis.

There are many theories for the proper sizing of battery capacity. We recommend your reading the articles by Richard Perez (Lead-Acid Batteries for Home Power Storage) and Eric Nashlund (Random Thoughts on Solar Dependence) following. For photovoltaic systems, we believe the most efficient battery capacity sizing is reached at C10 to C30: where C is the photovoltaic module output, 10 to 30 is the multiplier. For example: a single photovoltaic panel with a rated output of 3.0 amps should be connected to a battery rated at from 30 to 90 amphr. If your system generates 20 amps, the battery capacity would be 200 to 600 amphr. (C = 20 x 10 = 600 or C = 20 x 30 = 1,000). This happens to be the formula that allows our photovoltaic charge controllers to operate at maximum efficiency too.

For the small hydro-electric system we offer in this catalog, a battery capacity of C20 is recommended for proper voltage regulation.

Wind electric, on the other hand, will be sized according to the number of days in a row one is likely to be without wind power. If your electrical consumption is 2.0 kWh per day with a 24 VDC system and you can experience up to ten days without wind, you would require a battery storage capacity of at least 20 kWh or 833 amps. (20 kWh is 20,000 watthours divided by the system's voltage, which in this example is 24 = 833.3 amps).

For a hybrid system where wind, photovoltaics, a traditional generator, or photo-gen hybrid are combined, use the C30 to C50 formula for the average combined hourly output to determine the size of your battery storage capacity. As an example: if you employ a five module 12V photovoltaic system that collectively produces 12 amphr, a wind plant that contributes an average of another 2 amphr and a traditional generator/battery charger that inputs still another average 2 amphr, that totals 16. C30 to C50 of 16 amphr would dictate a battery storage of between 480 to 800 amphr and the best choice would be a deep cycle battery configuration.

The battery storage may be increased if it is not enough to sustain power during anticipated sunless or windless days. Most people prefer to prepare for the eventuality of three sunless days or "three days of autonomy."

Configuring Batteries

Batteries of similar types may be connected to each other in parallel or series. When you parallel two or more batteries, you increase the total amphour capacity. Placing two 220 amphr batteries in parallel stores 440 amphr. The voltages of all parallel batteries must match each other. So should the capacity. If not, the batteries will equalize to the smaller of the two capacities, causing the larger battery to rapidly sulfate.

Here is an example of what paralleled batteries look like.

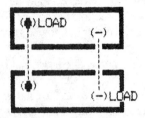

Batteries can also be connected in series. This combination gives you the total voltage of all the batteries in the series. Placing six 2V batteries in series produces a 12V battery (6 x 2 = 12). Be sure all series wired batteries match in amp/hr capacity, type and age.

Here is an example of what series wired batteries look like.

Finally, identical batteries may be connected in series and paralleled too. This increases the amphour capacity as well as the voltage.

Placing one group of two 220 amphr 12V batteries connected in series and paralleled with another group of two 220 amphr 12V batteries connected in series, produces a 440 amphr, 24VDC battery storage system.

Here is an example of what a series wired/parallel battery configuration looks like.

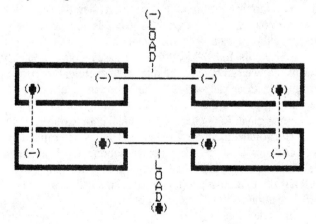

Helpful Hints

1. Batteries will self-discharge and can sulfate when left for long periods unattended and where no charge controller or trickle charge device is employed. Disconnect them in this case. However, self-discharge is not prevented by disconnecting batteries. Clean, dry battery tops is a must as a preventative measure.

2. Keep batteries warm. At 0° F, 50% of their rated capacity is lost.

3. All batteries must be vented to prevent potential explosion of hydrogen and other gases.

4. Do not, under any circumstances, locate any electrical equipment in a battery compartment.

5. Wear old clothes around electrolyte solution as they will soon be full of holes.

6. Be careful of metal or tools falling between battery terminals. The resulting spark can cause a battery to be destroyed or damaged.

7. Baking soda neutralizes battery acid, so keep a few boxes on hand.

8. Check the water level of your battery once a month. Excessive water loss indicates gassing and the need for a charge controller or voltage regulator. Use distilled water only.

9. Protect battery posts from corrosion. Use a professional spray or heavy-duty grease or vaseline. If charging has ceased or been reduced and the generator checks out, chances are corrosion has built up where the cable terminal contacts the battery post, preventing the current from entering the battery. This can occur even though the terminal and battery post look clean. Best yet, use our solid copper Pro-Start battery terminal protectors locating in the battery accessories section.

10. Never smoke or carry a lighted match near a battery, especially when charging.

11. Take frequent hydrometer readings. The battery manufacturer will provide instructions about the density of the electrolyte when their battery type is fully charged. The hydrometer is the best way to accurately monitor a battery.

12. Batteries gain a memory about how they are used. Large deviations from regular use after the memory has been established can adversely affect performance. That incurred memory can be erased by discharging the battery system 95%. Recharge it to 140% of its capacity at a slow rate, then rapidly discharge the battery again completely and recharge to normal capacity. A generator/battery charger combination should be used for this procedure.

13. Use fewer, larger cells in series rather than lots of small cells paralleled. For example use two 6V batteries in series rather than two 12V batteries in parallel. It is safer in the case of a shorted cell, you have half the chance of random cell failure, half the maintenance effort, and in general, thicker and more rugged plates - certainly more rugged than popular "ersatz" deep cycle batteries.

Lead-Acid Batteries for Home Power Storage

By Richard Perez, Home Power Magazine. (Note: All page references are to *Home Power Magazine*)

In 1970, we moved to the Mountains. The only desirable property we could afford was in the outback. Everything was many miles down a rough dirt road, far from civilized conveniences like electricity. We conquered the bad roads with a 4WD truck and countless hours of mechanical maintenance. The electrical power problem was not so easy to solve. We had to content ourselves with kerosene lighting and using hand tools. The best solution the marketplace could then offer was an engine driven generator. This required constant operation in order to supply power, in other words, expensive. It seemed that in America one either had power or one didn't.

We needed inexpensive home power. And we needed it to be there 24 hours a day without constantly running a noisy, gasoline eating, engine. At that time, NASA was about the only folks who could afford PVs. We started using lead-acid batteries to store the electricity produced by a small gas engine/generator. We'd withdraw energy from the batteries until they were empty and then refill them by running a lawnmower engine and car alternator. Since we stored enough energy to last about 4 days, we discharged and recharged the batteries about 100 times a year. Over years of this type service, we have learned much about lead-acid batteries— how they work and how to best use them. The following info has been hard won; we've made many expensive mistakes. We've also discovered how to efficiently and effectively coexist with the batteries that store our energy. Batteries are like many things in Life, mysterious until understood.

Before we can effectively communicate about batteries, we must share a common set of terms. Batteries and electricity, like many technical subjects, have their own particular jargon. Understanding these electrical terms is the first step to understanding your batteries.

Electrical Terms

Voltage. Voltage is electronic pressure. Electricity is electrons in motion. Voltage is the amount of pressure behind these electrons. Voltage is very similar to pressure in a water system. Consider a water hose. Water pressure forces the water through the hose. This situation is the same for an electron moving through a wire. A car uses 12 Volts, from a battery for starting. Commercial household power has a voltage of 120 volts. Batteries for renewable energy are usually assembled into packs of 12, 24, 32, or 48 volts.

Current. Current is the flow of electrons. The unit of electron flow in relation to time is called the Ampere. Consider the water hose analogy once again. If voltage is like water pressure, then current is like FLOW. Flow in water systems is measured in gallons per minute, while electron flow is measured in Amperes. A car tail light bulb consumes about 1 to 2 Amperes of electrical current. The headlights on a car consume about 8 Amperes each. The starter uses about 200 to 300 Amperes. Electrical current comes in two forms— direct current (DC) and alternating current (ac). In DC circuits the electrons flow in one direction ONLY. In ac circuits the electrons can flow in both directions. Regular household power is ac. Batteries store electrical power as direct current (DC).

Power. Power is the amount of energy that is being used or generated. The unit of power is the Watt. In the water hose analogy, power is can be compared to the total gallons of water transferred by the hose. Mathematically, power is the product of Voltage and Current. To find Power simply multiply Volts times Amperes. The amounts of power being used and generated determine the amount of energy that the battery must store.

Battery Terms

A Cell. The cell is the basic building block of all electrochemical batteries. The cell contains two active materials which react chemically to release free electrons (electrical energy). These active materials are usually solid and immersed in a liquid called the "electrolyte". The electrolyte is an electrically conductive liquid which acts as an electron transfer medium. In a lead acid cell, one of the active materials is lead dioxide (PbO_2) and forms the Positive pole (Anode) of the cell. The other active material is lead and forms the Negative pole (Cathode) of the cell. The lead acid cell uses an electrolyte composed of sulphuric acid (H_2SO_4).

During discharge, the cell's active materials undergo chemical reactions which release free electrons. These free electrons are available for our use at the cells electrical terminals or "poles". During discharge the actual chemical compositions of the active materials change. When all the active materials have undergone reaction, then the cell will produce no more free electrons. The cell is now completely discharged or in battery lingo, "dead".

Some cells, like the lead-acid cell, are rechargeable. This means that we can reverse the discharge chemical reaction by forcing electrons backwards through the cell. During the recharging process the active materials are gradually restored to their original, fully charged, chemical composition.

The voltage of an electrochemical cell is determined by the active materials used in its construction. The lead-acid cell develops a voltage of around 2 Volts DC. The voltage of a cell has no relationship to its physical size. All lead acid cells produce about 2 VDC regardless of size.

In the lead acid cell, the sulphuric acid electrolyte actually participates in the cell's electrochemical reaction. In most other battery technologies, like the nickel-cadmium cells, the electrolyte merely transfers electrons and does not change chemically as the cell discharges. In the lead-acid system, however, the electrolyte participates in the cell's reaction and the H_2SO_4 content of the electrolyte changes as the cell is discharged or charged. Typically the electrolyte in a fully charged cell is about 25% sulphuric acid with the remaining 75% being water. In the fully discharged lead-acid cell, the electrolyte is composed of less than 5% sulphuric acid with the remaining 95% being water. This happy fact allows us to determine how much energy a lead-acid cell contains by measuring the amount of acid remaining in its electrolyte.

A Battery. A battery is a group of electrochemical cells. Individual cells are collected into batteries to either increase the voltage or the electrical capacity of the resulting battery pack. For example, an automotive electrical system requires 12 VDC for operation. How is this accomplished with a basic 2 VDC lead-acid cell? The cells are wired together in series, this makes a battery that has the combined voltages of the cells. A 12 Volt lead-acid battery has six (6) cells, each wired anode to cathode (in series) to produce 12 VDC. Cells are combined in series for a voltage increase or in parallel for an electrical capacity increase. Figure 1 illustrates electrochemical cells assembled into batteries.

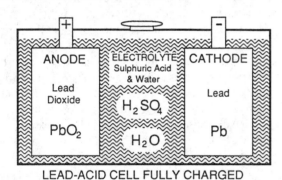

LEAD-ACID CELL FULLY CHARGED

$$PbO_2 + Pb + 2H_2SO_4 \underset{DISCHARGE}{\overset{CHARGE}{\rightleftarrows}} 2PbSO_4 + 2H_2O$$

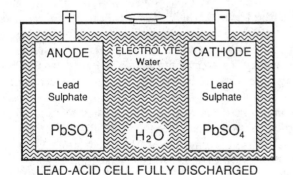

LEAD-ACID CELL FULLY DISCHARGED

FIGURE 1

Battery Capacity. Battery capacity is the amount of energy a battery contains. Battery capacity is usually rated in Ampere-hours (A-h) at a given voltage. Watt-hours (W-h) is another unit used to quantify battery capacity. While a single cell is limited in voltage by its materials, the electrical capacity of a cell is limited only by its size. The larger the cell, the more reactive materials contained within it, and the larger the electrical capacity of the cell in Ampere-hours.

A battery rated at 100 Ampere-hours will deliver 100 Amperes of current for 1 hour. It can also deliver 10 Amperes for 10 hours, or 1 Ampere for 100 hours. The average car battery has a capacity of about 60 Ampere-hours. Renewable energy battery packs contain from 350 to 4,900 Ampere-hours. The specified capacity of a battery pack is determined by two factors— how much energy is needed and how long the battery must supply this energy. Renewable energy systems work best with between 4 and 21 days of storage potential.

A battery is similar to a bucket. It will only contain so much electrical energy, just as the bucket will only contain so much water. The amount of capacity a battery has is roughly determined by its size and weight, just as a bucket's capacity is determined by its size. It is difficult to water a very large garden with one small bucket, it is also difficult to run a homestead on an undersized battery. If a battery based renewable energy system is to really work, it is essential that the battery have enough capacity to do the job. Undersized batteries are one of the major reasons that some folks are not happy with their renewable energy systems.

Battery capacity is a very important factor in sizing renewable energy systems. The size of the battery is determined by the amount of energy you need and how long you wish to go between battery rechargings. The capacity of the battery then determines the size of the charge source. Everything must be balanced if the system is to be efficient and long-lived.

State of Charge (SOC). A battery's state of charge is a percentage figure giving the amount of energy remaining in the battery. A 300 Ampere-hour battery at a 90% state of charge will contain 270 Amperes-hours of energy. At a 50% state of charge the same battery will contain 150 Amperehours. A battery which is discharged to a 20% or less state of charge is said to be "deep cycled". Shallow cycle service withdraws less than 10% of the battery's energy per cycle.

Lead-Acid Batteries. Lead-acid batteries are really the only type to consider for home energy storage at the present time. Other types of batteries, such as nickel-cadmium, are being made and sold, but they are simply too expensive to fit into low budget electrical schemes. We started out using car batteries.

Automotive Starting Batteries. The main thing we learned from using car batteries in deep cycle service is DON'T. Automotive starting batteries are not designed for deep cycle service; they don't last. Although they are cheap to buy, they are much more expensive to use over a period of several years. They wear out very quickly.

Car Battery Construction. The plates of a car battery are made from lead sponge. The idea is to expose the maximum plate surface area for chemical reaction. Using lead sponge makes the battery able to deliver high currents and still be as light and cheap as possible. The sponge type plates do not have the mechanical ruggedness necessary for repeated deep cycling over a period of many years. They simply crumble with age.

Car Battery Service. Car batteries are designed to provide up to 300 Amperes of current for very short periods of time (less than 10 seconds). After the car has started, the battery is then constantly trickle charged by the car's alternator. In car starting service, the battery is usually discharged less than 1% of its rated capacity. The car battery is designed for this very shallow cycle service.

Car Battery Life Expectancy & Cost. Our experience has shown us that automobile starting batteries last about 200 cycles in deep cycle service. This is a very short period of time, usually less than 2 years. Due to their short lifespan in home energy systems, they are more than 3 times as expensive to use as a true deep cycle battery. Car batteries cost around $60. for 100 Ampere-hours at 12 volts.

Beware of Ersatz "Deep Cycle" Batteries. After the failure of the car batteries we tried the so called "deep cycle" type offered to us by our local battery shop. These turned out to be warmed over car batteries and lasted about 400 cycles. They were slightly more expensive, $100. for 105 Ampere-hours at 12 volts. You can spot these imitation deep cycle batteries by their small size and light weight. They are cased with automotive type cases. Their plates are indeed more rugged than the car battery, but still not tough enough for the long haul.

True "Deep Cycle" Batteries. After many battery failures and much time in the dark, we finally tried a real deep cycle battery. These batteries were hard to find; we had to have them shipped in as they were not available locally. In fact, the local battery shops didn't seem to know they existed. Although deep cycle types use the same chemical reactions to store energy as the car battery, they are very differently made.

Deep Cycle Physical Construction. The plates of a real deep cycle battery are made of scored sheet lead. These plates are many times thicker than the plates in car batteries, and they are solid lead, not sponge lead. This lead is alloyed with up to 16% antimony to make the plates harder and more durable. The cell cases are large; a typical deep cycle battery is over 3 times the size of a car battery. Deep cycle batteries weigh between 120 and 400 pounds. We tried the Trojan L-16W. This is a 6 Volt 350 Ampere-hour battery, made by Trojan Batteries Inc., 1395 Evans Ave., San Francisco, CA (415) 826-2600. The L-16W weighs 125 pounds and contains over 9 quarts of sulphuric acid. The "W" designates a Wrapping of the plates with perforated nylon socks. Wrapping, in our experience, adds years to the battery's longevity. We wired 2 of the L-16Ws in series to give us 12 Volts at 350 Ampere-hours.

Deep Cycle Service. The deep cycle battery is designed to have 80% of its capacity withdrawn repeatedly over many cycles. They are optimized for longevity. If you are using battery stored energy for your home, this is the only type of lead-acid battery to use. Deep cycle batteries are also used for motive power. In fact, many more are used in forklifts than in renewable energy systems.

Deep Cycle Life Expectancy & Cost. A deep cycle battery will last at least 5 years. In many cases, batteries last over 10 years and give over 1,500 deep cycles. In order to get maximum longevity from the deep cycle battery, it must be cycled properly. All chemical batteries can be ruined very quickly if they are improperly used. A 12 Volt 350 Ampere-hour battery costs around $440. Shipping can be expensive on these batteries. They are corrosive and heavy, and must be shipped motor freight.

Deep Cycle Performance. The more we understood our batteries, the better use we made of them. This information applies to high antimony, lead-acid deep cycle batteries used in homestead renewable energy service. In order to relate to your system you will need a voltmeter. An accurate voltmeter is the best source of information about our battery's performance. It is essential for answering the two basic questions of battery operation— when to charge and when to stop charging.

Voltage vs. Current. The battery's voltage depends on many factors. One is the rate, in relation to the battery's capacity, that energy is either being withdrawn from or added to the battery. The faster we discharge the battery, the lower its voltage becomes. The faster we recharge it, the higher its voltage gets. Try an experiment- hook the voltmeter to a battery and measure its voltage. Turn on some lights or add other loads to the battery. You'll see the voltage of the battery is lowered by powering the loads. This is perfectly normal and is caused by the nature of the lead-sulphuric acid electrochemical reaction. In homestead service this factor means high powered loads need large batteries. Trying to run large loads on a small capacity battery will result in very low voltage. The low voltage can ruin motors and dim lights.

Voltage vs. State of Charge. The voltage of a lead-acid battery gives a readout of how much energy is available from the battery. Figure 2, on page 26, illustrates the relationship between the battery's state of charge and its voltage for various charge and discharge rates. This graph and its companion, Fig.3, are placed in the center of the magazine as a tearout so you can put them on your wall if you wish. This graph is based on a 12 Volt lead-acid battery at room temperature. Simply multiply the voltage figures by 2 for a 24 Volt system, and by 4 for a 48 Volt system. This graph assumes that the battery is at room temperature 78°F. Use the C/100 discharge rate curve for batteries at rest (i.e. not under charge or discharge).

Temperature. The lead-acid battery's chemical reaction is sensitive to temperature. See the graph, Figure 3 on page 25, which shows the same info as Figure 2, but for COLD lead-acid batteries. Note the voltage depression under discharge and the voltage elevation under charge. The chemical reaction is very sluggish at cold temperatures. Battery efficiency and usable capacity drop radically at temperatures below 40° F. At 40°F., a lead acid battery has effectively lost about 20% of its capacity at 78°F. At 0°F., the same battery will have effectively lost 45% of its room temperature capacity. We keep our batteries inside, where we can keep them warm in the winter. Batteries banished to the woodshed or unheated garage will not perform well in the

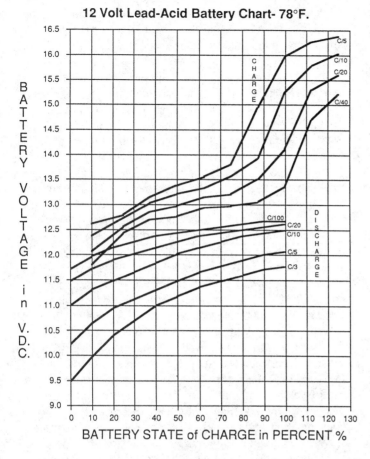

12 Volt Lead-Acid Battery Chart- 78°F.

BATTERY VOLTAGE in V.D.C.

BATTERY STATE of CHARGE in PERCENT %

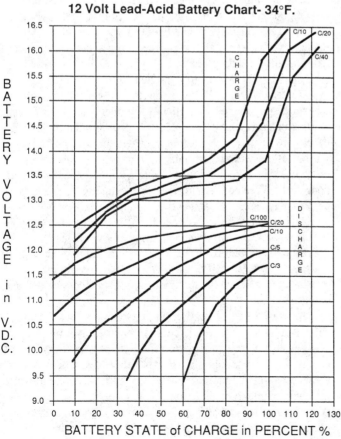

12 Volt Lead-Acid Battery Chart- 34°F.

BATTERY VOLTAGE in V.D.C.

BATTERY STATE of CHARGE in PERCENT %

winter. They will be more expensive to use and will not last as long. The best operating temperature is around 78° F..

The situation with temperature is further complicated by the lead-acid system's electrolyte. As the battery discharges the electrolyte loses its sulphuric acid and becomes mostly (≈95%) water. IT WILL FREEZE. Freezing usually ruptures the cell's cases and destroys the plates. Lead-acid batteries at < 20% SOC will freeze at around 18°F. If you're running lead acid batteries at low temperatures, then keep them fully charged to prevent freezing on very cold nights.

Determining State of Charge with a Hydrometer. A hydrometer is a device that measures the density of a liquid in comparison with the density of water. The density of the sulphuric acid electrolyte in the battery is an accurate indicator of the battery's state of charge. The electrolyte has greater density at greater states of charge. We prefer to use the battery's voltage as an indicator rather than opening the cells and measuring the electrolyte's specific gravity. Every time a cell is opened there is a chance for contamination of the cell's inards. Lead- acid batteries are chemical machines. If their cells are contaminated with dirt, dust, or other foreign material, then the cell's life and efficiency is greatly reduced. If you insist on using a hydrometer, make sure it is spotlessly clean and temperature compensated. Wash it in distilled water before and after measurements.

Rates of Charge/Discharge. Rates of charge and discharge are figures that tell us how fast we are either adding or removing energy from the battery. In actual use, this rate is a current measured in Amperes. Say we wish to use 50 Amperes of current to run a motor. This is quite a large load for a small 100 Ampere-hour battery. If the battery had a capacity of 2,000 Ampere-hours, then the load of 50 Amperes is a small load. It is difficult to talk about currents through batteries in terms of absolute Amperes of current. Battery people talk about these currents in relation to the battery's capacity.

Rates of charge and discharge are expressed as ratios of the battery's capacity in relation to time. Rate (of charge or discharge) is equal to the battery's capacity in Ampere-hours divided by the time in hours it takes to cycle the battery. If a completely discharged battery is totally filled in a 10 hour period, this is called a C/10 rate. C is the capacity of the battery in Ampere-hours and 10 is the number of hours it took for the complete cycle. This capacity figure is left unspecified so that we can use the information with any size battery pack.

For example, consider a 350 Ampere-hour battery. A C/10 rate of charge or discharge is 35 Amperes. A C/20 rate of charge or discharge is 17.5 Amperes. And so on... Now consider a 1,400 Ampere-hour battery. A C/10 rate here is 140 Amperes, while a C/20 rate is 70 Amperes. Note that the C/10 rate is different for the two different batteries; this is due to their different capacities. Battery people do this not to be confusing, but so we can all talk in the same terms, regardless of the capacity (size) of the battery under discussion.

Let's look at the charge rate first. For a number of technical reasons, it is most efficient to charge deep cycle lead-acid batteries at rates between C/10 and C/20. This means that the fully discharged battery pack is totally recharged in a 10 to 20 hour period. If the battery is recharged faster, say in 5 hours (C/5), then much more electrical energy will be lost as heat. Heating the battery's plates during charging causes them to undergo mechanical stress. This stress breaks down the plates. Deep cycle lead-acid batteries which are continually recharged at rates faster than C/10 will have shortened lifetimes. The best overall charging rate for deep cycle lead-acid batteries is the C/20 rate. The C/20 charge rate assures good efficiency and longevity by reducing plate stress. A battery should be completely refilled each time it is cycled. This yields maximum battery life by making **all** the active materials participate in the chemical reaction.

We often wish to determine a battery's state of charge while it is actually under charge or discharge. Figure 2, on page 25, illustrates the battery's state of charge in relation to its voltage for several charge/discharge rates. This graph is based on a 12 Volt battery pack at room temperature. For instance, if we are charging at the C/20 rate, then the battery is full when it reaches 14.0 volts. Once again the digital voltmeter is used to determine state of charge without opening the cells and risking contamination. Figure 3, on page 24, offers the same information as Figure 2, but in Figure 3 the information pertains to a lead-acid battery at 34°F. Note the depression of voltage under discharge and the voltage elevation under charge. This reflects an actual change in the batteries internal resistance to electrical flow. The colder the battery becomes, the higher its internal resistance gets, and the more radical the voltage swings under charge and discharge become.

The Equalizing Charge. After several months, the individual cells that make up the battery may differ in their states of charge. Voltage differences greater than 0.05 volts between the cells indicate it is time to equalize the individual cells. In order to do this, the battery is given an equalizing charge. **An equalizing charge is a controlled overcharge of an already full battery. Simply continue the recharging process at the C/20 rate for 5 to 7 hours after the battery is full.** Batteries should be equalized every 5 cycles or every 3 months, whichever comes first. Equalization is the best way to increase deep cycle lead-acid battery life. Battery voltage during the equalizing charge may go as high as 16.5 volts, especially if the battery's temperature is < 40°F.. This voltage is too high for many 12 Volt electronic appliances. Be sure to turn off all voltage sensitive gear while running an equalizing charge.

Wind machines and solar cells are not able to recharge the batteries at will. They are dependent on Mama Nature for energy input. We have found that most renewable energy systems need some form of backup power. The engine/generator can provide energy when the renewable energy source is not operating. The engine/generator can also supply the steady energy necessary for complete battery recharging and equalizing charges. The addition of an engine/generator also reduces the amount of battery capacity needed. Wind and solar sources need larger battery capacity to offset their intermittent nature. Home Power #2 discusses homebuilding a very efficient and supercheap 12 Volt DC source from a lawnmower motor and a car alternator.

Self-Discharge Rate vs. Temperature. All lead-acid batteries, regardless of type, will discharge themselves over a period of time. This energy is lost within the battery; it is not available for our use. The rate of self-discharge depends primarily on the battery's temperature. If the battery is stored at temperatures above 120° F., it will totally discharge itself in 4 weeks. At room temperatures, the battery will lose about 6% of its capacity weekly and be discharged in about 16 weeks. The rate of self-discharge increases with the battery's age. Due to self-discharge, it is not efficient to store energy in lead-acid batteries for periods longer than 3 weeks. Yes, it is possible to have too many batteries. If you're not cycling your batteries at least every 3 weeks, then you're wasting energy. If an active battery is to be stored, make sure it is first fully recharged and then put it in a cool place. Temperatures around 35° F. to 40° F. are ideal for inactive battery storage. The low temperature slows the rate of self-discharge. Be sure to warm the battery up and recharge it before using it.

Battery Capacity vs. Age. All batteries gradually lose some of their capacity as they age. When a battery manufacturer says his batteries are good for 5 years, he means that the battery will hold 80% of its original capacity after 5 years of proper service. Too rapid charging or discharging, cell contamination, and undercharging are examples of improper service which will greatly shorten any battery's life. Due to the delicate nature of chemical batteries most manufacturers do not guarantee them for long periods of time. On a brighter note, we have discovered that batteries which are treated with tender love and care can last twice as long as the manufacturer's claim they will. If you're using batteries, it really pays to know how to treat them.

Battery Cables. The size, length, and general condition of your battery cables are critical for proper performance. While the battery may have plenty of power to deliver, it can't deliver it effectively through undersized, too long or funky wiring. Battery (and especially inverter) cables should be made of large diameter copper cable with permanent soldered connectors. The acid environment surrounding lead-acid system plays hell with any and all connections. Connectors which are mechanically crimped to the wire are not acceptable for battery connection. The acid gradually works its way into the mechanical joint resulting in corrosion and high electrical resistance. See Home Power #7, page 36, for complete instructions on home made, low loss, soldered connectors and cables.

Battery Safety. Location plays a great part in battery safety. A battery room or shed, securely locked & properly ventilated, is a very good idea. Children, pets, and anyone not aware of the danger should never be allowed access to battery areas. Lead-acid batteries contain sulphuric acid, and lots of it. For example, a medium sized battery of 12 VDC at 1,400 A-h will contain some 18 gallons of nasty, corrosive, dangerous acid. Such a battery pack is capable of delivering over

4,000 Amperes of 12 VDC for short periods. Direct shorts across the battery can arc weld tools and instantly cause severe burns to anyone holding the tool. Be careful when handling wrenches or any metallic object around batteries. If tools make contact across the batteries electrical terminals, the results can be instantly disastrous.

When a lead-acid battery is almost full and undergoing recharging, the cell's produce gasses. These gasses are mostly oxygen and hydrogen- a potentially explosive mixture. Battery areas should be well ventilated during recharging and especially during equalizing charges to dissipate the gasses produced. If a blower is used in ventilation, make sure that it employs a "sparkless motor". See Home Power #6, page 31, for specific info on venting lead-acid batteries.

Battery Maintenance. There is more to battery care than keeping their tops clean. Maintenance begins with proper cycling. The two basic decisions are when to charge and when to stop charging. Begin to recharge the battery when it reaches a 20% state of charge or before. Recharge it until it is completely full. Both these decisions can be made via voltage measurement, amperage measurement and the information in Figures 2 and 3.

1. Don't discharge a deep cycle battery greater than 80% of its capacity.

2. When you recharge it, use a rate between C/10 and C/20.

3. When you recharge it, fill it all the way up.

4. Keep the battery as close to room temperature as possible.

5. Use only distilled water to replenish lost electrolyte.

6. Size the battery pack with enough capacity to last between 4 to 21 days. This assures proper rates of discharge.

7. Run an equalizing charge every 5 charges or every 3 months, whichever comes first.

8. Keep all batteries and their connections clean and corrision free.

Acid vs. Alkaline: the Electrochemical Shootout

Richard Perez & Chris Greacen

Ever wonder how different battery technologies stack up? Which is best? Which lives longest? How do different battery technologies compare in cost, weight or volume per amount of energy stored, temperature performance, self-discharge rate, and many other operating characteristics? Well, this is a comparison between the most used battery technologies. Included in these comparisons are three types of lead-acid cells and five types of nickel cells. All the info is in the table and the text here merely helps define how the info is categorized.

Electrochemical Cells- the contenders

In the table to the left, there is information about eight different types of cells. Three of these types are lead-acid cells: automotive, deep-cycle, and gel. Five of the types are nickel cells using alkaline electrolyte. Three of the nickel systems use cadmium as a cathode, while one uses a variety of metal hydrides, and the other iron. All of the cells mentioned are are secondary cells. Secondary cells can be recharged, while primary cells (like flashlight batteries) can only be discharged once and cannot be recharged.

Lead Acid Cells

All these cells use lead compounds as their anode and cathode material. The first type listed are automotive batter-

ies. These are the standard car battery with plates constructed out of lead sponge. The car battery is designed to do one thing, start your car at the minimum price. The deep-cycle cells mentioned are heavy-duty types whose plates are made from scored sheet lead alloyed with antimony. For example, the Trojan L-16W is such a deep-cycle battery. The third lead-acid type is the gel cell. Gel cells have a jellied electrolyte and are sealed cells. All these lead-acid systems use a dilute solution of sulphuric acid as their electrolyte.

Nickel Alkaline Cells

Sintered plate nicads are small sealed cells, like AA, C or D sized cells. The pocket plate types are vented wet cells, and two types are mentioned- new and reconditioned. As an example of the pocket plate nicad consider the Edison ED-160. The nickel-hydride cells are a new type of sealed cell being made by Ovonics. They use a variety of metal hydrides as their cathode material. The last type of alkaline secondary cells is the venerable nickel-iron type. All of these cells use a dilute solution of potassium hydroxide as their electrolyte.

The Comparisons

Here are the criteria and standards used in these battery comparisons.

Capacity. This row details the range of available cell

capacities, expressed in Ampere-hours, for each particular cell chemistry. The lead acid gel cells are becoming available in larger capacities, now up to over 100 Ampere-hours. The new nickel hydride types are not yet manufactured in sizes bigger than 3.8 Ampere-hours (a "C" sized cell). The other types are made in everything from tiny cells to ones you need a forklift to move.

Cell Voltages. Three types of voltages are covered for each cell at 78°F. The first called "Cell Operating Voltage" is a nominal voltage value for the cell under moderate discharge. All the lead acid types produce about 2 Volts. The nickel alkaline technologies all have cell voltages around 1.2 to 1.25 VDC under moderate discharge rates. "Cell Full Charge Voltage" indicates the voltage of a cell that is full and still undergoing recharging at a C/10 rate. "Cell Discharge Cut-off Voltage" is the voltage at which the cell is considered to be fully discharged and still under a moderate ($\approx$C/10) discharge rate.

Cost. For the cost figures, expressed in dollars per kiloWatt-hour of power stored, we used the following collections of cells (batteries). The lead acid automotive battery was a standard type from the local Les Schwab tire shop. The lead acid deep cycle battery is a Trojan L-16 W. The lead acid gel cell is a Panasonic 6.5 Ah model. The sintered plate nicad is a Panasonic "D" sized cell. The pocket plate nicad (both new

and reconditioned) is an Edison ED-160. The nickel hydride cell is an Ovonics "C" sized cell. The nickel iron cell is a Gould 35 Ah cell.

What follows in the table is an analysis of the cell's cost for its cycle lifetime and calendar lifetime. Cycle life is the average number of discharges (to 20% State of Charge) that the cell will undergo before failure. Calendar lifetime is rated for cells in float service. Float service means that the cell is continually under charge and only rarely sees a shallow (<10%) discharge cycle. All these figures are averages and are placed at extremes of service, i.e. regular deep cycling and virtually no cycling at all. As such, these lifetime figures cover the spectrum of longevity that an average user may expect.

Lifetimes on all electrochemical cells are greatly dependent on the cell's user. A careful user, one who follows the rules for that particular cell technology, will receive greater lifetimes than those on the chart. A slob will receive less. It's up to you to learn how to properly use your cells and then to do it.

Energy Density. We rated energy density in two fashions, one by the cell's weight in pounds, and the other by the cell's volume in cubic inches. One very notable feature here is the high powered Ovonics cells.

Cycle Rate. This row ranks the cells according to their

COMPARISON BETWEEN LEAD ACID AND NICKEL ALKALINE ELECTROCHEMICAL CELLS

Data complied by Richard Perez and Chris Greacen

	LEAD ACID SYSTEMS			NICKEL ALKALINE SYSTEMS CADMIUM			HYDRIDE	IRON
	Automotive	Deep-Cycle	Gel	Sintered	Pocket Plate	Reconditioned Pocket Plate	Sintered	Pocket Plate
Cell Capacity in Amp-hours/Cell at77°F.	33 to 340	60 to 2,000	1 to 120	0.5 to 4	20 to 1600	20 to 1600	0.5 to 3.8	20 to 1600
Cell Operating Voltage	2	2	2	1.25	1.25	1.25	1.25	1.25
Cell Full Charge Voltage	2.5	2.5	2.5	1.5	1.65	1.65	1.5	1.5
Cell Discharge Cut -off Voltage	1.75	1.8	1.8	1	1	1	1	1
Cost in $/kWh- Average	$69	$98	$375	$1,800	$668	$267	$3,372	$351
Cycle Life (discharge to 20% SOC)	200	1000	500	1000	3000	3000	500	3000
Cost in $/kWh/Cycle in Deep Cycle	$0.34	$0.10	$0.75	$1.80	$0.22	$0.09	$6.74	$0.12
Calendar Life in Float Service	3	15	10	10	40	40	10	40
Cost in $/kWh/year in Float	$22.84	$6.50	$37.46	$180.00	$16.70	$6.68	$337.17	$8.78
Energy Density by Weight (Wh./lb.)	16.40	17.22	16.48	12.50	9.14	9.14	23.75	7.03
Energy Density by Volume (Wh/cu. inch)	1.97	1.71	1.43	1.63	0.58	0.58	2.99	0.58
Self-Discharge Rate in %/Week- NEW Cells	6%	6%	6%	5%	2%	2%	10%	10%
Self-Discharge Rate in %/Week- OLD Cells	50%	50%	50%	10%	5%	5%	15%	10%
Cycle Rate -Cell's ability to transfer current	High	High	Medium	Very High	High	High	High	Low
Temperature Performance	Fair	Fair	Fair	Good	Good	Good	Good	Good
Capacity available at 104°F.	105%	105%	108%	98%	98%	98%	98%	98%
Capacity available at 32°F.	70%	70%	87%	90%	90%	90%	90%	90%
Capacity available at -20°F.	20%	20%	40%	78%	78%	78%	75%	65%
Ease of Use	Low	Medium	Medium	Medium	High	High	High	High
Cell Maintenance Requirements	High	High	None	None	Medium	Medium	None	Medium
CellDamaged by Total Discharge?	Yes	Yes	Maybe	No	No	No	No	No
Memory Effect?	No	No	No	Yes	No	No	No	No
Cell Requires Equalizing Charges?	Yes	Yes	Yes	No	No	No	No	No
Cell Portability: is it sealed?	No	No	Yes	Yes	No	No	Yes	No

ability to be rapidly cycled. It details the cell's ability to be discharged and recharged at fast rates (≥C/5). Electrochemical cells that power an inverter must deliver high rates of current in relation to their capacity.

Ease of Use. This category summarizes the characteristics listed below it into an overall rating of low, medium or high. "Ease of Use" gives the prospective battery user a look how easy it will be to keep the battery up and running. Many things, like physical maintenance, equalizing charges, etc. are all part of life on some electrochemical cells. This category summarizes the characteristics listed below it into an overall rating of low, medium or high.

Conclusions. Not in a minute. The table contains the data, you can make up your own mind about what type of electrochemical battery best suits your application.

We will, however, point out some interesting data on the table:

1) While lead acid automotive batteries are the cheapest to buy in $/kWh of storage, they are more expensive to operate than their lead acid cousins. Car batteries cost over three times more to use in either deep cycle or float service than do deep cycle types.

2) The self-discharge rate for all lead acid systems goes out of sight at the end of the cell's life. All nickel alkaline technologies maintain constant rates of self-discharge throughout their life.

3) Memory effect is limited to the sintered plate nicads and NOT shared by any other nickel alkaline types.

4) If you're running cells at low temperatures, then they had better be nickel alkaline types. Lead acid systems lose much of their capacity at low temperatures.

5) If you require maximum energy density, then the Ovonics nickel hydrides (NiH) are the only ones to consider. The energy density by weight of the Ovonics NiH is over 31% higher than any other type on the chart. The volumetric energy density is up over 50% from the nearest competitor. This, coupled with the NiH cell's sealed package, make it the best to use in portable service.

Nickel-Cadmium Storage Batteries

We've recently located a great supply of reconditioned nicad storage batteries. While we've always been enamored and fascinated with the potential of nickel-cadmium battery storage, the cost has always been way out of reach. Now there is a reasonable alternative to lead-acid!

There are a number of advantages to nickel-cadmium storage batteries over traditional lead-acid batteries. First, there is outstanding retention of charge while standing idle. At normal temperatures, only a small part of the charge is lost in a period of months, and this energy can be easily restored by recharging even after years of idleness. This phenomenon translates into a 100% depth of discharge for nicads! We recommend that a 50% discharge depth be employed with light duty batteries and a maximum of an 80% discharge rate be used with heavy duty industrial batteries (like chlorides). Nicads also have an incredibly low water consumption rate, allowing water additions only once every several years. Nicads have an excellent low-temperature performance and capacity with no bursting from exposure to sub-zero temperatures, regardless of charge state. Nicads can maintain a full charge in float service without periodic equalizing charges.

Nickel-cadmium batteries use a different electrolyte - potassium hydroxide, and a 12V system comes configured as 10 linked 1.2 volt cells. One disadvantage to Nicads is that their state-of-charge cannot be reliably measured by specific gravity or voltage. These reconditioned nicads are expected to last a full 20 years and are warranted on a pro-rata basis for 5 years. *There are usually 20-30 different sizes and styles of nicads available at any given time. If you are flexible, we can usually get you a battery quickly. The batteries listed below are only a few of the more common styles. All reconditioned nicads sell for $4.85 per amphour.*

■15-501	80 amphour Nicad Battery	$ 388
■15-502	120 amphour Nicad Battery	$ 582
■15-503	160 amphour Nicad Battery	$ 776
■15-504	240 amphour Nicad Battery	$1,164

Shipped freight collect from Oregon

Nicad batteries are subject to availability - check before ordering.

Trojan and Interstate Batteries

The battery spec'd out in our Remote Kits 1, 2, & 3 are 6 volt, 220 amphour batteries that are hooked up in series to provide 12V, and then paralleled for increased amperage. These batteries are either Trojan T-105's or a generic equivalent like the Interstate US-2200. Size 10.25"L x 7.25"W x 10.25"H. You can expect these batteries to last 3-5 years.

For a larger capacity battery with increased longevity (7-10 years if properly cared for) we recommend the Trojan (or Interstate) L-16 battery which is 6 volt at 350 Amphours. Size: 11-11/16"L x 7"W x 16"H.

We've been using Interstate batteries for many years with good success. There are more than 160,000 service centers around the United States to handle warranty and other services. The 24-SRM and the 27-SRM, 12V batteries, are of the lead-acid, high antimony (6%) deep cycle variety; same as the L-16's and the T-105's. Many people like the 24-SRM because it's interchangeable with car batteries (Size: 13"L x 6-5/8"W x 8.5"H) and has a handy carrying strap, and is rated at 85 amphours. The 27-SRM is slightly longer, (Size: 13"L x 7"W x 9"H) has the same carrying strap, and is rated at 95 amphours.

■15-101   Trojan T-105 (or equiv) $ 89
■15-102   Trojan L-16 (or equiv) $229
■15-103   Interstate 24-SRM $ 85
■15-104   Interstate 27-SRM $ 95
Batteries shipped freight collect from Northern California

Johnson Photovoltaic Batteries

The Johnson series of sealed, maintenance-free, deep-cycle, lead-acid batteries are designed for extremely long life in PV charging applications. These maintenance-free batteries are designed to give 100% recovery from deep discharge applications. With their uniquely thick lead-calcium grids and gelled electrolyte formula, these batteries far exceed the performance specs of their deep cycle competitors. These batteries are more expensive; when their cost is considered over time, they are well worth the extra outlay. Conservative test measurements show a minimum 1000 cycle life at 60% D.O.D. (depth of discharge), 1,500 cycles at 50% D.O.D. and 3,000 cycles at 40% D.O.D. Johnson batteries are a good middle ground between the less expensive, less long-lasting Trojan T-105 batteries and the far more expensive chloride industrial batteries in the under 500 amphour range. In applications using in excess of 500 amphours, generally the Chloride batteries are a better buy. Amphour rating is at 20 hour rate and all batteries listed are 12V.

Item Code	Battery	Amphours	Price
■15-301	12-31	31	$ 99
■15-302	12-55	55	$149
15-304	12-90	90	$215

We send all Johnson batteries UPS & our regular shipping charges apply to customers west of the Rockies. Additional charges apply to east of the rockies customers and HI/AK customers.

East of the Rockies Customers: *Add $25 additional to regular shipping charges for each battery ordered.*

Alaska & Hawaii Customers: *Because of weight and tracking considerations we must send these batteries either 1st class USPS or 2nd Day Air UPS. Therefore please add and additional $55 to our regular shipping charges*

15-308 East USA Add'l freight per battery $25
15-309 HI & AK Add'l freight per battery $55
* Amphour capacity is given at the 20 hr rate.

Prevailer Gel Cell Batteries

Prevailer gel cells, made by Sonnenschein, are truly 100% maintenance free. Prevailer batteries can be charged and discharged more frequently than any other battery because the lead-acid electrolyte is totally gelled. Gel cells are leakproof, sealed for life and can operate in any position (even upside down) with an extremely low self-discharge rate. These batteries can be taken down to a 100% depth of discharge and can sit totally discharged for 30 days with absolutely no damage. They are really the only batteries to use in extreme cold environments as they are rated at. To compare gel cell capacities with standard lead-acid batteries, multiply the amphour rating by 1.5. To summarize some of the advantages of gel cells (to justify their high cost!) over standard lead-acid batteries:

- They don't require acid checks or watering
- They can withstand shock and vibration better
- They can be stored up to 2 years without recharging
- They tolerate extreme cold and extreme heat
- There is virtually no gassing or emission of corrosive acid fumes

All Prevailer batteries come with a 5-year warranty and can be shipped via standard UPS (if 70 lb or less). There are numerous Prevailer gel cells available at voltages from 2V to 12V and capacities available from 1 to 110 amphr, but we'll only list the two most popular here as an introduction. Send SASE for more information.

■15-201 Gel Cell 85 amphour(12V) (A212/85G) $265
■15-202 Gel Cell 110 Amphour(12V) (A212/110A) $375
Shipped freight collect from the Midwest
Allow 4 to 6 weeks for delivery

Chloride Industrial Batteries

We used to shy away from selling batteries long distance due to the costliness of shipping them. Lately we've discovered that our chloride battery prices are so good that we can ship them to the East Coast and still beat local prices even with freight. We encourage you to request a freight quote if you're in the market for batteries even if you live far away.

Chloride batteries are by far the best batteries we sell. They come with a 5-year warranty, but with proper maintenance they can easily last 15 years and often 20 or more. They come in individual 2-volt cells hooked up in series to make your desired voltage. They can come either in individual metal cases or one case around a 12-volt battery. Write for a more detailed spec sheet. The prices given below are for a single can with six two-volt cells hooked in series to make a 12V battery.

Six or more cells in one case allows hinged battery covers to be provided. Covers must be specified. Case heights are 22.625" without cover and 23.25" with cover. Standard cable size is #2 for cells from 5-17 plates (75N5-75N17), 1/0 for cells from 19-23 plates, and 2/0 for cells with 25-33 plates.

All trays are coated with acid resistant paint. Jars are made of molded high-impact material. The cover is a high-impact material and lead terminal inserts have a leak tight coating and a floating bushing for the positive terminal. The separators are made of a micro-porous material and the vent cap is a bayonet type that is spray proof. The retainers are vertically wrapped glass mat with a folded perforated envelope and molded plastic bottom plate shield. Specific gravity is fully charged from 1.280 to 1.290 at 77 degrees F.

Chloride batteries are made of lead and lead is a commodity subject to price fluctuations. Check before purchasing batteries!

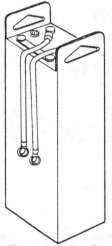

	Battery	Amphours	Weight	Price
■15-401	75N05	184	125#	$ 795
■15-402	75N07	277	228#	895
■15-403	75N09	389	300#	1,050
■15-404	75N11	461	396#	1,125
■15-405	75N13	554	468#	1,225
■15-406	75N15	646	546#	1,375
■15-407	75N17	738	618#	1,475
■15-408	75N19	830	690#	1,595
■15-409	75N21	923	768#	1,750
■15-410	75N23	1,015	846#	1,895
■15-411	75N25	1,107	918#	2,050
■15-412	75N27	1,199	996#	2,175
■15-413	75N29	1,292	1,068#	2,295
■15-414	75N31	1,384	1,146#	2,445
■15-415	75N33	1,476	1,218#	2,595

Shipped freight collect from Northern California
Allow 4-6 weeks for delivery

Nickel-Metal Hydride Battery Pack

Our 3.4 amphour Nickel-Metal Hydride battery pack represents a breakthrough in portable power. It can be clipped to a belt or placed in a leather/canvas portable case and used to power any number of 12-volt appliances or small AC loads through a small inverter. Portable cellular phones, radios and hobby toys are examples of applications, as well as diagnostic equipment requiring a reliable portable power source. Nickel-metal hydride is non-toxic and has no "memory effect," a characteristic of conventional nickel-cadmium rechargeable batteries to lose capacity if they are repeatedly recharged before fully discharging. It's almost three times the capacity of similar sized Nicad batteries. It measures 1.5" x 3.25" x 6.5", weighs 2.5 lbs. and has a female lighter plug attached. Like with nickel-cadmium cells, this battery uses nickel for the positive electrode, but it uses metal hydrides as the negative electrode. The disposal of these new rechargeables is much safer because they contain no cadmium.

15-551   Nickel-Metal Hydride Battery (3.4 AH)   **$195**

Battery Care Accessories

Charg-Chek Battery Tester

The Charg-Chek is a very simple hydrometer that accurately tests the state of charge within the temperature extremes of -40° to 130° F. It uses less than 1/4 oz. of acid, is easy to use, easy to read, and features an unbreakable rubber tip, leak-proof container, and pocket protector. You read the state of charge by the number of balls floating - one ball floating is 25% charged, four balls floating indicates a full charge.

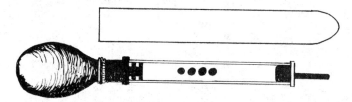

15-701   **Charg-Chek Mini Hydrometer**   **$5**

874-L Battery Hydrometer

This full size specific gravity tester is accurate and easy to use in all temperatures. Specific gravity levels are printed on the tough, see-through plastic body. It has a one-piece rubber bulb with a neoprene tip.

15-702   **Full Sized Hydrometer**   **$6**

Battery Service Kit

This kit has almost everything you need to service a battery. It contains an open end wrench to remove terminal bolts and nuts, a wire brush, a battery post and terminal cleaner, a safety grip battery lifter, a terminal clamp lifter (a must if you've ever tried to remove a corroded battery terminal,) an angle nose plier, a terminal clamp spreader, and a box wrench.

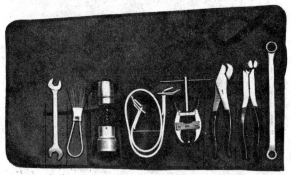

15-703   **Battery Service Kit**   **$49**

Post & Clamp Cleaner

The most basic battery tool for cleaning your posts and clamps.

15-704   **Post & Clamp Cleaner**   **$4.50**

Pro-Start Battery Terminals

Pro-Start solid brass non-corrosive battery terminals will last a lifetime, making a positive sealed connection between the battery and electrical system. The positive connection and higher conductivity allows more power solidly stored in the battery. Solid brass terminals have 3.5 times higher conductivity than lead connections. This elimination of corrosive oxides allows faster power flow back into the battery from any charging source. Pro-Start terminals can extend battery life potential by 50%, increase available power by 23% and forever end cable replacement and costly maintenance. Each terminal comes with a grease fitting which is filled with grease upon installation. When purchasing any battery system (cannot be used on Chlorides), we highly recommend Pro-Start. An **Arm Accessory Kit** fits into the terminal that can be used for multiple electrical hook-ups. Its application includes CBs, stereos, cellular phones, on-board computers, winches, multiple battery hook-ups, and generators.

15-781 Pro-Start Terminals (set of 2) $19
15-782 Pro-Start Arm Accessory Kit $ 6

Battery Treatment Kit

This kit includes a 1 oz. tube of corrosion preventative, an applicator brush, and two protection rings. It's a good idea to put a treatment kit on every new battery to prevent corrosion.

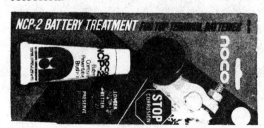

15-706 Battery Treatment Kit $4

Battery Maintenance Kit

This kit contains one 1-1/4 oz. aerosol can of NCP-2 battery corrosion preventative, one 2-1/2 oz. aerosol can of Noco battery bath cleaner, and one ST-11 side terminal pouch of two protectors. The protectors are impregnated with a corrosion preventative and slip over the battery post before the terminal is attached.

15-705 Battery Maintenance Kit $ 11

Insulated Battery Cables

For interconnecting batteries in series you need these insulated battery cables. Cables are 4 gauge stranded copper. Our most popular seller and what is used most frequently in putting 6V batteries in series are the 16" cables.

15-776 #4 Gauge Insulated Battery Cable - 16" $3
15-777 #4 Gauge Insulated Battery Cable - 24" $3.50

Battery Terminals

Wing Nut Battery Terminals

This is the terminal to use to convert conventional post-type batteries to a wing nut connection adapter with a 5/16" stud.

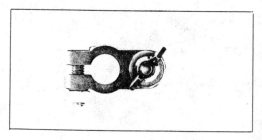

15-751 Wing Nut Battery Terminal   **$2**

Quick Connect Battery Terminal

This Quick Connect battery terminal converts conventional post-type batteries to a quick-disconnect adapter. The slide-apart feature allows disconnection when leaving the house or when current needs to be cut in a hurry.

15-752   Quick Connect Battery Terminal   **$6**

509   Bell Telephone Laboratories
Later Versions of This Solar Battery May Power Equipment in Satellites

Reprinted from National Geographic Magazine, circa 1954.

Large Battery Chargers

The battery charger you need to use with a generator in a hybrid system is not the same as those used to deliver surge power for starting a car or as part of the generator to maintain its starting battery. Those chargers are relatively inexpensive and are designed to deliver some high amperage for a short time to wake up a dead battery, after which they revert to their continuous rated output, usually no more than 10 to 30 amps until the battery is about 75% recharged. The charge rate thus tapers to as little as 2 amps for the last 25% of the recharging cycle. If you are using a 200 amphour battery, trying to put in the last 25% at 2 amps would be an expensive proposition in fuel cost, maintenance, and wear and tear on the generator.

The solution is the high-powered, industrial type battery chargers that follow. They plug directly into the generator and are activated when the generator is, thus automatically filling your battery system with stored power for later use while you pump water or use other high-powered appliances or tools that require the generator. Many units are equipped with timers and all can be connected to automatic load sensors through the generator that run on and shut off the generator/charger by remote control.

Selecting the Proper Charger

Selecting the proper charger is in some ways similar to the process used for choosing the proper size backup generator or inverter.

Look for a charger with input current that's not more than 90% of the generator's continuous duty rating at its available voltage level (120V or 240VAC). Also keep in mind that other tools or equipment will often be operating concurrently with the charger and if it's too powerful, the average charge rate you are planning on won't be achieved.

Another factor to check for is the charger's surge characteristics. The charger must not bog down the generator on start up, otherwise pumps and other equipment connected to the generator simply will not turn on or sometimes may be damaged by the resulting low-voltage, low-current condition.

If the charger's duty cycle is limited, i.e., you only intend to use it infrequently like once a week because you have a deep-cycle battery system, choose one with the highest ampere or kWh output possible while being careful to match the maximum charge rate to your battery bank's ability to tolerate it. Conversely, if you choose a shallow-cycle battery system where the discharge rate will not be more than 25%, choose a less powerful charger.

All the chargers we feature here are regulated to prevent overcharging the battery. The regulation is accomplished by tapering to a lower charge rate as the battery fills up with energy. This means that even though you may select a unit that delivers 100 amps, it will do so for only a short period of time like 10 or 15 minutes (except Todd which is much more efficient and will do it longer) and do it only if you battery is fully depleted. If your battery is 75% fully charged and the 100 amp charger is activated, the charge rate may only average 50 amps. This is important to realize, especially if you are depending on a given amount of power each hour the generator/battery charge runs to supplement photovoltaics or wind power output.

Based on our experience, a good general rule for matching charger-to-systems choice should be as follows:

- For shallow-cycle batteries where the charger will be activated most of the time at a 25-35% discharge level, select a 60 to 90 amp unit. The average delivery will be about 35-45 amphour.
- For deep-cycle batteries where the charger will be activated most of the time at a 50-75% level, select a 90 to 125 amp unit. The average delivery will be about 45-60 amphr.
- For deep-cycle or shallow-cycle batteries where the total amphour capacity exceeds 1,000, select a 125 to 270 amp unit. The average amphour delivery will vary depending on field conditions it should average near or slightly below 50% of the maximum rated output.

Battery Charger Types

The light-duty Marquette chargers used to be our economy line. Since Todd Engineering came up with its fine line of battery chargers, it no longer makes sense to deal with this type of "gas station" battery charger which does more "tapering" than it does hard charging.

The heavy-duty IBE industrial chargers are among the best made. We have sold these chargers sporadically over the last 12 years and have been generally quite pleased with their performance. However, once again, since Todd Engineering has come up with its great charger it makes no sense to spend the extra money on an IBE charger as the *Todd gives you everything and more for around half the price!*

Battery manufacturers can provide maximum rate data, but a good rule of thumb for deep-cycle lead-acid or lead-calcium batteries is not to deliver amperage of more than 80% of their rated capacity at the beginning of the charge cycle. Example: a fully depleted 100 amphr deep-cycle lead-acid battery could be charged at a rate of 80 amps.

Todd Battery Charger

The new battery charger manufactured by Todd Engineering (and also marketed and re-labeled by Heliotrope General) is the best battery charger we've found to date for a stand alone charger. Our most popular model is a 75 amp high-output, regulated voltage, pure DC output charger. It is available in three output voltages. It puts out 70-80% of rated output at just under 14V. It begins to taper only after 14V is achieved, allowing for battery charging in hours instead of days, **greatly reducing generator time and fuel costs.** Another big plus for this charger is its amazing efficiency, *using a scant 1050 watts for the 75 amp model!* It can deliver full output, on generators as small as 2000 Watts. With larger generators, this charger leaves you with enough reserve power to do other chores such as laundry, pumping, and vacuuming. It's a very versatile charger allowing series connection for 24V charging or unlimited paralleling to achieve as many amps as desired (making it far more reasonable than IBE). It is compact (15.5" x 7.25" x 4") and light (9#), can be wall mounted for added convenience and is warranted for one year. The low voltage model (14.0V) is designed for charging using standard AC power and can be left hooked up indefinitely for unattended operation. The medium voltage model (14.8V) is designed for faster and more efficient charging using generator power. The high voltage model (16.5V is designed for charging Nicad batteries regardless of the power source. All three models come with 2-speed fan. The charger is also available in a 45 amp model (using less than 1000 watts) and a 30 amp model (using less than 600 watts.)

15-631	75 amp Charger (14.0V)	$299
15-632	75 amp Charger (14.8V)	$299
15-633	75 amp Charger (16.5V)	$299
15-637	45 amp Charger (14.0V)	$249
15-634	45 amp Charger (14.8V)	$249
15-636	45 amp Charger (16.5V)	$249
15-639	30 amp Charger (14.0V)	$149
15-635	30 amp Charger (14.8V)	$149
16-638	30 amp Charger (16.5V)	$149

Balmar Diesel Battery Charger

The Balmar PC-100 is the original lightweight diesel charging unit. It uses a super efficient direct drive high output alternator to charge a full 114 amps @12-14V DC into your battery (or 65 amps at 24-28V DC for the 24V unit). It uses a 4.2 HP Yanmar engine with manual start (electric start optional) and uses less than 1 pint of diesel (#1 or #2 or Jet A) per hour. It is an ideal charger for emergency power, industrial charging, portable aircraft APU, alternative energy, or marine uses.

■15-621	Balmar Battery Charger (PC-100)	$2,495
■15-622	Mounting Base for Balmar	$ 195
■15-623	Electric Start for Balmar	$ 425

Shipped Freight Collect from Washington

Specific Gravity	Percentage of Full Charge
1.265	100%
1.225	75%
1.190	50%
1.155	25%
1.120	0%

Voltage	Percentage of Full Charge
12.7	100%
12.4	75%
12.2	50%
12.0	25%
11.9	0%

Charge Controllers, Metering, and Fuse Protection

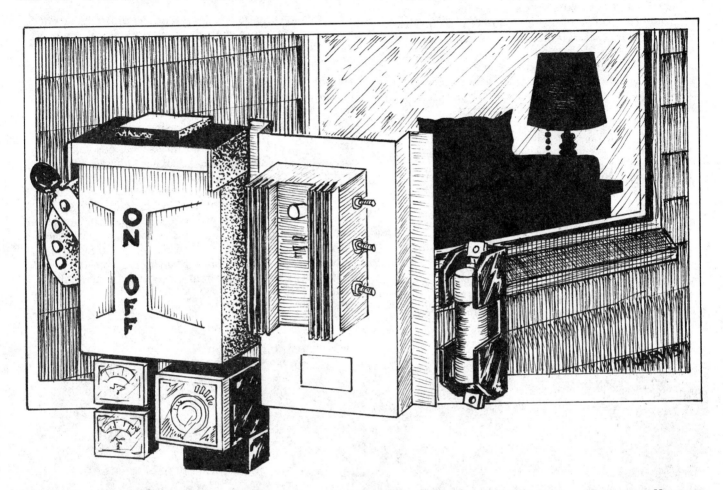

As a general rule, all systems should be equipped with a controller or regulator to protect the battery from overcharging and to maintain it properly for maximum life. Many of the controllers and meters we offer in this section may be interfaced with wind generators or hydro-electric plants; others can be used with PV only.

Meters are useful devices and necessary to indicate the status of your system. Some people have avoided installing them only to discover, sometimes up to a year later, that their system was not installed correctly or a malfunction had occurred without their knowledge. Protect your system and your warranties by knowing at a glance how everything is working. We offer compact, attractive, ready-to-connect monitoring panels and devices. Individual meters are available for those wishing to design and fabricate their own panel systems.

Fuses and circuit breakers are important to maintaining a safe low-voltage system. Use a safety disconnect switch and a main distribution fuse box or Suntronics fused receptacles and switches. In addition, fuse the line between your battery system and the main fuse box and fuse all generators, motors, equipment, or appliances connected directly to the battery.

Photovoltaic Charge Controllers

Photovoltaic power systems today often include batteries for storage of electricity generated during the day for use at night. Storage batteries are usually of the lead-acid type - which are electrically very sensitive to overcharging and undercharging. The battery charge control regulator, whose primary function is to control the magnitude and rate at which the PV array is allowed to charge the batteries, protects storage batteries from overcharge and undercharge.

Regulators today employ two basic components to electrically control the PV array: solid-state devices and electromechanical relays. Both connect and, upon detection of a full charge condition, disconnect the PV array output from the battery.

Solid-state regulators have no moving parts. Their usual method of controlling PV array output power provides two paths; one path leads to the battery (charging mode) while the other leads back to the PV array through a short circuit. (This is called the shunt method, and contrasts with the series method which provides an open path to halt all current flow.)

Upon detection of a full charge, a semiconductor transistor switch safely short-circuits the PV array. The majority of solid state regulators in 12 and 24VDC applications use the shunt method.

Electromechanical regulators use relays or contactors (moving parts) to switch the PV array current. This type of regulator is capable of operating to 120VDC or higher. Current capacity can exceed 200 amps, making this type applicable in large PV systems.

The newest hybrid type regulator employs the advantages of solid-state components, namely efficiency and reliability, with the high voltage and high current advantages of electromechanical components.

The primary consideration in choosing the proper charge controller for your system is in the amount of charging current you have to regulate. You also need to accommodate future potential needs by selecting a controller that will handle the extra power you may later add on.

The control center for the North Carolina Biodome project mentioned earlier consisting of 156 Sovonics 20 watt PV modules, six 2248 Trace inverters, and Stirling generator backup.

A List of PV Controller Absolutes

Do not use a PV charge controller to regulate any other charging device such as wind or hydro-electric. It will not work.

Do not connect more modules than the controller is rated for. However, more controllers of the same voltage may be connected together in parallel to increase the number of modules it will regulate.

Do not connect fluorescent lighting to the controller circuit without first going through the battery. Some fluorescent lighting will cause an RF (radio frequency) malfunction in the controller sensing ability.

Do use a charge controller for all installations unless you are planning a self-regulating system. Otherwise the battery will discharge into the PV module during sunless conditions at about 30 mA per module per hour. For a five module system or more, a blocking diode should be used for reverse current protection. No diode is needed with less than five modules because the line loss through the diode would be equal or more than the 30 mA loss through each of the modules. Many of our controllers are equipped with reverse-current protection, hence no diode is necessary.

Charge Controllers

Trace C-30A Charge Controller

The Trace Charge Controller has quickly become our best selling basic charge controller for systems up to 30 amps of charging current, and it is the charge controller we've chosen for both reliability and economy in our Remote Home Kits #2 and #3. It can be operated from either 12 or 24 volt battery systems. It comes factory preset for high voltage disconnect and low voltage return, but it's also user adjustable. The C-30A features a night-time disconnect eliminating the need for a diode and an "equalizing charge" switch that allows for occasional heavy charging of the battery bank.

| 25-103 | C30A Charge Controller | $95 |

M8 / M16 Sun Selector Charge Controller

If you don't need to regulate a full 30 amps of solar power, then the M8 or M16 is the way to go. The M8 and M16 are full-feature sophisticated charge controllers for a very reasonable price. Made by Sun Selector Electronics, who also makes the ingenious "Linear Current Boosters," these fine charge controllers eliminate the internal or external need for blocking diodes. They are specifically designed to eliminate battery gassing and maintenance. The M8 is rated at 8 amps maximum current, but realistically will handle up to 10 amps so it is the ideal controller for up to three 48-watt PV modules. Both the M8 and the M16 measure only 2" x 2" x 1-1/4" and are simple to install with just four wires to connect. The M16 will handle conservatively up to 16 amps of charging current. Both units feature four LED indicators which inform the user of a wide variety of system conditions. Sun Selector makes excellent charge controllers for a very reasonable price.

25-101	M8 8 amp Controller - 12V	$ 65
25-102	M16 16 amp Controller - 12V	$ 75
25-128	M8 8 amp Controller - 24V	$ 55
25-104	M16 16 amp Controller - 24V	$ 69

SCI Mark III 30 Amp Charge Controller

This 12 volt 30 amp charge controller is similar to the Trace C-30A. The two major differences are that the Mark III is nicer looking and it provides two step charging rather than the simple array connect/disconnect provided by the trace. The SCI Mark III will charge with 100% of available charging power up to a factory preset voltage of 14.8 V; here it switches to float charge mode charging at 3 Amps and tapering the batteries to 14.1 Volts. The 100% charging rate is resumed any time the battery voltage drops below 12.7 volts. A low voltage warning light indicates when the battery voltage falls below 11.5 Volts. The Mark III comes mounted on a 5-1/4" X 8-1/2" flush mount faceplate. The optional knock-out box will fully enclosed the unit. The Mark III has reverse-current protection eliminating the need for a diode.

25-113	Mark III (30A) Charge Controller	$109
25-114	Single SCI Knock-out Box	$ 25
25-115	Double SCI Knock-out Box	$ 35

Suntronics DV-15 Load Diverter

The DV-15 is a solar load diverter that allows you to switch from a charged battery to another battery or a load. It works with any charge controller. As soon as battery "A" is fully charged to 14.2 volts, it switches over to battery "B" to charge it, or to another load. The unit is fused for charging protection, and diode protected in case you hook it up backwards. Two 5 milliamp LED lights let you know which bank of batteries is being charged.

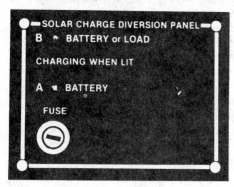

25-201   DV-15 Load Diverter $45

Outdoor Lighting Timer

This solid state timer is designed to sense darkness at the PV array or via an optional photo sensor and then automatically turn on the lights or other loads. The loads are then allowed to run for an adjustable period of time (10 min. - 15 hrs.).
This 12 volt timer can operate up to 8 amps. Enough for several low wattage walkway lights or two 50 watt porch lights. This timer can prevent deep battery discharging and because it senses darkness your outdoor lighting will automatically change with the seasons. Operates on only 12 mA.

25-402   Outdoor 12V Lighting Timer $ 89

Day/Night Switch

Another way to operate outdoor lighting is by sensing battery voltage and turning off the lights when the batteries discharge to a specific voltage. This does just that. This 12V 10 amp switch turns on the lights at dark and shuts them off at a user adjustable (11.5V to 13.5V) voltage level. It uses only 12 mA to operate.

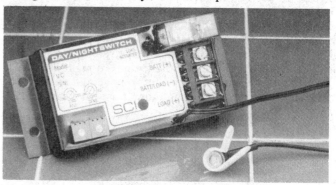

25-403   Day/Night Switch $ 89

Generator Auto-Start Controller

Our new GenMate is a super versatile computerized generator controller that can be customized to automatically start and stop nearly any electric start generator. Flipping tiny switches programs the GenMate - a one time operation. This controller can be told to automatically start the generator when the batteries get below a certain voltage or (for generator powered pumping) when the water tank is too low, then turn off when the task is complete. You can also select a pre-start warning beeper and the number of start cycle retrys to attempt before sounding a start failure alarm. To use this controller be sure that your generator is equipped with a low oil cutoff switch. All of the Makita and Onan generators in our catalog have this feature.

25-170   GenMate Auto-Start Controller $395

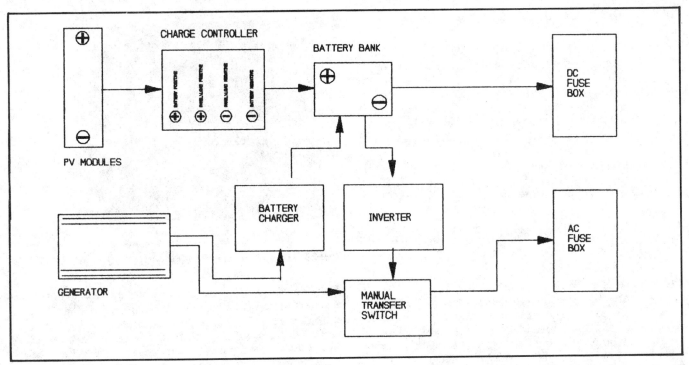

Courtesy The Solar Electric Independent Home Book

SCI Mark III 15 Amp Metered Charge Controller

For smaller systems this attractive full featured charge controller contains a blocking diode, circuit breaker, LED charging light and built-in volt and ammeters. It is available in 12 or 24 volts and will regulate up to 15 amps of incoming power. It uses only 28 mA of power. The Mark III contains everything you need in a small charge controller.

25-116	Mark III Charge Controller (12V)	$139
25-117	Mark III Charge Controller (24V)	$139

Heliotrope CC-20 Charge Controller

The CC-20 is a 20 amp, non-relay type series charge controller. It has "taper charge" circuitry, reverse polarity protection, selectable state-of-charge voltages, and a two battery selector switch. It also comes with LED bar graphs to monitor both the voltage and amperage of the system. The CC-20 is a PWM (pulse width modulation) controller which maintains the charge voltage by frequently switching the PV current source on and off, unlike series type regulators which simply disconnect the charging source once the upper voltage is achieved. The CC-20 thus ensures that the battery is kept at the fullest possible charge. An option board is available with temperature compensation, remote battery voltage sensing, low voltage warning, and generator start.

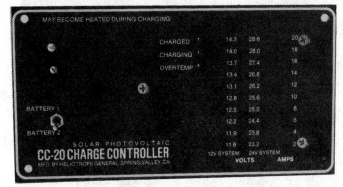

25-120	CC-20 Heliotrope PWM Controller	$225
25-129	Option Board for CC-20 (see above)	$ 99

Heliotrope CC-60 Charge Controller

The CC-60 is a larger version of the CC-20 handling 45 amps of charging current, and is easily upgradable to 60 amps. It can be user adjusted for 12 or 24 volts, includes a 60 amp Shottky blocking diode, and has adjustable set points. It has a low voltage warning light selectable from 10.5 to 11.0 volts, and indicator lights for charging and charged. It is a PWM type charge controller just like the CC-20.

25-105	CC-60 Heliotrope PWM Controller	$245
25-106	Fan-60 (to expand to 60A)	$ 39

PPC PV Power Controller

The PPC made by Specialty Concepts (SCI) is a professional grade charge controller system, with a full array of bells and whistles that brings systems together. It is available for 12, 24, 36, and 48 volt systems with models for 30 or 50 amps of charge current. It consists of a series-relay battery charge regulator with low voltage load disconnect, a load circuit breaker, system status lights, and full metering. The lights indicate "Charging" and "Load Disconnect" conditions, and meters monitor battery voltage and charging current, providing system status information at a glance. The PPC is housed in a waterproof NEMA enclosure and has a terminal block with adaptors for up to 6 gauge wire.

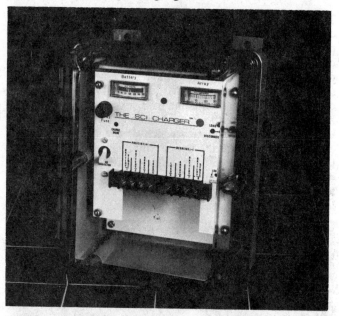

25-141	PPC Controller - 12V	$395
25-142	PPC Controller - 24V	$395
25-143	PPC Controller - 36V	$595
25-144	PPC Controller - 48V	$595
25-145	PPC-50amp option -12V & 24V	$115
25-146	PPC-50amp option -36V & 48V	$185
25-147	PPC Temperature Comp.	$ 25

Enermaxer Universal Voltage Controller

The Enermaxer Universal Battery Voltage Regulator is an excellent state-of-the-art voltage controller and has been extremely reliable. *It is the only voltage controller that will work with many different charging sources at once (Wind, Hydro, PV).* It is a parallel shunt regulator that uses the excess power of the charging source to perform useful work such as heating water, space heating, powering fans, pumps, and lights. It is made of all solid state components so that there are no relay contacts to wear out. Amperage is reduced gradually to provide a complete finishing charge and then accurately maintain a user selected float voltage. Float voltage of the battery is user adjustable. It will handle loads up to 50 amps. The 12 and 24 volt configurations will handle 50 amps, and the 32 and 36 volt configurations will handle up to 45 amps. Two-Year warranty. When ordering the Enermaxer, keep in mind that you must have an external load at least equal to your charging capacity. This is usually performed with either water-heating elements or air-heating elements which draw 15 amps each at 12V. *Always specify voltage when ordering.* Many of our customers order the water heating elements separately. They are 3/4" male iron pipe thread and measure 5.5" from threads to tip.

25-107	Enermaxer Controller - 12/24V	$249
25-108	Water Heating Element - 12V	$ 89
25-130	Water Heating Element - 24V	$ 69
25-109	Air Heating Element	$ 25

Linear Current Booster (LCB) by Sun Selector

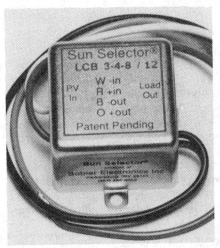

The LCB matches the electrical characteristics of the solar panel with the electrical characteristics of the device being powered and produces dramatically increased performance (up to 66%), especially in cloudy weather or in early morning and late afternoon hours. It converts the PV panel into a variable current source. A ceiling fan, for example, which is wired PV direct may deliver 2.4 amps in full sun. With an LCB between the fan and PV panel, you might now measure 4 amps going to the motor, resulting in a much better performing fan. LCB's are stackable both in parallel and in series to achieve any needed level of voltage or current.

The *new* LCB-3M and LCB-7M are designed specifically for "motorized" loads. Virtually every pumping system which is powered PV-direct (like the Flowlight Slowpump or any of the Flojet pumps) will benefit from the "M" series LCB. Water production can be increased as much as 40% to 66%. LCB's are stackable both in parallel and in series to achieve any needed level of voltage or current.

The "T" option indicates that the unit is "tuneable." Even better performance may be had with the T option since you can fine-tune the unit to match the load and PV array. The LCB can also be ordered for 20 amps of input (LCB20). All LCB20's are tunable.

An option to the LCB is the RC or Remote Control feature. An extra wire is brought out of the unit which when shorted to PV will turn the LCB and the motor load off. The remote control wire can be actuated by a float switch, or for more reliable operation we offer an electronic **water level sensor called the WLS-1**. This device can be virtually any distance from the LCB and only requires a pair of inexpensive 22 gauge telephone wires. The sensor may be mounted inside or outside of any tank. The sensor is waterproof and requires no power supply. When the water in the tank touches the probes, the LCB will shut off the pump!

25-126	WLS-1 (option)	$ 29
25-131	LCB 3-4-8/12	$ 65
25-131-RC	LCB 3-4-8-RC	$ 80
25-131-TR	LCB 3-4-8-TRC	$ 95
25-132	LCB 3-4-8/24	$ 65
25-132-RC	LCB 3-4-8/24 RC	$ 80
25-133	LCB 3M (12V)	$ 65
25-133-RC	LCB 3M - RC	$ 80
25-133-T	LCB 3M-T	$ 80
25-134	LCB 7M (12V)	$ 85
25-134-RC	LCB 7M-12V - RC	$100
25-134-T	LCB 7M-T	$100
25-134-TR	LCB 7M-TR	$115

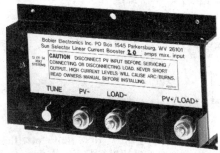

25-135	LCB 20-T	$279
25-138	LCB 7M/TR	$115
25-139	LCB 3M/24V-RC	$ 80

20 Amp Float Charge Controller

Our new fused 20 amp charge controller provides a constant float charge for a very reasonable price. It comes with a full charge indicator and charging status light.

25-111	20 amp Float Controller	$89

Are your nerves frayed from battling polluters and land-rapers? Have you been frustrated by the EPA? Do you feel wasted by years of doing all the right things to save the environment? Well then, follow these suggestions for a vacation from righteousness and **LET THE LITTLE BLUE PLANET TAKE CARE OF ITSELF FOR A WEEK OR TWO**

By Malcolm Wells

Metering

Emico Analog Meters

We have found the ammeters and voltmeters manufactured by Emico to be the best all-around meters that work in the 5% accuracy range. All ammeters operate without the need for external shunts. The meters measure 2-1/4" x 2-1/4". The expanded scale voltmeters are very easy to read and accurate. Rather than reading the low end of the scale (0-9 volts) which never registers in a 12V system or 0-18V in a 24V system, these expanded scale voltmeters show you only what you need to see and give a more accurate reading.

25-301	10-16V voltmeter	$ 18
25-302	20-32V voltmeter	$ 18
25-311	0-10A ammeter	$ 18
25-312	0-20A ammeter	$ 18
25-313	0-30A ammeter	$ 18
25-315	0-60A ammeter	$ 25

Equus Computer Voltmeter

This new voltmeter will give you your battery voltage with a digital LCD display to one decimal place at a glance. It comes with a programmable bar graph that shows the state of charge and an internal light that allows easy reading day or night. The meter mounts in a 2-1/16" hole. It runs on only 12 milliamps.

25-341   Equus Digital Voltmeter $ 49

SCI Mark III Battery Monitor

A selector switch on this combination volt/amp meter lets you change the large LED display between: incoming amperage up to 30 amps, battery voltage, and outgoing amperage up to 500 amps (with the addition of an optional shunt). A high voltage alarm light indicates when battery voltage exceeds 15.5 volts and a low voltage light turns on at 11.5 volts. The Mark III Monitor comes mounted on a 5-1/4" X 8-1/2" flush mount faceplate. The optional knock-out box will fully enclosed the unit. Two Mark III's, a monitor, and a charge controller, can be mounted in a double box to provide a very attractive and compact battery control center.

25-362	Mark III Battery Monitor	$149
25-363	100 Amp Shunt	$ 49
25-364	500 Amp Shunt	$ 79
25-114	Single SCI Knock-out Box	$ 25
25-115	Double SCI Knock-out Box	$ 35

Dual Battery Digital Volt-Amp Monitor

Our new digital volt-amp monitor operates from 8.1 volts to 35.1VDC using only 95 mA with backlight. It will read up to 500 amps (making it useful for inverter loads) and two NASA developed shunts are included in the price. It features 1% accuracy and measures only 2.7" x 2.8" x 2.4". It will read volts from two batteries, amps in and out from one battery; or amps in or amps out from two batteries.

25-361   Dual Battery Monitor $159

AHM-100 Sun Selector Digital Amphour Meter

This digital amphour meter is much heavier duty than the Accu-Slope listed above. It will operate unattended to monitor and calculate the total amphours that have flowed through a circuit. The unit is used to monitor PV array performance over time and load consumption over time, and the two resultant readings can indicate net system gain or loss over time. 12 and 24 volt units are available. The readout is a seven segment LED digital display. It is available for a maximum 100 amps or a maximum 400 amps in two different models.

All steel, wall mount cabinet.

Easy access to wiring terminals.

25-381	Sun Selector Amphour Meter - 100A	$295
25-382	Sun Selector Amphour Meter - 400A	$325
25-383	Sun Selector 24V Option	$ 15

SPM 2000 Amp Hour Accumulator

The SPM 2000 is ideal for monitoring 12 or 24 volt systems. It's two channels allow you to simultaneously measure and record power produced and power used. The digital LCD display will accurately tell you battery voltage, amps, watts, amphours, and watthours for two separate inputs or outputs. You can monitor two different inputs like your PV array on 1 and your wind generator on 2 or you can monitor two different loads, like your lights on 1 and your Sunfrost refrigerator on 2. This is a great educational device for us on our showroom floor! It will handle up to 199.9 amps (two shunts included), and 10-35 volts. It will accumulate up to 999,999 amphours. The SPM 2000 was designed to be virtually foolproof with regard to installation. It is accurate to 1% on amperage.

Power Meter 15

The **Power Meter 15** is the first meter that looks like it belongs in your battery-powered home, RV or boat, mounting just like your thermostat. It's designed to record and give you up-to-the-minute status on your entire electrical system. From across the room the **Power Meter's** LED's will tell you at a glance if your battery voltage is adequate, if the inverter or generator are on or off, and if your solar array is producing power (in three different colors).

For a more in-depth accounting, press one or both of the **Power Meter's** two keys and learn what the solar array current is (factory supplied 15 Amp shunt or optional 250 Amp shunt) and what the cumulative amphours or watthours are. It also will tell you how long the generator has been running since you last turned it on, or what the cumulative runtime has been.

The last feature is great for maintenance schedules and oil changes. Available on the system battery channel is a *battery voltage* reading and a *battery voltage high* (from last reset), *low,* and *average* reading. On the inverter and generator channels, both voltage and frequency are available. Inverter voltage is RMS calibrated and generator voltage is peak.

Standard equipment is a backup battery which is used by the **Power Meter** should the System battery be dead or disconnected. The backup battery channel will display the current voltage of the internal 9V alkaline battery, or the voltage of a second external battery should the user prefer to connect another battery to the **Power Meter**. It works on 12 or 24 volt systems.

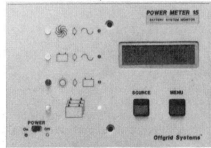

| 25-367 | Power Meter 15 | $295 |
| 25-369 | Power Meter 250 w/shunt | $345 |

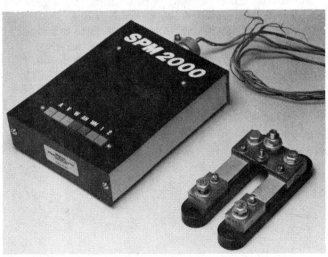

25-355 SPM 2000 AmpHour Accumulator $495

Accu-Slope Cumulative Amphour Meter

When connected to any brand charge controller the Accu-Slope will display both the instantaneous amp charging rate and the cumulative total amphours. It can be mounted between the solar panels and the battery to show the total charging amps, or between the battery and the load to show the total amps of usage at the load. The unit is reverse polarity damage protected, easy to install, and lifetime guaranteed. *The Accu-Slope must have a shunt to operate.* Match the maximum current output of the current source. Shunts are available in even numbers 10 through 30, 35, 40, 45, and 60 through 120 amps. You must specify the maximum array output in amps in order to size the shunt properly - available up to 120 amps. It can be used on systems up to 50 volts.

| 25-321 | Accu-Slope Amphour Meter | $120 |
| 25-322 | Shunt for Accu-Slope *(Specify Size)* | $ 45 |

Cruising Equipment AmpHour Meter

Our amphour meter made by Cruising Equipment allows you to moniter your energy system at a glance. When hooked up to a fully charged battery bank, the amphour meter (LCD w/backlighting standard) reads zero. As amphours are consumed, the meter counts and displays them with a "minus" sign. It automatically compensates for efficiency during charging: if one amphour is removed from a battery, it typically must be charged with 1.2 amphours to replace it. As the batteries are charged the meter counts forward to zero and on up the "plus" scale to an overcharged state. When discharging begins again amphours are automatically reset to zero and the meter begins counting the amphours used. Three shunts are available for different amperages: 100A, 200A, & 800A. The unit must have a shunt to operate.

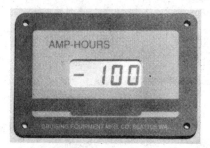

25-350	Amp-Hour Meter	$195
25-351	100 amp shunt	$ 29
25-352	200 amp shunt	$ 55
25-353	800 amp shunt	$ 89

Energy Monitor

This energy monitor measures volts and amps, and also tracks amphours into and out of the battery, through the use of a powerful microcomputer using an accurate crystal timebase. Because the Energy Monitor is built with precision circuits, no calibration is required, yet the instrument is the most accurate and stable amphour meter available. The recharge factor, which varies from battery to battery is even adjusted automatically by the microcomputer. The Energy Monitor operates from either 12 or 24 Volts, and measures 400 Amps full scale with a resolution of 0.1 Amps. The unit comes standard with equal distribution backlighting. Drawing less than 0.015 amps, it consumes a mere 0.36 Amphours per day. The Energy Monitor is similar to Cruising Equipments' meter but it includes DC Volts to one decimal place, and a 400 Amp shunt in the low price.

25-365   Energy Monitor $299.00

Timers, Photoswitches, Load Diverters & Rheostats

24 Hour 12V Programmable Timer

This timer will operate any appliance up to 20 amps, including lights, pumps, fans, and appliances. It will also start and stop remote generators. It comes complete with a 1.2V nicad battery giving it a 200-hour power reserve in case your battery is dead. It allows for up to three separate on/off operations per day. The timer is run by a 12V quartz movement and best of all it draws only 10 mA (1/100 of an amp).

25-401 Programmable 12V Timer $ 99

12V Photoswitch

Our new 12V photoswitch will automatically turn on your lights at sunset and off at sunrise! It uses a simple two wire installation: you simply cut the positive lead to the light to install. It will operate for appliances of up to one amp of current.

25-601 12V Photoswitch for 1 amp $ 49

Conserve Switches

Conserve switches are variable speed control switches (rheostats) for 12V applications such as lights, fans and motors, using less than 1/2 watt of power to control them. They are available in both a 4 amp and 8 amp configuration. These work great on incandescent and halogen lamps and also on our 12V fans. The 4 amp switch comes either with or without the forward/reverse switch. The 8 amp switch comes only without the forward/reverse switch. The forward/reverse feature is useful for summer/winter reversal of air direction in fans.

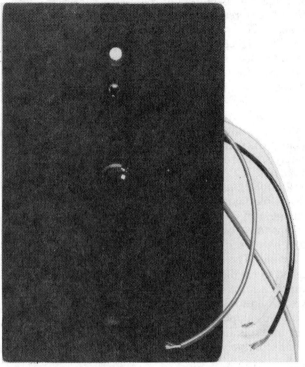

24-101 4 amp Switch $26
24-102 4 amp Switch (w/forward/reverse) $28
24-103 8 amp Switch $42

Spring Wound Timer

Our timers are the ultimate in energy conservation - they use absolutely no electricity to operate. Turning the knob to the desired timing interval winds the timer. Timing duration can be from one to twelve hours with a hold feature that allows for continuous operation. This is the perfect solution for automatic shutoff of fans, lights, pumps, stereos, VCRs, and Saturday morning cartoons! It's a single pole timer, good for 10 Amps, and it mounts in a standard switch box. A brushed aluminum faceplate is included.

25-400 Spring Wound Timer $29

Load Centers
& Transfer Switches

Code Approved DC Load
& Distribution Centers

Being on the leading edge of the PV industry we at Real Goods think it's time to move towards safer, code-approved equipment. We pledge to continue to work with responsible manufacturers to comply with the new electrical code requirements for DC systems. Although code-approved equipment can often be more costly, we feel when weighed together with safety factors it is a small price to pay for insurance. In this same spirit, we are in the process of a thorough re-write of our Remote Home Kit Owners' Manual to comply with the 1990 National Electrical Code.

We have worked with a manufacturer to engineer and build this new low voltage DC distribution center designed to meet NEC code for solar photovoltaic systems. It meets all safety standards and utilizes only UL listed components. *All current handling devices have been UL approved for 12 and 24 volt DC applications.* Enclosures are available that will handle up to 16 branch circuits with a maximum of 70 amps on an individual branch and up to 200 amps total for all branches. All overcurrent handling devices have been UL tested for 12 & 24V DC applications. The enclosure and wiring lugs have been designed to handle the larger wires used in low voltage DC applications. Extra wire bending room & multiple knockouts up to 2.5" are standard features. All load centers are shipped with a 100A main fuse and two 15A circuit breakers installed and extra breakers must be ordered separately.

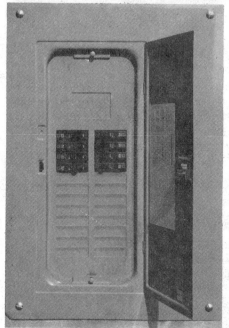

23-123	8 Circuit Distribution Center	$250
23-124	12 Circuit Distribution Center	$350
23-125	16 Circuit Distribution Center	$375
23-131	10 Amp Circuit Breaker	$ 9
23-132	15 Amp Circuit Breaker	$ 9
23-133	20 Amp Circuit Breaker	$ 9
23-134	30 Amp Circuit Breaker	$ 9
23-135	50 Amp Circuit Breaker	$14
23-136	40 Amp Circuit Breaker	$14
23-137	70 Amp Circuit Breaker	$29
23-139	200 Amp Main Fuse	$45

Low Voltage DC Load Center

The low voltage DC load center is much more than just a fuse box. It's a full-sized load center designed for safe and convenient whole-house wiring. It comes ready to use with all terminals, bus bar, fuse blocks, and meters wired in. The roomy 11" x 15" x 3.5" box has ample knockouts and quick access is provided with a hinged door with a quick-twist latch.

To install the unit you simply bring the positive lead from the solar array to the "C+" terminal and the negative lead to the bus bar. The positive lead from the battery bank then is attached to the "B+" terminal, the negative to the bus bar, and you're in business. The two ammeters allow reading of charging amperage as well as load amperage, and the voltmeter shows battery status.

Specifications

Meters:	Voltmeter: 10-16 V
	In/Out meters: 0-30 A or 0-60 A
Terminal connections:	Twelve 20A circuits, each fused Beryllium-copper fuseclips Large negative bus bar
Charging & battery terminals accommodate up to #4 wire	
Fuses	Uses standard AGC glass fuses
Mounting	Flush or surface mount

23-101 MB3012 Low Voltage DC Load Center $175
Unit will be shipped with 2 ea. 30A ammeters unless you specify 60A.

Automatic Transfer Switch

Transfer switches are designed as a safety device to prevent two different types of voltage from traveling down the same line to the same appliances. This transfer switch made by Todd Engineering will safely connect an inverter and an AC generator to the same AC house wiring. If the generator is not running, the inverter is connected to the house wiring. When the generator is started, the house wiring is automatically disconnected from the inverter and connected to the generator. A time delay feature allows the generator to start under a no-load condition. They come in both a 110V model and a 240V model. Each will handle up to a maximum of 30 amps.

23-121	Transfer Switch - 30A, 110V	$ 95
23-122	Transfer Switch - 30A, 240V	$145

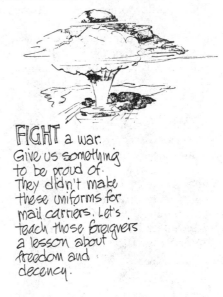

FIGHT a war. Give us something to be proud of. They didn't make these uniforms for mail carriers. Let's teach those foreigners a lesson about freedom and decency.

Safety Disconnects & Fuseboxes

Low Voltage Disconnect

This new fused unit is designed to disconnect loads up to 15 amps to prevent damage to batteries and loads in the event of an over-discharge condition. It automatically resets when there is sufficient charging capacity to bring the battery voltage above 11.5V. We use these often in water pumping applications to prevent batteries from over-draining while pumping.

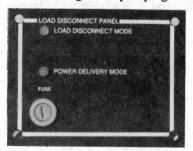

24-205   Low Voltage Disconnect   **$59**

Safety Disconnects

Every professional PV system should employ a fused safety disconnect for safety. The function of this component is to disconnect all power generating sources and all loads from the battery so that the system can be safely maintained and disconnected in emergency situations with the flick of a switch. We carry several different safety disconnects for different sized applications. Refer to our Remote Home Owner's Manual for exact sizing - fuses should be sized according to the wire size you are using.

The 30 amp two-pole fused disconnect should be used for small systems where neither the load nor the charging capacity will ever exceed 30 amps. This could easily be exceeded even with a small inverter like the Trace 612: 360 watts will pull 30 amps out of your battery - so size your fusing accordingly! (Better to use the 60A disconnect for the Trace 612.)

The 60 amp two-pole fused disconnect can be used for medium-sized systems where neither the load nor the charging capacity will ever exceed 60 amps. Of course, when employing large inverters you must wire directly from the inverter to the battery or you'll quickly exceed the 60A capacity.

The 60 amp three-pole fused disconnect should be employed when the Enermaxer voltage regulator is used which requires the third pole for disconnection.

24-201	30 amp 2-pole Safety Disconnect	$ 59
24-204	30 amp 3-pole Safety Disconnect	$ 95
24-202	60 amp 2-pole Safety Disconnect	$105
24-203	60 amp 3-pole Safety Disconnect	$139
24-209	100 amp 3-pole Safety Disconnect	$249

Fuses sold separately (see below)

Fuses & Fuseholders

Fuses and circuit breakers are circuit protectors. Their only function is to break an electrical circuit if the current (amps) flowing in that circuit exceeds the rating of the device. Any size fuse may be used safely as long as its rating is lower than the maximum ampacity of the smallest wire in the circuit.

Most fuses and breakers will pass three times their rated current for a few seconds, but they will open the circuit immediately in the event of a short circuit, which draws hundreds or even thousands of amps. This is necessary so that the fuse will melt or the breaker will open before the wire catches fire in the event of a short circuit. Fuses or breakers should be installed as close to the batteries as possible, and definitely before wires pass through a flammable wall. The chart below gives the maximum ampacity for various sizes of copper wire.

Some types of wire can handle slightly more current, and all types of wire can handle much more current for very short amounts of time. For example, a 2,000-watt inverter can surge to 6,000 watts for a few seconds when starting large motors. During this few seconds, a 12V inverter is drawing 600 or more amps from the battery, but 4/0 wire and a 250-amp fuse will work fine.

Wire gauge	Maximum ampacity
14	15
12	20
10	30
8	45
6	65
4	85
2	115
0	150
2/0	175
4/0	250

If you choose not to use our highly recommended Safety Disconnects listed above, you must at a bare minimum fuse your system both between your solar array and your battery, and between your battery system and the distribution panel or load. This can be done cheaply and easily with our Bakelite Fuse Blocks, available for 30A fuses or for 60A fuses. Don't forget to order fuses for your Disconnects!

24-401	30 amp Fuseholder	$5
24-402	60 amp Fuseholder	$9
24-404	100 amp Fuseholder	$25

24-501	30A Fuse	$4
24-502	45A Fuse	$7
24-503	60A Fuse	$7
24-506	100A Fuse	$8

15 Position Buss Bar

You can use this 15 position buss bar as a common negative for any fuse block.

24-431   15 position buss bar $ 8

AGC Glass Fuses

These glass fuses go into the Newmark fuse box. The most common size for systems is the 20A fuses. They are available in the following amperages: 2A, 3A, 5A, 10A, 15A, 20A, 25A & 30A. They come in a box of five, and the fuses measure 1.25" long x 0.25" in diameter and are good up to 32V maximum.

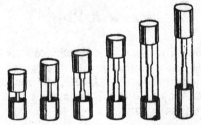

24-521	AGC glass fuses - 2 amp, 5/box	$2
24-522	AGC glass fuses - 3 amp, 5/box	$2
24-523	AGC glass fuses - 5 amp, 5/box	$2
24-524	AGC glass fuses - 10 amp, 5/box	$2
24-525	AGC glass fuses - 15 amp, 5/box	$2
24-526	AGC glass fuses - 20 amp, 5/box	$2
24-527	AGC glass fuses - 25 amp, 5/box	$2
24-528	AGC glass fuses - 30 amp, 5/box	$2

MIX some wastes.
Let the people at the dump
segregate them if it's so
important to them. You didn't
pay top dollar for all those
products just so you could
mess with yucky wrappers.
Besides, isn't this whole
"environment" thing about
over? I think I heard
that, somewhere.

Inline Fuse Holder

There is no excuse for not fusing your equipment with a fuseholder that installs this easily. This Bayonet type fuse holder has 15" of #14 wire in a loop that can be cut anywhere along the length for versatility. This is a "universal" type with 3 different springs for any length AGC style fuse. Always keep a couple of these on hand.

25-408   Inline Fuse Holder $3

Newmark Fuse Box

This is our bare bones basic fuse box. It consists of an eight-position fuse block riveted into a hinged metal box with three knock-outs in place for wiring. It's made to accept AGC type fuses up to 20 amps. It has a screw terminal for positive input from battery and one screw terminal for each of the six fused outputs.

24-411   8-position fuse box (Newmark) $14

Adapters, Plugs & Receptacles

Use these adapters as replacement equipment or to convert lighting and other low-voltage equipment to operate from the special low-voltage receptacle. Regular AC plugs and receptacles should *never* be used in a low-voltage system. Reverse polarity and confusion over AC/DC results because the plugs and receptacles are subject to "Murphy's Law" when used in low-voltage systems. (Murphy's Law says that anything that can go wrong, will.) Many low-voltage TV's, stereo systems, and answering machines have been fried by accidentally plugging into an AC inverter circuit because a friend couldn't tell the difference. Be safe rather than sorry. Use the proper low-voltage adapter, receptacle, and plug.

Receptacles and Switches

In this section we feature three lines of low-voltage receptacles and outlets: Safco, Suntronics, and a new line of UL listed DC rated units. The Safco line is made from plastic and metal and carries ampere ratings up to 10 amps. It is a good economical line and has been used very successfully in RV's for years. We felt that with the advent of low-voltage home power systems, a good decorator line of receptacles (outlets) and switches were needed with higher ampere ratings too. We spoke with Suntronics and worked with them to develop a new line of 12V receptacles rated at 20 amps and switches rated at a full 15 amps.

Besides their high current rating (the highest commercially available), some model receptacles and switches are individually fused, thus eliminating the need for a fuse box in small systems. They come attractively packaged and are made from .035 solid brass. Options include marine quality solid brass, stainless steel, chrome, or copper-brushed finish.

Fused Replacement Plug

The 80KF is a fused adapter plug kit with a unique polarity reversing feature. It comes complete with a 2-amp fuse and four different sizes of snap-in strain relief wire connectors to accommodate gauges from 24 to 16. It is rated at 8 amps continuous duty and can be fused to 15 amps to protect any electronic device.

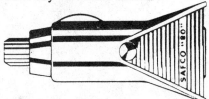

| 26-103 | Fused Replacement Plug (80KF) | $2.25 |

Replacement Plug

The SP-36 is the simplest 12V plug on the market. The plug is supplied with two solderless terminals for easy attaching of terminals to wire. The terminals are inserted into the rear of the plug and lock into place.

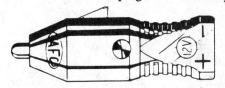

| 26-101 | Replacement Plug (SP-36) | $1 |

Heavy Duty Replacement Plug

The SP-20 is designed for heavier duty applications than the SP-36. It is supplied complete with three styles of end caps to accommodate coil cords, SJ, SJO, and round wire. The plug is unbreakable and can accommodate currents of 8 amps under continuous duty.

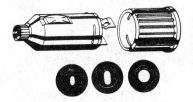

| 26-102 | Heavy Duty Plug (SP20) | $2 |

Wall Plate Receptacle

Our most popular basic 12V receptacle. It is made of a break-resistant plastic housing and all-brass metal parts and fits a standard junction box. SPECIFY BROWN OR IVORY.

| 26-201-B | Wall Plate Receptacle (Brown) | $4 |
| 26-201-I | Wall Plate Receptacle (Ivory) | $4 |

Adapter (#70)

This simple durable plastic adapter plug produces an elegant conversion from any standard 110V fixture. Simply plug the 110V plug into the base on the #70 adapter (No cutting!), insert a 12V bulb and you're in 12-volt heaven for cheap!

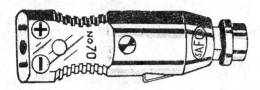

26-104 Adapter (#70) $2

Triple Outlet Box

The triple outlet box permits the use of three plugs at one time. The plug is rated at 15 amps, comes with heavy duty 16 gauge wire, and a self-adhesive pad in back.

26-105 Triple Outlet Box $8

Double Outlet

The double power outlet permits the use of two 12V products at once out of one outlet. The two receptacles are connected to the adapter plug with short lengths of heavy duty 16 gauge wire.

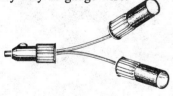

26-107 Double Outlet $5

Extension Cord Receptacle

This receptacle made of all brass parts in a break-resistant plastic housing, mates with all 12V cigarette lighter type plugs. It comes with a 6" leadwire. (20SGLB)

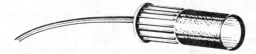

26-108 Extension Cord Receptacle $3

Non-Fused Cordset

This replacement cord set is handy where the existing cord set has been damaged or worn. It consists of a male cigarette lighter plug on one end and is attached to 8 feet of coiled 20 gauge, polarity coded wire, the free end of which is pre-stripped.

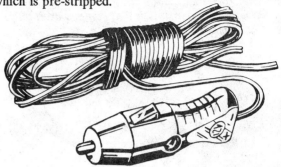

26-109 Non-Fused Cordset $2.25

Fused Cordset

The fused cordset is identical to the 26-109 listed above except that it comes with a fused male plug on one end.

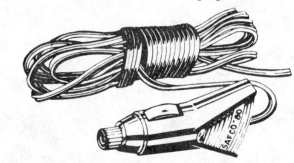

26-110 Fused Cordset $3.25

Extension Cords

These low-voltage extension cords come with a cigarette lighter plug on one end and a 12V male plug on the other. They are available with either a 10' cord or a 25' cord. The maximum current recommended is 4 amps for the 10' cord and 2 amps for the 25' cord.

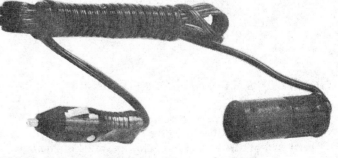

26-113 Extension Cord - 10 ft. $5.00
26-114 Extension Cord - 25 ft. $6.00

Extension Cords with Battery Clips

These unique cords have heavy clips for attachment directly to a battery. They increase the use and value of 12 volt products like lights, TV sets, radios or appliances. 1B comes with 1' cord and 10B comes with 10' cord. *The maximum amperage that can be put through the one foot cord is 4 amps and the maximum that can be put through the 1' cord is 10 amps.*

| 26-111 | Extension Cord w/Battery Clips - 1 ft. | $6.00 |
| 26-112 | Extension Cord w/Battery Clips - 10 ft. | $7.50 |

UL Listed Switches and Receptacles

In keeping with our trend of offering code approved equipment for all of our systems, we now have a line of U.L. listed DC rated 15A switches and 20A recepticles. All are heavy duty commercial grade. Although expensive, these rugged components may be the only way to keep your local building inspector happy (and maybe even give you better sleep at night). They are light years ahead of conventional RV grade plugs and switches. All are ivory colored and mount in standard electrical boxes. *Read the article on page 129 for a full discussion of these receptacles.*

26-310	Single pole 15A Switch	$ 5
26-311	3-Way 15A Switch	$ 7
26-312	Single 20A Receptacle	$ 6
26-313	Duplex 20A Receptacle	$14
26-314	Switch Plate	$ 1
26-315	Single Receptacle Plate	$ 1
26-316	Duplex Receptacle Plate	$ 1
26-317	Heavy Duty 15A Plug	$14

Suntronics 12-Volt Brass Outlets

Suntronics makes a very attractive line of solid brass outlets and switches designed for the 12V market. Constructed of .036 solid brass, they are rated at 20 amps for the outlets and 15 amps for the switches.

26-221	One Plug	$ 6
26-222	Two Plugs	$ 9
26-223	One Switch	$ 5.50
26-224	Two Switches	$10
26-225	Four Switches	$14
26-226	1 Switch w/1 Plug	$10
26-227	1 Switch w/1 Fuse	$ 9
26-228	1 Switch, 1 Fuse, 1 Plug	$11
26-229	1 Plug w/1Fuse	$ 9
26-230	2 Switches w/2 Fuses	$14
26-231	2 Plugs w/2 Fuses	$15

System Standards:

House Wiring, Standards & the Electrical Code

The following article contains hard to find information about low voltage wiring and receptacles. It is reprinted from Home Power Magazine and written by Windy Dankoff of Flowlight Solar Power (who makes many pumps that we sell) with help from Mike Mooney.

WARNING! "Cigarette lighter" type sockets are a de-facto standard for 12 volts, only because there is not yet an official standard for DC home wiring. They are light duty, all of them, and are questionable even for the 15 Amps that some of them are rated for (the plugs only handle skimpy #18 lamp cord!). Use them at your "entertainment center" for your 12V stereo and TV that came with cig. lighter plugs (their current draw is very low). DO NOT USE THEM for DC lights and appliances in general! NEVER mount them within reach of children. A paper clip inserted into one of these sockets can turn red hot!

What To Use

It will probably be a long time before a true standard will emerge. Meanwhile, there is a much better system that many of us have been using for years. It is safe, child-resistant, easy to wire, locally available, and compatible with ordinary wiring hardware and cover plates! Go to your local electrical parts supplier and order "240 volt 15 amp horizontal-prong DUAL receptacles". They look like ordinary sockets except for the position of the prongs. Suppliers generally stock only single receptacles, but will get the duals if you order them. Plugs can be found in most hardware stores when you run out.

Power Access for the AE Home

An important part of power distribution in any home is the method used to gain access to the system. The plugs and wall sockets to be used are critical.
* **120/240 vac:** The standard of access for alternating current has long been established and should be used for the A.C. current developed by the inventor in the AE home. All established electrical codes should be strictly observed.
* **12/24 VDC:** There is not yet a standard for low voltage D.C. power access, and it will probably be some time before one will emerge. Unfortunately, the automotive cigarette lighter type plug and socket are being used.

Sockets and plugs of this type have been adapted to conduit boxes for installation in motor homes and PV powered homes. THOSE NOW ON THE MARKET ARE FLIMSILY CONSTRUCTED, ELECTRICALLY UNSAFE, AND WE DON'T WANT ANY!

Described here, for your consideration, is an alternative method of access to the D.C. system which we have used for several years. It has proved to be both safe and child proof. As well as safety and convenience, we wanted a method which was durable, pleasing to the eye, and which would preclude any chance of cross-plugging an A.C. appliance in D.C., or vice versa.

We have found the 250 volt/15 amp straight blade plug and receptacle shown below to be quite workable. The receptacles are manufactured by many in both single and duplex units, and are available in ivory, white and brown. We use the Leviton "Spec-Master" variety.

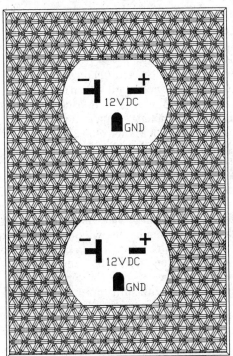

For the mating plug, we have found the 250 volt/15 amp Leviton "Spec-Master" to be a real jewel. It is very durable, looks good, provides excellent strain relief for the cord, and is very easy to assemble. Since we do use cigarette lighter plugs on occasion, we have made up a few "pigtail" pendants using the Leviton plug and Safeco automotive adaptors (Radio Shack, #RS270-1535A).

Wire & Connectors

Wire and wiring are very important. When selected and installed properly, it will ensure a successful and efficient system. The rules governing wire and accessories selection and installation are simple and uncomplicated but when ignored (wrong size wire and sloppy connections) will cause major problems and equipment malfunction in your low or high-voltage system.

The first thing to notice about wire is the color. The low-voltage part of your system will require the use of two wires to each plug, receptacle, switch and light; one black (or red) for positive (+) and one white for negative (-) or ground. This is not an earth ground as required for AC wiring. The black-for-positive and white-for-negative rule holds true for nearly all household and other electrical and appliance wiring, but there are a few exceptions. AC wiring uses a third, earth ground wire either bare copper or colored green. Also, automotive wiring usually uses red for positive (+) and black for negative (-). Keep this variance in mind when installing a house to car power system.

In a low-voltage DC system, the current flows in only one direction, hence the positive (+) and negative (-) distinction, making the color code extremely important. This distinction is called polarity. If you cross the wires (reverse polarity), DC appliances and lighting (all except incandescents and heating devices) will either not operate, short out, or blow a fuse. Motors will run backwards too. If a light won't light or an appliance won't operate, polarity is the thing to check in a low-voltage system.

In AC wiring, the current alternates, usually at a rate of 60 cycles per second in the U.S. and 50 cycles per second in Europe. This is known as hertz and is usually designated as Hz following the AC voltage (example 120VAC/60Hz). Because the current alternatives in AC, polarity is not of major importance and will not stop appliances or lighting from coming on, will not blow fuses, and will not reverse motor direction. It does contribute to the efficiency of appliances, so more U.S. manufacturers are using appliance plugs where two of three prongs are shaped differently, making it difficult to reverse AC polarity.

Another wiring distinction to keep in mind is the difference between a low-voltage system and the usual 120VAC system. The two operate on completely separate principles and carry different kinds of power loads. Even if your home is already equipped with AC wiring, you must add new wiring for the low voltage system. In some instances, like overhead lighting circuits equipped with #12 romex wire, those circuits can be disconnected entirely from the AC side and switched to the DC fuse box. *However, you absolutely cannot use the same wiring to carry low-voltage, 12VDC or AC current at the same time. Serious damage to the entire system will occur and even fire may result.*

Always use copper wire. Copper has a high conductivity value, much more than aluminum. Large aluminum wire can be used for long outdoor wire runs and some people do so because it's usually cheaper but you can expect 33% more resistance or voltage drop in aluminum wire over copper. Aluminum wire should never be used inside the house, as several house fires have occurred in homes using it and most codes ban it.

There is no difference in the conductivity of solid wire over stranded wire. However, stiff, solid wire is hard to work with in low-voltage systems where large wire is often necessary.

The following illustrates various wire sizes. Note that the smaller the wire number, the larger the wire size.

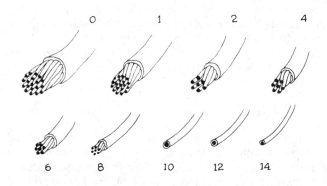

Voltage Drop

Voltage drop is caused by resistance to current flowing through a particular size wire. Think of voltage as pressure, like water pressure. The lower the voltage or pressure, the higher the resistance. If the wire in a system is not sized properly the result is a loss of power, like a 20-watt light receiving only 10 watts at the end of a long wire run not properly sized. This loss of power can and will shorten the life expectancy of fluorescent lighting tubes, and electronic equipment. So you can see that sizing wire is most important.

Here is a chart to help select the proper wire sizes for your low-voltage project. In the Voltage Loss column, it shows the voltage you can expect to lose up to a 200 foot wire distance with 12 volts at 10 amps using various sizes of wire. Keep in mind that when the distance between the battery or fuse box and a receptacle is 75 feet, the wire distance is actually 150 feet. If your amphour draw is only 5 at 12V, cut the voltage loss listed in the Volt Loss column in half. Conversely if the amphour draw is 30 at 12V, multiply the voltage loss in the Volt Loss column by 3. A different voltage will change the chart figures too. To calculate losses for 24V at 10 amps, cut the Volt Loss column figure in half, for 36V cut it by two-thirds, and for 48VDC cut it by three-fourths.

Helpful Hints

- Always earth ground AC wiring and equipment with an 8 ft solid copper grounding rod driven into the ground.
- Use romex wire for AC and follow the local electrical codes.
- Low-voltage systems should be wired with #12 or larger wire.
- In a low-voltage system, the heavier the wire, the greater the cost in dollars, but the fewer your problems.
- Use a good wire crimping tool to make wire terminal connections. Don't use pliers. Bad connections will result in voltage drop.
- Use a volt/ohm meter and check the wiring. Under load, you should have no more than a 4% drop in voltage at the appliance or from the charting system to the battery. If you have sized the wire right and you have voltage drop, it may be in the wire connections.
- If you can, solder connections with electrical solder, not plumber's solder. Use common sense. Sometimes soldering is impractical.
- Fuse all circuits and motors. Be sure to match fuses to the load, i.e., don't put a 20- amp fuse in a 2-amp lighting circuit. Use a 3- amp fuse.

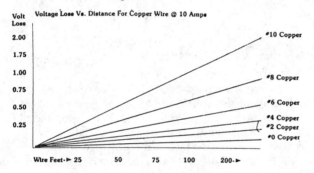

Plastic Coated Single Conductor Wire

This is your basic single conductor wire for connecting the discreet components of your alternative energy system. **We stock it in black only.** As mentioned before we recommend you use red tape on the ends for positive and white tape on the ends for negative. Wire is stranded copper. **The minimum order for #16, #14, and #10 is a 500 foot roll. Wire gauge size of #8 and larger can be ordered in any length.**

	Wire Gauge	Price/ ft
26-531	#16 THHN Primary Wire	$0.07
26-532	#14 THHN Primary Wire	$0.13
26-533	#10 THHN Primary Wire	$0.19
26-534	# 8 THHN Primary Wire	$0.35
26-535	# 6 THHN Primary Wire	$0.45
26-536	# 4 THHN Primary Wire	$0.75
26-537	# 2 THHN Primary Wire	$1.30
26-538	# 0 THHN Primary Wire	$1.95
26-539	#00 THHN Primary Wire	$2.50

Remember: #16, #14, & #10, must be ordered in 500' rolls!

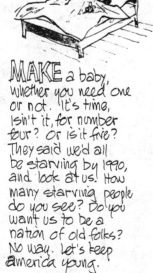

MAKE a baby, whether you need one or not. It's time, isn't it, for number four? Or is it five? They said we'd all be starving by 1990, and look at us! How many starving people do you see? Do you want us to be a nation of old folks? No way. Let's keep america young.

FELL a tree. Take an axe to it and you'll have seasoned logs before you know it. Trees get too mature, you know, and need to be thinned.

Type USE Direct Burial Cable

Type "USE" (underground service entrance) cable is moisture proof and sunlight resistant. It is recognized as underground feeder cable for direct earth burial in branch circuits by the National Electrical Code and UL. It is resistant to acids, chemicals, lubricants, and ground water. Our USE cable is a stranded single conductor with a sunlight resistant jacket. It is much more durable than standard romex.

	USE Gauge	Price/Foot
26-521	USE #10 Wire	$ .35
26-522	USE # 8 Wire	$ .40

Split Bolt Kerneys

Split bolt kerneys are used to connect very large wires to smaller wires. You must always wrap the kerney with black electrical tape to prevent corrosion and the potential for short-circuiting.

26-631	Split Bolt Kerney - #6	$ 5
26-632	Split Bolt Kerney - #4	$ 6
26-633	Split Bolt Kerney - #2	$ 7
26-634	Split Bolt Kerney - #0	$ 9
26-635	Split Bolt Kerney - #00	$14

Solderless Lugs

These solderless lugs are ideal for connecting large wire to small connections or to batteries.

26-622	Solderless Lug (#8,6,4,2)	$2
26-623	Solderless Lug (#2 thru 4/0)	$ 6

Copper is a commodity subject to rapid price fluctuations. Check prices before ordering large quantities.

Copper Lugs

We carry very heavy duty copper lugs for connecting to the end of your large wire from #4 gauge to 4/0 gauge (#0000). The hole in the end of the lug is 3/8" diameter. *Wire must be soldered to copper lugs.*

26-601	Copper Lug - #4	$2
26-602	Copper Lug - #2	$4
26-603	Copper Lug - #1	$4
26-604	Copper Lug - #0	$4
26-605	Copper Lug - #00	$5
26-606	Copper Lug - #0000	$6

Crimp-on Terminals

While of course it's always better to solder your connections or to use split bolt connectors and electrical tape, many of you are still going to fall back upon the old RV and vehicle type crimp-on connectors. They come in various sizes to accommodate different gauge wires. They also come in several different connection configurations.

26-640	Ring Terminal - #8, Stud 10	$ 1
26-642	Ring Terminal - #8, Stud 1/4"	$ 1
26-645	Ring Terminal - #10/12, Stud 10	$ .50
26-646	Ring Terminal - #10/12, Stud 1/4"	$ .50

Crimping Tool

This crimping tool cuts wire and strips insulation. It also crimps wire terminals.

26-501	Crimping Tool	$ 7

Also see our Electrician's Tool Kits in the Appliances and Tools Section on page 295.

Wiring and Alternative Energy

This article was written by Jon Vara, a customer from Vermont who has a flair for writing and is often published in New England journals. We welcome articles from customers!

Alternative-energy householders tend to be creative and resourceful. When it comes to electrical wiring, though, that's not always a good thing. I've seen houses where the wiring showed a degree of creativity that would give an electrical inspector a blood clot of the brain. But while you *can* run extension cords under rugs, thumbtack them to walls, splice circuits together with electrical tape, and tie the whole mess to the battery bank with alligator clips, it's safer - ultimately, easier and less costly - to do the job right. The following pointers should help you keep your house from turning into an extension-cord jungle.

I've seen houses where the wiring showed a degree of creativity that would give an electrical inspector a blood clot of the brain.

First of all, use standard wiring devices and techniques wherever possible. Low-voltage DC wiring (12 or 24 volt) is not very different from ordinary 110 AC wiring. With a few important exceptions, in fact, the two are virtually identical. (And, of course, in AC circuits powered by an inverter or generator, they *are* identical.) The basics of household wiring have been covered thoroughly in any number of well-illustrated, clearly written books. Time-Life's *Basic Wiring* is one good one; another is published by Ortho. It's a good idea to buy your own copy, rather than borrowing one from a friend or your local library, since you will be referring to it again and again over the years.

What are the exceptions that pertain to DC wiring? One of the most important has to do with the wire used. Most AC circuits are wired with 12- or 14-gauge wire, while higher-amperage DC circuits require heavier wire. As a rule of thumb, 12-volt lighting circuits will require 10-gauge wire, while 12-gauge is ordinarily adequate in 24-volt systems.

Most AC wiring uses non-metallic sheathed cable - often called Romex, after the name one of the most popular brands - in which a white neutral wire, a black hot wire, and an uninsulated ground wire are bundled together in an outer plastic sheath. That is convenient, but the solid, single-strand wire used in all non-metallic cable is very stiff and difficult to work with in larger sizes. Solid 10-gauge wire, in particular, is an installer's nightmare. A better option is primary type wire (sometimes called

THHN), which consists of a number of fine strands, and is far more flexible than single-strand wire of the same gauge. The only important difference is that it's difficult to wrap stranded wire neatly around the screw terminals of switches and receptacles. [**ed note**: *THHN must be installed in conduit according to code!*] Instead, make connections to screw terminals with solderless spade connectors (illustration), which are easily crimped onto the stripped ends of wires.

Because THHN wire does not come in cable form, however, you will need to buy separate spools of black, white, and green wire. (It's much cheaper in full 500-foot spools than in cut lengths). The black will serve as the positive wire, the white wire as the negative, and the green as a ground wire.

Strictly speaking, the separate ground wire is not required in a DC system, and may be omitted if you want to save a few dollars on wire. [**ed. note**: *the code requires all receptacles to be grounded - any voltage!*] However, by adding the ground wire (secured to electrical boxes, receptacles, and switches as in an AC system) you have the option of eventually reconnecting the entire system to run on AC, rather than DC. Given the dramatic improvements we've seen in inverters during the last few years, that day may arrive sooner than you think.

If your house will be equipped with both AC and DC circuits for the time being, however, you will need to guard against accidental interconnection between the two. Plugging a 12-volt appliance into a 110-volt receptacle - or vice versa - will yield results discouraging to any but the pure scientific mind.

Your best bet is to fit AC boxes with standard AC receptacles, and DC boxes with 15 amp, 250 volt receptacles. The latter look very much like standard AC receptacles. They are designed to fit standard electrical boxes, and accept conventional cover plates. The key difference is that the prongs on the 250-volt receptacle are turned 90 degrees from those on a conventional outlet (see illustration), which means that the plugs which fit one system will not fit the other. The 250-volt receptacles will be perfectly happy with a diet of low-voltage DC, and unless a house guest shows up lugging an electric hot water heater or portable air conditioner - to name the two most common patrons of 250-volt AC receptacles - there is no chance of an AC-to-DC mixup.

One frustrating problem that sometimes arises when working with DC wiring (especially when bulk 10-gauge wire is called for) is electrical box overcrowding. An uncomplicated duplex outlet and its attendant wires will fit neatly into a standard 3" deep electrical box. But if a branch circuit enters the box as well, or if the receptacle is to be controlled by a switch (two common configurations explained in detail in books on wiring), the additional wires and wire connectors may stuff the box so full that it is impossible to attach the receptacle.

If the thickness of the wall doesn't allow you to substitute a deeper box, a neat solution is to use a two-gang box of the same depth, which is a double-wide box that ordinarily houses two switches or two duplex receptacles. By installing only one duplex receptacle, however, you leave twice as much room for the network of wires that feed it. Cover the box with a duplex/blank outlet cover, which is wide enough to cover the entire box, but has holes for the receptacle face on only one side. These, again, are available from most good-sized electrical suppliers.

A variation of that trick can save the day when you need to add DC wiring to an existing house - the usual case, since relatively few homes are constructed with low-voltage DC wiring in mind. The solution is a **wiring chase** concealed in the baseboard. Remove the existing baseboards, and screw shallow, 1-1/2" deep two-gang boxes to the wall studs, an inch or two above the floor. (Use metal boxes, which have holes in the back that will accept screws, rather than plastic ones.) Next, nail a pair of 1" wide strips of 1/4" plywood around the perimeter of the room, one at floor level and one just above the tops of the electrical boxes. Wire one receptacle to each two-gang box. (If you are using 12-gauge wire, you may be able to use single-gang boxes, but with bulkier 10-gauge, the extra-wide boxes are necessary to compensate for their lack of depth.)

Cut new baseboards of the required width and length, position them temporarily, and mark the locations of the cutouts for the electrical boxes. Cut the holes for the boxes with a reciprocating saw, and after making sure that the runs of wiring between boxes fit neatly in the channel formed by the plywood strips, secure the baseboards to the wall with finishing nails.

If you have used 3/4" stock for the baseboards, the electrical boxes will protrude 1/2" beyond the surface of the baseboards. Place a 1/2" shim over each box - these can be very attractive if made from scraps of nice hardwood - and screw a duplex/blank cover plate on over it.

If your house will be equipped with both AC and DC circuits for the time being, you will need to guard against accidental interconnection between the two.

Finally, use a strip of 3/8" quarter-round (or other molding, if you prefer) to conceal the raw edge of the plywood spacer that remains visible along the top edge of the baseboard. [**ed. note:** *this is legal only with Romex!*]

And you're done. The finished product is neat, good looking, and quite inexpensive. Could you call it creative? I guess you could. -**Jon Vara**

Santa Fe, NM. This privately financed PV powered home uses 78 Arco ASI 16-2300 modules. The system has a daily power output of 12-14 kWhs. Courtesy Arco Solar.

Wire Sizing and Voltage Drop in Low Voltage Power Systems

John Davey and Windy Dankoff

Properly sized wire can make the difference between inadequate and full charging of your energy system, between dim and bright lights, and between feeble and full blast performance of your tools and appliances. Even wiring that is slightly undersized can cheat you out of a major portion of your system's energy.

Designers of low voltage systems are often confused by the implications of voltage drop and wire size. In conventional home electrical systems (120/240 volts ac), wire is sized according to its safe amperage carrying capacity know as "ampacity". The overriding concern here is fire safety. However in low voltage (12/24/48 volts DC) systems, sizing for larger wire is usually necessary to minimize power loss due to voltage drop before increased wire size is required for amperage safety.

Typically, low voltage systems are seen in Alternative Energy (AE) home systems and Recreational Vehicle (RV) systems. The heart of these systems is DC power because DC electrical power can be stored in batteries. With photovoltaic systems, the electrical power produced is also DC. DC systems are primarily low voltage because most of the DC lights and appliances have traditionally been built for the vehicular market, which is typically 12 or 24 volts. There is also increased fire danger with high voltage DC because of the high potential for arcing in switches and poor electrical connections. High voltage DC also has a high shock hazard (more than at an equivalent ac voltage).

Voltage Drop is caused by a conductor's electrical resistance (Ohms) and may be calculated according to Ohm's Law—

(1) Voltage Drop (Volts) = Electrical Resistance (Ohms) X Current(Amps)

Power Loss is calculated by—

(2) Power Loss (Watts) = Voltage Drop (Volts) X Current(Amps)

By substituting the Voltage Drop Equivalence from equation (1) into equation (2), we find—

Power Loss (Watts) = Ohms X Amps^2

If we have a 12V system with a 100 ft. wire run of 12 gauge wire (0.33 Ohms) and a 72 watt load, there will be a 6 amp current (Amps = Watts/Volts) and a power loss of 12 watts (0.33 Ohms X [6 Amps]^2). If we converted this system to 24V, we would have a current of 3 amps and a power loss of 3 watts. The significance here is that by DOUBLING the system voltage, power loss is reduced by a FACTOR OF FOUR. Or for no increase in power loss, we can use ONE FOURTH the wire size by doubling the voltage. This is why the trend in AE full home systems with DC circuits is towards 24V instead 12V systems. It is also why it is important to reduce the current by using efficient loads and putting fewer loads on the same circuit. Likewise, reducing wire resistance by using large wire and shorter wire runs is important. All of these are particularly critical with AE systems, where cost per kilowatt of electrical power may be several times that of "Grid" supplied electrical power.

Wire Size Chart

Because of the significance of voltage drop in low voltage electrical systems, we have developed an easy-to-use wire sizing chart. Most previous charts published assume a 2 or 5% voltage drop for 12 and 24 volt systems and result in pages of numbers. This new chart works for any voltage and accommodates your choice of % voltage drop. You'll find it the handiest chart available. The chart applies to typical DC circuits and simple ac circuits (refer to footnote on Wire Size Chart).

We recommend sizing for a 2-3% voltage drop where efficiency is important. We shall discuss this as it applies to specific loads in greater detail in Part II of the article.

Wire Size	Copper Wire		Aluminum Wire	
AWG	VDI	Ampacity	VDI	Ampacity
OOOO	99	260	62	205
OOO	78	225	49	175
OO	62	195	39	150
O	49	170	31	135
2	31	130	20	100
4	20	95	12	75
6	12	75	•	•
8	8	55	•	•
10	5	30	•	•
12	3	20	•	•
14	2	15	•	•
16	1	•	•	•

Practical Applications and Considerations

Here, we will consider voltage drop and wire sizing for different types of electrical loads, alternatives to the use of large wire and long wire runs, and some recommended wiring techniques. Different electrical loads (power-consuming devices) have different tolerances for voltage drop. These guidelines will help you determine how much drop is acceptable.

Lighting Circuits

Incandescent and Quartz Halogen. A voltage drop below appropriate levels results in a disproportionate loss in performance. A 10% voltage drop causes an approximate 25% loss in light output. This is because the bulb not only receives less power, but the cooler filament drops from white-hot towards red-hot, emitting far less visible light.

Fluorescent. Voltage drop here is less critical, causing a proportional drop in light output. A 10% voltage drop results in an approximate 10% loss in light output. Because fluorescents are more efficient, they use 1/2 to 1/3 the current of incandescent or QH bulbs and therefore many be used with smaller wire (including most pre-existing ac wiring). We strongly advocate use of fluorescent lights. The unpleasant qualities of flicker and poor color rendition may be eliminated by using the more advanced 12, 24, and 120 volt fluorescents now available. See our "Efficient Lighting" article in HP#9 for details. We suggest using a 2-3% voltage drop for sizing wire in lighting circuits. If several lights are on the same circuit but are rarely all on at once, see the Part-Time Loads section for an economical approach.

Motor Circuits

DC Motors. DC motors operate at 10-15% higher efficiencies than ac motors and eliminate the costs and losses associated with DC/ac inverters. DC motors have minimal surge demands when starting, unlike ac induction motors. Voltage drop results in the motor running at a proportionally slower speed and starting more gradually. We suggest using a 2-5% voltage drop under normal operating conditions for DC wire sizing.

DC motors used for hard-starting loads, particularly deep-well piston pump jacks and compressors, may have high surge demands when starting. High power demands are also seen in DC power tools when overloaded. DC refrigerators (e.g. Sun Frost) with electronically controlled (brushless) motors will fail to start if the voltage drops to 10.5 volts, in a 12V system, during the starting surge. This is due to a low voltage shut-down device in the refrigerator intended to protect your batteries from damage. We suggest sizing wire here for a 5% voltage drop at surge current (use 3X operating current).

ac Motors

Alternating Current (ac) induction motors are commonly found in large power tools, appliances and well pumps. They exhibit very high surge when starting. Significant voltage drop ion these circuits may cause failure to start and possible motor damage.

Universal Motors. Brush type ac motors ("Universal Motors") are found in smaller appliances and portable tools. As with DC motors, they do not have large sure demands when staring. However, wire should still be generously sized sized to allow for overload and hard-starting conditions. Consult an electrician or the *National Electrical Code* for wiring standards in ac tool and appliance circuits.

Photovoltaic Battery-Charging Circuits

In PV battery charging a voltage drop can cause a disproportionately higher loss in power transfer. To charge a battery, a generating device must apply a higher voltage than exists in the battery. That's why most PV modules are designed for 16 volts or more. A voltage drop of 1 or 2 volts in wiring will negate this necessary voltage difference, and greatly reduce charge current to the battery. A 10% voltage drop in a wire run may cause a power loss of as much as 50% in extreme cases. Our general recommendation here is to size for a 2-3% voltage drop.

PV array voltage also drops in response to high temperatures. Use high voltage modules (over 17 volts peak power) in very hot climates (where module temperatures commonly exceed 117°F./47°C). In moderate climates, high voltage modules allow for more line voltage drop, but they cost more per Amp delivered to the battery bank. You may, therefore, size wire for a somewhat larger voltage drop, e.g. 5%, when high voltage modules in a moderate climate.

If you think you might expand your array in the future, install wire appropriately sized for your future needs NOW, while it is easier and less costly. It never does any harm to oversize your wire.

Number Of Circuits. If circuits are designed with numerous loads requiring large wire, overall wire cost may be adding additional circuits and putting fewer loads on each circuit. Fewer loads per circuit reduces circuit current which in turn allows for the use of smaller wire.

More Than One Size Of Wire In A DC Circuit

If you size wire for the loads on "End Branches" of a circuit, smaller wire may be used. For instance, voltage drop sizing may specify 10 gauge wire for a circuit but a light on an "End Branch" of the circuit, when sized separately, may allow for the use of 12 gauge wire from the switch to the light. Using smaller wire for "End Branches", may also make your electrical connections faster and easier because it is physically difficult to make connections to standard household switches, receptacles, and fixtures with wire larger than 12 gauge. BE SURE THAT THE AMPACITY RATING OF ALL WIRE IN A CIRCUIT MEETS OR EXCEEDS THE FUSE PROTECTION RATING OF THE CIRCUIT.

Part-Time Loads

If a number of loads are on the same circuit but are rarely all operating at the same time, you can size the wire for voltage drop according to the TYPICAL load demand. AGAIN, BE CERTAIN THAT THE AMPACITY RATING OF ALL WIRE IN THE CIRCUIT MEETS OR EXCEEDS THE FUSE PROTECTION RATING OF THE CIRCUIT.

System Voltage. Consider 24 volt DC instead of 12 volt where feasible. Use 120 volt ac from inverter to loads where 10-20% conversion loss is not a major comprise. See our article "Selecting System Voltage" in HP#14.

Location Of System Components. Locate batteries, inverter, ac battery charger, and distribution panel near each other. Also, locate the distribution panel as close as possible to very large loads and as central as possible to all other loads. This will shorten wire runs and for some circuits, reduce the wire size required.

Water Well Pumps. Consider a slow-pumping, low power system with a storage tank to accumulate water. This reduces both wire and pipe sizes where long lifts or runs are involved. An ARRAY-DIRECT pumping system may eliminate a long wire run by using a separate PV array located close to the pump. (For more about water system design, see our article "Solar Powered Pumping", HP#11.)

Soldering vs. Mechanical Connections. Soldering is recommended around battery and inverter terminals (see "Build Your Own Battery/Inverter Cables" in HP#7) and in other corrosive, high-current environments OR at the discretion of the installer. Soldering requires skill and has numerous pitfalls—too much or too little heat, oxidized or dirty metal, the wrong solder or flux, or just lack of experience will GUARANTEE poor solder joints. Do not attempt to solder connections in your system unless you have learned do it properly. A tight mechanical joint is far safer than a questionable solder joint.

Grounding And Lightning Protection. We've seen thousands of dollars of damage to electrical equipment from lightning. In one PV home a lightning bolt entered the house via the PV wiring and exited the other side of the house, popping plaster and light bulbs, and burning wire along the way. Proper grounding PREVENTS nearly all such occurrences. For a more thorough discussion, see our article "Grounding and Lightning Protection", HP#6.

Audio Signal Wires. Wires that carry audio signals (telephones, intercom, speakers) may pick up buzzing noise if run alongside ac wiring. This is especially true when the ac power is from an inverter. Avoid this problem by running audio wires along a separate path (or in a separate trench) from the ac wires. Keep then as far apart as possible, especially on long runs. Proper grounding also helps. Audio wires will NOT pick up noise from DC lines.

LIGHT a smoke. Find a tiny restaurant, wait till it's full, then start puffing. Revenge is sweet.

ADD a lawn. Don't go all teary-eyed about the glory of wildflowers. They're only weeds. Grass is the ideal ground cover.

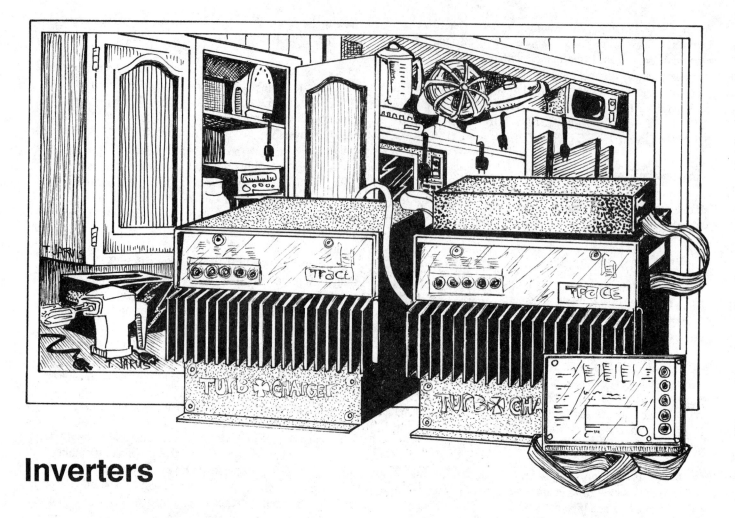

Inverters

An inverter is the device that interfaces between a low-voltage storage system and a 110V appliance. Low-voltage systems alone will take care of many basic needs, but in the average U.S. home, there are numerous appliances, tools, motors, and equipment that use standard 120-240VAC 60 Hz electricity and cannot be easily converted.

In Europe and many foreign countries, 220-240VAC 50 Hz is commonly used to run appliances. We feature inverters for both markets. While inverters at first glance appear to be a panacea (running all your old 110V appliances from your new 12V system), they do have some drawbacks, although nothing like the inefficient inverters of the ancient past (about 8 years ago!)

Inverters take energy to run themselves, from as low as 0.36 watt for the newest F.E.T. equipped solid state units (on standby) up to 10% of the rated maximum power output; plus the power an appliance draws when turned on. Rotary inverters (which we'll explain later) take up to 25% of their rated power output to run themselves and as a consequence have all but disappeared from the market place.

Basically, the inverter takes nominal DC (direct current) from a battery storage system and changes the wave form to AC (alternating current) at a particular voltage and frequency. There are technical names for the different forms, square wave and sine wave and there are synchronous and non-synchronous types of inverters.

We will not delve too deeply into the physics of inverter operation here but will explain the points necessary to help make an intelligent choice.

Synchronous Inverters

Synchronous inverters are used with alternative energy systems designed to feed the public utility grid with AC electricity generated from a DC source, be it photovoltaics, wind, or hydro-electric. It sounds almost too good to be true, selling power back to the power companies, and it nearly is because the cost of synchronous utility-intertie inverters is virtually prohibitive.

Static Inverters

Square and modified sine wave non-synchronous static inverters feature solid state, space-age electronic technology. They have come a long way in the past few years and are now used successfully all over the world to provide 120-240VAC, both 50 and 60 Hz current, from various battery sources. Models are available to run the most sensitive electronic equipment and powerful induction or capacitor start motors as well. Some inverters without the newest F.E.T. technology (which can result in a high standby power draw) can be equipped with an automatic load demand, a device that turns it on and off only when an appliance is activated, and some models may be ordered with a remote control switch option.

Inverters are not intrinsically efficient and, if left on to run appliances continuously, will quickly deplete a battery system. (As an example, one wouldn't operate an electric clock or clock radio from an inverter not equipped with the F.E.T. technology because the standby power draw would be prohibitive. Battery powered clocks are the solution in this case.) But if one uses an inverter correctly, that is, intermittently or only when needed, the user can justify the inefficiency by the flexibility gained in a low-voltage system.

There are several price ranges for inverters that reflect their ability to perform. The higher priced inverters will operate a wide range of appliances; from those requiring less than 100 watts, to microwave ovens (some models take longer to cook using an inverter output) to induction motors that require substantial wattage (up to 6 times their continuous rating) to start. They are excellent for heavy-duty service in the home (especially where an all AC-system is desired), and for light industrial use with some exceptions.

Water pumps are often troublesome to operate from inverters, especially deep well submersibles, because of the highly inefficient AC motors most pump manufacturers use. If some model inverters can deliver an impressive 48 kW surge, why can't they run a water pump? The answer is time. Often the inverter is incapable of delivering the surge requirement for the time it takes the motor to get started. The more inefficient the AC motor, the longer it takes to start, and most inverters, except for the heavy-duty, high wattage units, can't handle it. The Trace and Heart line of inverters will handle some kinds of heavy duty water pumping with their 2 kW and above models, but we strongly recommend you choose one of our DC water pumping systems for the sake of efficiency. (As an example, not counting surge requirements, a water pump that runs on 500 watts DC will require up to 800 watts to run through an inverter.) There are several DC deep well submersible pumps to choose from in this Sourcebook.

Air conditioners and electric heaters are not practically run on inverters. Not that most of our top-of-

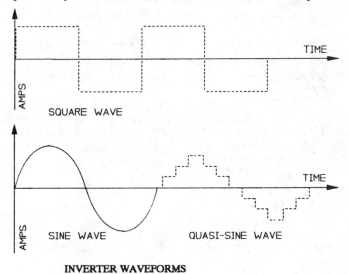

SQUARE WAVE

SINE WAVE   QUASI-SINE WAVE

INVERTER WAVEFORMS

the-line inverters wouldn't run them easily, but the power consumption for both is horrendous. It would be the same were they available as DC equipment and would require an expensive generating system to keep up with the demand.

Instead, check our cooling section and read about solar-powered evaporative cooling. Proper design of your home to take advantage of passive cooling, along with the fans and evaporative cooling systems we offer in this catalog, can be combined to realize great energy savings

---

Harmonic distortion is what gives sensitive electronic equipment and certain motors a fit.

---

while maintaining a comfortable pleasant and natural environment.

Thermal solar and gas are recommended for water heating, and wood heat is still the most efficient and inexpensive means to heat space.

Moderately priced inverters do not have the high wattage or surge load capability of the larger models. However, on some models their wave form shape does approximate a sine wave through pulse width modulation. Harmonic distortion is what gives sensitive electronic equipment and certain motors a fit.

The economy line of inverters has limits on the number and type of appliances or equipment they will operate. As an example, they will not support wattage more than 10% above their power rating. They will not operate induction or capacitor start motors. They produce a completely square wave form which means these inverters will not operate sensitive video (VCR) or other electronic equipment that requires a sine wave or suppression of third harmonic distortion. (We promised to stay away from inverter physics, but some of you need to know this information; however, this is as far as we'll go.) Surprisingly though, the inexpensive inverters will operate many personal computers, because of the recent advances in chip design, and a host of small appliances.

Even if you select a larger inverter, having a few low-wattage economy models on hand can be a good, practical idea; for instance, when you want to use a small portable sewing machine, tool, personal computer or a 19 inch color TV that uses less than 200W, the smaller inverter can be plugged right into a nearby low-voltage receptacle and run efficiently for hours on end while eliminating the need to activate the larger inverter. (By the way, numerous people have bought 120VAC television and FM signal boosters to be run on an inverter, not realizing that we have a 12VDC booster available!)

Computers on Inverters

Although many personal computers will operate satisfactorily on the economy line of inverters, computer video monitors do not. The square wave form causes some annoying screen interference as does the lack of frequency control on some models. Printers like Epson, Okidata, and the daisy wheel models will not operate satisfactorily either. Use the Trace or Heart model inverters rated at 500 watts or more for computers with external video monitors, hard disk drives, and printers. 12VDC USI 9 and 12 inch external video monitors are available and carried by many computer dealers. Even if you have public utility power, Heart and Trace model inverters are a good investment. A small car battery and inexpensive battery charger can be connected to the inverter which in turn will operate the computer and peripheral hardware. Utility company power spikes, surges, brown-outs and failures won't affect your computer programs.

Small personal computers with up to two disk drives can be operated from a non-frequency controlled 200 watt economy model PV-200 Tripp-Lite inverter or from one of our smaller Powerstar or Statpower inverters.

Power Requirements - What Size Inverter?

To properly select an inverter, you will need to know how much current a particular unit will use when operating all the AC equipment and appliances you have chosen for your house. Many people do not follow this rule and end up buying an inadequate, expensive piece of equipment and then try to tailor the system to their demands. The beginning part of this Sourcebook shows you how to figure your power needs and also explains the relationship between watts and amps.

Knowing how much power you need will affect the wire size you need, the amount of battery storage required, etc. It is not enough to look at an AC appliance and use the wattage or amperes listed. Because your inverter will operate from a DC battery system, the nominal voltage of your system will be used to figure your power requirements. Because batteries are rated in amphour capacity, this is important to know. Regular appliances or equipment that run on 120-240VAC household electricity will have a specification plate listing their wattage. The formula for figuring amps from wattage is watts divided by DC volts equals amps. (Watts = Amps x Volts.) As an example, let us say the household appliance draws 700 watts and your alternative energy system's nominal voltage is 12: 700 divided by 12 equals 58.3 amps. That, plus the energy it takes to run the inverter, is the current you will be pulling from the battery system for the time the appliance is on.

Our economy-model inverter uses about 10% of its rated capacity to run itself, and if it is a 1,000-watt model (necessary to operate a 700-watt non-inductive load) you would add 100 watts, or approximately 8 amps, to the 58.3 amps the appliance uses for a total power draw of 66.3 amps. Sixty-six amps is a lot of current and requires large gauge wire to conduct it properly. For equipment with induction, capacitor start, or split phase motors, the surge of electricity required to start them can reach astronomical heights, up to six times the running watts for some loaded motors. Your inverter must be sized to handle these surge loads.

Once you have estimated your AC surge load requirements, the next step is to estimate your peak

demands on the AC side of your system. Domestic peak loads usually occur at mealtimes and early morning and evening, when most appliances, lights and equipment are in use. Look at the electrical usage curves taken from a typical rural American homeowner, as provided by Independent Power Development, Sand Point, Idaho.

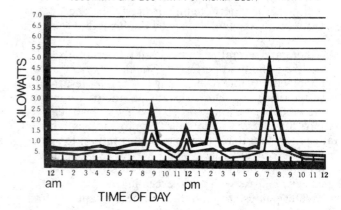

TYPICAL USAGE PATTERN FOR RURAL USER OF ELECTRICITY
(500 KwH and 200 KwH Per Month User)

Two things are apparent; first, the peak loads are substantially higher than the average continuous load, and second, the need for electricity is 24 hours a day. The system you put together must be able to handle the peaks and supply power all the time. The well-planned alternative energy system can be a combination of AC/DC or all AC and will use a lot less electricity than shown in this chart!

A rule of thumb for figuring AC loads is: peak loads will usually be 6 to 12 times the average hourly load, although momentary surge peaks, when large motors are turned on, may be 12 to 24 times the average hourly load. So, if you use 5 kWh (5000 watthours) per day (.208 Kwh per hour), your peak AC usage is likely to be from 1.25 to 2.5 kWh with motor starting peaks from 2.50 to 5.0 kW.

Even if you have public utility power, Heart and Trace inverters are a good investment.

Here your choice could be a Trace 2524 inverter with a surge capability of over 7 kW operated from a 24V system.

The generating system you plan will most likely not match your electrical usage pattern. If you need electricity available 24 hours a day (and that's the objective of the system), that means batteries. They will play a major role in your system, and the subject of choice and sizing is covered in Chapter 3.

Life Expectancy

The solid state, modular electronics used in the top-of-the-line and moderately priced inverters contribute to their two-decade life expectancy. That's not to say you can expect 20 years without the need for service, but it should be minimal. Like your television set, solid state electronic devices sort of "fade out" after that many years.

The economy lines of inverters usually ail earlier because they are more often subjected to use beyond their rated specifications, but we have units still operating in intermittent service after more than 9 years in the field.

Helpful Hints

The most common inverter problems stem from placement of the unit too far away from the battery, undersized wiring, and exceeding the rated capacity. If the distance is long to the appliance, use the high-voltage AC side of the inverter to cover the distance. Place the inverter as close as possible to the battery *but not in the same compartment*. You don't want the inverter sparking in a battery compartment. Use the largest wire possible from the battery to the inverter.

All inverters will cause some radio frequency (rf) interference which will show up as static noise in some stereo systems, especially on the AM radio band. FM television and radio bands stay fairly clear but sometimes scratchy white lines may appear on your TV screen. Usually the cheaper the inverter, the more interference, especially when battery voltage is low. There are noise filters which can eliminate most of these problems. How much noise is eliminated will be determined in part by the quality of the entertainment system you own.

Non-frequency controlled inverters will not operate tape decks or turntables at the proper speed under varying battery voltages. A "Cinemascope" type black line will also appear at the top of an AC television screen. Frequency controlled models will eliminate this problem.

How to wire an inverter to a 120 vac mains/ breaker panel

Richard Perez

Getting the inverter's power output into a conventional 120 vac mains panel can be a problem. Unless the wiring is properly connected and sized, the inverter's power will not be effectively transferred. So here's the straight dope about wiring your inverter to a mains panel that was designed for conventional grid power input. These panels are found in all "electrically standard" homes and manufactured housing.

The Marriage of Inverter to Mains

If you're reading Home Power, then you probably are already familiar with inverters. These marvelous devices change the low voltage DC stored in our batteries into 120 vac, 60 cycle, power. They allow us to use PV produced and battery stored energy in conventional appliances. The inverter's power output, while not an exact replica of that supplied by the power company, is close enough to run almost all conventional 120 vac appliances. Just like downtown.

Now the mains panel is a different matter. This piece of electrical equipment lurks in basements, closets, and other dark, unfrequented places. The function of the mains panel is to connect your building with the conventional commercial power grid. It provides a terminus for your building's wiring.

Within the mains panel each 120 vac circuit, via its individual circuit breaker, connects with the main power input. Hence its name, mains panel.

Our mission is to wed the inverter, from the world of renewable energy, with the mains panel, from the world of costly, pollution ridden, commercial grid electricity. Maybe not a marriage made in heaven, but certainly one made in the sunshine. Consider yourself an ecological/electrical match maker.

Getting the Power out of the Inverter

All high quality inverters offer us two ways to connect to their output- via a plug or via hardwired terminals. Let's look at plugs first. The male plugs are a standard 3 prong, grounding, 20 Ampere plugs known in electrical jargon as "cord caps" (don't ask me why). You can use just about any male plug, but get one that is of high quality. This means strong prongs, anti-corrosion plating, and a solid case. A high quality cord cap will cost around $5 to $7 and is worth it. The inverter's entire output is passing through this plug, so it's not the place to save a buck. Connect the plug as follows. The GOLD colored terminal of the plug is HOT and connected to the BLACK wire in the output cable. The SILVER colored terminal of the plug is COMMON and connected to the WHITE wire in the output cable. The GREEN colored terminal of the plug is GROUND and connected to the BARE copper wire in the output cable.

You may also have a hardwire output for your inverter. This output consists of three electrical terminals that will accept either bare wire ends, or ring connectors. Wire these according to the manufacturer's instructions on your particular inverter. Here's some info on two of the most common types of inverters. The Trace inverters offer their hardwired output via a barrier strip under their Plexiglas window. The terminus is located in the upper right hand corner of the window just below the standby input line. Trace supplies ring connectors and an Allen wrench with every inverter. These allow you to install lightweight wire into the barrier strip. The Heliotrope inverter supplies three large & easily used connectors on the lower left hand side of their main PC board. These connectors will accept 10 gauge wire ends directly. Once again the wiring scheme is the same: HOT to BLACK wire, COMMON to WHITE wire, and GROUND to BARE wire.

Inverter to Mains Panel Wiring

The wire transferring the inverter's power to the mains panel must be of sufficient size to handle the current over the distance without excessive losses. If the inverter to mains panel wiring distance is less than 70 feet, then 12 gauge copper will will do the job at 98% or better efficiency. If the inverter to mains panel wiring distance is about 120 feet, then 10 gauge copper wire will be 98% efficient. These facts are computed on the round trip wiring distance (two conductors) and half of these distances is the actual physical distance between the inverter and the mains panel. Use conventional ROMEX cable for this purpose, like NM12/2 with Ground. This cable contains three solid 12 gauge copper wires: one black insulation, one with white insulation, and one without insulation. For longer distances, use the 10 gauge equivalent, NM10/2 with Ground. If the cable is exposed to sunlight or buried, then use cable with USE (Underground Service Entrance) insulation. The USE insulation on the outside of the cable will not photodegrade in sunlight, or rot in moisture.

Connecting to the Mains Panel

Route the wire into the mains panel from its top. Connect the hot (BLACK) to the main input breaker(s). Connect the Common (WHITE) and the Ground (BARE) to the wiring terminal beside the rows of circuit breakers. See the diagram below.

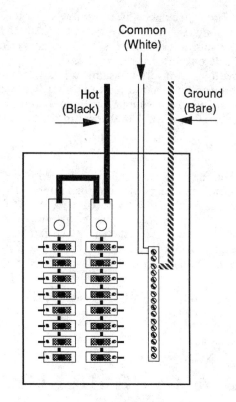

The mains panel is designed for commercial power input. Each row of breakers (and there are two), is connected to a 120 vac leg of the grid input power. Together, these two 120 vac legs make 240 vac. Well, the inverter just makes 120 vac. In order to energize the second set of breakers, we must add a jumper between the two main input terminals as shown in the diagram. This effectively converts the mains panel from 120/240 vac operation to just 120 vac operation.

The terminals where the common and ground wires are all connected should be grounded. This is the main system ground for the 120 vac distribution system. This terminus should be connected, with 6 gauge bare copper wire, to a metallic rod driven at least six feet into the ground.

And they lived happily ever after...

The inverter is now wedded to the mains panel, and all 120 vac circuits are energized. As with many weddings there are leftovers. Most weddings produce an excess of toasters and cuisinarts, but in this case the leftovers are circuit breakers. The inverter contains its own output circuit breaker, there are the main breakers at the top of the panel as well as the individual breakers for each ac circuit. There are actually three circuit breakers in series with every circuit. More than enough to please even the fussiest building inspector.

Report on the Inverter Shootout at SEER '90

Richard Perez

Seer '90 at Willits, CA was probably the very first time that this industry had just about everyone in the same place at the same time. A perfect opportunity to place different brands of inverters in exactly the same system and compare their performance under a variety of loads.

The Test Inverters & People

Just about every inverter manufacturer got into the act. Inverter manufacturers present were (listed alphabetically) Heart, Heliotrope, and Photocomm, Statpower, and Trace. We were only able to test the 600 Watt inverters (Heart 600, Statpower 600, & Trace 600) because of system limitations. So the larger (>1 kW.) inverters made by Heart, Heliotrope, Photocomm and Trace were not able to be tested, but in all fairness they were ready and willing. The reasons why we couldn't test the larger inverters was voltage loss through the system's cables, fuses, switches, circuit breakers, shunts and connectors. More on this problem below.

The testing was organized by the fine fellows from ATA, Johnny Weiss and Ken Olsen. The testing was conducted on Sunday August 12, 1990 in front of a live audience of more than 50 fairgoers and the tech reps from the aforementioned inverter manufacturers. The whole show was video taped by Paul Wilkins of The Photovoltaic Network News (PVNN).

The Test System

The test system contained eight Trojan L-16 lead acid batteries configured as a 1,400 Ampere-hour battery at 12 Volts DC. The system contained lots of other gear like eight PV models on a Zomeworks tracker, regulators, controls and instrumentation. We hunted through the crowd and were able to find three Fluke 87 Digital Multimeters to take accurate test data. All inverters used the same set of heavy weight copper cables for connection to the system. A large board of 100 Watt incandescent lightbulbs served as loads. Other loads tried were an approximately 650 Watt Microwave oven and a medium sized (about 400 Watt) circular saw. These last two loads were used to measure the inverter's performance under inductive loads.

The Data

The table and chart below give the data just as it was taken. All inverters were run into exactly the same loads. The most meaningful data was the output voltage of the inverter under a variety of loads. We measured RMS voltage and peak voltage of the inverter's output. We also measured battery

Richard Perez (safari hat) and company testing inverters at the SEER '90 fair.

voltage, amperage, and inverter frequency. In terms of battery voltage and amperage, it became apparent early on in the testing that the instrumentation was not accurate, so I have omitted this data from the table and chart. In terms of frequency, all the inverters were so stable and close to 60 cycles that the data was trivial. Copies of the data were supplied to all the manufacturers of the inverters immediately after testing.

In order to match the output of the commercial electric grid, the inverter should have an RMS voltage output of 117 volts ac. RMS voltage on the grid commonly varies by about six volts RMS or about ±5%. Peak voltage of the commercial power grid is 162 volts peak. Since inverters don't really make sine wave power, their peak voltage is different from that of the sine wave grid power. The peak data is, however, accurate and provides a basis for relative comparison of inverter performance. What really counts in the inverter test data is how close the inverter was able to keep its output voltage to 117 volts RMS under a variety of loads and within its specified operating range of 600 Watts.

Conclusions from the data

I am content to let the data speak for itself.

Now this test system was set up according to Code. This means that it had all the fuses, circuit breakers, fused disconnects, and other paraphernalia required by the National Electric Code (NEC) in addition to the cables and connectors necessary to move the power from the batteries to the inverter.

The major problem we had testing the larger (over 1,000 Watt) inverters was voltage loss. By the time all the code required safety devices added their individual voltage losses, we couldn't move much more than 100 Amperes of current into the inverter. At amperages higher than this, the accumulated voltage loss of all the components in the inverter's low voltage supply lines was about 2 Volts. This meant that the larger inverters were shutting themselves off because of low voltage at their terminals.

And this is perhaps the most important thing we learned from this testing. Large inverters are capable of drawing surges of well over 1,000 Amperes from the batteries. They are capable of consuming over 200 Amperes during normal operation at their limits. In order for a low voltage line to move this much current without excessive voltage loss the line much have very, very low resistance. The inverter lines at SEER '90 had a resistance of about 0.02 Ohms. This was too much resistance to operate an inverter larger than 600 Watts. Today's inverters commonly put out over 2,000 Watts. In order to have these larger inverter works well, the electrical lines feeding them must have very low resistance (less than 0.0015 Ohms). This means heavy copper cables of between 0

gauge for cables totaling less than 6 feet, to 0000 gauge for longer cable lengths. Every series connection and device in this low voltage line adds some resistance. Every fuse, fuse holder, mechanical connector, circuit breaker, switch, and disconnect adds some resistance.

I appreciate that the NEC is concerned for our safety and the safety of our systems. My concern is that by the time they've made us safe enough, our system will be crippled by the accumulated voltage losses in all the protective devices. Please understand that I am all for safety and agree that we need protection in low voltage lines. I respectfully submit that the NEC needs to spearhead the development of safety devices (like fuses, fuse holders, disconnects and circuit breakers) that have about ten times less loss than those they are now proposing. The NEC and the electrical products industries are used to working with 120 vac where a volt or two loss doesn't make much difference to performance. In 12 Volt systems, however a volt or two loss is the difference between working and not working. If low resistance protection devices are not developed, then we are faced with two choices: 1) running an outlaw system, or 2) sitting safely in the dark.

600 Watt Inverter Shootout, Willits, CA on 12 August 1990

120 Vac Loads	TRACE 600			STATPOWER 600			HEART 600		
	Vac RMS	% High or Low	Vac PEAK	Vac RMS	% High or Low	Vac PEAK	Vac RMS	% High or Low	Vac PEAK
100 Watt Lamp	118.9	2%	141.2	115.2	-2%	148.0	124.4	6%	148.0
200 Watt Lamp	118.1	1%	136.8	115.0	-2%	148.8	124.2	6%	140.8
500 Watt Lamp	116.7	0%	125.2	104.6	-11%	137.6	115.8	-1%	122.8
600 Watt Lamp	115.8	-1%	123.2	99.3	-15%	130.8	109.2	-7%	115.6
800 Watt Lamp	108.9	-7%	116.8	91.5	-22%	122.4	98.0	-16%	104.0
Microwave	118.0	1%	140.8	104.7	-11%	141.2	122.6	5%	221.6
Microwave & Saw	116.0	-1%	181.2	97.0	-17%	142.4	110.1	-6%	191.2

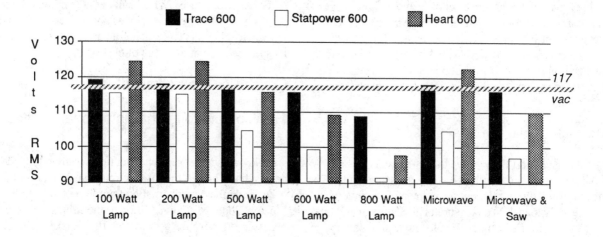

Statpower Pocket Power Inverter - PC100

The Statpower inverter represents a revolution in small inverter technology. Now you can run most 19" color TV's and even some 25" color TV's off of this 90% efficient modified sine wave inverter that fits in your coat pocket and weighs only 14 ounces. Simply plug it into your vehicle's cigarette lighter or your 12V home outlet and you'll have instant AC power up to 100 watts continuous and 200 watts peak. It will operate standard and fluorescent lights, sound and video equipment, data and communications equipment, small appliances, and other electronic products.

The no-load current draw is only 0.06 amps. It comes with a low battery alarm set at 10.7 volts to avoid surprise shutdowns, and an auto shutdown will activate at 10.0 volts to prevent harm to the battery. One year replacement warranty and two additional years limited warranty.

27-101   Statpower Inverter - 100 watt $149

Statpower ProWatt 250 300-Watt Inverter

This brand new Statpower inverter fills the gap left since *Heart Interface* discontinued its 300 watt inverter. This inverter will surge to 500 watts, produce 300 watts for ten minutes, 250 watts for thirty minutes and 200 watts continuously. It has a low battery cut-out with audible alarm at 10V. It is voltage regulated and frequency controlled and features all of the protection features of its big brother, the ProWatt 600. It comes with a 33" long cigarette lighter cord and measures only 5.5" x 4.5" x 1.5" and fits in the palm of your hand! The Prowatt 250's precisely regulated output allows you to run your sensitive electronic equipment. With an efficiency of 90%, almost all of your battery's current is converted into useable AC power. One year replacement warranty and six additional months limited warranty.

27-109   Statpower ProWatt 250 $195

Statpower ProWatt 600 600-Watt Inverter

The ProWatt 600 is a 600-watt inverter that will continuously operate TV's, VCR's, stereo systems, power tools, lighting, kitchen appliances, refrigerators, and a wide range of other electrical and electronic equipment. The ProWatt 600 is almost totally silent. It is a smaller and more compact unit than either the Trace 612 or the Heart HF-600X. It weighs in at 5 pounds and occupies the space of a small car stereo amplifier (2.5" x 8" x 10"). The Prowatt 600 will produce 800 watts of power for 15 minutes so that it can run microwaves and hair-dryers. It will produce 500 watts continuously and will surge to 1,500 watts. It operates at 90% efficiency, with a no-load current draw of 0.12 amps.

Unique to Statpower inverters is an excellent visual representation of the power being drawn and the state of the battery charge via an LED bar graph and a voltage LED bar graph. One year replacement warranty and six additional months limited warranty.

27-103   600W Statpower Inverter $495

PowerStar Upgradable Inverters (400W-700W-1,300W)

PowerStar has recently introduced the very first low-wattage upgradable inverters. The three models, which all use the same case are rated in terms of their continuous true RMS power capability. The *continuous* power rating on the smallest inverter is 380 watts; the first upgrade takes you up to 700 watts continuous; and the final upgrade takes you up to 1,300 watts continuous. The actual upgrade can be performed by the factory in a few days. The price for the upgrade is simply the difference between the two units. *Therefore, there is no cost penalty for buying a smaller unit and later upgrading it, making these inverters extremely versatile!*

The small 380 watt unit is suitable for computer systems, power tools and small appliances. The manufacturer claims that its 3,000 watt surge capacity can start and run 1/4 horsepower motors.

The first upgrade - the 700 watt unit - will run a 500 watt microwave, a vacuum cleaner, a hair dryer (on medium), or a small coffee pot or hotplate. It is preferred for power tools or appliances that run heavily loaded for long periods of time. Like the smaller unit, the surge capacity is 3,000 watts.

The second upgrade - the 1300 watt unit - will continuously operate a full size microwave, any 1,500 watt appliance or a worm-drive circular saw.

The following specifications apply to all three units:

Input Voltage:	10.5 to 16.5 volt DC
Output Voltage:	115 volts AC true RMS +- 5%
Output Frequency:	60Hz, modified sinewave
Idle Current:	60mA DC
Low Battery Warning:	Audible alarm below 10.9V input
Over Voltage Shutoff:	16.8 Volts
Indicator:	A green micropower lamp shows AC output present
Over Temp. Protection:	Proportional power reduction
Overload Protection:	Dual mode limiter enables operation of appliance rated higher in power than inverter. A moderate overload lowers the output voltage. A severe one causes shutdown. Reset by cycling the power switch.
Remote Control:	A built-in connector socket has provisions for a remote switch as well as a hard-wired AC output.
Fusing:	Replaceable fuses on both input & output.
Efficiency:	Over 90% at half rated power
Size:	3.15" x 3.3" x 11"
Weight:	Less than five lbs.
Warranty:	Two years.

27-105 **PowerStar 380 Watt Inverter (UPG400)** **$399**
27-106 **PowerStar 700 Watt Inverter (UPG700)** **$499**
27-107 **P'Star 1300 Watt Inverter (UPG1300)** **$799**

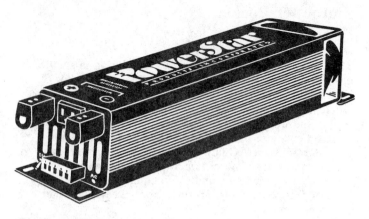

Pre-Press Update:

We've just completed thorough testing of these new upgradeable Powerstar inverters, and are very impressed. First of all, one is immediately taken by the fact that they are totally silent under load - unlike any other large inverter that we offer. Secondly, even with their incredibly low standing power loss of 65mA (.065A), they are continually on rather than in a "seek" or "search" mode like many other inverters, thus providing instant power. Futher we were surprised and delighted to find that even the smallest unit (400W) will indeed (as the manufacturer claims) start and run a 1/4 HP motor without delay!

PowerStar 200 Pocket Inverter

This new compact inverter has quickly become our best seller. It's the best mini-inverter in the industry. It is ideal for powering most 19" color TVs, drills, computers, VCRs, guitar amplifiers, water pumps, curling irons, sanders, jigsaws, video games, stereos, and lots of other small appliances *directly from your car or 12V system.* It's great for travellers to carry in their cars to power these standard appliances. It comes with a cigarette lighter plug that plugs into any cigarette lighter and will deliver 140 watts of 110V AC power continuously from your 12V battery. It will provide 400 peak watts and 200 watts for over two minutes. Amp draw at idle is 0.25 amps (3 watts). PowerStar has recently made an improvement to the 200 watt inverter. In case of overload, the unit now safely delivers as much power as it can, into that overload. This means a hair dryer, which would shut down the older unit, can now be used on the LOW setting. One year parts and labor warranty.

27-104 **PowerStar 200 Inverter** **$149**

Trace Inverters

Trace inverters have literally turned the inverter industry upside down in the past four years. They have been consistently one of the best selling items in our catalog. Customer satisfaction on Trace is close to 100%, and the folks at Trace are a delight to deal with, always willing to lend a patient ear to a confused customer. Just a short few years ago Heart Interface was the big name in innovative inverter technology, but it seems now that Heart has all but bowed out of the remote home market, choosing instead to concentrate on marine applications. Trace has proved itself again and again to be the most efficient, durable, and reasonable (in dollars per watt) inverter on the market.

Trace inverters are designed with a wide range of user installable options - allowing them to grow in sophistication and power as your needs change. The "Standby" option provides a powerful, internal, programmable, automatic battery charger, that works well with standby generators allowing backup power for your solar array. If your power requirements increase over time, the "Stacking Interface" option allows an additional unit to be operated in parallel, doubling your output power capability.

Perhaps the greatest feature of Trace inverters is their microscopic "No-Load Power Drain." The 2012, Trace's 2,000-watt inverter, has a no-load current of only 0.03 amps or 1/3 of a watt! This means you can leave the inverter turned on for an entire day and only use 8 watt hours of power or 2/3 amphour! (It's even less for the 612 inverter - 0.02 amphours.)

Trace's protection circuits are automatically resetting. If an overload occurs, the inverter shuts down until the condition is corrected and then resumes operation. It does not need to be manually reset. Trace inverters employ a unique "impulse phase correction" circuit allowing the inverter to closely duplicate the characteristics of standard 117V grid power. With this design approach the limitations of the modified sine wave format are largely overcome. Protection circuitry carefully monitors the following conditions: over current, over temperature, short circuit, reverse output voltage application (output connected to grid), induced electrostatic charge (lightning), high battery and low battery.

Trace uses a very large number of output transistors (44 FET's) and specially wound transformers which results in exceptional high power performance. The 2012 will surge to well over 6,000 watts of power.

Here is a partial list of appliances typically used with the Trace inverter:

Here is a detailed photograph of the six Trace 2248, 2200 watt, 48 volt inverters interfaced together in series/parallel configuration for the North Carolina Biodome project that Real Goods designed in 1987.

Stereos	Sewing machines
Flood lights	Electric typewriters
Ceiling fans	Electronic musical instruments
Night lights	Vacuum cleaners
Shavers	Compact disc players
Hair dryers	Trash compactors
Blenders	Satellite dishes
Ice makers	Color TV's
Toasters	VCR's
Microwave ovens	Computers & peripherals
Coffee makers	Electric blankets
Fluorescent lights	All power tools
Refrigerators	1/2 hp deep well pumps
Washing machines	Radial arm saws & table saws

Trace 612 Inverter

The 612 is a welcome addition to the Trace family. It will produce 600 watts for 30 minutes, 500 watts for 60 minutes, or 400 watts continuously. It uses only 0.24 watt (0.02 amps) on standby. It will power TV's, VCR's, computers, and test equipment, and it will also power many vacuum cleaners, Champion juicers, and some microwave ovens. For an extra $100 you can order an optional 25-amp battery charger in the "Standby Option". This internal 25-amp battery charger comes with a 30-amp transfer switch that switches back and forth between battery power or grid/generator power. The battery charger requires 2,000 watts of generator power at a minimum to operate it. The current is not adjustable. Also available is the RC/3 Remote Control option allowing you to turn the unit on and off from a remote location (although with the extremely low no-load power drain you can let the unit stay on all day for around 0.5 amps.) The Trace 612 comes with our very highest recommendation. TWO YEAR GUARANTEE. Weight: 22#.

We now offer a 100 amp fuse with holder for installing between the battery and the Trace 612. This will help you comply with state and local electrical codes.

27-211	Trace 612 Inverter	$550
27-212	612 Inverter w/Battery charger	$650
27-214	Remote control for 612	$ 50
27-215	100 amp fuse w/holder for 612	$ 32

Trace 724 Inverter

The new Trace 724 is the newest addition to the Trace line. It was created as a direct response to the demands of our customers who prefer to base their systems upon a 24V battery bank. Power ratings show a 450 watt continuous operation with a very strong 2500 watt surge available, for starting inductive loads. The 724 will produce 1000 watts for 7 minutes, 600 watts for 55 minutes, and 500 watts for 90 minutes. Idle current remains at an incredibly low .025 amps. A 12 amp battery charger is available as is a remote control (RC/3) option that will allow remote monitoring via an LED and will also allow the user to turn the 724 on and off. Warranty is 2 years and the weight is 16#.

27-221	Trace 724 Inverter	$625
27-222	Trace 724SB (w/battery charger)	$725
27-214	Remote Control for 724	$ 50

"Since we've brought home the Trace 612 inverter it has constantly held us in awe. First it was so quiet when we turned it on that it was a good thing there was a light on it. Then the testing of one electric tool or appliance after another: drills, palm sanders, industrial high speed grinder, sewing machine, mixer, blender, all worked better than they ever had before when I had been running them with an Onan generator. The electricity is so clean that our Hi-Fi sounds brand new. The computer and printer work better and faster than B.T. (Before Trace)

Yesterday I looked at our old but functional Kirby, an antique legacy from grandparents. Why not give it a try?, I thought. The plate on it's shiny bottom read 4 amps. I plugged the Kirby in, hit the switch and it roared to life like a new machine and vacuumed the whole house. And the batteries didn't even seem to be touched. The efficiency is amazing.

When we used the inverter to power our 19" TV and video player we were blown away by the beautiful color and full sized picture. Again better than before using the generator.

If you think I'm enthusiastic that's an understatement. This certainly has been one of the most life enhancing purchases that we have ever made. Thank you for the "offer I couldn't refuse" and thank you for providing such a valuable service which is allowing us to live an abundant life in closer harmony with the world around us. We will be staunch supporters of Real Goods and AE in the years to come. And if you talk to the Trace people tell them thanks for that little box of magic that quietly hums away, the Trace 612!"

Laird Sutton & Gloria Molica,   **Bodega, CA.**

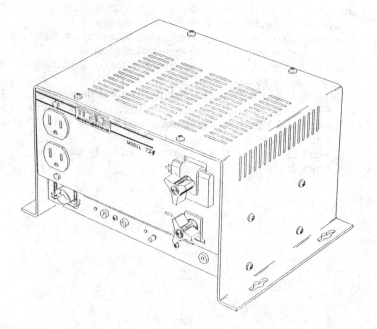

Trace 2012 Inverter

The Trace 2012 is by far our best selling inverter (and probably the best selling item in our entire catalog). It will produce 2,000 watts of power at 110 VAC from a 12VDC battery source for up to 1/2 hour with a 6,000+ watt surge for starting induction and other high surge motors. The no-load power drain is the best in the industry - a scant 0.03 amps (0.36 watts). The unit measures 6.25" high x 10" wide x 12.4" deep. We strongly recommend you go over the options available very carefully to tailor the Trace inverter to your needs. The 2012 is also available with 234VAC output at 50 Hz for export use for $1,150 (order the **2012/E**). As with all Trace products, a 2-year warranty is included.

Late flash at presstime: Trace has just added something new to all domestic series 2000 inverters. They are all now being shipped with a easy to use hardwire access. The improved hardwire system utilizes an externally accessible 30 amp terminal block and wire clamps for strain relief that accomodate wire sizes up to #10 gauge. All units with the Standby option will also include an externally accessible 30 amp circuit breaker on the AC input. An AC outlet will no longer come as standard equipment, but is available as an option.

The new Model 2012 requires Stacking Interface Module SI/B. The SI/B is backward compatible with all units now in the field, however, the old SI/1 Stacking Interface is NOT compatible with inverters featuring the new hardwire access. All new inverters will require the SI/B for stacking applications. The SI/B will place the master unit on the left simplifying the installation of options to a "Stacked Pair."

27-201	2012 Trace inverter	$1,090
27-202	2012SB inverter w/battery charger	$1,310

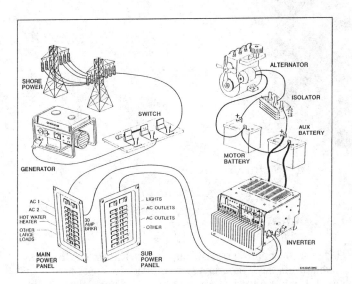

Power vs. Efficiency

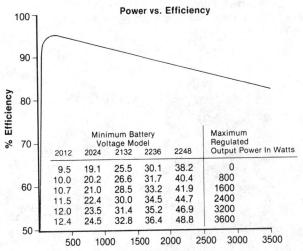

Minimum Battery Voltage Model					Maximum Regulated Output Power In Watts
2012	2024	2132	2236	2248	
9.5	19.1	25.5	30.1	38.2	0
10.0	20.2	26.6	31.7	40.4	800
10.7	21.0	28.5	33.2	41.9	1600
11.5	22.4	30.0	34.5	44.7	2400
12.0	23.5	31.4	35.2	46.9	3200
12.4	24.5	32.8	36.4	48.8	3600

Power vs. Time

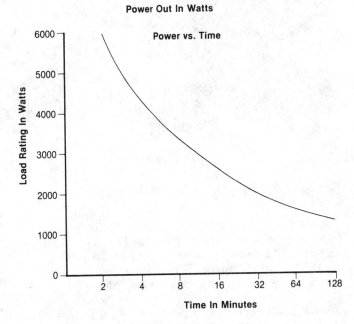

Trace 2524 Inverter

The Trace 2524 will provide 2,500 watts of power at 110 VAC for 1/2 hour from a 24V battery source. The unit was recently upgraded (July 1989) from the old 2024. Once again the price didn't increase. The no-load current is 0.018 amps. The unit will provide several seconds of surge power to 6,000+ watts. It's available with 234 VAC output at 50 Hz for export use for $1,410 (order the 2524/E).

27-203   2524 Trace inverter $1,350
27-204   2524 inverter w/battery charger $1,570

Trace 2236 Inverter

The Trace 2236 will provide 2,200 watts of output power at 110 VAC from a 36VDC input for 30 minutes. It has a surge capacity of 6,200 watts and a no-load current of 0.025 amps.

27-209   2236 Trace inverter $1,400
27-210   2236 inverter w/battery charger $1,620

Trace 2248 Inverter

The Trace 2248 will provide 2,200 watts of output power at 110 VAC from a 48VDC input for 30 minutes. It has a surge capacity of 6,200 watts for a few seconds. The no-load current is only 0.022 amps.

27-205   2248 Trace Inverter $1,500
27-206   2248 inverter w/battery charger $1,720

Trace 2132 Inverter

The Trace 2132 will provide 2100 watts of output power at 110 VAC from a 32VDC input for 30 minutes. It has a surge capacity of 6,000 watts and a no-load current of 0.016 amps.

27-207   2132 Trace inverter $1,400

WE PROMISE TO BEAT ANY PRICE IN THE USA (PUBLISHED OR VERBAL!) ON ANY TRACE INVERTER AT ANY TIME - DON'T HESITATE TO ASK.

Options for Trace Inverters

Stacking Interface Module (SI/1)

The stacking interface allows two inverters to be paralleled for double output power at 117 VAC. If both units have the battery charger option, the charging capability is also doubled. This is the option that Trace pioneered to give its units maximum flexibility.

27-304   Stacking interface (SI/1) $200
27-308   Stacking Interface (New) SI/B $200

Turbocharger (ACTC)

This is a fan cooling option that increases the continuous power of all the Trace inverter models. It senses the heatsink temperatures and automatically operates the fan. The continuous power rating is increased by 400 watts. It is user installable on all units but not available on the 612.

27-302   Turbocharger (ACTC) $120

Standby Battery Charger

A programmable battery charger (the Standby Option) is available so that you can plug your inverter into your generator for back-up power. The battery charger is a constant current voltage limited design (16.5V maximum). This means that it charges at a set current until a set voltage is approached and then tapers to zero charge rate. Both the maximum charge rate and maximum charge voltage can be programmed by the user. Time delay transfer circuitry (a 1 hp, 30 amp transfer relay) detects the presence of grid or generator power and automatically changes the inverter to and from the battery charger mode. A 30-amp automatic transfer switch is standard with the battery charger that switches the unit from battery power to grid/generator power. Many of our customers purchase the battery charger option and use it in conjunction with their generators. Typically a generator is used for very large loads (water pumping, washing machine, etc.) If you leave the generator on for a couple hours per day for that load, you can charge your batteries at the same time supplementing your solar panels.

The standby battery charger puts out the following maximum charge rates for the various sizes of Trace Inverters:

Inverter	Maximum charge rate
2012	110 amps
2524	50 amps
2132	39 amps
2236	35 amps
2248	25 amps

One important note on the Standby Option: *You need 6,000 watts of power (a 6-kW generator) to utilize the full power of the battery charger.* If you have that much input power you'll get 110 amps (with the 2012) of charging capacity when your batteries are fully drained. However, don't fret if you only have a 4-kW generator (as most of us do) because the Trace battery charger will still charge at approximately two-thirds capacity or 75 amps.

If you choose to add the standby option to an inverter at a later date there is a $75 installation charge.

Battery Cables for Trace Hook-Ups

As we mentioned earlier the wire sizing between the battery and the inverter is of critical importance and many people tend to undersize. There is a tremendous current draw in this area and it's always better to err on the side of overkill. For this reason, Trace has decided to make up very heavy duty 4,000- strand, 4/0 (that's four-ought!) welding cable. They're available in either a 5-foot or 10-foot pair of terminated and color coded battery cables. Highly recommended to complete your Trace installation with security. Our 2012 fuse & cable assembly contains a DC rated 300 amp fuse, fuse holder, and enclosure assembled on a pair of 10 foot 4/0 cables to help you comply with state and local electrical codes.

27-311	2 each 5-foot 4/0 cables	$ 79
27-312	2 each 10-foot 4/0 cables	$125

Remote Control

The remote control option is for use in installations where the inverter is not easily accessible. The old option RC1 has been discontinued in favor of the new and improved RC-2000 which comes complete with a digital voltmeter (DVM). It provides a duplicate set of control switches and indicator lamps and is only available in models with the standby battery charger option. Option RC2 is a more basic remote with on/off control and an LED that indicates on/off and search mode states.

27-306	RC2 Remote Control	$ 75
27-309	RC-2000 Remote w/DVM for 2012	$250
27-310	RC-2000 Remote w/DVM for 2524	$250
27-316	RC-2000 Remote w/DVM for 2132	$250
27-317	RC-2000 Remote w/DVM for 2236	$250
27-318	RC-2000 Remote w/DVM for 2248	$250

Digital Volt Meter (DVM)

The Digital Volt Meter (DVM) is a must option for purchasers of the Standby charger. The meter monitors battery voltage to tenths of a volt, the charge rate of the battery charger, the frequency of the generator in hertz (Hz) and the peak voltage of the charging source. Without this option on your battery charger, you're stuck with a lot of guesswork! The DVM can be easily installed in 5 minutes by the customer either in the base unit or in the Remote Control (RC1). The DVM is available as an option to all Trace inverters except the 612.

27-301-12	DVM (Digital volt meter) -12V	$130
27-301-24	DVM (Digital volt meter) -24V	$130

110V to 220V Transformer (T-220)

The T-220 is a 2,500 watt 1-hour transformer that may be configured by the user to function as either a step-up or step-down autoformer (120 to 240 or 480 VAC), or as an isolation transformer or a generator balancing transformer. Many alternative energy homesteads have everything on 12V or 110V with the exception of the oddball 220V submersible pump. By installing the T-220 between the submersible pump and the pressure tank the transformer will only come on when the pump is activated. 5,000 watt maximum surge.

27-307 T-220 Step-up transformer (110V-220V) $250

Low Battery Cut Out (LBCO)

The LBCO protects against over-discharging the batteries. The circuitry evaluates battery voltage and current to shut down the inverter in a low battery condition. User programmable.

27-303-12	Low battery cut-out - 12V	$75
27-303-24	Low battery cut-out - 24V	$75

Heart Inverters

For several years Trace has dominated the inverter market for remote home applications. Heart chose instead to concentrate on the marine and RV markets. Last year, with its introduction of the Energy Management System (EMS), Heart made some strong design changes and engineering breakthroughs in an attempt to recapture some of its lost market-share to Trace. We enjoy this friendly competition because it keeps both manufacturers on their toes and hastens the development of new technologies to the end-user! (Note that our inverter test was done prior to the introduction of the new Heart EMS line.)

Heart Interface has just introduced its all new 1991 version of the EMS line. These new inverter/battery chargers make use of microprocessor based logic, running on software developed at Heart Interface. This new design offers several departures from traditional inverter performance. For one thing, Heart has rated these inverters for their continuous power output, no intermittent ratings are used. Furthermore, these new inverters produce a multi-stepped output waveform, which more closely approximates a true sine wave. This unique waveform will provide more consistent peak-to-peak AC voltage values, and, among other characteristics, provides exceptional motor-starting power for high-torque appliances such as washing machines and refrigerators.

These new EMS systems from Heart Interface are UL listed. They are designed to conform to National Electrical Code standards and feature an easy to access hard-wiring compartment with input and output strain reliefs. Heavy duty battery cable lugs and an AC input/output terminal strip designed for 30 amp 10 AWG wire are provided. The Heart warranty covers parts, labor and return shipping for 30 months. A cascade option will be available in mid-1991 which will allow connecting multiple units together for increased inverter output and charging capacity. The EMS Spectator will work with either of the 2 EMS units.

EMS Competitor Heart Inverter

The new EMS Competitor produces 1,500 watts of silent AC power continuously. It comes standard with a 65 amp 12 volt battery charger and features sealed, nitrogen filled 30 amp relays for AC power source switching. It will operate sensitive electronics such as computers, TV's, and VCR's. The load detection circuit reduces power consumption in a no-load condition and responds almost instantly when an appliance is turned on. The unique waveform provides extremely high motor-starting torque. The EMS battery charger uses two-step regulation to initially provide constant current for a rapid recharge and then a constant, lower voltage float charge for safe long-term battery maintenance. The charging circuitry will compensate for low line voltage, maximizing the charging current when using small generators. Use of the EMS Spectator allows adjustment of charging parameters and digital monitoring of the system. *30 month warranty.*

27-401 EMS Competitor Heart Inverter $1,595

EMS Olympian Heart Inverter

The new EMS Olympian will produce 3,000 watts of silent AC power continuously. It comes standard with a 100 amp 12 volt battery charger and the same relay switching as the EMS Competitor. Like the EMS Competitor, the Olympian is UL listed and provided with AC reverse polarity, thermal, overload and circuit breaker protection. These two new models also share the same waveform, load detection, and battery charging circuitry, with the EMS Olympian having twice the motor-starting power. *30 month warranty.*

27-403 EMS Olympian Heart Inverter $2,395

EMS Spectator

The new EMS Spectator is a flush mounted remote monitor that is available for both the EMS Competitor and the EMS Olympian. It features separate inverter and battery charger on/off switches. It allows digital metering of the following functions: True RMS AC input and output voltage, AC output current, AC input current and DC battery voltage. It has digital monitoring of faults including low battery, over-temperature, overload, and AC reverse polarity. It features an audible alarm for all the above fault conditions. It allows for programming the following functions: Inverter load detection threshold, charger AC input current, charger bulk and float voltages and inverter low battery cutout voltage.

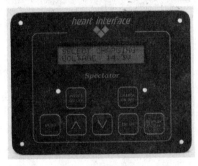

27-402   EMS Spectator Remote Monitor $200

Heart HF 12-600 Inverter

Heart Interface has discontinued its 300-watt unit (HF 12-300X) and replaced it with this 600-watt unit. It will run 600 watts for 25 minutes and surge to 1,200 watts. It uses only 0.7 watt (0.06 amps) on standby (idle loss) and reaches 90% efficiency at only 50 watts. It is an ideal unit for computers, printers, TV's, VCR's, and test equipment. Modified sine wave output. Guaranteed one year.

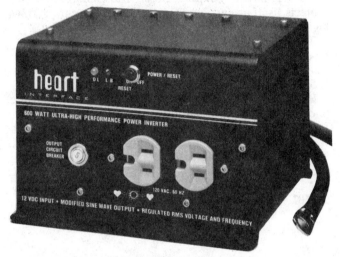

27-404   Heart HF 12-600X 600 Watt Inverter $595

EFFICIENCY VS OUTPUT POWER
FOR EMS 1800-12 & HD 1800-12 INVERTERS

CHARGING CURRENT VS BATTERY VOLTAGE
FOR EMS 1800-12

LOAD SIZE VS TIME UNTIL SHUTDOWN
FOR EMS 1800-12 & HD 1800-12 INVERTERS

1100 VA continuous
@ 13 VDC, TA = 25 C

EFFICIENCY VS OUTPUT POWER
FOR EMS 2800-12 & HD 2800-12 INVERTERS

CHARGING CURRENT VS BATTERY VOLTAGE
FOR EMS 2800-12 INVERTERS

LOAD SIZE VS TIME UNTIL SHUTDOWN
FOR EMS 2800-12 & HD 2800-12 INVERTERS

2000 VA continuous
@ 13 VDC, TA = 25 C

PSST 12-2500 Heliotrope Inverter

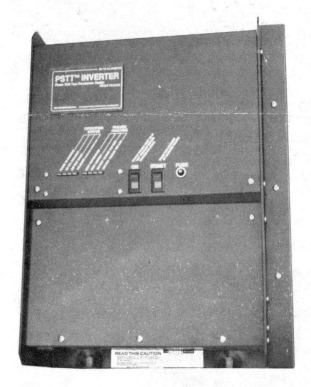

Heliotrope, makers of solar thermal panels, has recently introduced a very solid, heavy duty 2,500-watt inverter. One is immediately taken with the construction of the unit, as it's obvious that materials have not been scrimped upon. Unlike other inverters it mounts to a wall, leaving more usable horizontal space. It comes with four AC outlets, instead of the usual two, and is reverse polarity protected, another unique feature. The Heliotrope protects itself against over temperature, over current, and high and low battery voltage.

The Heliotrope has two operating modes: "standard mode" and "battery saver mode." In standard mode, very small appliances can be run and the no-load power drain is 5 watts, while in battery saver mode, the no-load drain is 0.4 watts. The inverter uses two transformers instead of one and has a built-in fan to cool them down. The wave form is "quasi-sinewave" and the rated output is 2,500 watts continuous, with a surge rating of 6,800 watts, and is 95% efficient. It comes with a 10-year limited warranty, the longest in the industry.

27-501 PSST Heliotrope 12-2500 Inverter $1,895

The Great Real Goods Inverter Test

We got tired of using our built-in bias on recommending inverters to customers and decided it was time to do a bonafide controlled experiment to compare the efficiency, cost-effectiveness, and power economy (no-load power drain) of the various inverters that we sell. In fairness, this test was run over two years ago. We recommend that you also read the more recent inverter test (included in this Sourcebook) that was conducted by Richard Perez, of HomePower Magazine at the SEER fair in Willits, CA in August 1990.

We tested three of the smaller inverters: Heart HF-600X, Trace 612, and the Westec W600-12; as well as three of our best selling large inverters: Heliotrope PSST 12-2300, Heart 12-1200U, and the Trace 2012. The test was conducted using a 100 amp-100 millivolt shunt, two 6V, 220-amphour batteries, #2/0 wire between battery and inverters, and a GC Electronics digital multi-meter #20-200. The load used was a resistive load consisting of a bank of 110V incandescent light bulbs.

The purpose of the inverter test was to determine the relative efficiencies of the inverters at various resistive loads and to calculate the idling current or "no-load power drain" of the units. Thanks to Steve Taylor of Northwest Energy Center for helping to graph the results.

The Heart HF 12-600 and the Trace 612 both performed famously, with the Heart being slightly more efficient at 100 watts but the Trace showing higher efficiency at 250, 500, and 1,000 watts, although not significant enough to make a major difference. The Trace, as expected, displayed its greatest attribute of a minuscule no-load power drain, using 1/7 the power of the Heart!

All three large inverters performed famously with good efficiency. The Heliotrope has several advantages over the Trace and the Heart with four AC outlets instead of two, with wall mounting capability, with a cooling fan as standard equipment to cool down the two transformers, and with two idling modes for more versatility.

Again the Trace demonstrated its sole possession of first place in the ultra-low no-load power drain category. At a scant 0.03 idling amps it draws half what the Heliotrope does in "battery saver" mode and less than one-third the drain of the Heart.

We're happy to report that all inverters tested represent a vast improvement over the older style inverters, but we must conclude that all things considered we still recommend the Trace inverters in both sizes in all categories. The Trace is as efficient or more efficient than any inverter on the market, is by far the most energy economical with the lowest no-load power drain, and is the least expensive in dollars per peak watt.

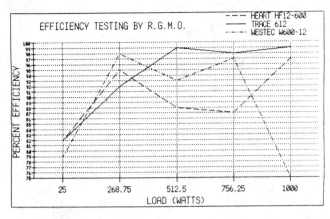

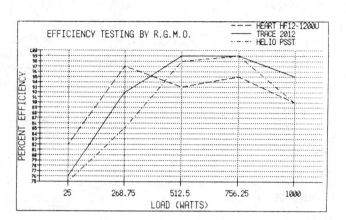

Computer graphics courtesy Steve Taylor, Steamco Solar.

Tripp-Lite Inverters

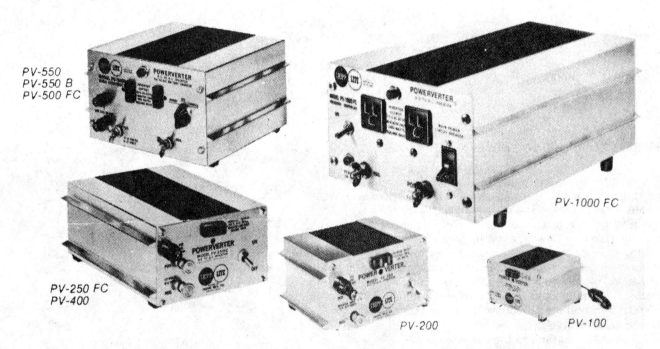

PV-550
PV-550 B
PV-500 FC

PV-1000 FC

PV-250 FC
PV-400

PV-200

PV-100

Tripp-Lite is our economy line of light duty inverters. It produces 117VAC square wave current. Some models have frequency control and some do not. Their 100-watt model square-wave inverter furnished much of the power for the old Osborne 1 computer on which many pages of the first 1984 Sourcebook were written. It would be hard to find a better inverter-cost-per-watt buy; however, this inverter will not operate induction, or capacitor-start motors or any other reactive-load equipment. The surge load capability is limited to 10% above each unit's rated power output. In case you are connected to a public utility and it should fail, Trip-Lite also manufactures a good line of economy standby power systems.

Tripp-Lite inverters will operate AC equipment from 12, 24, or 32-volt battery systems: TV's, stereos, small hand tools, appliances, lights, minicomputers, and terminals. (Check wattage requirements on your appliances to determine the correct model for the application.)

Regulated output maintains voltage and frequency within a specified range across a wide variation of battery voltages. Output is compatible with most AC appliances and electronics. The input is protected against DC input line voltage spikes and transients. A sensor automatically turns off the inverter when battery voltage is low on most models. Reverse polarity protection (most models) guards the inverters against improper connection to batteries.

Tripp-Lites are encased in rustproof anodized aluminum chassis with heat sink fins for convection cooling. Integral shock mounts reduce component vibration.

All units come with a one-year limited warranty.

Model#	Input Volts DC	Output Volts AC	Frequency	Maximum Watts Continuous Output	Maximum Watts Intermittant Output	Input Amps No Load	Polarity Protection and Overload Indicator	Size HxWxD
PV-100	12	117	60 Hz	100	110	2.5	No	3½x6½x5½
PV-200	12	117	60 Hz	200	220	3.5	No	3½x6½x5½
PV-400	12	117	60 Hz	400	440	3.5	No	3-5/8x7¼x6
PV-550	12	117	60 Hz	550	600	4.0	Yes	5x9½x9
PV-550/B	12	117	60 Hz	550	600	4.0	Yes	5x9½x9
PV-250/FC	12	117	60 Hz	250	275	1.5	No	3½x6½x11½
PV-500/FC	12	117	60 Hz	500	550	3.5	Yes	5x9½x9
PV-1000/FC	12	117	60 Hz	1000	1100	3.5	Yes	5x10x14
PV-1000FC/24	24	117	60 Hz	1000	1100	1.8	Yes	5x10x14
PV-1000FC/220	12	220	50 Hz	1000	1100	3.5	Yes	5x10x14

Item #	Model	Price	Item #	Model	Price
27-601	PV100	$ 95	27-606	PV250/FC	$245
27-602	PV200	$125	27-607	PV500/FC	$395
27-603	PV400	$195	27-608	PV1000/FC	$575
27-604	PV550	$245	27-609	PV1000/FC/24	$625
27-605	PV550/B	$325	27-610	PV1000/FC/220	$625
			27-611	SB1000G	$895

Vanner Voltmaster Battery Equalizer

The Vanner Voltmaster was designed to eliminate the overcharging of one battery in a split 24/12-volt system. All batteries will charge and discharge equally even when there is an unequal load. The Vanner Voltmaster allows you to upgrade to a larger 24-volt inverter system, but still use all your 12V appliances. The device electronically monitors voltages of both batteries and transfers current whenever one battery discharges at a rate different than the other. By maintaining equalization down to 0.01 volt, the Voltmaster will extend battery life by preventing both over and under charging.

All units are guaranteed for one year.

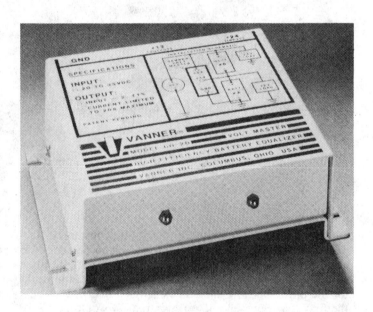

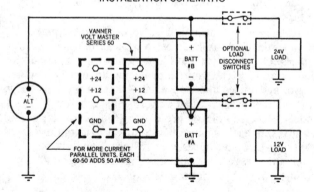

INSTALLATION SCHEMATIC

Item #	Model	Voltages	Max cont. draw	Price
27-801	60-10A	24/12	8 amps	$340
27-802	60-20A	24/12	20 amps	$365
27-803	60-50A	24/12	50 amps	$495

Chapter 4
Alternative Power Loads

Energy Efficient Lighting

"Astounding progress has been made in the efficiency of lighting technology since the candle. Take, for example, a state-of-the-art 20 watt compact fluorescent lamp (light bulb) which screws into a standard household light fixture socket and produces the same appealing, warm-quality light as an incandescent light bulb: This innovative technology will produce the same amount of light as a standard 75 watt incandescent light bulb and, compared to the candle, it emits 100 times more light, is about 1,333 times longer lived, and over its 10,000 hour lifetime, will produce the same amount of light as that from about 133,000 candles."
Robert Sardinsky, Rising Sun Enterprises, Inc.

Lighting is one of the *most important* and *potentially most significant* electrical loads in PV-powered homes. In comparison, lighting in utility-powered homes typically accounts for only 7% of total energy usage. In either case, using state-of-art, energy-efficient technologies such as the one mentioned above can enable you to wring far more "work" out of each precious watt of energy you use, whether it be from renewable or nonrenewable resources.

Although various types of lights illuminate our lives, we usually take light for granted - unless it's not there when we need it. As winter gets deeper and darker and more time is spent indoors, lighting can become a significant part of total power consumption.

Because lighting is such a basic need and so often utilized, it is wise when designing a system to become familiar with the different kinds of electric lights available and carefully select those types which will meet your needs best, both in terms of esthetics and energy efficiency.

Since our 1982 AE Sourcebook went to press nearly eight years ago, leaps and bounds of technological improvement have occurred in low-voltage lighting. It used to be that all the manufacturers of low-voltage lighting catered exclusively to the RV industry, and churned out an endless supply of ugly, white plastic, nondescript light fixtures.

Real Goods is actively continuing to forge new partnerships with a number of innovative manufacturers to offer you attractive, high-quality, energy-efficient lighting products. Over the coming year, we intend to greatly expand our DC lighting product offerings. And because many of you are building alternative-energy-driven AC homes powered with inverters or are striving to live as efficiently as possible on the utility grid, we will also be making a great effort to bring you a variety of new AC lighting products that can illuminate your homes with high-quality lighting without guzzling electricity.

At the heart of many of our new lighting products is the compact fluorescent lamp. These lamps ("light bulbs") provide a wonderful, warm quality light just like from standard incandescent light bulbs, but they operate much more energy-efficiently and are far longer lived. To help you understand the application and limitations of this state-of-the-art technology, we have asked one of the leading distributors of these extraordinary products in the country, Rising Sun Enterprises, Inc., to prepare some basic consumer information for you. Please note, if you already have light fixtures in place, we offer a variety of compact fluorescent lighting products which you can screw right into these fixtures; or if you are shopping for new fixtures, a select variety of these is offered as well.

Although you may pay more for most high-quality energy-efficient lighting products than for conventional lighting, they can offer you - and the world around you - *paybacks* in many ways, from money savings from lower utility bills to a lower cost for your alternative energy system to less fossil fuels polluting the atmosphere to provide for our energy needs.

The Basics of Lighting

Lamps

There are four basic types of electric lamps, or "light bulbs", which are most suitable for use around the home: 1) incandescent, 2) halogen, 3) compact fluorescent, and 4) standard fluorescent. A fifth family, high-intensity discharge lamps, such as high-pressure sodium and metal halide, has some practical outdoor residential applications as well, and will be discussed in an upcoming Real Goods News along with announcements for dozens of new lighting products!

Incandescents

The "all-American" incandescent light bulb is, in reality, an electric space heater which produces a little light. The electric current passing through an incandescent lamp's tungsten filament heats it until it glows white, emitting light that we can see; approximately 90% of the electricity that passes through a standard incandescent lamp is converted to heat, and only 10% into visible light. An inert gas fill, such as nitrogen or argon, is used inside the glass bulb to slow the oxidation of the filament which, over its lifetime, slowly evaporates to blacken the inside of the lamp envelope and typically reduces light output by + or - 20% .

Eventually, this evaporation/oxidation weakens the filament and it breaks, causing the lamp to burn out.

The simple construction of incandescent lamps minimizes their initial cost, but, on a life-style basis, they represent the most energy *inefficient* and *shortest* lived of all lighting sources. Lamp *efficacy* (light output divided by power input) ranges from 8 to 20 lumens per watt and lamp service from 750 to 2,000 hours. While longer lamp life is possible with extended-service, rough-duty, and long-life lamps, these operate the least efficiently of all incandescents, producing fewer lumens per watt, because their filaments are run at lower temperatures. Note, the more red/orange an incandescent lamp appears, the lower its operating efficiency.

Hundreds of different incandescent lamp styles (general service, reflector, decorative, etc.) have been developed both for stationary and mobile applications. Many of these are available for low-voltage applications and can simply be screwed into regular light fixtures. Two of the most versatile low-voltage options include the "A" lamp, conventional-looking incandescents with standard screw in medium edison bases, and automotive brake light bulbs, miniature lamps with a bayonet-to-edison base adapter to screw them into conventional lamp sockets.

Operating most types of incandescent lamps at low voltage produces significantly more light than higher voltage lamps of the same wattage. Low-voltage lamps incorporate a stouter filament to withstand the higher current, at which they operate (ten times more current passes through the filament of a 12V incandescent lamp than that for a comparable wattage 120V model.) This raises filament operating temperature, and boosts lamp efficiency. Approximately 40% more light is produced for each watt of power consumed.

Operating incandescent lamps above or below their rated input voltage dramatically affects their performance as illustrated in Table 1 (the effect of voltage on standard incandescents is nearly identical to that for the illustration of tungsten-halogen incandescents shown). At lower than rated input voltages, for example, though lamp life may be significantly longer, light output (lumens), power consumption (current), and lighting energy efficiency (lumens per watt) all decrease. Furthermore, the appearance of the light is warmer (that is richer in the red and orange portion of the visible spectrum, - the rainbow of colors we can see) versus cooler (richer in the blue and violet portion of the spectrum). At higher than rated input voltages, the opposite effects occur.

From an energy-efficiency standpoint, incandescent lamps should ideally only be used where 1) no more energy-efficient alternative is available, or 2) they will be operated for short periods, relatively infrequently, and energy consumption is not an issue, 3) the light source must turn on instantly and operate at full light output regardless of temperature, or 4) full range dimming is desired. Where incandescents are used, it is most efficient to use smaller numbers of higher wattage lamps than greater numbers of lower wattage ones. The higher wattage models operate more efficiently. Also, note that incandescents will operate on AC or DC; just make sure that the voltage of the lamp you need is the same as that for the system it will be operated on.

Tungsten-Halogens or Quartz-Halogens

Tungsten-halogen (or quartz) lamps are turbocharged incandescents. Compared to standard incandescents, these produce a brighter, whiter light and are more energy-efficient by operating their tungsten filaments at higher temperatures. In addition, unlike the standard incandescent light bulb which loses approximately 25% of its light output before it burns out, halogen lights' output depreciates very little over their life, typically less than 10%.

To make these gains, lamp manufacturers enclose the tungsten filament in a relatively small, quartz-glass envelope filled with halogen gas. During normal operation, the particles which evaporate from the filament combine with the halogen gas and are eventually redeposited back on the tungsten filament, minimizing bulb blackening and extending the lamp life. Where halogen lamps are used on dimmers, they need to occasionally be operated at full output to allow this "regenerative" process to take place.

Tungsten-halogen lamps produce about 10% more light per watt input than standard incandescents and last longer, having useful service lives ranging from 2,000 to 2,500 hours, depending on the model. Tungsten-halogen lamps are also very sensitive to operating voltage, as shown in Table 1.

There are many styles of halogens available for both 12V and 120V applications. The most popular low-voltage models for residential applications include the miniature multifaceted reflector lamp (MR 16) and the halogen version of the conventional A-type incandescent lamp, which incorporates a small halogen light capsule within a protective outer glass globe. The reflector lamp offers a light source having very precise light beam control, allowing you to distribute light exactly where you need it without wasteful spillover, while the latter option is intended for general ambient-lighting applications. A new family of 120VAC voltage, sealed-beam reflector halogen lamps are now coming on the market to replace standard incandescent reflectors.

The higher operating temperatures used in halogen lamps produce a whiter light, which eliminates the yellow-reddish tinge associated with standard incandescents. This makes them an excellent light source for applications where good color rendition is important or fine-detail work is performed. Because tungsten-halogen lamps are relatively expensive compared to standard incandescents, they are best suited for applications where the optical precision possible with the compact reflector models can be effectively utilized.

Never touch the quartz-glass envelope of a halogen lamp with your bare hands. The natural oils in your skin will react with the quartz glass and cause it to fail prematurely. Because of this phenomenon, and for safety reasons, many manufacturers incorporate the halogen lamp capsule

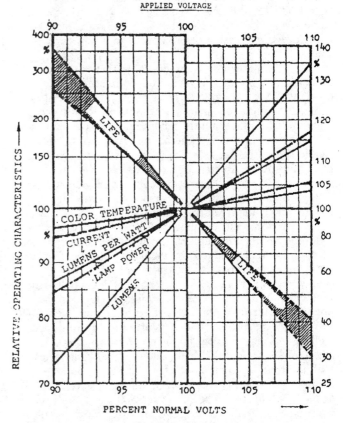

TYPICAL TUNGSTEN HALOGEN LAMP OPERATING CHARACTERISTICS
VERSUS
APPLIED VOLTAGE

PERCENT NORMAL VOLTS

NOTE: Data only valid for operation between + 10% rated lamp voltage. Characteristics for lamps operated at voltages outside that range can not be predicted.

(generally about the size of a large flashlight bulb) within a larger outer globe. Such is the case with the "A" type halogens, but most of the bulbs used in the MR family of lamps are exposed.

Compact Fluorescents and Standard Fluorescents

Please note, a more in-depth discussion on *using* compact fluorescents is presented in the following section, "All You Need to Know About Compact Fluorescent Lamps."

There are several types of fluorescent lamps. The two best for residential application are single-ended compact fluorescents (commonly called "PL" lamps) and double-ended standard fluorescents. Both are low-pressure mercury-vapor gas-discharge light sources. When an arc is struck between the lamps' electrodes, the electrons collide with mercury atoms which emit ultraviolet radiation. This radiation in turn excites the phosphors (the white powder coating the inside of fluorescent tubes) to fluoresce, or emit light at visible wavelengths.

The quality of the light fluorescents produce depends largely on the blend of chemical ingredients used in making the phosphors; there are dozens of different phosphor blends available. The most common and least expensive are "cool white" and "warm white". These, however, provide a light of relatively poor color rendering capabilities, making colors appear washed out, lacking luster and richness. On a color rendering index scale (CRI) of 1 to 100, with 100 being best, they rate 69 and 52 respectively. The higher the number, the more accurately a light source renders colors compared to a reference source.

Phosphor blends are available that not only render colors better, but also produce light more efficiently. Most notable of these are the fluorescent lamps using tri-stimulus phosphors, which have CRI's in the eighties. These incorporate relatively expensive phosphors that peak out in the blue, green, and red portions of the visible spectrum (those which people are most sensitive to), and produce about 15% more visible light than standard phosphors. Wherever people spend much time around fluorescent lighting, specify lamps with higher (80+ CRI) color rendering ratings.

Fluorescent lamps are also available in several different color temperatures, which relates to how warm or cool the light produced by the lamp appears. The lower the color temperature, the warmer or more orange/red/yellow the light appears. The higher the color temperature, the cooler or more violet/blue/green the light appears. Color temperature is rated in degrees Kelvin (K). For reference, a candle has a color temperature of around 1,800K and incandescent lamps range from 2,500K for standard lamps to 3,000K for halogens. Sunlight varies with the time of day and weather conditions; on a clear summer day, the sky's color temperature is about 5,500K at noon, while it may be between 6,500 and 7,500K on an overcast day. Fluorescent lamps are available with color temperatures between 3,500 and 5,500K.

We recommend selecting fluorescent lamps with the color temperature that will best blend with the other electric lights you are using, and that complements the decor/ambiance of your home. For residential applications, most people will prefer a 3,000K lamp. Where a space is illuminated primarily with daylight, a cooler temperature lamp such as a 3,500K lamp would be more complimentary, or 4,000 or 5,000+K lamps where northern-exposure light is received.

The rated service life of fluorescents varies greatly. Most compact-fluorescents have an average rated life of 9,000-10,000 hours, depending on model. The rated life for standard double-ended fluorescents range from 7,500 hours for standard 15-watt models to 20,000 hours for 40-watt models. These are based on the lamps being operated on a 3-hour duty cycle, on for 3 hours, off for 20 minutes. Though turning fluorescents off and on more frequently will shorten their life, the cost for lamp replacements is far less than that for the electricity used to operate them when they are not needed. It is most cost-effective to turn off fluorescents when they will not be needed for five minutes or more.

To get the greatest lamp life possible, fluorescents must be properly operated. All fluorescent lamps require a ballast to regulate voltage and current to the lamps to ensure that they are properly driven during start-up and normal operation. There are two types of ballasts for AC applications, electromagnetic and electronic, and one type for DC applications, the electronic ballast inverter which convert low-voltage DC to the higher-voltage AC that fluorescents need. The electronic units are intrinsically

about 75% more efficient than the electromagnetic technology and operate the lamps at high frequency, which both raises lamp efficiency by about 17% and eliminates perceptible lamp flicker (even at 60 Hz line frequency, the staying power of the better fluorescent phosphor blends minimizes flicker).

There are many grades of ballasts available. Each type of fluorescent lamp has its own unique electrical requirements. Lower quality or poorly designed ballasts may not properly drive the fluorescents they are designed for. When lamps are operated at higher than their rated voltage, for example, the arc operates at a higher current loading, and the lamps fail prematurely; excessive current also overheats the ballast and will cause it to fail prematurely too. At lower than rated voltage, inadequate current may be delivered to the cathodes to properly heat them, which results in delayed starting, premature end-blackening, and shortened lamp life.

Many ballasts do not operate fluorescent lamps at

The quality of the light depends largely on the blend of chemicals used in making the phosphors.

---

their full rated light output. To fairly compare the relative efficiencies of different fluorescent ballasts, you must know, at minimum, what their ballast factor is, and input watts. This indicates what percentage of a fluorescent lamp's rated light output the product is designed to deliver. This phenomenon makes it difficult for the uninformed consumer to evaluate different fluorescent lighting options against each other. Most descriptions in retailers' product catalogs suggest that their fluorescent lamps will delivery whatever lumen output they are rated for, without acknowledging that most ballasts are not designed to deliver this.

Also, it is important to realize that fluorescent lighting systems are sensitive to temperature. Manufacturers rate the ability of lamps and ballasts to both start and operate at various temperatures. Light output and system efficiency both fall off significantly when lamps are operated above or below the temperatures for which they are designed to operate. The temperature at which any given fluorescent system will effectively work varies greatly, depending on the specific lamp and ballast combination. The optimum operating temperature for most fluorescents is between 60 and 70°F, though most standard double-ended types will operate satisfactorily between 50 and 120°F. Special lamps and ballasts must be specified for applications outside of these temperature ranges such as low-ambient ballasts for cold applications.

When comparing the efficiency of fluorescent lighting systems, you should check out the system's efficacy (that is, how many lumens of light are produced per watt of total power consumed, including what both the lamp and the ballast draw). In general, fluorescent lighting operates far more efficiently than incandescents - two to three times more efficiently than low-voltage incandescents and four to six times more efficiently than 120V incandescents - and is much longer lived.

When installed in appropriate fixtures and properly situated, fluorescent lighting is suitable for a wide variety of ambient lighting applications as well as task lighting. Because fluorescent lamps are linear-light sources, versus a point source, they produce a relatively diffuse light which can be effectively used to provide direct or indirect lighting with minimal shadowing. Incorporating them into task lights equipped with good specular reflectors controls their diffuse-light output by directing it where it is desired.

Fluorescent lighting used effectively is one of the keys to illuminating your home well and energy-efficiently.

Future product updates in the *Real Goods Catalog* will feature many new, hard-to-find fluorescent lamps and fixtures.

The following table shows the relative efficiencies of fluorescent lamps/ballasts compared to incandescents. It illustrates the dramatic differences in efficacy, and rated service life between various lamp types.

Please note that when using <u>AC ballasted fluorescents</u>, there are some other losses associated with the "power factor" of the ballast that occur in the generation and distribution systems (house wiring, generators, inverters, and transformers) which are not accounted for here. Power factor (pf) is the ratio of watts ("real power") to volt amperes (total "apparent" power), and measures how much of the power drawn by the load is "real" (in phase with voltage) and thus able to do work. Simply divide the real watts drawn by a fluorescent lighting system by the ballast's power factor to determine the volt-amps. Unless you properly allow for this demand, your generator or inverter may not have ample capacity to power all of your lighting loads, or your wiring may be overloaded. Because most compact fluorescents generally draw such small amounts of current at 120V, the line losses will be minimal in most residential applications (assuming standard wire gauge sizes are used, #12 or #14), even with low-pf ballasts.

This concern does not affect DC fluorescent ballasts (current and voltage are always in phase in DC systems) or any incandescents (they have a pf of 1). Power factor is, however, a concern with some electronic devices, and *all* AC *inductive* loads, including not only ballasts but also transformers and any appliance or tool equipped with an inductive motor.

The "Watts" column listed in the Lamp Comparison Table for AC ballasts indicates only the real power used, versus the total or apparent power (in volt-amperes), which each fluorescent lamp/ballast uses. It also assumes a ballast pf of 1. In reality, however, most ballasts have lower power factors, as indicated by the pf ratings for the AC fluorescents listed. There are presently no screw-in, compact fluorescent high-pf ballasts available; the typical power factors for existing ballasts range from 0.5 for core-coil electromagnetic ballasts to 0.7 for electronic ballasts.

Ideally, we would recommend using only ballasts with a pf greater than 0.9. Some special AC high-pf

electromagnetic and electronic ballasts are available, but only to operate standard, double-ended 20 or 40 watt fluorescent lamps, and for compact fluorescents in fixtures where the ballast is hardwired separately. Whenever possible, choose the ballast with the highest power factor available for your application.

Energy-Efficient Lighting Primer Updates

Providing lighting as energy-efficiently as possible requires that you take a systematic approach. Energy-efficient lighting involves not only the careful selection of hardware, but also the application of it. Lighting is delivered through a system of components, including lamps, ballasts, reflectors, lenses/louvers, controls, and so forth, which together deliver light into a space. From an energy-efficiency standpoint, it is equally important (or more so) to incorporate daylighting into your home wherever practical, and to finish the interiors with light-colored sheathings, paints, and wallpapers as it is to install efficient hardware. Future updates to this primer will focus on energy-efficient light fixtures, lighting design and applications, and lighting-energy economics.

Fluorescents and Health?

By carefully assembling the best fluorescent-lighting-system components for your application, it is possible to provide high-quality lighting very energy-efficiently. Many people ask, however, "is this lighting good for me?" This question broaches what is perhaps the most controversial and complex issue in all of the lighting profession today. Electric lighting can profoundly affect our well-being, especially for people who spend much time indoors. There are many facets to the subject, including full-spectrum lighting, cathode-radiation shielding, radio-frequency shielding, long-wave ultraviolet supplements, and ultraviolet transmitting lenses. Many of these concerns involve not only fluorescent lighting, but also electrical household appliances such as televisions, microwave ovens, computers, smoke alarms, etc., which emit electromagnetic radiation.

Full Spectrum Fluorescents

Many customers have inquired as to why we don't sell full spectrum fluorescents. First of all, we're not convinced that they perform as promised. The Ultraviolet (UV) end of the lighting spectrum is the first part of the lamp's spectrum to disappear lasting only a fraction of the overall lamp life. Therefore the purported benefits of full spectrum lighting are extremely short-lived and thus overrated and not justifying the high price of these lamps. Further it's very difficult for us to ship anything as fragile as fluorescent tubes through the mail. That's also why we recommend that you purchase *all* your lamp tubes locally - there is no difference between a 110V tube and a 12V tube - only the ballast is different.

LAMP COMPARISON TABLE

LAMP	LUMENS Light Output -1-	WATTS Power Consumption -2-	LUMENS PER WATT Efficacy -3-	LIFE Average Rated Hours -4-	LUMENS HOURS -5-	PURCHASE PRICE -6-	OPERATING COST -7-
5 Watt QHI*, 12V	60 L	5 W	12 L/W	2,000hrs	120,000 LH	med	high
PL5, CF, 12V, EB	250L	6.6W	38 L/W	10,000 hrs	2,500,000 LH	high	med
PL5, CF, 120V, CCB PF.52	250L	8.8W	28 L/W	10,000 hrs	2,500,000 LH	high	med
PL7, CF, 12V, EB	400L	8.4W	48 L/W	10,000 hrs	4,000,000 LH	high	low
PL7, CF, 120V, CCB PF.52	400L	10.8W	37 L/W	10,000 hrs	4,000,00 LH	high	low
EL7, CF, 120V, CCB PF .70	400L	7W	57 L/W	10,000 hrs	4,000,000 LH	high	low
PL9, CF, 12V, EB	600L	10.2W	59 L/W	10,000 hrs	600,000 LH	high	low
PL9, CF, 120V, CCB PF.52	600L	12.8W	47 L/W	10,000 hrs	6,000,000 LH	high	low
10W QHI*, 12V	140L	10W	14 L/W	2,000 hrs	280,000 LH	med	low
EL11, CF, 120V, EB PF .70	600L	11W	55 L/W	10,000 hrs	6,000,000 LH	high	low
PL13, CF, 12V, EB	860L	15W	57 L/W	10,000 hrs	8,600,000 LH	high	low
PL13, CF, 120V, CCB PF .50	860L	16W	54 L/W	10,000 hrs	8,600,000 LH	high	low
EL15, CF, 120V, EB PF .70	900L	15W	60 L/W	10,000 hrs	9,000,000 LH	high	low
15W, Incand. A Lamp, 12V	187L	15W	12.5 L/W	1,000 hrs	187,000 LH	low	high
15W Incand. A Lamp, 120V	126L	15W	8.4 L/W	2,500 hrs	315,000 LH	low	high
Lt Cap, CF, 120V, CCB PF.52	720L	15W	48 L/W	9,000 hrs	6,480,000 LH	med	low
SL18, CF,120V, EB PF .60	1,100L	18W	61.1 L/W	10,000 hrs	11,000,000 LH	high	low
20W, QHI*, 12V	320L	20W	16 L/W	2,000 hrs	640,00 LH	med	high
EL20, CF, 120V, EB PF .70	1,200L	20W	60 L/W	10,000 hrs	1,200,000 LH	high	low
25W Incand A Lamp 12v	385L	25W	15.4 L/W	1,000 hrs	385,000 LH	low	high
25W Incand A Lamp, 120V	235L	25W	9.4 L/W	2,500 hrs	587,000 LH	low	high
PL36, CF, 12V, EB	2,900L	38.4W	76 L/W	10,000 hrs	29,000,000 LH	high	low
40W Incand A Lamp, 120V	480L	40W	12 L/W	1,500 hrs	720,000 LH	low	high
40W St Fluor, 12V, EB	3,150L	40W	78.7 L/W	20,000 hrs	63,000,000 LH	high	low
40W St Fluor, 120V, CCB PF.90+	3,150L	46W	68 L/W	20,000 hrs	63,000,000 LH	high	low
40W St Flr 120V ESCCB PF.90+	3,150L	44W	71.5 L/W	20,000 hrs	63,000,000 LH	high	low
40W St Fluor, 120V, EB PF.90+	3,150L	36W	87 L/W	20,000 hrs	63,000,000 LH	high	low
40W HE Fluor, 120V, EB PF .90+	3700L	44W	84 L/W	24,000 hrs	88,800,000 LH	high	low
50W QHI*, 12V	900L	50W	18 L/W	2,000 hrs	1,800,000 LH	med	high
50W Incand A Lamp 12V	790L	50W	16 L/W	1,000 hrs	790,000 LH	low	high
51 QHI, 120V	870L	51W	17 L/W	2,000 hrs	1,740,000 LH	med	high
60W Incand A Lamp, 120V	890L	60W	14.8 L/W	1,000 hrs	890,000 LH	low	high
75 Incand A Lamp, 120V	1,220L	75W	16.3 L/W	750 hrs	915,000 LH	low	high
100W Incand, A Lamp 120V	1,750L	100W	17.5 L/W	750 hrs	1,312,500 LH	low	high
200W Incand A Lamp, 120V	3,900L	200W	19.5 L/W	750 hrs	2,925,000 LH	low	high

Source: Rising Sun Enterprises, Inc.

*Lumen output shown for bare bulb, (deduct 3% for addition of outer, protective, frosted globe).

(1) Lamp manufacturers rated light output. Actual light output will vary with voltage input for incandescents and ballast factor for fluorescents.

(2) For fluorescents, this includes both lamp and ballast power consumption and assumes a power factor of 1 (see "Power Factor and AC Fluorescents" discussion)

(3) "Lumens per Watt" indicates how efficient any given light source is at converting power to light. The higher the number, the better. Note: For fluorescents it is important that ballast power consumption be included.

(4) For fluorescents, based on 3 hour duty cycle: on for 3 hours, off for 20 minutes. Assumes ballast properly drives fluor. lamps per manufacturer specs.

(5) Indicates the total quantity of light a lamp will produce over its lifetime. Divide Lumen-Hours by the cost of the lamp to find out what your cost is per LH.

(6) This index factors in the cost for the lamp and, for fluorescents, the ballast.

(7) This index factors in the life-cycle operating cost for each option, including energy cost ($/kWh) and lamp replacement costs.

KEY

QHI: Quartz Halogen Incandescent	ESCCB: Energy Saving Core-coil Ballast
CF: Compact Fluorescent	PF: Power Factor
EB: Electronic Ballast	EL: Osram Dulux EL
CCB: Core-coil Electromagnetic Ballast	SL: Philips SL
	Lt. Cap: Panasonic Light Capsule
St Fluor: 40W lamp with standard phosphors	HE Fluor: 40W "high-efficiency" lamp w/tri-stimulus phosphors

All You Need to Know About Compact Fluorescent Lamps

Our sincere thanks to Robert Sardinsky and Christopher Myers of Rising Sun Enterprises, Inc. for preparing the following concise and easy-to-understand summary of the new compact fluorescent technology.

1. **Lighting Color Quality:** Fluorescent lighting has a deservedly poor reputation for the quality of light it provides. Typically, it is known for its "cool white" appearance which makes people look pale and interior decors "washed out." Standard "warm white" fluorescent lamps (or tubes) tend to "soften" the light somewhat, but still provide relatively poor rendition. The fault lies in the phosphors - the white powder coating the inner lamp wall - which determines the quality of light emitted. Both the appearance of the light (how "warm" or "cool" it is) and the lamp's ability to render color (show colors accurately) depend on the blend of phosphors used.

Most compact fluorescent lamps (commonly referred to as "light bulbs") use an advanced blend of tri-stimulus, rare-earth phosphors which emit a light of greater spectral balance than that found in standard fluorescents (i.e., they are richer in the three visible wavelengths that people are most sensitive to: red, green, and blue). The most common variety (and all those featured in this Sourcebook) appear "warm" in color like standard incandescents and render colors much more accurately than conventional fluorescents, making them well-suited for residential application. (A new generation of full-sized fluorescent tubes have also recently been introduced which use these advanced phosphors.)

2. **Modular vs. Integral Compact Fluorescents:** Compact fluorescent lamps are available in modular and integral models (see Figure 1).

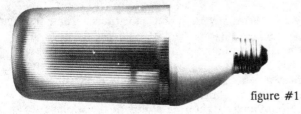

figure #1

The modular variety (e.g., the 12VDC PB/S series pictured above) consists of a replaceable, plug-in compact fluorescent lamp; a reusable, screw-in ballast adapter base; and, for some product lines, a series of reflectors

and diffusers. Integral compact fluorescent systems (e.g., the 110-volt AC SL*18 pictured above), on the other hand, incorporate the same components, but have no replaceable parts and thus must be thrown away when they burn out. Therefore, when a replacement is needed, it is less costly to buy a new lamp for a modular unit than it is to purchase an entire integral one. Integral units, however, offer features which may not be available in some modular lamps, such as electronic ballasts, more compact size, and certain light outputs.

3. **The Ballast:** A ballast regulates the voltage and current delivered to a compact fluorescent lamp and is essential for proper lamp operation. The electrical input requirements for each compact fluorescent lamp varies and hence, each type/wattage requires a ballast specifically designed to drive it. There are two types of ballasts which operate on AC: **core-coil** and **electronic**. The **core-coil ballast**, the standard since fluorescent lighting was first developed, uses **electromagnetic** technology. The **electronic ballast**, only recently developed, uses **solid-state** technology. All DC ballasts are electronic devices.

Relative to the core-coil ballast, electronic ballasts weigh less; operate lamps at a higher frequency (20,000+ cycles per second vs. 60 cycles); are silent; generate less heat; and are more energy-efficient. However, electronic ballasts cost significantly more, particularly DC ones because they are manufactured only in very small numbers. With few exceptions, electronic ballasts for AC lamps are presently only available as a part of integral units (the entire ballast/lamp must be discarded when the lamp burns out after approximately 9,000-10,000 hours of use). Modular compact fluorescents, on the other hand, have replaceable 10,000-hour lamps and reusable ballasts, good for approximately 50,000+ hours of use (*there are 8,760 hours in a year*).

All of the compact fluorescent lamp assemblies and pre-wired ballasts featured in the AE Sourcebook are equipped with standard medium, screw-in bases (like normal household incandescent lamps.) The ballast portion, however, is wider than an incandescent light bulb *just above the screw-in base* (see Figure 2). Therefore, fixtures having constricted necks or deeply recessed sockets may require a socket "extender" (to extend the lamp beyond the constrictions). These are readily available in several sizes at most hardware stores.

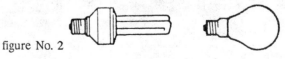

figure No. 2

4. **Compact Fluorescents on Inverters:** Most inverters will operate compact fluorescent lamps satisfactorily. However, because all but a few specialized inverters produce an alternating current having a modified sine wave (versus a pure sinusoidal waveform), they will not

drive compact fluorescents which use electromagnetic, core-coil ballasts as efficiently or "cleanly" as possible and may emit an annoying buzz. Electronic ballasted compact fluorescents, on the other hand, are more tolerant of the modified sine wave input, and will provide better performance, silently.

5. **PL Lamps vs. "Quad" Lamps:** There are two popular styles of compact fluorescent lamps being used with screw-in adapters: the "PL" type (as designated by Philips Lighting Corp.) and the "quad" type (see Figure 3).

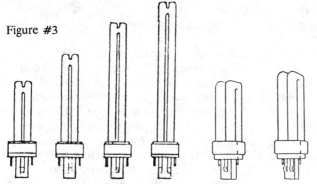

Figure #3

The PL lamp consists of two straight, parallel, interconnected glass fluorescent tubes mounted side-by-side on a plug-in base. These are most readily available in 5, 7, 9, and 13-watt models. The 9 and 13-watt models are more than 6" in length, making them too long to fit inside some fixtures or reflectors (discussed in the next section). To reduce the overall length of these lamps and of higher wattage models, lamp manufacturers developed "quad" lamps, which incorporate four short, interconnected glass fluorescent tubes, rather than the two longer tubes characteristic of the PL lamps. At roughly half of the overall length of PLs, "quad" lamps provide approximately the same light output (as PLs of equivalent wattages) and are available in a variety of models from 9 to 28 watts, the most common being the 9 and 13-watt models.

6. **Reflectors:** Several manufacturers offer a variety of reflectors for modular and integral compact fluorescent lamps to help you satisfy many lighting needs (see Figure 4). These plastic or metal units clip on (e.g., the PL reflector), or may be built into the housings (e.g., the SL*18 R40) of compact fluorescent lamps.

figure No. 4

If a directional light source is needed, such as in recessed downlights, tracklights, and "bullet"/floodlight fixtures, a lamp with a reflector should be used to direct the light where it is desired. In wall- and ceiling-mounted fixtures having only white-painted surfaces

serving as reflectors, a clip-on reflector which fits most "PL" and "quad" compact fluorescent lamps can effectively double the light output from the fixture.

7. **Lamp Size:** Compact fluorescent lamps are generally taller/longer and wider (at their base) than incandescent light bulbs (see Figure 2). Therefore, when selecting a "screw-in" compact fluorescent lamp, it is *essential* to measure the maximum width and length of a light bulb that your fixture can accommodate *to make sure you get one that fits*. Please take note of any constrictions in your fixtures, particularly around their light sockets, and for fixtures with lamp shades, within their lamp shade harps (the metal framework which forks over a light bulb and supports the fixture's lampshade). Where you have a lampshade which "clips" on to the light bulb itself, it may be difficult, though not impossible, to make the lampshade securely fit on the new compact fluorescent lamp. CAUTION: Before taking any measurements around light sockets, be sure to turn off the power to the fixture (or unplug it) and, as an extra precaution, use a plastic or wooden ruler.

8. **Starting Time:** The start-up time for compact fluorescent lamps varies. It is normal for most core-coil compact fluorescents to flicker for up to several seconds when first turned on while they attempt to strike an arc. Grounding the fixture greatly eases the starting process. If your compact fluorescent lamp flickers for more than 3 seconds before coming on, it is probably not adequately grounded. One way to overcome this is to touch or stroke the bare bulb while it is attempting to come on. If this is frequently the case, we suggest you loosely install a copper grounding piece (available from us) between the lamp's tubes and wire it to a grounded point.

Most electronically ballasted units start their lamps instantly, though, depending on the ambient temperature, it may take from several seconds to several minutes for the lamps to come to full brightness. The start-up times for lamps driven by core-coil ballasts vary from almost instantaneous to several seconds (they flicker during this brief period). These, too, may take from several seconds to several minutes to "warm up" and attain full brightness, depending on the ambient temperature.

Caution: Several vendors are offering customized PL lamps which turn on instantly. They do this by disconnecting part of the circuit built-into the lamp base. This voids any lamp and ballast warranty and may cause premature failure of the ballast.

9. **Cold Weather/Outdoor Applications:** Temperatures below freezing inhibit compact fluorescent lamps in starting and in attaining full brightness (normally reached between 50-70°F). In general, the lower the ambient temperature, the greater the difficulty in starting and attaining full brightness.

Manufacturers rate the ability of their lamps to reliably start at lower temperatures (from 0-50°F depending on lamp wattage). We have found that most lamps will start at temperatures 10-20°F lower than those stated by manufacturers. The lower wattage PL models

(5/7/9 watts) function better in colder temperatures than the 13 watt units, and PL9 and PL13 lamps have lower starting temperatures than their "quad" lamp equivalents. At present, all electronically ballasted lamps (including the SL*18) are rated to start in temperatures down to 0°F.

To help improve the operation of compact fluorescents used in cold temperatures, we recommend that they be installed in enclosed fixtures. This insulates them, improving their ability to both start and attain full brightness. It is also very important that the fixture be well grounded; a grounded metal fixture or reflector near the lamp, ideally within 1/4" of the lamp, helps it to start (i.e., strike an arc of electrons between the electrodes). If this is not possible or if the lamp flickers for more than several seconds before coming on, a thin strip of copper can be loosely inserted between the tubes of the PL lamp and attached to a grounded point. We offer a simple copper grounding piece to do this.

Most compact fluorescent lamps are not designed for use in wet applications (e.g., in showers or in open outdoor fixtures). In such environments, the lamp should be installed in a fixture rated for wet use.

10. **Service Life:** Most manufacturers rate lamp life based on a three-hour duty cycle, meaning that the lamps are tested by turning them off-and-on once every three hours until they burn out. *Turning lamps off-and-on more frequently will decrease lamp life while keeping them on longer will increase lamp life.* Generally, it is more cost-effective to turn fluorescent lights off whenever you leave a room for more than a few minutes, since the greatest cost associated with operating a light is for the electricity it uses versus the cost to replace it. Please note that the rated lamp life represents the *average life* of a lamp; as a result, some will have longer lives and some shorter. As illustrated below (Figure 5), a typical compact fluorescent lamp will last as long as (or longer than) 10 standard incandescent AC light bulbs or 5 standard incandescent AC floodlights, saving you the cost of numerous bulbs, in addition to much electricity from the greater efficiency.

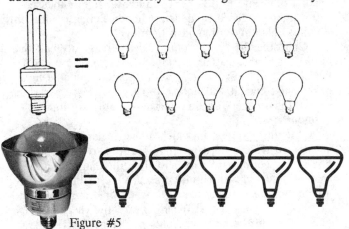

Figure #5

11. **Lamp Efficiency:** To determine how energy-efficient a lamp is, divide its light output (rated in lumens) by its power consumption (rated in watts). The greater the lumens per watt, the greater a lamp's efficacy ("efficiency"). Since the ballast for a compact fluorescent lamp consumes power, this, too, must be factored in to determine a compact fluorescent lamp's true energy-efficiency (see note 12 for details). Electronically ballasted compact fluorescents are more efficient than core-coil ballasted ones, and in general, all compact fluorescent lamps have substantially better efficiencies than incandescent lamps.

12. **Power Consumption:** The power consumption rating for a compact fluorescent lamp should indicate the total watts that both the lamp and the ballast use when operating. Most product literature for *modular* compact fluorescents is misleading due to the fact that it only notes power consumption of the lamp, without including ballast dissipation. The ballast, however, can increase system power consumption significantly, up to 75% with some models. Wattage ratings for *integral* compact fluorescent lamps, in contrast, *do account* for the unit's complete power consumption.

13. **Light Output/Wattage Range:** Screw-in compact fluorescent lamps are available in wattages ranging from 5 to 28 watts, with light outputs approximately equivalent to incandescent lamps ranging from 25 to 100 watts. The higher wattage compact fluorescents (greater than 20 watts), however, are considerably bulkier and longer than their incandescent equivalents and, consequently, do not fit into most existing household fixtures. As with incandescent light bulbs, compact fluorescent lamps' light output drops 10-20% as they age.

14. **Fixture Stability:** Please note that compact fluorescent lamps with core-coil ballasts (e.g., the Light Capsule) are substantially heavier than standard incandescent lamps, and thus should not be used in table or floor lamps which they would make top-heavy. This should not be a problem with most electronically ballasted compact fluorescents, which are relatively lightweight.

15. **Dimming:** At present, there are no dimmable, screw-in compact fluorescent lamps on the market. **DO NOT** install compact fluorescent lamps on any circuits controlled with a dimmer (even with the dimmer switch turned to full brightness) as they will overheat and become a fire hazard.

16. **Three-Way Light Sockets:** Compact fluorescent lamps can be screwed into any three-way light socket. However, the lamp will only operate at full light output, in two of the three "on" settings.

17. **Payback:** The payback period is the length of time it takes for your investment in energy-efficient lighting to be returned to you in savings. Compared to incandescent lamps, compact fluorescent lamps, in effect, pay for themselves through the substantial amount of electricity they save over their lifetime and the reduced number of light bulbs you will need to buy. Beyond the payback point, the savings you realize represent your profit on this investment. The length of the payback period will vary, depending on the amount of time a light is used, the efficiency of the existing light bulb and its replacement, and your cost for electricity.

In homes powered with alternative energy systems, the wise purchase of compact fluorescent lamps can dramatically reduce household energy requirements and hence, the cost to purchase and operate your independent power system.

Fluorescents & Harmful Rays

Fluorescents have often been accused of emitting harmful rays and of adversely affecting people's behavior. This is true to some extent for 110VAC fluorescents, because many people can subconsciously pick up the 60 cycles per second frequency that units with conventional electromagnetic core-coil ballasts operate the lamp at. This 60 cycle frequency can cause negative physical or emotional reactions, particularly in children. High-quality fluorescent lighting fixtures which use electronic ballasts (either AC or DC powered) operate the lights at 25,000 to 35,000 cycles per second - this is far too rapid to be detectable by the conscious or subconscious mind.

Lighting Fixtures

When electricity first began to appear in homes, in the late 19th century, there were no television sets, electric can openers, or telephone answering machines. The only domestic use for electricity was to power incandescent light bulbs. Electric generating stations--or "electric light plants," as they were then called--started producing power at sunset, and shut down at sunrise.

Things have changed a great deal since then. In grid-connected households, lighting typically accounts for only 3% of the total power consumption. In an alternative-energy supplied home, however, where wasteful or frivolous power use is held to an absolute minimum, lighting generally accounts for a much larger slice of the energy budget. As a result, it pays to devote careful thought to designing an efficient home lighting system.

If you are already involved in alternative energy, or have done any reading on the subject, chances are that you already know some important facts about lighting technology. Here's a capsule summary:

Incandescent bulbs (or "lamps", in lighting jargon) are remarkably inefficient devices. A standard AC lamp uses about 90% of the power it consumes to produce heat, turning only about 10% into visible light. The efficiency of low-voltage DC lamps is typically somewhat higher, but still poor. In practical terms, incandescent lamps are electric heaters that produce a small amount of light as a by-product. However, they produce a crisp, pleasant light, are low in cost, and fit almost all standard lighting fixtures.

Quartz-halogen lamps are also a form of incandescent lamp. They operate at higher temperatures than standard incandescents, and produce somewhat more light per watt of energy consumed. On the negative side, quartz halogens are three to four times as costly as standard incandescents. They last considerably longer, but are subject to premature failure if the system voltage does not remain very close to the level at which the lamp is designed to run. That can be a problem in many home systems, where the battery voltage may rise to more than 14 volts during on a sunny day, and fall to 12 volts or less at other times.

Fluorescent lamps produce light by generating a high-voltage, high-frequency arc within a sealed glass tube, which in turn causes a coating within the tube to give off visible light. Fluorescent lamps produce much more light per watt than standard incandescents or quartz-halogens. Although they are expensive, they last for thousands of hours, and are very tolerant of fluctuations in voltage. The quality of the light produced has improved a great deal in recent years, but it still falls a bit short of the white, natural-appearing light produced by incandescent lamps.

Despite their differences, however, all three types have their place in the home. The important thing is to choose the right lamp for the application. Energy-efficient fluorescent lamps, for example, are a good choice for a general background illumination, in room lights that are switched on at dusk and left aglow until everyone has gone to bed. Standard incandescent lamps are an obvious choice in closets, hallways, and other areas where light is needed only intermittently. The intense, tightly-focused white light produced by quartz halogen lamps is ideal for reading lights and other task lighting.

Choosing the proper lamp, however, is only half the battle. The other half is in selecting--or, in some cases, devising--a light fixture that will efficiently put the light produced where it will do the most good. Here are some suggestions to get you thinking about utilizing light efficiently:

Hanging fixtures, in which a shaded lamp is suspended from a cord or chain secured to an electrical box in the ceiling are excellent for providing direct, shadow-free light to a limited area. One of the best applications is directly over a kitchen or dining-room table. That tends to cast the light directly where it is needed, while the table prevents passersby from walking directly beneath the light and banging their heads. Hanging fixtures can also work well in an area with a very high ceiling, such as a stairwell. Avoid hanging fixtures in areas where there is foot traffic, particularly in confined spaces, such as bathrooms.

Many commercially available hanging fixtures will

accommodate modern compact fluorescents, such as those made by Osram or Phillips. The long tube of a high-wattage lamp, however--such as the popular 13-watt PL lamp--may protrude beyond the mouth of the shade, resulting in an awkward appearance. One option is to select a more compact quad-tube lamp in place of the twin-tube type. Another approach is to screw a Y-shaped two-socket adapter to the fixture--available for a few dollars at any electrical supply shop--and substitute a pair of 7-watt lamps for the longer 13-watt lamp. That involves the expense of an added lamp, but not an additional ballast, since the paired 7-watt lamps will operate on a single ballast.

Recessed ceiling lights are more versatile than hanging lamps as a source of background light, since they are above the reach of passing heads, and cast their light over a wider area. A special fixture described as a "can" is installed in the ceiling framing, so that the lamp itself is concealed. Various types of lenses, reflectors, and baffles are available for recessed lights, although many are quite inefficient. The best choice for general area lighting is fitted with a type of lens described as a "drop opal diffuser".

Some recessed light fixtures will accommodate a 13-watt PL tube, while others will not take anything larger than a 7 or 9-watt tube. To avoid disappointment, it's a good idea to bring a tube with you when shopping for fixtures. That precaution is especially important if you intend to use the convenient, but bulky, pre-wired lamp kits, in which lamp and ballast are combined as a single unit. All recessed lights incorporate an electrical wiring box that serves as a handy place to stash the ballast if a separate ballast and lamp is to be used.

Surface-mounted fixtures, as the name implies, sit on the surface of the ceiling, instead of being recessed within it. They are comparable to recessed lights in function. However, they are far more visible, and must be selected with care if they are to look good. Cheap surface-mounted lights, unfortunately, generally look cheap.

A good budget choice is to assemble your own ceiling lights, by mounting a fluorescent ballast in a ceiling electrical box, and wiring a simple porcelain lamp holder onto it. Instead of screwing the compact fluorescent tube directly into the lamp holder socket, however, first install a metal UNO thread socket extension, such as the #346 socket extension manufactured by Eagle Electrical Manufacturing Co. That will lengthen the socket by an inch. More important, the fine threads on the outside of the extension will accept a spun aluminum reflector, of the sort often supplied with clip-on utility lights. (Any well-stocked lighting shop will carry the shades alone, in a variety of sizes.) Screw the reflector onto the end of the adapter, twist in a compact fluorescent tube, and you're in business. You can leave the aluminum reflector unfinished, or paint it with any good enamel paint. The resulting fixture is quite

attractive in a plain, straightforward way.

The same approach, minus the ballast, can be used with incandescent lamps as well. Where an incandescent lamp is used, it's possible to switch the lamp with a pull chain on the lamp holder. With a fluorescent lamp, however, the fixture must be controlled by a wall switch, or the ballast will continue to draw power from the batteries even when the lamp is off. Finally, if the fixture is to be controlled by a three-way switch, the necessary wire connections within the ceiling box may not leave room for a fluorescent ballast. In that case, it will be necessary to mount the ballast in a box of its own, in an inconspicuous but accessible location.

Task lighting should be thoughtfully placed so that a strong, bright light source can be directed exactly where it

---

A standard AC lamp uses about 90% of the power it consumes to produce heat, turning only about 10% into visible light.

---

is needed--onto the kitchen sink or cutting board, into an open book, or over the surface of a desk. Accuracy, rather than sheer wattage is the key to success here. A tiny 5-watt quartz halogen lamp that shines squarely on your fly-tying vise is more effective than hundreds of watts of unfocused background light, and saves energy in the bargain.

For best results, have plenty of individual task lights, and make a habit of turning them on and off as needed. It makes sense to use highly efficient fluorescent lamps or quartz halogens for task lights that will be left on for long periods--such as a desk lamp in a home office--but ordinary incandescents are fine for lights that are used less frequently.

One versatile and inexpensive type of task light is a swingarm-type desk light, which can be mounted on a wall or clamped to a table or desktop. Gooseneck-type quartz halogens can be used in the same manner.

Low voltage quartz-halogen track lighting is superb for spotlighting specific areas. Tracks and fixtures are available at all good lighting shops. When connected to the power grid--as is usually the case--such fixtures require a small transformer to step the AC line voltage down to 12 volt DC. In an alternative energy household, however, they will operate perfectly directly from the battery. By running a strip of track lighting beneath upper level kitchen cabinets, it's possible to fine-tune the countertop illumination exactly the way you want it.

With some thought and a little imagination, you'll find that it's possible to achieve a pleasant balance of task lighting and background illumination without breaking the household budget--or the battery bank.

-Jon Vara

AC High-Efficiency Lighting

Because many of you are living in or building alternative-energy-driven AC homes powered with inverters or are striving to live as efficiently as possible on the utility grid, we are making an effort to bring you a variety of new AC lighting products that can illuminate your homes with high-quality lighting without guzzling electricity.

We're always reminded of Amory Lovins' (of the Rocky Mountain Institute) energy analysis of the far reaching effects of compact fluorescent light bulbs. Amory points out that if every American household converted just one incandescent lamp to an energy efficient compact fluorescent bulb, *we could immediately shut down one large nuclear power plant. If every American household (there are currently 100 million) converted all lamps to compact fluorescents, America would instantly become an energy exporting nation! Think about it - that's five lamps per household or $80 each. Multiply that by 100 million and you get $8 billion - that's less than one third of what President Bush is spending to defend the oil fields of Saudi Arabia! If every human on earth replaced their incandescents with compact fluorescents we could shut down 50 nuclear power plants!*

We've had great success with the Panasonic light capsules up to now and are happy to introduce lots more efficient AC lighting in this Sourcebook. If you haven't already done so we strongly recommend that you regress for a moment and read the entire "consumer-oriented" section on compact fluorescent technology in the *DC lighting introduction*.

Dulux "EL" Compact Fluorescents

These AC lamps are one of the most efficient and light-weight compact fluorescent units available to screw into standard 110VAC light fixtures. They incorporate a built-in electronic ballast (absolutely no hum on inverters!), which operates the lamp at high frequency (35kHz), starts it instantly, works in temperatures from 0 to 140°F, and are completely silent. Many of our customers prefer these bulbs to the popular Panasonic light capsules as the Osrams contain absolutely no radioactivity of any kind. These lights are best installed inside table lamps or other light fixtures, which will diffuse their bright light and keep dust from settling on them. The Dulux EL's rated lifetime is 10,000 hours. Four wattages are available; the 11 watt and 15 watt are our best sellers.

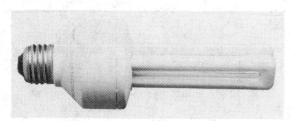

		Approx. incand. equi.	Fixture watts AC		
36-111	Dulux EL 7	25	7	5-11/16"x2-1/4"	$24
36-112	Dulux EL 11	40	11	5-11/16"x2-1/4"	$24
36-113	Dulux EL 15	60	15	6-7/8"x2-1/4"	$25
36-114	Dulux EL 20	75	20	8-3/16x2-1/4"	$25

Dulux EL Reflector Lamps

The Dulux EL Reflector combines the convenience and efficiency of the Dulux EL lamp with a high performance parabolic reflector. With ten times longer lamp life and up to 75% less power consumption, the Dulux EL Reflector is the energy saving alternative for R30 and R40 incandescent lamps. These lamps give a warm, pleasant incandescent-like light with excellent color rendering. They are ideal for track lighting and down lighting in offices or homes. Two models are available: the Dulux EL-R-11 Watt runs on 11 watts and is equivalent to a 50 watt incandescent, and the Dulux EL-R-15 is 15 watts and equivalent to a 75 watt incandescent. The EL-R-11 has a 70 degree beam spread, puts out 600 lumens, is 5-7/8" high x 2-1/4" in diameter and weighs 4.1 oz. The EL-R-15 has an 80 degree beam spread, puts out 900 lumens, and is 7-1/4" high x 4-7/8" in diameter.

As an interesting anecdote: *the city of San Rafael, CA recently converted from 150 watt incandescents to Osram Dulux 11 watt reflector bulbs. The payback period for break-even with the savings in electricity was only 8 months!*

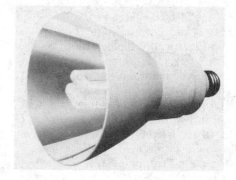

36-131	Dulux 11 Watt w/Reflector	$29
36-132	Dulux 15 Watt w/Reflector	$29

Panasonic Light Capsules

These are one of the most efficient 110V bulbs on the market and a best seller for us. Many of our customers with Trace inverters choose to wire their homes completely 110VAC and for that you can't beat these light capsules for efficiency, quality, and price. The technology is still young - first there was the Mitsubishi Marathon bulb, but they cancelled their marketing agreement with the USA. Next came GE, which we discontinued in early 1989 because of the boycott. Now we have Panasonics, which are working out as well or better than any previous product.

These attractive, core-coil ballasted, compact fluorescent lamps offer a high-quality, economical option for providing light from 110VAC fixtures. Available in either globe- or tubular-shaped, white-frosted glass bulbs, the 15-watt Light Capsule produces a "soft" light comparable to that from a standard 60-watt incandescent. But a 60-watt incandescent typically lasts only 500-1,000 hours while these light capsules last up to 10,000 hours! and only use 1/4 of the electricity! These one-piece units operate at line frequency (60 Hz) and screw into standard light bulb sockets. While they at first glance seem expensive, they pay for themselves in less than three years and over their life save more than $40 when compared to standard incandescent light bulbs.

Note: Because these units are relatively heavy, they may make some fixtures top-heavy. When installed in unsheltered outdoor areas, use only in weatherproof fixtures. *These units may not be used with dimmers.* Check Light Capsule dimensions to be sure it will fit your fixtures. Not suitable for use inside globed or sealed fixtures.

The ballast is included inside the bulb and a standard edison screw base makes it simple to use with any fixture. When mounted with the optional reflector, they are excellent for recessed lighting - "Halo Cans" are available for different recessed mounts. The lights come in two configurations: the G15 is the globe shaped bulb which is 3.7" in diameter and 5.7" long and the T15 is the tube shaped bulb which is 3.1" in diameter and 3.1" long and able to fit in narrower fixtures. Minimum starting temperature is 32 degrees Farenheit. Both bulbs use only 15 watts but put out the equivalent of 60 watts of incandescent lighting. Their efficacy is 48 lumens per watt.

36-101	Light Capsule - Globe (one)	$ 17
36-103-KT	Light Capsule - Globe (ten)	$155
36-102	Light Capsule - Tube (one)	$ 17
36-104-KT	Light Capsule - Tube (ten)	$155

A Note on Radioactivity of Some Compact Fluorescents:

Panasonic Light Capsules employ a rare-earth element called Prometheum that in its gaseous form aids in starting the lamp. Prometheum puts out 330 nanoCuries of radioactivity. Chemically, Prometheum is almost identical to the very elements which are used in tristimulus phosphors. This extremely small quantity is of even less *chemical toxicity* concern than the phosphor itself. We believe that the miniscule amount of radioactivity involved is of little concern considering the tremendous energy savings brought about by this wonderful technology. If the radioactivity concerns you, we recommend you consider the electronically ballasted lamps.

TRUST a politician. He wouldn't be running for office if he weren't trustworthy, would he? Once he's in office he'll take care of the environment. You can be sure of that.

Panasonic Electronic Light Capsules

These brand new electronic light capsules represent a technological breakthrough in energy-efficient lighting. The 18 watt compact fluorescent light is equivalent to a 75 watt standard incandescent light bulb and is the brightest of any equivalent compact fluorescent presently available with *frosted diffuser/globe*. It lasts, on average, eleven times longer than standard incandescent light bulbs, minimizing the need to purchase eleven replacement lamps. Its solid state circuitry allows it to start instantaneously and operate completely flicker free. The diffuser provides a "soft" even illumination, which is particularly desirable in bare bulb applications. The light fits most table lamps, when installed with our specially-matched extended lampshade harp (the metal framework supporting the lampshade), as well as in many ceiling, wall, hanging pendant, or pole mounted fixtures. 7.4" long x 3.0" in diameter, 1100 lumens. *No hum on inverters & no radioactivity!* **May not be used with dimmers.**

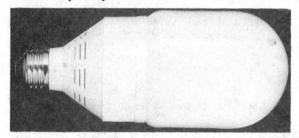

| 36-141 | 18 Watt Electronic Light Capsule | $27 |
| 36-403 | "10" Lampshade Harp | $ 3 |

SL18 Compact Fluorescents

Introduced to the U.S. market in 1983 and since much refined, the SL18 was the first electronic-ballasted 110VAC compact fluorescent and it produces the same amount of light as a 75W standard incandescent; the reflector version - the SL18R40 - will replace a 75-watt incandescent floodlight. These screw-in, one-piece units operate at high frequency, work in temperatures from 0 to 140°F, are silent, and screw right into standard light bulb sockets. The SL18's are rated to last for 10,000 hours. Their lightweight polycarbonate housing and prismatic diffusor/lens keep dirt from settling on the bulb, making them an excellent choice for use in applications where a "bare bulb" is presently used on most regular fixtures. The Reflector version is designed for use anywhere a directional light source is desired and may be installed in recessed-can or track fixtures or simple surface mounted, porcelain-type lampholders. The SL18's are one of the most energy-efficient compact fluorescent lamps presently available.

Note: when installed in unsheltered outdoor areas, use only in weatherproof fixtures. Caution: May not be used with dimmers. Check SL18's dimensions to make sure it will fit your fixture.

Electronic Twin Light Capsules

Panasonic has recently come out with this brand new energy saving twin tube light capsule. Just like other compact fluorescents these lamps last up to 13 times longer than a conventional incandescent bulb. Because they are electronically ballasted they have instant start, more lumens and less weight. They fit into standard incandescent sockets. These are twin-tube bulbs and do not include the diffuser like the standard light capsules or the other Panasonic electronic light capsules listed above. The 27 watt bulb is the *brightest compact fluorescent on the market and puts out the equivalent of a full 100-watt incandescent*. It is 7.8" long x 2.1" in diameter and puts out 1,550 lumens. **May not be used with dimmers.**

36-143   27 Watt Twin-Tube (100 watt equiv.) $29

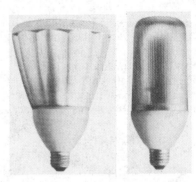

		Approx. incand. equiv.	Fixture watts		
36-121	SL18	75	18	6 7-3/16"x2.5"	$27
36-122	SL18R40	75	18	7-1/4"x5"	$32

Osram Dulux Combi Lamp

This is an extremely attractive and compact energy-saving lamp. It was developed to provide supplementary light in small places, as built-in fitting for furniture, to brighten up kitchens and bathrooms or for small areas such as desks, tables, and bookshelves, where it can be table-mounted or wall mounted (brackets included for easy fastening). It includes a PL9 compact fluorescent putting out the equivalent of 50 watts of incandescent lighting. Operable on 110V AC only. Featured in the Sharper Image catalog for $89.95, we're introducing it at a very attractive price. Available in black or white.

36-315-W	Dulux Combi (White)	$39
36-315-B	Dulux Combi (Black)	$39

Outdoor Compact Fluorescent Floodlight

This unique 110VAC compact fluorescent floodlight delivers approximately the same amount of light as a 50-watt incandescent floodlight, yet uses only 10 watts! It is designed for general area, landscape, and security lighting. This well-designed floodlight incorporates a core-coil ballast in the aluminum base, has a replaceable 10,000 hour life PL7 lamp and a PAR 38 glass floodlight reflector, comes with a glare shield, and has a swivel mount which screws right into a standard 1/2" IPS threaded junction box cover. A "standard" floodlight can't come close to matching the long-term economy and performance of this unit, which operates in temperatures down to -20°F. Finish in flat black. Dimensions: 9" L x 6".

36-311	J-910-7 Outdoor floodlight	$45
36-312	J-910-GS with glare shield	$59

Reflect-A-Star

This is the most durable, high quality, modular compact fluorescent floodlight available. Its 13-watt replaceable bulb produces the same amount of light as a 75-watt incandescent floodlight and will typically last the life of five of them. Engineered to provide many years of reliable service, this product incorporates a torroidal ballast, ALZAK finished aluminum reflector, and clear acrylic prismatic lens into its design. It screws into most recessed downlight can fixtures and track fixtures which presently accommodate 4.5" and larger diameter incandescent floodlights. The fixture is rated for use in damp locations and may be used outdoors in locations protected from the weather. Operates down to 0° F. Choose between a 5.25" diameter reflector and 4.5" - the larger one produces more light. Both units are 6" long and 2" wide at screw-in base end. Available in 110 VAC only and it uses 16 watts.

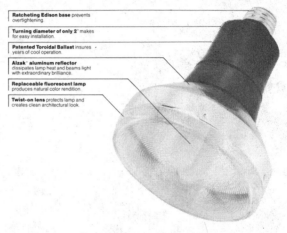

Ratcheting Edison base prevents overtightening.

Turning diameter of only 2" makes for easy installation.

Patented Toroidal Ballast insures years of cool operation.

Alzak aluminum reflector dissipates lamp heat and beams light with extraordinary brilliance.

Replaceable fluorescent lamp produces natural color rendition.

Twist-on lens protects lamp and creates clean architectural look.

36-313	Reflect-A-Star (5.25")	$46
36-314	Reflect-A-Star (4.5")	$46

Ultra-Short Compact Fluorescent System

This new versatile, modular system is great for illuminating areas where lower light levels are desired. The system is available in four versions: 1) as a bare bulb package, 2) with glass PAR 38 (4.75" diameter) floodlight reflector, 3) with white frosted glass (3.75" diameter) decorative globe diffuser, and 4) with ornate frosted glass candle flame diffuser. All use the same ballast adapter base and replaceable PL7 lamp. The screw-on reflector and diffuser options offer additional utility for a variety of lighting applications. The low-power factor, core-coil ballast and PL lamp draw 10.8 watts and replace comparable 40 to 60 watt incandescents. Operates down to -20° F and can be installed in outdoor locations protected from the weather. Adapter base diameter is 2-5/8". Screws into conventional light fixture socket. Available in 120V AC only.

36-151	Bare Bulb Package	$21
36-152	w/PAR 38 Reflector	$35
36-153	w/White Frosted Glass	$32
36-154	w/Ornate Frosted Diffuser	$31

Etta Electronic Ballast

Operate standare 4', double-ended, 40 watt fluorescent tubes in the most energy-efficient manner possible with this new Etta Sinusoidal ballast. This product is a true electronic ballast, incorporating integrated circuit technology throughout its design. It produces and drives the tubes with an extraordinarily clean sinusoidal waveform. Drawing only 1-2 additional watts per tube, it is a great energy-saver compared to standard electromagnetic ballasts, which typically draw 8 additional watts per lamp. In addition, the Etta ballast operates fluorescents at high frequency and silently gives you flicker-free lighting without the annoying hum (at 110VAC) associated with many conventional ballasts. It is designed to be installed into any 120V AC fluorescent fixture and will operate one or two conventional 40 watt fluorescent tubes. *Our offline customers have reported to us that the Etta will cause significant hum in inverters, so be aware.*

36-201 Etta Electronic Ballast $59

Dazor Asymmetria PL Task Lamp

This Finnish-made, 120VAC, swing-arm lamp is one of the most versatile and energy-efficient of its kind. Its sleek, contemporary design, articulating arm, and adjustable reflector head enable you to provide good quality task lighting where you need it. The built-in specular aluminum reflector is asymmetrically shaped to bias light distribution in one direction, lowering reflective glare on work surfaces and computer screens. The Asymmetria has a 34" arm with three swiveling joints, each with its own tension control, eliminating the temperamental behavior of spring-tensioned lamps. The fixture's 13-watt compact fluorescent lamp delivers a soft, warm colored light, without the excessive heat that makes working by comparable incandescent desk lights uncomfortable. Available in both black and white models with a clamp-on or weighted base. Total fixture power consumption, including ballast, is 15 watts.

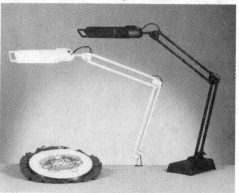

Note: The plug-in ballast, located at the end of the power cord, sometimes emits a slight audible hum which can be easily muffled by plugging it into an outlet situated underneath a desk or behind furniture. Also beware, there are numerous imitations of the Dazor now on the market which, though lesser priced, are also not nearly so well made.

36-301-B	PL task lamp w/clamp (black)	$59
36-301-W	PL task lamp w/clamp (white)	$59
36-302-B	Task lamp w/weight base (black)	$69
36-302-W	Task lamp w/weight base (white)	$69

Energy Efficient Lighting- Compact Fluorescents on 120 VAC

George Patterson

Compact fluorescent lights are one of the most energy efficient lamps available in the market today. They produce 3 1/2 times more lumens per watt than incandescent lights and 7 to 13 times the lamp life as a standard "A" type incandescent. The lamps use 70% less power than standard incandescent. Modern types use high frequency electronic ballast and produce silent, flicker-free light. These lamps are color correct. They produce light that is a very good imitation of daylight. We are seeing a revolution in lighting!

Compact Fluorescent Lamp Data

Lamp Type	Initial Lumens	Lumens per Watt	Lifetime in hours	Min. Start Temp.	Color Temp.	Color Index
7w. twin tube	400	57	10000	0 °F.	2700 °K.	82
9w. twin tube	600	67	10000	25 °F.	2700 °K.	82
13w. twin tube	900	69	10000	32 °F.	2700 °K.	82
13w. quad tube	860	67	10000	32 °F.	2700 °K.	82
18w. quad tube	1250	69	10000	32 °F.	2700 °K.	86
26w. quad tube	1800	69	10000	32 °F.	2700 °K.	86
25w. Incandescent	260	10	1500		2500 °K.	91

The data in the table shows performance data for six compact fluorescent lamps and two types of incandescent lamps. Lumens is a unit of light intensity and ranks the lamps by brightness (the higher the lumen value the more light the lamp produces). Lumens per watt shows how efficient the lamp is, note that the compact fluorescents are about six times more efficient than incandescents. The lifetime (in hours) is rated by the manufacturer assuming that the lamp remains burning for three hours when switched on. Minimum starting temperature is just that, the lowest temperature at which the lamp will reliably start. Color temperature is a scientific system for measuring the spectral output of a light producing object. In the color temperature scheme, the object color is related to a black body at a certain temperature in degrees Kelvin (°K.). The color rendition index is more easily understood. The color rendition index of daylight is 100 by definition. The closer a lamp's color rendition index is to 100, the closer its color is to daylight.

OK! Are all of these deals real? How do they apply to real life?

The fluorescent light as a system

The lighting fixture is truly an energy system with four elements - 1) Input power, 2) Ballast, 3) Starter, and 4) Fluorescent tube. The efficiency and performance of the system is dependent on the interaction of all four elements. Change any one element and the light's performance and efficiency changes.

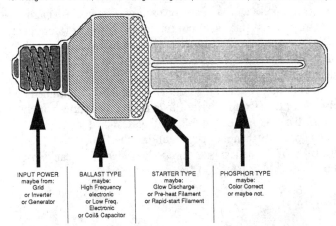

Fluorescent Lighting is a System
for it to give color correct, efficient & long-lived light all parts must be in proportion and harmony

INPUT POWER
maybe from:
Grid
or Inverter
or Generator

BALLAST TYPE
maybe:
High Frequency
electronic
or Low Freq.
Electronic
or Coil& Capacitor

STARTER TYPE
maybe:
Glow Discharge
or Pre-heat Filament
or Rapid-start Filament

PHOSPHOR TYPE
maybe:
Color Correct
or maybe not.

The reality of lighting is that we are not going to get something for nothing. Of course, in trying to do so we are likely to take ourselves to the cleaners. There is no substitute for doing our homework and making decisions based upon actual experiences. The 10,000 hour life figure quoted for most compact fluorescent tubes is just a starting point. The truth is that we may get anywhere from 2,000 to 20,000 hours from the same tube depending on the ballast type and operating environment. The light output from a 13 watt compact fluorescent tube may be 900 lumens at 75° F (100%), 720 lumens at 120° F. and 450 lumens at 40° F. This is especially a problem where housings and lighting fixtures trap heat inside, or they are used outdoors in the cold.. A typical graph of the operating temperature characteristics is shown in figure 1. Note that the efficiency we seek so dearly is effected by the position of the base.

Temperature characteristics
DULUX® S 5W; 7W; 9W, 11W, 13W
DULUX® S/E 5W, 7W, 9W, 11W
DULUX® L 18W, 24W, 36W

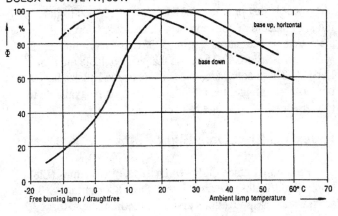

Ballast and tube life on inverters (square wave) may be cut in half compared to use on true sine wave for 120 VAC applications. On modified sine wave inverters there is not known to be a problem, but the jury is still out.

Tube Life and Starting

The electronic ballast may deliver promised efficiency, but the design of the starting circuit is critical. Some compact fluorescent tubes have built in glow discharge starters, while others use pre-heat filaments for starting. Pre-heat filaments require external starting circuitry. Life of compact fluorescent tubes designed for use with external starting circuits (rapid start, pre-heat, and electronic ballasts) is determined by the design of the starting circuit. The life of a fluorescent tube with built in glow discharge starter is primarily determined by the life of the starter. Starter life in these tubes varies widely with ballast design. If the fellows that designed the ballasts did a good job, then the starter will last the 10,000 hour life of the tube. If the ballast is not properly designed we can expect life times as short as 2,000 hours.

Ballasts

The newly developed high power factor coil capacitor ballasts for 120 VAC have energy efficiencies similar to electronic ballasts. When operated at normal AC line frequency (60 Hz.) the color temperature is 2700° K. By operating compact fluorescent lamps on an electronic ballast at high frequency, 25 kHz to 35 kHz, the lamps phosphors are about 14% to 17% more efficient at producing light and flicker is eliminated. The color temperature drops from 2700°K to about 2300°K..

Very few residential ballast designs address the power factor requirements imposed by fluorescent lamps. Power factor relates to the lag between current and voltage and values less than 1.0 translate into wasted energy. Some ballasts have power factors as high as 0.9, but many fall short with values as poor as 0.2. Normally, coil capacitor ballasts have a power factor of 0.2 to 0.4, however, high power factor (HPF) designs achieve values as high as 0.9. Electronic ballasts usually have power factors above 0.6 and the more expensive and bulky designs above 0.9.

The OSRAM Corporation, A Siemens Company (the same people that purchased ARCO Solar!), is one of the industry leaders in both compact fluorescent lamp and electronic ballast manufacturing. OSRAM has a line of 12 VDC and 120VAC/DC ballasts that are available only in Europe. These commercial grade electronic ballasts have a power factor greater than 0.9., and will soon be available in the USA in 5 to 26 watt sizes.

Dulux EL Electronic Light Bulbs

Residential grade OSRAM DULUX™ EL lamps (Electronic Light Bulbs) are available in the USA right now. These are for retrofit applications and have medium bases that replace incandescent light bulbs. These lamps may be used on inverters at 120VAC and there is no hum! The power factor for these DULUX™ Electronic Lightbulbs is 0.6 to 0.7. Its built in ballast is designed with a full wave bridge rectifier capacitor input filter followed by a 35 kHz oscillator to drive the fluorescent tube. All of this is integrated and the expected tube life and ballast life is well matched. As a result of this design, these electronic light bulbs may be operated on DC or 120VAC. Since the capacitor acts as a peak detector of the 120 V RMS AC, the DC required would be around 165 V. This may be only interesting, but I thought that I would mention it. Also, these electronic light bulbs have received FCC Part 18C certification for residential use. This means that they aren't going to interfere with radios or TVs. Most magnetic ballasts have never been tested by the FCC, they can be very noisy and interfere with radios and TVs.

We have learned that coil capacitor ballasts produce much more heat than electronic ballasts. In fact, the manufacturers of compact fluorescent fixtures, that get them UL approved, must keep the internal temperatures of the fixtures below 120 °F. The entire industry is waiting for the low temperatures from electronic ballasts. They cost more, but have advantages. If we let the fixture manufacturers know that the trade off is worth it to them, they are with us.

Conclusions:

Compact fluorescent lighting systems are much more efficient than incandescent lighting. We see about four times the lumens per watt as compared to incandescent. There are more efficient systems than the compact fluorescent, but they usually aren't suitable for indoor use. Recently, fluorescent lighting has become much better at color rendition and can start almost as rapidly as incandescent lamps. With the emergence of electronic ballasts, heat dissipated in the ballast has been reduced and the performance of fluorescent lamp starting improved.

Why bother?

Energy savings!!! & $$$

Don't forget that your local power utility (maybe even you!) don't have to produce as much energy to feed your lighting needs.

Ecological Benefits!!!

CO_2 from burning fossil fuels adds to "greenhouse effect" and global warming..

Acid rain kills trees and fish in lakes.

EMR envelope diagram: width (2nd number), height (1st number), EMR envelope

Legend: ● BEST ◐ Okay NO Unsuitable

EMR envelope	Incandescent equivalent / actual wattage drawn	Size height/diameter	EMR 1 milligaus envelope	Temperature range °F / °K	Color Temperature °K	Coil or Electronic ballast	Replaceable bulb	Radioactive starter	Shaded lamp	Enclosed indoor	Open indoor	Bare bulb	Track lighting	Recessed can	Protected flood	Enclosed outdoor
Panasonic G-15	52w / 15w	6.3" / 3.7"	25" / 12"	32 / 95	2800	C	NO	YES	NO	●	●	◐	◐	NO	◐	◐
Panasonic T-15	52w / 15w	6.3" / 3.0"	18" / 16"	32 / 95	2800	C	NO	YES	◐	●	●	◐	◐	NO	◐	◐
Panasonic Electronic 18 watt capsule	75w / 18w	7.4" / 3.0"	5" / 3"	32 / 95	2800	E	NO	NO	◐	●	●	◐	◐	NO	◐	◐
Panasonic Electronic Twin 18 watt tube	75w / 18w	7.4" / 2.2"	na	32 / 95	2800	E	NO	NO	●	◐	◐	◐	◐	NO	◐	◐
Panasonic Electronic Twin 27 watt tube	100w / 27w	7.8" / 2.2"	7" / 7"	32 / 95	2800	E	NO	NO	●	◐	◐	◐	◐	NO	◐	◐
Osram Dulux EL 7 watt	25w / 7w	5.7" / 2.3"	3" / 3"	0 / 140	2700	E	NO	NO	◐	●	◐	◐	◐	NO	◐	◐
Osram Dulux EL 11 watt	40w / 11w	5.7" / 2.3"	5" / 5"	0 / 140	2700	E	NO	NO	◐	●	◐	◐	◐	NO	◐	◐
Osram Dulux EL 15 watt	60w / 15w	6.8" / 2.3"	5" / 5"	0 / 140	2700	E	NO	NO	◐	●	◐	◐	◐	NO	◐	◐
Osram Dulux EL 20 watt	75w / 20w	8.2" / 2.3"	5" / 5"	0 / 140	2700	E	NO	NO	◐	●	◐	◐	◐	NO	◐	◐
Dulux EL 11 R with reflector	50w / 11w	5.8" / 4.8"	6" / 1"	0 / 140	2700	E	NO	NO	NO	NO	◐	NO	●	●	◐	NO
Dulux EL 15 R with reflector	75w / 15w	7.3" / 4.8"	4" / 1"	0 / 140	2700	E	NO	NO	NO	NO	◐	NO	●	●	◐	NO
Phillips SL 18	75w / 18w	7.2" / 2.5"	4" / 0"	0 / 140	2700	E	NO	NO	◐	●	●	◐	◐	NO	◐	◐
Phillips SL18 with reflector	75w / 18w	7.3" / 5.0"	4" / 0"	0 / 140	2700	E	NO	NO	NO	NO	◐	NO	●	●	◐	NO
Reflect-A-Star 4.5"	75w / 14w	6.0" / 4.5"	32" / 30"	0 / 140	2700	C	YES	NO	NO	NO	NO	NO	●	●	●	NO
Reflect-A-Star 5.25"	75w / 14w	6.0" / 5.3"	32" / 30"	0 / 140	2700	C	YES	NO	NO	NO	NO	NO	●	●	●	NO
Outdoor Compact J-910	50w / 17w	9.0" / 6.0"	12" / 30"	-20 / 140	2700	C	YES	NO	NO	NO	NO	NO	NO	NO	●	NO
Ultra Short System	40w / 8w	5.5" / 2.6"	18" / 8"	-20 / 95	2700	C	YES	NO	◐	●	●	◐	◐	NO	◐	◐
Dulux Combi Lamp	50w / 8w	8.8" / 2.8"	18" / 16"	32 / 95	na	C	YES	NO	NO	NO	●	NO	NO	NO	NO	NO

chart by Michael Potts based on data developed by Doug Pratt and the Real Goods Applications Engineers

6Feb91

DC Energy Efficient Lighting

PL Lamps

PL bulbs are more sensitive to voltage than other lightbulbs, and will not fire or will fire slowly at voltages below 12.0 volts. It also helps to ground them by touching the bulb, making them fire more rapidly (known as "stroking the lamp"). Some of our customers have suggested placing a 1/2" piece of aluminum foil around the two tubes of the lamp for quicker start. We have a copper grounding piece available which will solve the problem of firing at low voltage for $3.

Steve Willey of Backwoods Solar has come up with a great solution to the slow PL firing problem. (However keep in mind that it will void your warranty.) Steve claims the solution will result in allowing every PL bulb to start every

time without failure with any ballast! Here is how it's done with the two major PL bulb manufacturers, Phillips and Osram:

Osram: Cut off the "skin" from the bottom of the PL bulb's base, using a band saw, hawksaw, or grinding wheel. This is a very thin layer at the bottom - be very careful not to catch any wiring when cutting. You will expose a glass cylinder (starting tube) and a grayish plastic cylinder (the capacitor). Cut one of the wires going to the capacitor and bend it out of the way so it doesn't touch other wires. Use glue to seal if you want.

Phillips: Cut as with the Osram, but in the Phillips PL bulbs, the capacitor is encased in a blue plastic unit. Snip one wire (any of the three going to the capacitor) and bend it back as with the Osram so that it doesn't touch any other wires. Use glue to seal if you want.

Pre-wired Screw-In Compact Fluorescent 12V PL Lamp Kits

These modular compact fluorescent lamp assemblies screw into standard light bulb sockets. They incorporate a 12V solid-state ballast, a standard edison base, a compact fluorescent lamp socket and a long-lived 10,000 hour life PL lamp, which can be replaced when it burns out. *Remember, these lamps are for 12V use only!* **For all PBS series lights, the center must be positive.**

ATTN: ALL PL LAMPS REQUIRE BALLASTS TO OPERATE

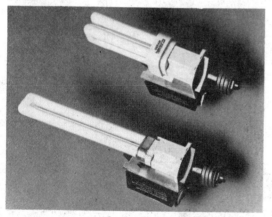

		Equivalent AC incan. watts		Price
31-661-KT	PBS-PL5	25W	5.25"x2"	$59
31-662-KT	PBS-PL7	40W	6.5"x2"	$59
31-663-KT	PBS-PL9	50W	7.75"x2"	$59
31-664-KT	PBS-PL13	60W	8.25"x2"	$59
31-665-KT	PBS-Quad9	50W	5.5"x2"	$65
31-666-KT	PBS-Quad13	60W	5.75"x2"	$65

12 Volt Ballast/PL Lamp Kits

Use this kit to install energy-efficient DC **powered** compact fluorescent lighting into fixtures which cannot accommodate the bulkier size of the (above) PBS series or where the bare lamp may be visible and a neater look than that of the PBS series is desired. The RK-PL kits come with the appropriate hard-wire ballast, compact fluorescent lamp, and PL lamp/edison screw-in base adapter. The ballast measures 2.5" long x 2" wide x 1" deep and may be installed in the base of adequately sized floor or table light fixtures, or in a junction box mounted in the wall/ceiling, a fixture, or on a fixture power cord. Please note: switching of power to the lamp must be done on the positive input lead to the ballast. The height of the PL lamp/edison adapter base is the same as for the above comparable PBS series units and the width is 1.25".

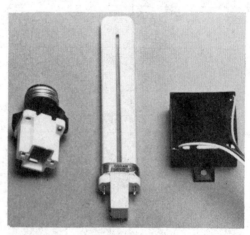

31-671-KT	RK-PL5	$43
31-672-KT	RK-PL7	$43
31-673-KT	RK-PL9	$43
31-674-KT	RK-PL13	$43
31-675-KT	RK-Quad9	$49
31-676-KT	RK-Quad13	$49

DC Ballasts

The Sunalex Corporation, with 11 years manufacturing experience, produces one of the best lines of DC ballasts made. Their solid-state ballasts are very energy-efficient, operate at high frequency (eliminating any perceptible flicker), have built-in polarity protection, and are totally encapsulated, offering protection against high humidity, insects, and corrosive elements such as salt air. No external heat sinking is required, though anything you can do to facilitate its cooler operation (i.e., mounting it on a metal surface or installing it in a ventilated fixture or junction box) will likely extend its useful life. The 12V models will operate with input voltages between 10 and 14V and the 24V models between 20 and 28V. Operating these ballasts within their lower voltage range will slightly decrease the lamp's light output and increase its life, while operating it at a higher voltage, within its specified range, will slightly increase light output and decrease lamp life. (Please note: input voltages in excess of those specified will irreparable damage the unit.) The compact size of these products offers you freedom in installing these in a variety of fixtures. Sunalex offers the only low-voltage ballast that will operate more than one lamp on a single ballast, as noted in the following table. All ballasts are 2"W x 2.5"L x 1"D except for the SXE 1012, which is 2"W x 6"L x 0.25"D.

Item #	Voltage	Will operate	Amp draw	Price
31-200	12V	1-PL5	0.55A	$34
31-200	12V	2-PL5 (series)	0.95A	$34
31-200	12V	3-PL5 (series)	1.35A	$34
31-200	12V	1-PL7	0.70A	$34
31-200	12V	2-PL7 (series)	1.3A	$34
31-200	12V	1-PL9	0.85A	$34
31-201	12V	1-PL13	1.25A	$34
31-202	12V	1-F40	3.5A	$65
SXE 1012 (standard 4'-40W fluorescent)				
31-203	24V	1-PL7	0.35A	$55
31-203	24V	1-PL9	0.40A	$55
31-204	24V	1-PL13	0.60A	$55

Copper Grounding Strip/Starting Aid

We have discussed at length the uses of the copper grounding strip. This is simply a piece of 1/2" copper that when placed between the tubes of a PL bulb and grounded will greatly enhance its starting speed. It is normal for most compact fluorescents to flicker for up to several seconds when first turned on while they attempt to strike an arc. Grounding the fixture greatly eases the starting process. If your lamp flickers for more than 3 seconds before coming on, it is probably not adequately grounded.

31-401	Copper Grounding Strip	$3

PL Twin Tube Fluorescent Bulbs

The PL bulbs come in 5, 7, 9, and 13-watt versions. *All are designed to be used with ballasts - not on 110V house current!* They can be purchased in three different configurations. First, and least expensive is the bulb with a two pin base as a replacement bulb for the RK or PBS series to fit into an edison adapter. Next you can purchase the PL bulb with an edison base (either attached for a shorter overall length or with a separate edison base adaptor which makes the bulb longer overall but replaceable), and lastly you can purchase the bulb with an edison base AND with a frosted globe on the outside, which many people consider more attractive, although it cuts the light output down marginally. The quad tube bulbs are the answer to the common complaint that the bulbs are too long. They come with and without frosted globes. The Quad-9 consists of two 5-watt PL bulbs and the Quad-13 consists of two 7-watt PL bulbs. You can also order the edison base adapter separately that accepts a pin-based PL bulb on one end and screws into a standard edison base fixture. Also available is a hardwire PL socket for those of you who choose not to employ the edison base. **All PL bulbs must use a ballast!**

31-101	PL-5 w/pin base	$ 7
31-102	PL-7 w/pin base	$ 7
31-103	PL-9 w/pin base	$ 7
31-104	PL-13 w/pin base	$ 7
31-121	Quad 9 w/pin	$11
31-122	Quad 13 w/pin	$12
31-151-KT	PL-5 w/separate edison base	$15
31-152-KT	PL-7 w/separate edison base	$15
31-153-KT	PL-9 w/separate edison base	$15
31-154-KT	PL-13 w/separate edison base	$15
31-161-KT	Quad 9 w/separate edison base	$18
31-161-KT	Quad 13 w/separate ed. base	$18
31-156	PL-5 w/attached edison base	$13
31-157	PL-7 w/attached edison base	$13
31-158	PL-9 w/attached edison base	$13
31-159	PL-13 w/attached edison base	$13
31-165	Quad 9 w/attached ed. base	$19
31-166	Quad 13 w/attached ed. base	$19
31-111	PL5 w/ed. base & globe	$18
31-112	PL7 w/ed. base & globe	$18
31-113	PL9 w/ed. base & globe	$18
31-114	PL13 w/ed. base & globe	$18
31-131-KT	Quad 9 w/ed. base & globe	$25
31-132-KT	Quad 13 ed. base & globe	$25
31-402	Edison to PL adapter	$5
31-403	Hardwire PL Socket	$3

PL Reflector

This highly polished, specular aluminum reflector clips onto compact fluorescent lamps to improve light output where directional light control is desired, such as in ceiling-mounted or wall-mounted fixtures. Because many fixtures have poorly reflective interior finishes, the reflector can provide nearly twice the light from a compact fluorescent lamp out of such fixtures. Being both bendable and cuttable, it can be adapted to fit most 5-13 watt PL and "quad" compact fluorescent lamps.
Dimensions: 3.5" W x 4-3/8"H x 3/4" D

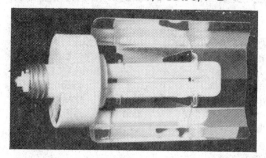

| 31-404 | PL Reflector | $5 |

Edgelight Wall Sconce

This decorative wall sconce can serve as an ambient or accent lighting source. The edges of its three illuminated glass discs have a rich green sheen when the fixture is turned off and glow clear when it is turned on. The polished brass finish compliments the glassware creating an elegant wall piece. The 12V version is equipped with two 7-watt compact fluorescent lamps comparable to a 75 watt incandescent. The 110V version has two 13 watt compact fluorescent lamps comparable to 100 watts of incandescent lighting. Most of the fixture's light output is distributed through its open top and bottom. Great for dining room, hallway, living room and bedroom applications. Available in 12V DC or 120V AC. Specify voltage. Total power consumption is 18 watts for the 12V and 30 watts for the 110V.

| 31-381 | Edgelight Wall Sconce (12V-brass) | $159 |
| 31-382 | Edgelight Wall Sconce (110V-brass) | $115 |

Sundancer

This handsome fixture is designed for indoor or outdoor areas where a clean, low-profile, surface mounted light is required. It may be mounted on the wall, ceiling, or underside of cabinets or shelves. The fixture incorporates a specular aluminum reflector, a 12-volt solid-state ballast, and a rugged, clear prismatic, polycarbonate lens, which provides a sparkling appearance and maximizes light distribution. Corrosion resistant materials are used throughout - Outdoor rated. A gasketed lens keeps dirt and insects out. An elegant wood trim (T), which frames the lens, is available. Compact fluorescent lamp included. **12VDC only!**

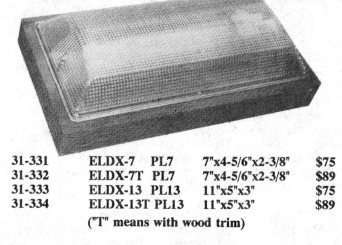

31-331	ELDX-7	PL7	7"x4-5/6"x2-3/8"	$75
31-332	ELDX-7T	PL7	7"x4-5/6"x2-3/8"	$89
31-333	ELDX-13	PL13	11"x5"x3"	$75
31-334	ELDX-13T	PL13	11"x5"x3"	$89

("T" means with wood trim)

Pagoda

These traditionally styled fixtures use a compact fluorescent lamp, mounted inside a gasketed, clear prismatic, glass jar to provide energy-efficient lighting for pathway, garden, and other landscape lighting needs. Non-corrosive cast metal construction is available in either powder coat black (BK) or forest green (GR) finish. Multi-tier design distributes light downward, minimizing direct glare. Comes with threaded coupling for mounting to standard 1/2" (IPS) threaded pipe. Compact fluorescent lamp included. 12-volt solid-state ballast. **12VDC only!**

| 31-376 | W1000-7GR | (PL7) | 9"x6" | $95 |
| 31-375 | W1000-7BK | (PL7) | 9"x6" | $95 |

Quartz Halogen Lighting

Quartz halogen 12-volt lights burn hotter and brighter than incandescent lights. They can be up to 50% more efficient and lead a longer life. While not as efficient as PL or other fluorescent lighting, they lend themselves to situations where direct, bright lighting is called for, like desks, sewing tables, workshops, and spot lighting.

Quartz Halogens Inside Frosted Globes

These are one of the best-selling light bulbs in our entire catalog. A small cottage industry in the Northeast manufactures these incredibly ingenious 12-volt light bulbs. On the outside they appear identical to an incandescent, but they have the increased efficiency and longevity of quartz halogens. The three-way bulbs are especially suited for situations where versatility of ambiance is called for. Multiply the watts on halogens by 1.5 and you'll get an idea of their equivalent light output compared to an incandescent lamp. *Warning: these bulbs are often backordered due to the manufacturer's inability to make them fast enough for us!* **12V DC only!**

Item #	Stock no.	Watts	Amps	Price
33-101	601F5	5	0.42	$14
33-102	601F10	10	0.8	$14
33-103	601F21	21	1.7	$14
33-104	601F35	35	2.9	$14
33-105	601F50	50	4.2	$14
33-106	601F10/20/30 - 3 way			$21
33-107	601F35/60/95 - 3 way			$21

MR-16 Halogen Flood Lamp

The MR-16 is a very bright flood lamp with a faceted reflector and a quartz halogen bulb in the center. It is ideal for situations where a bright light is desired but not overly directed to a pinpoint. The bulb is a 20-Watt halogen which draws 1.7 amps but is far brighter than you'd expect. **12V DC only!**

33-108   MR-16 12V Halogen bulb $22

Halogen Co-Pilot Tasklights

These rugged, compact, flexible 12V tasklights can be positioned precisely to deliver light where you need it, without getting in your way. Great for use by your bedside, kitchen sink, sewing machine, desk, or in your car, boat, or RV. Its 5-watt halogen lamps deliver a bright white light and operates approximately 30% more efficiently than standard incandescents. On-off switch and reflector is built into the cap of the lamp housing. Gooseneck stem is offered in three lengths for the hardwire mounted model and one length for the cigarette lighter plug-in model M. Matte black finish. The L-20 is our best seller with the K-30 running a close second.

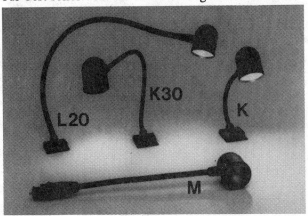

Item #	Lamp(s)	Fixture Amps 12V	Dimension (Length)	Price
33-301	L20 (5W Halogen)	0.42	20"	$26
33-302	K30 (5W Halogen)	0.42	20"	$24
33-303	K (5W Halogen)	0.42	4"	$24
33-304	M (5W Halogen)	0.42	6"	$24
33-109	5W Replacement Halogen			$9

Hi-Intensity Bulbs

High-intensity bulbs are really automotive brake lights. We offer a special adaptor (different from the one used for quartz halogen), so these hi-intensity lights can be used in standard 110VAC fixtures. The bulbs are rated in candle-power (CP) and are about 50% brighter than a standard incandescent counterpart (A 21-CP hi-intensity bulb uses 16.8 watts or 1.4 amps but delivers near the lumen equivalent of a 25-watt incandescent which uses 2.08 amps at 12 volts.)

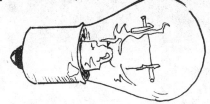

Item #	12V bulbs	Amps	Price
33-112	#1141	2.25	$1.50

Littlite High Intensity Lamp

We've just lowered the prices on these wonderful 12-volt quartz halogen lamp. The Littlite is a great gooseneck lamp for your desk, stereo, headboard, or worktable. It's available in a 12" or an 18" length. The "A" series comes with base and dimmer, 6-foot cord, gooseneck, hood and bulb. Two pieces of snap mount are included for permanent mounting. Available options include a weighted base (WB) for a movable light source, plastic snap mount with adhesive pads (SM), and an adjustable mounting clip (CL) that adjusts from 1/16" to 3/4" wide. Also available is a replacement halogen bulb (RHB). The WXF power transformer is available for 110V users to convert the Littlite from 12V to standard house current.

"These are incredible little lamps. My wife used to always get upset when I read in bed because the big overhead lamp kept her awake. Five months ago I installed the Littlite with the easily movable snap mount which attaches to just about anything. You can put the light up at night and take it down in the morning. With the halogen bulb, it casts a very directed light which enables me to read clearly at night while my wife sleeps on..." J.S., Ukiah, CA

33-306	12" Littlite (L-3-12A)	$39
33-307	18" Littlite (L-3-18A)	$45
33-401	Weighted Base (WB)	$11
33-402	Snap Mount (SM)	$ 3
33-403	Mounting Clip (CL)	$12
33-109	Replacement Bulb (RHB)	$ 9
33-501	110V Transformer (WXF)	$14

Navigator Lite

Our new rechargeable Navigator Lite is a high tech, pure bright red light source that greatly reduces night blindness. When the light is turned off, the eyes rapidly return to their full nighttime abilities. This is a must for sailors, astronomers, and pilots, and also very handy for campers and outdoorspeople. It will produce continuous illumination from its rechargeable batteries for over 25 hours. The recharger plugs into a standard cigarette lighter plug or 12V socket, where it will recharge in six hours. The light measures 4" x 2" x 1" thick and comes with a 4" velcro strip for attachment to clothing, table, or wherever. The light will last in excess of 100,000 hours!

37-307 Navigator Lite $35

12V Night Light

This Super Efficient night light is ideal for the childrens room or for shining directly down on the toilet seat at night. The Red MT 5000 light source is the most efficient light source on the market today. It won't destroy your night vision because red light doesn't washout night vision like white light does. It also works well to light door locks or to mark trees, posts, or other areas on your driveway that you don't want to back into at night. The 12 volt Nite Lite draws only 2/10ths of an amp so it can be left on continuously with little effect on your power system. If operated for 24 hours a day it should last for 10 years.

37-315   12 Volt Nite Lite $29

12-Volt Christmas Tree Lights

It just doesn't seem like Christmas without lights on the tree and these 12-volt lights will dazzle any old fir bush. These lights are actually great all year 'round for decorating porches, decks, and accenting homes and showrooms. The light strand is 20 feet long and consists of 35 mini, colored, non-blinking lights (amp draw = 1.2A), with a 12-volt cigarette lighter socket on the end. Many of you enjoyed the story (Spring '88 *Real Goods News*) of the California Highway Patrol pulling Nancy over last Christmas for being an illegal distraction with her gold Cadillac's interior bedecked with these lights, only to exonerate her after discovering that no code section applied. Needless to say, you'll crank many heads on the freeway Christmas Eve when you plug a set of these into your car's cigarette lighter. The optional installation kit is a roll of electrical tape. Installation not included. *Note: Some of the boxes say "Flashing Lights" - but don't believe it!* **Our lights do not flash!**

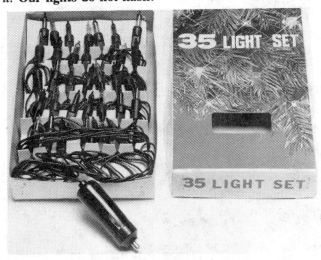

37-301   12V Christmas Tree Lights $12

2003 Tail Light Bulb Adapters

This simple adapter has a standard medium Edison base and accepts a standard automotive type bulb like the #1141 listed above. One of our best selling items. It's a very easy way to convert pole lamps to 12V. 1/2" long. *Light pictured but not included - see page 179 for light.*

33-404   Tail light bulb adapter $7.50

Bullet Lights

Though not as efficient for general illumination, bullet lights are excellent for areas where a focused light is desired (over your bed or in a workshop or kitchen). The high-intensity bulb makes it a perfect reading light, yet it can be directed to give area lighting as well. A positive lock swivel assures illumination where desired. It comes in four different configurations. It's available as a single or double tube-shaped bullet light, or as a single or double cone-shaped bullet light. The lights come in satin brass on a wood grain base and have a positive lock swivel. The single lamps draw 1.4 amps at 12V and the doubles draw 2.8 amps using 18 watt bayonet bulbs.

38-401	Bullet light - Double tube-shaped	$18
38-402	Bullet light - Single tube-shaped	$12
38-403	Bullet light - Single cone-shaped	$12
38-404	Bullet light - Double cone-shaped	$18

Incandescent Lights

Clearly the undisputed energy hog of the lighting industry, incandescent lighting nevertheless captures a necessary niche in the low-voltage lighting market. They are very cheap, wonderfully easy to install, and if you have power to spare (hydro users take note) are the right tool for the job. Most people use them in the beginning of their off-line transformation or in rarely used spots like closets, bathrooms, and alcoves.

Item #	Watts	Amps	Price
38-101	15W	1.2	$2
38-102	25W	2.1	$2
38-103	50W	4.2	$2
38-104	75W	6.2	$2
38-105	100W	8.3	$2

Standard Fluorescent Lamps

While not as efficient as the new "compact fluorescent" lamps featured above, standard fluorescent lamps are nonetheless far more efficient than incandescents and halogens. We are featuring the Thin-Lites here which are very reasonably priced and well built.

Thin-Lites

Thin-Lites are made by REC Specialties, Inc. Their 12-volt DC fluorescents are built to last, and are all U.L. listed. Easy to install, they have one-piece metal construction, non-yellowing acrylic lenses, and computer grade rocker switches. A baked white enamel finish, along with attractive wood grain trim completes the long-lasting fixtures, warrantied for two years. All use easy-to-find, standard fluorescent tubes, powered by REC's highly efficient inverter ballast. For those that didn't know, bulbs used in DC fixtures are exactly the same as those used in AC fixtures and can be purchased at any local hardware store - only the ballast is different. Because they are so delicate and inexpensive, bulbs are usually not included with lamp fixtures.

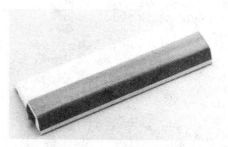

8 watt 12VDC light. 12"x4"x1" thin. 0.7 amps. Uses one F8T5/CW fluorescent tube. Light output: 400 lumens.

32-111   #111 8 Watt Thinlite $39

30 watt 12VDC light. 18" x 5-1/2" x 1-3/8" deep. 1.9 amps. Uses two F15T8/CW fluorescent tubes. Light output: 1,760 lumens.

32-116   #116 30 Watt Thinlite $44

15 watt 12VDC light. 18"x4"x1-3/8" deep. 1.26 amps. Uses one F15T8/CW fluorescent tube. Light output: 800 lumens.

32-115   #115 15 watt Thinlite $35

22 watt 12VDC circline. 9-1/2" diameter x 1-1/2" deep. 1.9 amps. Uses one FC8T9/CW fluorescent tube. Light output: 1,100 lumens.

32-109   109C 22 Watt Thinlite $43

32 watt 12VDC circline, 13-1/4" diameter x 2-1/4" deep. 2.6 amps. Uses one FC12T9/CW fluorescent tube. Light output: 1,900 lumens.

32-110   #110 32 watt Thinlite $58

Weatherproof Outdoor Area Lights

Thin-Lite weatherproof outdoor area lights feature anodized aluminum housings. They provide excellent nighttime security and area lighting. When attached to the rear of a vehicle, they provide clear visibility for parking, hitching and backing up. They offer the option of energy efficient outdoor area and security lighting for remote sites where standard grid power is unavailable. An amber diffuser lens is standard (will not attract night flying insects). Clear lens optional.

8 watt 12VDC light. 12-1/2"x3-3/4"x2-3/4". 0.9 amps. Uses one F8T5/CW fluorescent tube. 400 lumens.

32-161   #161 8 Watt Thinlite $45

Weatherproof outdoor area light with amber lens (will not attract night flying insects such as mosquitoes). 16 watt 12VDC light. 12-1/2"x3-3/4"x2-3/4". 1.9 amps. Uses two F8T5/CW fluorescent tubes (white lens optional). Light output: 800 lumens.

32-162   #162 16 Watt Thinlite $49

15 watt 12VDC light. 18-1/2"x3-3/4"x2-3/4". 1.3 amps. Uses one F15T8/CW fluorescent tube. 870 design lumens.

32-163   #163 15 watt Thinlite $57

Commercial and Industrial Fixtures: 180 Series

Thin-Lite commercial, industrial and residential surface mount fluorescents are designed for practical applications where decorative styling is desirable. The 180 series feature anodized aluminum housings and clear diffuser lenses for maximum light output. They are designed for remote switching.

40 watt 12VDC light. 48"x3-5/8"x3-1/2" deep. 2.9 amps. Uses one F40T12/CW fluorescent tube. 3,150 design lumens.

32-183   #183 40 Watt Thinlite $65

Commercial and Industrial Lights: 150 Series

Thin-Lite 20 and 40-watt surface mount fluorescents are designed for practical lighting in commercial, industrial, and remote area site applications. They feature anodized aluminum housings in two and four foot lengths. They are shipped without the fluorescent tubes. These fixtures utilize standard AC 20 and 40-watt fluorescent tubes readily available worldwide. They are designed for remote switching, and are also available with a diffuser lens (see 180 series).

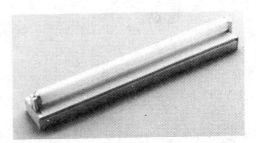

20 watt 12VDC light. Housing dimensions: 24"x3-3/8"x1-5/8". 1.6 amps. Uses one F20T12/CW fluorescent tube. 1,250 design lumens.

32-151   #151 20 Watt Thinlite $37

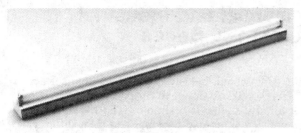

40 watt 12VDC light. Housing dimensions: 48"x3-3/8"x1-5/8". 2.9 amps. Uses one F40T12/CW fluorescent tube. 3,150 design lumens.

32-153 #153 40 Watt Thinlite $49

30 watt 12VDC light. 36-1/8"x2-1/4"x2-7/16". 2.1 amps. Uses one F30T8/CW fluorescent tube. 2,200 design lumens.

32-197 #197 30 Watt Thinlite $55

Hi-Tech Styles

Thin-Lite Hi-tech styles were developed for both efficient and attractive lighting where maximum light is required. Anodized aluminum housings and clear acrylic diffuser lenses provide high light output on three sides. They are designed for commercial and industrial vehicles, and for use in remote are a housing, schools, and medical facilities in conjunction with alternative sources of energy. Almond end caps standard. Black end caps optional.

Surface Mount Lights

The ST 130 series are our most popular Thin-Lites. They are economically priced, practical lights that feature prepainted aluminum housings and acrylic diffuser lenses. They are available as large as 5 feet long with two standard 40-watt AC fluorescent tubes to meet maximum lighting requirements. Where practicality is the principal consideration, the ST 130 series suit the requirement perfectly.

8 watt 12VDC light. 12-3/8"x2-1/4"x2-7/16". 0.9 amps. Uses one F8T5/CW fluorescent tube. 400 design lumens.

32-191 #191 8 Watt Thinlite $36

16 watt 12VDC light. 12"x5-3/8"x1-3/4". 1.5 amps. Uses two F8T5/CW fluorescent tubes. 800 design lumens.

32-130 #130 16 Watt Thinlite $36

15 watt 12VDC light. 18-1/8"x2-1/4"x2-7/16". 1.3 amps. Uses one F15T8/CW fluorescent tube. 870 design lumens.

32-193 #193 15 Watt Thinlite $37

30 watt 12VDC light. 18"x5-3/8"x1-3/4". 2.1 amps. Uses two F15T8/CW fluorescent tubes. 1,760 design lumens.

32-134 #134 30 Watt Thinlite $38

40 watt 12VDC light. 24"x5-3/8"x1-3/4". 2.5 amps. Uses two F20T12/CW fluorescent tubes. 2,500 design lumens.

32-138   #138 40 watt Thinlite $49

40 watt 12VDC light. 48"x5-3/8"x1-3/4". 2.9 amps. Uses one F40T12/CW fluorescent tube. 3,150 design lumens.

32-139   #139 40 watt Thinlite $59

80 watt 12VDC light. 48"x5-3/8"x1-3/4". 5.8 amps. Uses two F40T12/CW fluorescent tubes. 6,300 design lumens.

32-136   #139-2 80 Watt Thinlite $85

Thin-Lite Ballasts

Thin-Lite inverter ballasts convert a wide range of standard 110-120VAC fluorescent fixtures to 12V, 24V, 32V, and 48VDC operation. One ballast is required for each fluorescent tube adaptation in a fixture unless otherwise noted.

As a general rule, to find the amp draw of a particular light or inverter ballast, divide the watts by volts (W/V = A.) As an example, the 12VDC circline table lamp adapter Model 107 is rated at 22 watts. 22 watts divided by 12 volts = 1.8 amps. *Note: Thin-Lite ballasts are available by special order only; allow 2-4 weeks for delivery.*

Thin-Lite Inverter Ballasts, 12VDC and Other Low-Voltage Options

32-201	4 watt, single lamp	$29
32-202	4 watt, dual lamp	$29
32-203	22 watt, single lamp	$29
32-209	32 watt, single lamp	$29
32-210	8 watt, single lamp	$29
32-211	8 watt, dual lamp	$29
32-212	14 watt, single lamp	$29
32-213	14 watt, dual lamp	$29
32-214	15 watt, single lamp	$29
32-216	15 watt, dual lamp	$29
32-225	13 watt, single lamp	$29
32-226	13 watt, dual lamp	$29
32-247	30 watt, single lamp	$29
32-251	20 watt, single lamp	$29
32-252	20 watt, dual lamp	$29
32-253	40 watt, single lamp	$35

Note: Lamps and sockets not included.

To order different DC voltage, add 24VDC, 32VDC or 48VDC after the model number and add $8.50 to the price of each ballast so ordered.

PAVE a garden. Kill two birds with one stone by expanding your driveway into the vegetables. Just think: no more cabbage worms to squish by hand. No more messy compost.

Solar Outdoor Lights

Siemens (Arco) Solar Lights

All Siemens solar lights employ single-crystalline solar cells - the same that are used in the M55 industrial solar module - these new outdoor solar lights are strictly top-of-the-line. The solar cells are 16-17% efficient compared to Nordic's poly-crystalline cells which are 10-12% efficient. These cells provide brighter and longer running solar lights. Siemens employs an exclusive type of nicad battery that is extremely tolerant of high temperatures. Unlike conventional nicad batteries, these high temperature nicads will continue to accept a full charge from the solar cell even in hot temperatures in excess of 50° C. While most competitive solar lights use either butyl tape or an EVA backing to encapsulate their cells, Siemens uses an exclusive silicon based encapsulate designed to correctly encapsulate the cells. All lights have a 1-year warranty except the Pathway Light Plus and the Anywhere Light which have a 2-year warranty.

Mini Coach Light

The Mini Coach Light is an economical lantern-style light which mounts atop a two-piece 12" stake. An incandescent bulb, provides a soft glow within a unique diffraction grated lens. The bulb is driven by high-temperature Nicad batteries through environmentally-sealed circuitry. The batteries are charged by a powerful single crystalline solar cell. The light's body and stake are injection molded from black ABS, treated with UV inhibitors to prevent the fading and brittleness associated with untreated plastics in longterm sunlight exposure. The Mini Coach Light will run up to six hours per night. It decoratively accents driveways, walkways and patios and comes with a full one year warranty. The Nicad batteries have a three year life and the bulb has a nine month life.

34-322 Mini Coach Light $49

Pathway Light

The Pathway Light is an economical pagoda-style light which mounts atop a two-piece 12" stake. An incandescent bulb provides a soft glow within a unique prismatic lens. The bulb is driven by high temperature Nicad batteries through environmentally-sealed circuitry. The batteries are charged by a powerful single crystalline solar cell which is encapsulated in a waterproof silicone compound. The Pathway Light decoratively accents driveways, walkways, patios and gardens. Up to six hours nightly run time. Full one year warranty. Three-year battery life. Nine-month bulb life.

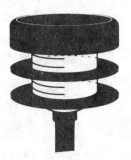

34-323 Pathway Light $59

Pathway Light Plus

The Pathway Light Plus is our premium pagoda style light which mounts atop a two-piece 16" stake. Incandescent bulbs provide a high/low option. Utilizing a 3M designed prismatic fresnel lens, the Pathway Plus evenly distributes light out and down. Bulbs have approximately 1,000 hours life (9 months). The bulbs are driven by high temperature full-sized C Nicad batteries through environmentally sealed circuitry. Batteries have a three year life. The batteries are charged by a powerful single crystalline Siemens solar cell. The panel has a 1.6 watt output. The Pathway Light Plus decoratively accents driveways, walkways, patios, and gardens. 2-year warranty. It will light up to 8 hours per night.

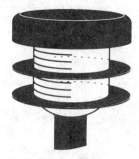

34-324 Pathway Light Plus $79

The Pathmarker

The Pathmarker is used to outline or mark driveways, paths, walkways, patios, or steps. It comes with an 8" stake for easy mounting. A large textured acrylic lens casts a soft red glow by way of a specially-designed high brightness LED. The LED is driven by premium quality AA Nicad batteries through environmentally-sealed circuitry. The batteries are charged by a Siemens Thin Film Silicon (TFS) industrial solar panel which is fully protected by a polycarbonate lens cover. The Pathmarker's body and stake are injection molded from premium quality black ABS which has been specially treated with UV inhibitors to prevent the fading and brittleness associated with untreated plastics in longterm sunlight exposure. It will operate up to 12 hours every night, because of the low amp-draw of the red light. It is ideal for defining or outlining walkways, patios, paths, driveways or steps. The soft red glow portrays a feeling of safety and convenience. 1-year warranty and 3-year battery life.

34-321   Pathmarker Solar Light (set of 2) $45

Anywhere Light

The Anywhere Light is Siemens' most powerful light. It has a bright (15 watt) fluorescent light behind an optically superior prismatic lens that allows it to illuminate a large area with an even shower of light. A high quality industrial 12-volt lead acid battery powers the lamp through environmentally sealed electronics. The electronics are designed for fast cold weather starts and provide a unique 20 second time-delayed shut-off feature giving one plenty of time to leave the area while the light remains on. The battery is charged by a Siemens Solar industrial solar module. With its 14 ft. cord, it plugs into the end of the Anywhere Light, and is designed to be remotely mounted on any rooftop or other surface which gets optimum sunlight exposure. The unit is ideal for patios, garages, sheds, barns or spas. It has up to four hours of run time capacity nightly and comes with a two year warranty. The sealed lead-acid battery has a three-year life.

34-325   Anywhere Light $169

Mailbox Light

The Mailbox Light is a house address light which attaches to a mailbox or post. An incandescent bulb and reflector system, specifically designed for Siemens Solar, provides a soft background glow which illuminates up to five digits from either direction. The bulb is driven by industrial quality, high temperature Nicad batteries. The batteries are charged by a high output Thin Film Silicon (TFS) module which is protected by a durable polycarbonate cover. The Mailbox Light's framework is injection molded from black ABS, specially treated with UV inhibitors to prevent the fading and brittleness associated with untreated plastics in longterm sunlight exposure. The Mailbox Light is an ideal location finder for friends, deliveries or emergencies. Address digits are visible day or night. Up to 8 hours of nightly run time. 3-year battery life. Full 1-year warranty.

34-329   Mailbox Light $79

Conserve Switches

Conserve switches are variable speed control switches (rheostats) that are extremely useful for dimming 12V halogen and incandescent lights, (*don't attempt to use on ballasted fluorescent lamps!*) using less than 1/2 watt of power to control them. They are available in both a 4 amp and 8 amp configuration.

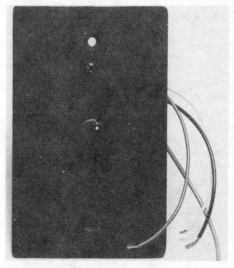

24-101   4 amp Conserve Switch $25
24-103   8 amp Conserve Switch $42

Nordic Solar Lighting

Nordic's solid state circuitry offers many features not found in other solar lights. It is one of the few solar lights to use crystalline solar cells (instead of amorphous), nicad batteries (instead of lead-acid) and dusk to dawn operation (many others go dead in 3-5 hours). Nordic backs up all its solar lights with a 5-year warranty.

Nordic Solar Post Lite

This full-sized carriage-type lamp provides great accent lighting for yards, patios, decks, driveways, and gardens. It will provide up to 12 hours of continuous illumination after being actived at dusk with a photoelectric eye. It can also be turned off manually or set to run for four hours only. The 5-volt, 60mA bulb provides an 8-foot radius of diffused light. The 3-section, 18-gauge mounting poles, constructed of a powder-coated urethane weather-resistant finish allow the lamp to be positioned at a 12", 39", or 68" height. Two bulbs are included, each of which will last three years. The lamp measures 12.5" x 9.25" x 9.25" and weighs 5 lbs.

34-308   Nordic Solar Post Lite $159

Nordic Solar Security Lite

The Security Lite has all the great features of the Post Lite listed above plus a built in high-beam security light. When its infrared sensing circuit detects a person approaching, a brilliant krypton light suddenly illuminates the area. After 30 seconds, if no one is in the vicinity, the security light shuts off automatically and normal low level operation is resumed. The circuitry is tuned to detect human frequencies so wind, small animals and trees will not cause "false alarms." It's an ideal light for providing bright temporary illumination on walkways and around steps or other stumbling points at night. Like the Post Lite it comes with a three section steel post.

34-309 Nordic Solar Security Lite $159

Nordic Walkway Lites

Nordic Walkway Lites are unique in the industry. Nordic features the only set-up we know of that allows for remote mounting of the solar panel. This enables you to put the PV in a sunny spot and spread out your four lights elsewhere. Included are 4 light fixtures, ground mounting stakes, surface mounting hardware for decks and patios, plus 50 feet of hookup wire.

34-310 Nordic Walkway Lites $159

1st Star Solar Outdoor Light

This outdoor light comes with a 10-watt Solec module and a 5-watt halogen beam lamp that provides 800 candlepower. It can provide from 4-8 hours of lighting per night and is ideal for lighting up patios, driveways, sidewalks, gardens, or backyards. It is much heavier duty than the Chronar (Walklites) units and puts out much more light.

34-301 First Star Outdoor Light $350

1st Watch Solar Security Light

The 1st Watch does everything that the 1st Star does except it has two lights instead of one, and a passive infrared sensor to detect heat and motion. When a person, automobile, or other large moving object passes within 40 feet in an 80 degree arc of view, the lamps automatically turn on. It's available with different size halogen bulbs- 5 or 20 watts.

"We've had a 1st Watch in our driveway for over a year now. It comes on whenever you drive in or walk in and stays on for about 5 minutes. The two 20 watt halogens are plenty bright and the beam lights up about 200 feet of driveway, with the adjustable lights."   JS

34-302 First Watch $425

Flashlights

MagLite Flashlights

Many of you are familiar with MagLites, the best flashlights in the world. We've decided to introduce MagLites into our catalog because they are clearly superior to anything on the market and they will last you a lifetime. Here are a few of MagLite's features:
- Adjustable beam (from spot to flood)
- Triple action self-cleaning internal switch
- On, off, and blink switch capability
- O-Ring sealed for water resistance
- Spare bulb in the tail cap (except mini series)

The Adjustable Beam!

Item #	Stock no.	Batteries	Price
37-341	M2A010	2 AA	$14
37-342	S2C010	2 C	$24
37-343	S3C010	3 C	$25
37-344	S4C010	4 C	$25
37-345	S5C010	5 C	$27
37-346	S2D010	2 D	$27
37-347	S3D010	3 D	$27
37-348	S4D010	4 D	$28
37-349	S5D010	5 D	$29

All-In-One Solar Flashlight and Battery Charger

Our solar flashlight/battery charger serves two very useful functions. It recharges and operates on multiple battery sizes: "AA", "C", and "D". It incorporates a heavy-duty, high-quality case that is water and weather-proof. The built-in encapsulated solar panel is made of high efficiency solar cells. The unit provides two sets of "battery size adapters" which allow you to operate and recharge multiple battery sizes. A LED charge indicator light shows when a solar charge is occurring. In addition to being extremely useful around the homestead, this product is also ideal for use in disasters, power emergencies, outdoor adventures, and yes even wars! *Nicad batteries recommended but not included.*

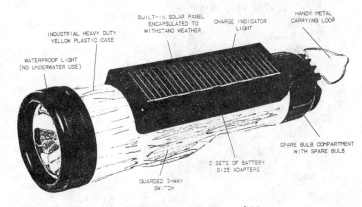

37-309 All-in-One Solar Flashlight $39

Solar Lantern

If you're like us, you're tired of all the Hong Kong produced throwaway flashlight lanterns. We've finally located the state-of-the-art rechargeable solar lantern and it's made by Heliopower (an American Standard & Hoxan PV joint venture). This lantern is the epitome of quality. It provides up to 5 hours of continuous lighting on a single charge with its bright 6-watt fluorescent bulb and broad 180-degree field of illumination. It is lightweight (3.3 lb) and very durable (made of corrosion resistant, high impact plastic). The solar cells consist of 14 single crystal individual cells with a maximum output of 2.5 watts. The power storage unit is a compact lead-sealed battery with 4 volts and 6.0 amphours and will last year after year (minimum 1,000 hours life) with solar recharging. The lantern measures 5.9" long by 2.2" wide by 12.1" high. Although the price seems high, we find this lantern to be well worth it.

37-302 Solar Lantern $109

Dear Real Goods,

Your Solar Lantern just arrived and me and the boys are blown away! Life is tough here in the Saudi desert. It's bad enough not getting to drink beer or take regular showers but until you guys came along it was even getting hard to see at night! It seems there is a shortage of D-Cell batteries and the ones we can get don't last very long. Now that we have your lantern that's charged by the sun (no shortage of that over here!) we can even do reconnaisance at night - kind of amazing that our government didn't think of that - Huh? - PFC Larry Harbiter, Operation Desert Shield, APO NY.

Bright Eyes

This simple invention makes you wonder how you lived without them for so long. It's a hands-free flashlight that can be worn over standard eyeglasses and is so lightweight that you hardly know they're on. We wired up over 30 of our Marathon light fixtures in the new Real Goods office tower using a pair of these in the attic and they worked famously. The Bright Eyes glasses operate on 4 AAA batteries which are included! The quality and durability exceeds the simple Tops Light.

37-305 Bright Eyes Head Lights $14

The Forever Light

Here is a miniature sized (1-7/8") rechargeable light that will give you years of use and you'll never have to change a battery or a bulb. It is energized by solar cells, and entirely self-contained. Simply expose it to sunlight for 2-3 hours and it's ready to give weeks of intermittent use. The semiconductor LED emits a brilliant yellow light, turned on by a simple squeeze. It's great for key chains.

37-306 Forever Light (FL-100) $14

Tops Light Forehead Lamp

Here is a simple but extremely useful product that we've been years searching for. It answers the problem of chopping wood in the darkness on cold nights with an awkward flashlight lodged in your mouth. It's a high-tech version of the old coal miner's lamps and has a multitude of uses. It's also ideal for under the hood or dash of a car, camping and hiking, bicycling, or around the house.

The Tops Light is made of ABS plastic and requires 4 AA batteries to operate. The lighting angle is fully adjustable as is the length of the head strap, which is secured down to the proper size with velcro. A spare light bulb is included with the lamp. We used the Tops Light for our Nicad battery testing and discovered that it will operate for four hours on our Panasonic Nicads.

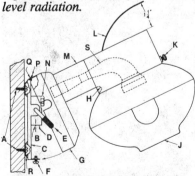

37-304 Tops Light Forehead Lamp   **$8**

Humphrey Propane Lights

Propane lighting is very bright and a good alternative if you don't have an electrical system. One propane lamp puts out the equivalent of 100 watts of incandescent light. The Humphrey 9T contains a burner nose, a tie-on mantle, and a #4 Pyrex globe. The color is "silver grey." The mantles seem to last around three months and replacements are cheap. *Note: Propane mantles emit low level radiation.*

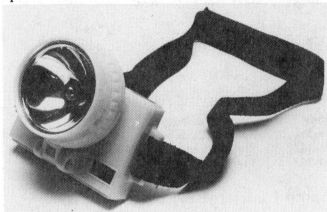

A Mounting Screws (4), **B** Gas Inlet, **C** Wall Bracket Assembly, **D** Valve Assembly, **E** On/Off Lever, **F** Valve Cover Lock Screw (2), **G** Valve Cover, **H** Globe Holder Tab (2), **J** Globe, **K** Globe Lock Screw, **L** Heat Deflector, **M** Globe Holder, **N** Nozzle, **P** Wall Bracket Tab (2), **Q** Valve Cover Slot (2), **R** Wall Bracket Boss (4), **S** Bunsen

35-101 Humphrey Propane Lamp (9T)   **$39**
35-102 Mantle for Propane Lamp (each)   **$ 1**

Elegant Brass Kerosene Lamps

We've recently located an importer of very high quality brass kerosene lamps from France. Our importer has been in the lighting field for fifteen years and every lamp is time tested. Spare parts and service are readily available on all lamps.

The Technology

The 1800's in Europe saw a period of revolution in indoor living after dark. The "state of the art" peaked with the invention of the Kosmos Burner System and its derivative the Matador (flame spreader) Burner featuring symmetrical central drafting and an area-maximizing circular burning surface. Each of our lamps is equipped with these burners. With only the occasional replacement of wick or chimney these lamps will provide decades of daily illumination. All use kerosene or lamp oil.

Lamp Maitresse (LA012-B)

Equipped with an elegant "bombe" chimney, this little lamp maintains an unobtrusive presence until it is needed; then it provides a light source exceeding conventional hardware store kerosene lamps. The overall height is 12".

35-330	Lamp Maitresse	$42
35-331	Spare Maitresse Chimney	$ 6
35-332	Spare Maitresse Wick	$ 2
35-333	Replacement Burner	$15

Lamp Patronne

Fitted with a Kosmos burner and chimney, our tallest lamp offers the optional elegance of a ball shade. The overall height is 19".

35-325	Lamp Patronne complete	$44
35-321	Spare Patronne Chimney	$ 6
35-322	Spare Wick	$ 2
35-323	Replacement Patronne Burner	$15

Lamp Patronne with Etched Ball

This is the same lamp as above but with the beautiful 6" etched glass ball that fits over the glass chimney.

35-320	Lamp Patronne w/Ball	$69
35-324	Spare Etched Ball	$28

Lamp Concierge (LL004-14)

With a Kosmos burner system and ball shade, this lamp has a pleasing profile and a very low center of gravity. The carrying handle is slotted for wall mounting. The overall height is 15".

35-345	Lamp Concierge	$49
35-341	Spare Chimney	$ 6
35-342	Spare Wick	$ 2
35-343	Replacement Burner	$18

Lamp Concierge w/Etched Ball

This is the Lamp Concierge lamp listed above including the beautiful 6" diameter etched glass ball.

35-340	Concierge w/Etched Ball	$76
35-344	Spare Etched Ball	$28

Water Pumping & Storage

For many people, water pumping will influence or actually dictate the type of power system needed. Water weighs 8.34 pounds per gallon, so it can take a large amount of power to deliver it. This is especially true if long distances, high pressure, or many gallons per minute are involved. The following list of terms will be very helpful in understanding pump specifications and recommendations.

Glossary of Terms

Flow: The measure of the liquid volume capacity of the pump expressed in gph for gallons per hour, or gpm for gallons per minute.

Pressure: The measure of force exerted on the walls of piping, tanks, etc. by the liquid being pumped, multiply psi by 2.31 to get feet in head.

Head: Another measure of pressure, expressed in feet. Indicates the height of a column of water being lifted, neglecting friction losses in the piping. For water, divide head in feet by 2.31 to get pressure in pounds per square inch (psi). Example: If there are 100 feet of head to be pumped, 43.29 psi would be required (100 divided by 2.31 equals 43.29). Conversely, if you know the pressure a pump will deliver, multiply the psi by 2.31 to arrive at the total head it will pump.

Flooded Section: Water source is higher than the pump and liquid flows to the pump by gravity. Preferable for centrifugal pump installations and strongly advised for positive displacement pump installations.

Suction Lift: Water source is lower than the pump. Pumping action creates a partial vacuum and atmospheric pressure forces water up to the pump. Theoretical limit of suction lift is 33 feet; practical limit is around 20 feet or less depending on the pump type. Some pumps have a 12 to 24 inch limitation. This lift is calculated at sea level. Generally the suction lift capability of a pump decreases by 1 foot for every 1,000 ft above sea level.

Total Lift: Total lift means the entire vertical distance from the water source to the tank.

Pump Types

Pump types are important, too. Here is a list of the different types available.

Centrifugal: Consists of a fan-shaped impeller rotating in a circular housing pushing liquid towards a discharge opening. Usually used where flow is more important at low pressure (head, lift). Centrifugals are used to deliver large amounts of water over relatively small lifts. Some centrifugals are self-priming and some require a foot valve in the suction pipe.

Positive Displacement: Pumping action created by moving chambers, gears, or pistons. The flow rate of this pump is almost the same at any pressure level. Generally, self-priming. Should never be operated dry because of internal wearing of sliding parts.

Diaphragm: Consists of a flexible diaphragm which moves back and forth in a chamber, creating suction and pressure. Diaphragm pumps are the only pumps that can run dry because their working parts are made of a rubber-like material. Our Flojet line of pumps and our Hypro pumps are an example of the diaphragm pump.

Roller or Vane: Rollers or vanes in a rotor, rotating in an eccentric housing like a flexible impeller pump. For pressures up to 200 psi. Intake must be well filtered, and the pump should never run dry.

Piston: Fluid is drawn in and forced out by pistons moving within cylinders. Used where pressures up to 500 psi are required.

Jet: A type of centrifugal pump utilizing water flow through a narrow opening or nozzle (jet ejector) to bring water from a well.

Deep Well Submersible: A centrifugal pump in which several impeller assemblies in a housing are mounted on a shaft directly coupled to a submersible motor. This entire assembly is located at the bottom of the well. Power is brought to the motor by a waterproof cable. Our Solar Jack series deep well pumps are diaphragm pump.

Shallow Source Water Systems

Here is another in a series of articles by Jon Vara, a customer from Vermont. This edition of the Real Goods AE Sourcebook features his articles on "Shallow Well Pumping Systems," "Basic Electrical Wiring," and a follow-up on "Deep Well Pumping Systems." We think you'll find the articles on water pumping extremely educational and an excellent prelude to our pumping products.

Consider the ultimate low-energy water system: A house sits on a flat bench in a sunny, south-facing slope. Behind it, a forested hillside rises and steepens. About an eighth of a mile from the house, and a hundred vertical feet above it, a clear, cold spring bubbles out of the ground. The homeowners have enclosed the spring in a concrete casing, and covered it with a concrete cap to keep out frogs, snakes, mice, and other surprises. One-inch black polyethylene pipe carries the spring water down the hill to the house; ample water pressure is provided by the hundred-foot drop, with no need for a pump. The system is quiet, inexpensive, and as reliable as gravity.

Unfortunately, even in hilly, well-watered areas - such as my home state of Vermont - such ideal water systems

are the exception. Most gravity-powered water systems lack sufficient vertical drop, or head, to dispense with pumping altogether.

Here are some significant numbers: Typical household water pressure runs between 20 and 50 psi. At 50 psi, water comes blasting out of the tap. Hold a glass loosely under the faucet, turn it on rapidly, and the force of the stream may knock it out of your hand. At 20 psi, the flow is much gentler; water pressure that falls much below 20 psi will seem sluggish and inadequate to most people. [ed. note: here are some comments from Windy Dankoff

For each psi of outlet pressure in a gravity system, 2.3 feet of head is necessary.

(Flowlight Solar Power) after reading Jon Vara's article: *"Under 20 psi can deliver plenty of flow. It is flow that you observe and benefit from at the water spout, not pressure. But you must have oversized plumbing to deliver satisfactory flow. I recommend low-pressure gravity flow systems use one or two sizes larger than "normal" piping throughout the system, and possibly drilling out larger holes in some shower heads. (Do not use water conserving shower heads). Toilets and washers will fill slowly but will work fine. Most tankless water heaters will not work properly, without a booster pump. Some people use an inexpensive booster pump just on the line to the heater." -* **Windy Dankoff**]

For each psi of outlet pressure in a gravity system, 2.3 feet of head is necessary. The hypothetical water system described above, with its hundred-foot head, would have water pressure of 100 divided by 2.3, or about 43 psi. (For the sake of simplicity, I'm ignoring head loss - the reduction in effective head that results from friction between the piping and the water it contains. The smaller the pipe, the greater the flow, and the longer the run, the more severe the head loss. If large enough pipe is used, however, head losses will be insignificant. Most books on plumbing contain head-loss tables for various sizes and types of pipe.)

STORAGE TANK

PIPE

GROUND SURFACE

STATIC WATER LEVEL

WELL

DRAWDOWN LEVEL

PUMP

Total Static Head (TSH) = A + B
Total Dynamic Head (TDH) = A + B + C + Pipe Friction

Because relatively few homes can tap into a hundred feet of head, though, partial gravity systems are more common. Let's relocate our hypothetical spring further down the hill, so that it sits only 10 vertical feet about the house. That will reduce the pressure of the incoming water to about 4 psi - enough to provide usable pressure at the tap. The most economical method of stepping up the pressure, in most cases, is to install some sort of booster pump.

In a booster pump system, the low-pressure, gravity-supplied water is piped into the inlet of a small pump, which forces it out under higher pressure. This high-pressure water may be piped directly from the pump to the faucets and other outlets, but ordinarily it first enters a pressure tank.

The pressure tank enables the pump to remain off for much of the time

As the high-pressure water is forced into an inlet fitting in the bottom of a steel tank - which may have a capacity of anywhere from 2 or 3 gallons to 80 gallons or more - the air trapped in the open space above is compressed into a smaller and smaller volume. When the pressure has risen to a pre-set level - typically 40 psi or so - an inexpensive pressure switch cuts off the power to the pump. A one-way check valve - which may be built into the pump, or may be installed in the pressure pipe between the pump and the tank - prevents the stored water from flowing backward through the pump.

When you turn on a faucet to fill the tea kettle, the pressurized water emerges from the outlet - also located somewhere in the bottom of the tank - and, pushed by the expanding compressed air, travels through the household plumbing, and out the open tap. The water will continue to flow, without any help from the pump, until the tank pressure falls to 20 psi or so, at which point the pressure switch will cause the pump to kick on again, recharging the tank.

The pressure tank, in other words, enables the pump to remain off for much of the time, and, when it *does* cycle on, to run for at least as long as it takes to recharge the pressure tank. In the absence of a pressure tank, the pump would have to switch on every time you ran a glass of water, increasing wear and tear on the pump and pressure switch and wasting energy in the process, since electric pumps require a substantial starting surge each time they cycle on.

Several types of low-voltage pumps will work well in booster pump-systems. The top-quality option is the Flowlight Booster Pump, available in either 12 or 24 volts. The Flowlight pumps are beautifully made, and will last forever if not abused. They are very quiet - an important consideration in a pump that is to be located indoors, as booster pumps usually are. Their only disadvantages are their relatively high cost, and their susceptibility to damage from particles of sand, grit, or sediment. The incoming water must be drawn through a cartridge-type

sediment filter before it enters the pump, and the filters replaced several times a year. **[Windy Dankoff, of Flowlight responds:** "*It is not correct to generalize that a Flowlight Booster Pump or Solar Slowpump requires several filter changes per year. If there is no sediment, it will never require changing. I haven't changed my booster pump cartridge in over a year. I would emphasize that filter maintenance is absolutely dependent on water quality! When the pump starts to make noise, it is time to change the cartridge, so it is not necessary to pull pump from a casing to check filter, either -- just listen!*"]

A cheaper option is a marine or RV-type pump, such as the Shurflo or Flojet. Both are diaphragm pumps, and far more tolerant of particulates than the vane-type Flowlight pumps. A filter is usually not necessary. Where large chunks of foreign matter may be present - if water is to be drawn directly from a stream, for example - a 40-mesh intake strainer is recommended.

The diaphragm pumps typically put out somewhat less than 2 gallons per minute - far less than the Flowlight Booster. When the pressure tank is fully charged, that makes little difference, but when the tank is depleted and the pump is supplying water directly to the house - as when someone is taking a long shower - turning on the kitchen faucet will cause the flow in the bathroom to fall to a dribble.

A second drawback to the diaphragm pumps is the

Children using hand pump in Burkina Faso. Photo by Sean Sprague.

loud, irritating buzz they emit when operating. If located indoors, they should be placed in a location that can be effectively soundproofed.

In the pressure-tank booster system just described, water is delivered to the booster pump by gravity. But what if - as is more than likely - your water source is located below the house, rather than above it?

If it is not too far below the house, no important changes are necessary. Both the vane-type Flowlight and the lighter-duty RV pumps have the ability to pull water uphill, as you would sip iced tea through a straw, and force it into a pressure tank at the same time. That makes it possible to draw water directly from a shallow well, or a buried rainwater cistern.

The catch is something called the "suction limit," which is the maximum height to which a given pump will pull water. The suction limit of the Flowlight Booster, for example, is about 17 feet. The diaphragm Flojet and Shurflo pumps will pull water to about half that height. (Those are the figures at sea level; for arcane reasons I won't go into here, the suction limit of any pump decreases by approximately one foot with each thousand-foot increase in elevation.)

In theory, the horizontal distance between the water source and the house is far less important, since water is much easier to move sideways than it is to lift (assuming, again, that the piping is large enough to keep head losses reasonably low). In practice, however, a pump cannot be expected to move water more than, say, a hundred feet or so by suction, even if the lift involved is well below its suction limit.

The problem is that any air bubbles in the intake line will stop a centrifigul pump dead in its tracks. Unless the suction line to the pump can be sloped uniformly uphill, enabling trapped air to rise out of the line of its own accord, the bubbles will congregate at any localized high points and cut off the flow. Where water need not be lifted far, but must be moved a great distance horizontally, it's best to locate the pump at the water supply, and have it force the water by pressure, rather than pulling it by suction, since air bubbles in a pressure line do no harm.

That, however, brings up a different problem, which is the difficulty of supplying low-voltage electricity to a remote pump. My own water, for example, comes from a shallow well only a few feet lower than the house, but nearly 800 feet west of it. Transmitting 12- or 24-volt power to a pump at the well would have required thousands of dollars worth of heavy-gauge wire.

The solution was to choose a 110-volt AC pump run through an inverter. Standard 110-volt centrifugal pumps, however, are inefficient (that is, they move relatively little water for the amount of power they consume), and can only be run with a large, expensive inverter, such as the Trace 2012. Instead, I chose an inexpensive 110-volt Flojet - identical to the usual low-voltage Flojets, except for the motor - that is easily powered by my small 300-watt Heart inverter. That made it possible to run ordinary, inexpensive #10 direct-burial wire from the house to the pump, in the same three-foot deep trench that protects the plastic water line from the frost. The pump itself sits on a shelf within the well casing, a few feet below the surface of the ground - and hence safe from freezing - but well above the high-water level.

Water pump powered by wind and phototovoltaic modules in Mendocino County, California. Photo by Sean Sprague.

The higher-quality Flowlight pumps are also available in 110-volt. I chose the cheaper, less durable Flojet because I use relatively little water, and because I suspected that - given the remoteness of the pump - I might not get around to changing the inlet filter on a Flowlight as often as necessary.

Those are the basics of shallow-source water pumping. If you are lucky enough to live where good water lies within easy reach of the surface, I hope this information will enable you to divert some of it into your kitchen and bathroom.   -Jon Vara

How Many Watts Do I Need For Solar Pumping

Here is a handy formula for estimating watts required for a given pumping task. It allows you to predict power requirement even if you don't know what type of pump to use:

$$\text{Watts} = \frac{\text{Feet X GPM}}{.053 \text{ X \% Pump Efficiency}}$$

Pump efficiency (wire-to-water) is 30 - 50% for most solar pumps and generally highest for the higher volume varieties.

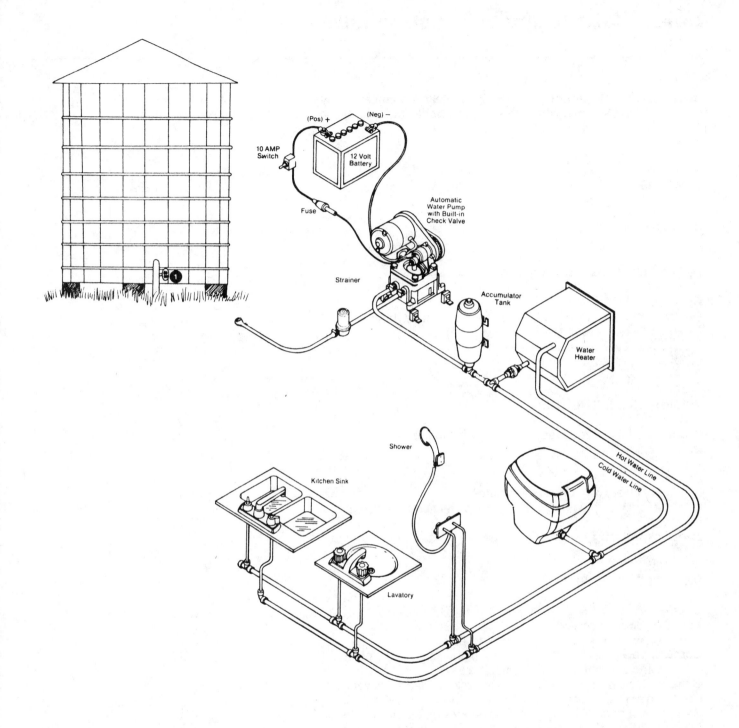

The Automatic Multi-fixture (Diaphragm) Water Pump System

The automatic multi-fixture pump system (shown above) delivers water the instant a faucet is opened, just like home. Pump starts automatically when a faucet is opened or a toilet is flushed. When all water outlets are closed, pressure in the discharge side of the pump rises to shut the pump off automatically.

Pump draws water from a non-pressurized water tank. Standard household fixtures are used throughout. The heart of the system is the automatic water system pump:

Solar Water Supply Questionnaire

In order for us to thoroughly and accurately recommend a water supply system for you, we need to know the following information about your system to the best of your knowledge.

NAME:

ADDRESS:

PHONE:

DESCRIBE YOUR WATER SOURCE:

Depth of well:
Depth to water surface: If water level varies, how much?
Estimated yield of well (gallons per minute):
Size of well casing (inside diameter):
Problems? (silty water, corrosive, etc.):

WATER REQUIREMENTS:

Irrigation: gal/day required: which months of year:
Is gravity flow/low pressure OK?

Domestic: gal/day required:
Is this your year-round full time home?
Is house already plumbed?   Conventional? - describe.

Livestock watering: gal/day required: which months of year?

DESCRIBE YOUR SITE:

Distance from well to point of use:
Elevation above sea level:
Vertical rise or drop from top of well to point of use:
Can you easily locate a storage tank higher than point of use?
    How much higher?   How far away?
Complex terrain, or multiple usage?--Enclose map to describe site.
Do you have utility power at site?   How far away?
Can you connect well pump to nearby PV home/battery system?
Is home PV system present or proposed?   Describe(voltage, etc.):
Distance from home system to well:

DESCRIBE EXISTING EQUIPMENT - Energy system, pumping, distribution, storage, etc.

How effective is your present system?
Do you have a specific budget in mind?
Do you have a deadline for completion?

Solar Slowpump

Since 1983, the Solar Slowpump has set the standard for efficiency and reliability in solar water pumping where water demand is in the range of 50-3,000 gallons per day. Solar pumps are more reliable than windmills, and eliminate the fuel and maintenance requirements of engine-powered systems. The Slowpump is a positive displacement pump that efficiently operates all through the solar day at the slow, varying speeds that result from variable light conditions. It is designed to draw water from wells, springs, cisterns, tanks, ponds, rivers, or streams and to push it as high as 450 vertical feet (about 200 psi), for storage, pressurization, or irrigation. Suction lift capacity is 20 feet at sea level with free discharge (subtract 1 foot for every 1,000 ft. of elevation.)

The Solar Slowpump can be powered PV direct or from batteries. They are available in eight 1/4 hp models and seven 1/2 hp models. The 1/4 hp models come in 12 or 24VDC and the 1/2 hp models come in 24 or 48VDC. They are also available in 115 VAC upon special request.

Slowpumps are precision built for years of dependable service. They outlast dozens of cheaper plastic pumps.

The precision Slowpump cannot tolerate sand or abrasive silt and therefore requires filtration (a range of filtration accessories follow) 1 yr. warranty.

Slowpumps have now endured 9 years of service both in irrigation and home use. Slowpumps are in use worldwide, from the tropics of Micronesia to the sub-arctic Yukon. Their positive displacement rotary vane pump has working parts of hard carbon-graphite and stainless steel - no plastic! The pump body is solid forged brass, so tough it survives most hard freezes. The pump

starts easily even in low sunlight conditions and produces a smooth, non-pulsating flow making it quiet and easy on your piping. They are self-priming and require no lubrication, and are rebuildable should damage or wear occur. Brushes last 5-10 years and are easy to inspect and replace.

Slowpumps must not be used without filtration. Their high precision parts are damaged by sand, rust, or abrasive silt. If your water is consistently clear, a very fine intake strainer will provide sufficient protection. Otherwise, use either an inline filter or our high-capacity intake filter. If in doubt, use a filter - the warranty does not cover damage from dirty water. Our filters use replaceable cartridges, which may last for months or years, depending on water quality. Commonly available drinking water cartridges may be used.

Power Sources & Storage for Slowpump

Slowpumps are powered by DC electrical current from solar panels or other power sources. They use a permanent magnet motor, which provides high efficiency even with varying voltages (typically encountered with direct solar power). They can also be powered by storage batteries including backup power from a vehicle battery. They are available in 110V DC or AC as well.

Slowpumps can run directly off solar panels with no storage batteries. Typically storage of water in a tank is more efficient and economical than battery storage. A float switch can be used to turn off the pump when the storage tank fills. We require employing our linear current boosters for solar direct applications, as it optimizes low light performance and allows for early morning start-up.

Solar Water Pumping Booklet

This booklet is a good place to start if you know nothing about solar water pumping. It's an 8-page booklet that introduces the concepts and the technology of solar pumping to potential users. It is written for general readership, and assumes little technical background.

80-613   Solar Water Pumping Booklet $1

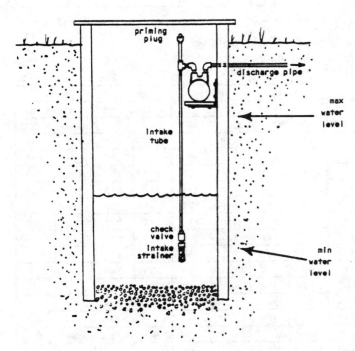

Typical Slowpump Installation For a Dug Well

SOLAR SLOWPUMP™ SPECIFICATIONS

VERTICAL LIFT = Vertical distance (FT) from water source to top of storage tank or pipe outlet

FLOW RATE = Gallons Per Hour (GPH). To estimate GALLONS PER DAY, multiply GPH by "Peak Sun Hours" per day for your locale (consult your dealer or a solar manual)

-- As voltage varies, flow rate will vary proportionally. Amperage stays nearly constant --

-- Use higher voltage models (24 V / 48 V) to reduce wire size and controller costs if possible --

WATTS = VOLTS X AMPS   FEET of lift = PSI X 2.31

1/4 HP SLOWPUMPS
– Available In 12 V or 24 V Models –
Max. Amp draw 16 A (12 V models), 8 A (24 V models)
Max. Working Voltage 18 V (12 V models), 36 V (24 V models)

VERTICAL	Model 1321-15		Model 1322-15		Model 1305-15		Model 1308-15		Model 1304-15		Model 2503-15		Model 2505-15		Model 2507-15	
LIFT (FT)	GPH	WATTS	GPH	WATTS	GPH	WATTS	GPH	WATTS	GPH	WATTS	GPH	WATTS	GPH	WATTS	GPH	WATTS
20							85	35	114	42	163	55	210	62	258	74
40					44	39	83	45	111	57	159	74	205	84	252	99
60			32	43	43	49	82	58	109	73	165	98	198	114	244	133
80	20	56	31	50	43	58	81	70	107	88	151	118	192	140	234	163
100	19	41	30	59	42	67	80	81	105	104	147	140	187	168	225	199
120	19	52	30	67	42	81	79	95	103	119	143	165	181	198	216	228
140	18	58	29	73	41	93	78	108	101	135	139	189	176	230		
160	18	62	29	81	40	105	77	122	99	151	135	216				
180	17	67	28	90	40	122	76	137	97	167	FLOW DETERMINED					
200	17	74	28	97	39	129	75	151	95	182	at 14 V for 12 V Model					
240	16	87	27	115	38	159	73	180	91	213	at 28 V for 24 V Model					
280	15	97	26	135	37	190	71	213								
320	14	115	26	160	36	227										
360	13	135	34	185												
400	12	161			-- ALLOW FOR POSSIBLE 10% VARIATIONS FROM SPECIFIED PERFORMANCE --											
440	10	199														

1/2 HP SLOWPUMPS
– Available In 24 V or 48 V Models –
Max. Amp draw 16 A (24 V models), 8 A (48 V models)
Max. Working Voltage 36 V (24 V models), 72 V (48 V models)

VERTICAL	Model 1321-30		Model 1322-30		Model 1305-30		Model 1308-30		Model 1304-30		Model 2503-30		Model 2505-30		Model 2507-30	
LIFT (FT)	GPH	WATTS	GPH	WATTS	GPH	WATTS	GPH	WATTS	GPH	WATTS	GPH	WATTS	GPH	WATTS	GPH	WATTS
100							88	101	120	136	163	163	201	193		
120							87	110	118	153	160	179	198	209		
140							86	116	116	168	157	197	195	228		
160					30	100	60	116	85	124	114	184	154	215	192	246
180					30	106	60	123	85	132	112	198	151	232	189	263
200	13	92	21	103	29	113	59	142	84	142	110	215	149	251	186	283
240	12	105	21	118	28	127	59	147	84	160	107	250	143	291	181	324
280	12	118	20	132	28	142	58	162	83	179	103	286	138	335	176	375
320	11	133	19	147	27	158	57	179	82	199	100	328	133	395		
360	10	148	18	162	26	174	56	197	82	218	97	386				
400	10	162	18	178	26	190	56	214	81	241	FLOW DETERMINED					
440	9	183	17	199	25	212	55	239	80	269	at 28 V for 24 V Models					
											at 56 V for 48 V Models					
	-- ALLOW FOR POSSIBLE 10% VARIATIONS FROM SPECIFIED PERFORMANCE --															

Sizing Your Slowpump

To determine which Slowpump model meets your needs, examine the specifications charts and find the **vertical lift** (feet) you need to raise the water from your water source. Move across this row and find the **flow** (gph) that you require. To the right of this flow rate is the **power** in watts needed for this lift/flow combination. The top of this column indicates the **pump model** to be used.

● **Flow Rate:** Water requirements are best estimated in gallons per day. When you have determined your daily requirement, divide this by the number of **"peak sun hours"** per day for your location to find the flow rate in gallons per hour. (*As an example, for the Central-Western USA, April through September, assume an average of 7 peak sun hours per day. If using a solar tracker, assume 10 peak sun hours.*) **Always size your system generously if you have no back-up source of water or energy.**

● **Solar Panel Sizing:** For PV-direct pumping applications the total wattage of the solar array should exceed the watts specified in the chart by at least 15% for better overall performance. A Linear Current Booster will prevent your pump from stalling in low light conditions.

● **Pump Voltage:** To determine what voltage to use (12V, 24V, or 48V) consider the available voltage options of your power source. The 1/4 hp Slowpumps are available in 12V or 24V. *A 24V system requires one-fourth the wire size and half the number of linear current boosters as a 12V system. It, however, requires an even number of solar panels, and if you jump to a 12V vehicle battery for back-up power, the pump will run at only half the rated flow.* The larger 1/2 hp Slowpumps are available in 24V and 48V. 90-120V DC and AC Slowpumps are available by special order to minimize wire size for very long wire runs.

● **Deep Well Sizes:** A minimum 6" inside diameter casing is required for the 2500 model pumps (see specs). The 1300 model pumps can be adapted to fit in a 5" casing by special order.

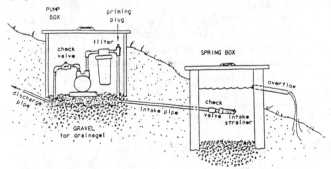

Typical Installation of Slowpump Pumping From a Spring

■41-101   **1/4 hp Slowpump (12V) $525**
■41-102   **1/4 hp Slowpump (24V) $525**
■41-111   **1/2 hp Slowpump (24V) $745**
■41-112   **1/2 hp Slowpump (48V) $745**

Please specify the complete model number from the Solar Slowpump chart. Send SASE for detailed spec sheet and illustrated sample installations.

To clarify what a total Slowpump system will cost you, let us give you an example of a typical system:

Example:
   Surface spring or shallow well; 200 feet vertical lift, 120 to 225 gallons per day.

   This system uses a non-submersible "surface pump" designed for high lift at low volume. It may be used to supply a mountain home from surface water in a valley or canyon below. Photovoltaic modules are located near the pump at the water source or within a few hundred feet to gain full sun exposure. Because of the low flow rate, a pipe run of over 1,000 feet requires only 1/2" pipe. A storage tank is required at the house, with either gravity-flow delivery or a separate pressurizing pump.

The Typical System:
   Flowlight Solar Slowpump - 24V,
   Water filter, foot valve, check valve
   2 each 48 watt photovoltaic modules
   Stationary mounting rack (with seasonal angle adjustment)
   Linear current booster (pump controller) - 3 amp model.
   Tank capacity required: 1,000 gallons minimum

   SYSTEM PRICE ESTIMATE: $1,300

Accessories for Booster Pump and Slowpump

(Refer to the Booster Pump section ahead for information on the Flowlight Booster pump.)

Adaptors: (included with the pump at **no charge**)
1300 Model pumps have a 1/2" threaded female adaptor.
2500 Model pumps have a 3/4" threaded male adaptor.

If your pump is to be placed in a well casing or other close quarters, request **adapting elbows** instead. They fit flexible 1/2" polyethylene pipe or 5/8" hose.

Dry Run Switch:
Slowpumps can be damaged from overheating of the pump head due to running dry. There is now a thermal shut-off switch available that can be clamped onto any Slowpump, old or new. This provides better, simpler protection than float switch or low-pressure cut-off systems. The switch shuts off the motor when the pump begins to warm up, and turns it back on again after about a 20 minute cooling-off period, OR specify "manual reset" (for water sources that do not reliably recharge, like tanks or cisterns.) Specify model number.

■41-135   **Dry Run Switch (Manual Reset)** $40
■41-134   **Dry Run Switch (Auto Reset)** $40
You must specify the pump model number!

Fine Intake Strainer/Foot Valve:

A fine screen strainer is the minimum protection required for water that is free of suspended silt and rust. Water filtration is extremely important with the Slowpump as one grain of sand can damage the pump's precision surfaces. The new strainer has been upgraded and is made of all metal with an extra-fine 100 mesh metal screen.

■41-136   Fine Intake Strainer/Foot Valve $75

30" Intake Filter/Foot Valve:

Replaces both the intake strainer and inline filter with a single high capacity submersed cartridge unit. Good for silty streams, drilled well casings, and other problem applications. A spare 30", 5 micron cartridge is included. It also accepts 10" cartridges. Comes with 3/4" female fitting and 3/4" to 1/2" reducer.

■41-427   30" Intake Filter/Footvalve $85

30" Replacement Cartridges
These replacement cartridges come three to a pack.

■41-428   30" Replacement Cartridges (3-pack) $45

Sumps and Filters

Inline Sediment Filter:

Sometimes your water isn't dirty enough to mess with fancy and expensive filtration systems and all you need is a simple filter. Our inline sediment filters accept standard 10" filters with one inch center holes. They are designed for cold water lines only and meet National Sanitation Foundation (NSF) standards. Easily installed on any new or existing cold water line (don't forget the shutoff valve), they feature a sump head of fatigue resistant Celcon plastic. This head is equipped with a manually operated pressure release button to relieve internal pressure and simplify cartridge replacement. They're rated for 125 psi maximum and 100 degrees F. They come with a 3/4 inch FNPT inlet and outlet and measure 14" high x 4-9/16" in diameter. It accepts a 10" cartridge and comes with a 5 micron high density fiber cartridge.

41-137   Inline Sediment Filter $45

Replacement Filter Cartridges

Our rust and dirt cartridge is made of white cellulose fibers with a graduated density. These filters collect particles as small as 5 microns (two ten-thousands of an inch). These are NSF listed components that take a maximum flow of 6 gpm. Our taste and odor filters are made with granular activated carbon. These filters effectively remove chlorine, sulfur, and iron taste and odor. These are NSF listed components. Maximum flow is 3 gpm. Note: filters should be replaced every six months to prevent bacterial growth or as needed. This is the cartridge to use with the inline sediment filter.

41-138   Rust & Dirt Cartridge (2) $14
41-436   Taste & Odor Filter (2) $34

Pressure Switch:
You must use a pressure switch for pressure tank systems. It will turn the pump off when the tank pressure reaches 40 psi, and then on again when the pressure drops to 20 psi. It also turns the pump off when pressure drops below 10 psi to prevent damage from running the pump dry. The settings are fully adjustable in the range of 5 - 65 psi. (This pressure switch is **included** in the Easy Installation Kit for the booster pump.)

41-140   Pressure Switch (6X535) $18

SOLAR/AUTO PREHEATED SAUNA
Drive the car into the garage, poking the engine compartment into the Sauna Preheating Cabinet. Take a shower and jump into the sauna, which has been preheated at no cost!

Temperature-responsive automatic insulated shade

© 1991 STEVEN M. JOHNSON

Engine heat radiates through metal panel

Insulating panel withdraws as engine heat is sensed

Sauna heater

Thermal mass

Solar Force Piston Pump

The Solar Force is a high flow, medium to high lift positive displacement pump, designed for water lift or pressurizing. It utilizes solar-electric power to draw surface water from a shallow well, spring, pond, river or tank. It can push water uphill and over long distances for home, village, livestock, irrigation and fire protection. It is capable of producing from 4 to 9 gpm and it can lift up to 230 ft. (or pressurize up to 100 psi). Its suction capacity is 25 vertical feet. (Subtract one foot per 1000 feet of elevation above sea level.) *It is by far the most efficient pump available in its performance range*, coming in 6 models to meet your specific pumping needs.

It works efficiently at all speeds, even in low light conditions. This very durable cast iron & polished brass pump is designed to last decades. Features include a non-slip gear belt drive that allows *mechanical back-up or hand power*, a pressure relief valve, and a circuit breaker/switch enclosed in a weatherproof box. This is an extremely low maintenance pump designed for a 2-6 year maintenance interval. No special tools or skills are required. A crankcase viewing window allows easy oil inspection. A repair kit is included with the pump which contains all the seals and gaskets, valve parts, gearbelt, motor burshes, brass cylinder, oil additive, and assembly diagram. This is enough supplies to maintain the pump for 10-20 years of typical use. The pump is available in DC voltages of 12, 24, or 48 volts or in 115 VAC. 2 yr. warranty.

Accessories for Solar Force Pump

Sediment Trap
Protects the pump from fine sand and debris. It is washable and comes with 1" male fittings. This is a one inch Arkal drip irrigation filter.

41-192	Sediment Trap for Solar Force Pump	$65

Pressure Switch for Solar Force
These are DC-rated, adjustable pressure switches. Use the 1/4 HP switch for the A-Series and the 1/2 HP switch for the B-series.

41-140	Pressure Switch - 1/4 HP	$18
41-194	Pressure Switch - 1/2 HP	$65

The dimensions of all Solar Force pumps are 22" x 13" x 16". The A1 and A# weigh 70 lbs. and the B2 and B# weigh 90 lbs. On all pumps the inlet is 1-1/4" and the outlet is 1" pipe thread.

■41-181	Solar Force - A1 - 12V	$1,350
■41-182	Solar Force - A1 - 24V	$1,350
■41-183	Solar Force - A3 - 12V	$1,350
■41-184	Solar Force - A3 - 24V	$1,350
■41-185	Solar Force - B1 - 12V	$1,485
■41-186	Solar Force - B1 - 24V	$1,485
■41-187	Solar Force - B3 - 12V	$1,485
■41-188	Solar Force - B3 - 24V	$1,485
■41-189	Solar Force - B1 - 48V	$1,510
■41-190	Solar Force - B3 - 115VAC	$1,510*

All Solar Force pumps are shipped UPS freight collect from New Mexico. The motor is packed separately.

*The 115 VAC model uses a DC motor with rectifier. It uses AC power most efficiently, with 1/5 the starting surge of the common induction motor. This reduces energy use, wire cost and load on power inverters.

Here's an example of a typical system we sell for the Solar Force Piston Pump, to give you a better idea of total costs:

Example: Sprinkler Irrigation, Domestic Pressurizing, and Fire Protection: 9 gallons per minute at 30 - 50 psi for seven hours per day (3,500 gallons per day, seven days per week). This system draws water from a storage tank or shallow water source. On-demand pressurizing necessitates a storage battery power system, which may be enlarged to power lights, tools and appliances. The DC piston pump uses *far* less energy than conventional AC jet pumps, thus requiring a smaller, less expensive power system.

The Typical System:
Solar Force Piston Pump (24V or 48V)
8 each 48 watt photovoltaic modules
Solar tracker / mounting rack
8 each storage batteries
Battery charge controller
System wiring
Pressure tank, etc.
SYSTEM PRICE ESTIMATE: $6,800

Here is a chart for Solar Force pumps with performance measured at 14V (12V nominal), 28V (24V nominal) or 56V (48V nominal). Remember for calculating the number of solar modules necessary: Watts = Amps X Volts. For pressurizing: Total Head = Vertical Feet + (PSI X 2.31)

Head in Feet	PSI	Model A1 - GPM	Model A1 - Watts	Model A3 - GPM	Model A3 - Watts	Model B1 - GPM	Model B1 - Watts	Model B3 - GPM	Model B3 - Watts
0-20	9	4.3	56	8.0	116	5.2	110	9.3	168
40	17	4.1	76	7.2	146	5.2	132	9.3	207
60	26	3.9	90	6.4	184	5.1	154	9.2	252
80	35	3.8	111	6.2	218	5.1	182	9.2	286
100	43	3.7	125			5.0	202	9.1	322
120	52	3.6	146			5.0	224	9.1	364
140	61	3.6	165			5.0	252	9.1	403
160	70	3.5	190			4.9	269		
180	78	3.5	190			4.9	280		
200	86					4.8	308		
220	95					4.7	314		

Solaram Surface Pump

The Solaram Surface Pump is the most efficient photovoltaic powered pump available for high head (in excess of 200 ft.) and medium to high flows (3 to 9 GPM). It is used to pump from a surface water source such as a shallow well, spring, pond, river, reservoir or holding tank. It is well suited to pressurizing as well as lifting. It may be used to lift water and pressurize a home water system at the same time. It may also be used for sprinkler and drip irrigation. A PV/battery system is needed for such on-demand pumping.. With a PV/battery system, the pump's flow output and wattage requirement will be 20% lower than specifications.

The Solaram is a high efficiency positive displacement diaphragm pump that is built like a tank for full time stock watering, home pressurizing, and irrigation use. It can pump as high as 960 vertical feet and produce flows as high as 9 GPM. It is a multiple diaphragm (2 in 200 series and 3 in 400 series) industrial pump driven by a permanent magnet DC motor. It uses a cogged gear belt for high efficiency and long term reliability with minimal attention. It comes with a Linear Current Booster (LCB), pre-wired in a weatherproof box. The LCB allows optimum pumping performance over a wide range of sunlight conditions. The entire unit is supported on a heavy galvanized steel base and is protected from the weather by a galvanized steel hinged cover. A brass strainer / foot valve is included. The pump is protected from excessive pressure damage by a pressure relief valve.

The motor is protected from overload by a DC circuit breaker and from overheating by an automatic reset, thermal cut-out switch.

The pump is filled with a non-toxic, vegetable base lubricating oil which should be replaced every 12 months of full time use (6-9 peak sun hours/day). The easy to replace pump diaphragms should also be replaced every 12 months of full time use. A replacement diaphragm kit is included with the pump.

The pump may be turned on and off for the filling of a storage tank with Water Level Sensor 1 and turned off with the depletion of a water source with Water Level Sensor 2. The two sensors may be used in conjunction with each other. For a pressurizing system, use a pressure switch. A special pressure switch system can be used to switch the pump on and off for extreme long distance pumping.

The Solaram Surface Pump is designed for decades of use. It comes in four basic models, powered by a 3/4, 1, 1-1/2, 2, or 3 hp motor depending on your needs. One example: Model 403, at 1 hp will deliver 8.2 gpm at 360 vertical feet, using 927 watts at 24V. The pump is highly

tolerant of dirty water and dry running and it comes with a one year warranty.

Installation of the Solaram is easy. Power is connected directly from the PV array to the terminals in the weatherproof LCB/breaker box. Water Level Sensors are easily connected with inexpensive telephone type wire. All pump models are 28" long x 16.5" high x 16" wide. The weight varies between 110 and 150 lbs. depending upon the model.

PERFORMANCE CHART

HEAD FEET	MODEL 201 GPM	WATTS	MODEL 202 GPM	WATTS	MODEL 203 GPM	WATTS	MODEL 401 GPM	WATTS	MODEL 402 GPM	WATTS	MODEL 403 GPM	WATTS	MOTOR HP / VOLTS
0-80	3.0	170	3.7	207	4.6	285	6.2	258	7.5	339	9.4	465	
120	2.9	197	3.7	238	4.5	319	6.0	305	7.3	396	9.1	539	3/4 HP/24 V
160	2.9	225	3.6	268	4.5	352	5.8	354	7.2	453	8.9	619	
200	2.9	247	3.6	296	4.5	388	5.7	400	7.1	513	8.9	693	
240	2.8	265	3.6	327	4.5	427	5.6	453	7.0	572	8.6	724	
280	2.8	286	3.6	356	4.4	466	5.5	499	6.9	628	8.4	801	1 HP/24 V
320	2.8	315	3.5	388	4.4	496	5.4	548	6.8	686	8.3	869	
360	2.8	342	3.5	416	4.4	536	5.4	592	6.6	733	8.2	927	
400	2.7	363	3.4	450	4.4	572	5.3	649	6.5	782	8.7	1122	
480	2.7	416	3.4	505	4.3	649	5.3	717	6.5	900	8.5	1265	1.5 HP/180 V
560	2.7	456	3.3	570	4.3	693	5.2	800	6.6	1045	8.4	1397	
640	2.7	502	3.3	623	4.2	774	5.1	893	6.5	1166	8.2	1540	
720	2.6	551	3.2	690	4.0	856	5.1	1031	6.4	1287	8.1	1683	2 HP/180 V
800	2.6	589	3.2	715	4.0	931	5.1	1114	6.4	1408	8.0	1815	
880	2.6	647	3.2	774	4.1	1082	5.1	1206	6.3	1529	8.0	1958	3 HP/180 V
960	2.5	705	3.1	838	4.1	1190	5.0	1289	6.1	1650	8.0	2145	

GPM and **Wattage** levels based on "Array Direct" voltages. With PV/Battery systems, they will be 20% lower.

PV Array Sizing

Our technicians will be glad to help you size your system. If you'd prefer to calculate it yourself follow these directions: From the pump's performance chart, identify the pump *wattage* needed for the desired *flow* (GPM) at the required *head*, where *head* is calculated by vertical lift + pipe friction head + pressure head (if pressurizing system) in feet (feet = PSI x 2.31). Also identify the appropriate pump motor *voltage*. Obtain specifications for the PV modules to be used. With this information, determine the total number of PV modules needed and the series/parallel module arrangement as follows:

Total number of PV modules = modules in series x modules in parallel

Where: Modules in series = Pump motor voltage

$$\overline{\qquad 12V \qquad}$$

Modules in Parallel = Pump wattage

$$\overline{\text{Modules in series x Module wattage}}$$

The Pressure System Kit includes all the parts you need to build a pressurizing system: Accessory Tee, Adjustable Pressure Switch, Pressure Gauge, Check Valve, Drain Valve, and Shut-Off Valve.

■41-286	Solaram #200 - 3/4 HP	$2,100
■41-287	Solaram #200 - 1 HP	$2,150
■41-288	Solaram #200 - 1.5 HP	$2,400
■41-289	Solaram #400 - 3/4 HP	$2,150
■41-290	Solaram #400 - 1 HP	$2,200
■41-291	Solaram #400 - 1.5 HP	$2,700
■41-292	Solaram #400 - 2 HP	$2,800
■41-293	Solaram #400 - 3 HP	$2,900
■41-294	Water Level Sensor 1	$ 30
■41-295	Water Level Sensor 2	$ 30
■41-296	Pulsation Damper	$ 90
■41-297	Pressure System Kit	$ 95
■41-298	Check Valve - Brass - 3/4"	$ 19
■41-299	Flow Gauge (1-10 GPM)	$ 85

All Solaram Pumps are shipped freight collect from New Mexico.

Flojet Diaphragm Pumps

Flojet 2100-12

The Flojet 2100-12 *(formerly called the 2000-12-30)* is by far our best selling 12-volt pump, and has proven itself durable and efficient against the test of time. It comes standard with a 30-psi pressure switch. It has automatic demand control which allows the pump to come on when flow is required (turn on a faucet) and shut off when the faucet is closed. It turns on at 20 psi and off at 30 psi, making it ideal for pressurizing a house system when insufficient gravity feed exists.

The pump is self-priming and may be located up to 11 feet above the water level. It has a powerful permanent magnet motor that operates at low speed and current draw. The pump may be pumped dry indefinitely with no damage if the water supply is lost. The motor has ball bearings, replaceable brushes, two 11" lead wires, and rubber mounting feet that reduce noise and vibration. The motor draws 3.5 amps at 0 psi and up to 7.0 amps at 30 psi. It comes with a 3/8" inlet and 3/8" outlet. Not recommended for more than a 15 minute duty cycle. One year warranty.

Performance

0 psi	10 psi	20 psi	30psi
156 gph	129 gph	120 gph	111 gph

41-201   Flojet 2100-12 $95

Flojet 2100-732

The 2100-732 *(formerly called the 2000-732)* is heavier duty than the 2100-12. It has the same motor as the 2000-12-30 but has a cooling fan, heavier duty ball bearings, a sealed motor, and a better seal on the shaft. It comes standard with a 60 psi pressure switch. If your use is very intermittent, stick with the less expensive 2100-12. If you have higher lift applications or you'll be using the pump more than 1 hour per day this pump is well worth the extra few dollars. It will pump up to a maximum head of 130' giving 1.9 gallons per minute drawing a maximum of 7.0 amps. It has a 3/8" inlet and 3/8" outlet. Duty cycle is maximum one hour. One year warranty.

41-202   Flojet 2100-732 $105

Flojet 2100-637

The 2100-637 *(formerly called the 2000-637)* is exactly the same pump as the 2100-732 except that it comes with a 35 psi pressure switch for those of you who don't need 60 psi. Duty cycle maximum one hour. One year warranty.

41-203   Flojet 2100-637 $105

Flojet 2130-132

The 2130-132 *(formerly called the 2030-132)* is designed for pumping as high as 210' vertically, or for situations where very high pressure is required, but with very little volume, and comes equipped with a 95 psi pressure switch. It will pump a maximum of 1.7 gallons per minute at a maximum draw of 10.0 amps at 12V. It comes with a cooling fan and a 1/4" inlet and outlet. The pressure switch cuts on at 70 psi and cuts off at 95 psi. Duty cycle is fifteen minutes maximum.

41-204   Flojet 2130-132 $115

Flojet Pressure Switches

All Flojet pressure switches are adjustable; a little known fact that makes these gems much more versatile! Open the top lid off the little square box and you'll see a small allen wrench socket within. If you turn it with an allen wrench clockwise one turn the pressure settings will increase by 5 lb (a 30 psi pressure switch is factory set to go off at 40 psi and on at 30 psi - by turning one turn clockwise it changes to 35 psi on and 45 psi off). Two turns will increase it by 10 psi, three turns by 15 psi, and four turns by 20 psi. By turning clockwise the procedure reverses and the pressure is decreased by 5 lb with each turn. Maximum of three turns either direction.

41-263   30 psi Pressure switch $15
41-264   40 psi Pressure switch $15
41-265   60 psi Pressure switch $15

Flojet 4100-143

The 4100-143 is Flojet's highest volume pump. Like the 4300-142 it is a quad pump with 1/2" inlet and outlet. It does not come with a pressure switch but does have a thermal overload cut-off to protect it. It will pump a full 4.8 gpm at open discharge and 2 gpm at 25 psi. If used in conjunction with a pressure system we suggest you use one of our pressure switches. Duty cycle is one hour maximum. One year warranty.

| 41-242 | Flojet 4100-143 | $109 |

Flojet 4300-142

The 4300-142 *(formerly called the 4000-142)* Quad Pump has a special four chamber design for higher capacity and smoother flow. It's a 12V pump that comes with a 45 psi pressure switch and a 1/2" inlet and outlet. It will pump a maximum of 3.4 gallons per minute, and pump to a maximum of 102' vertically. It draws a maximum of 9.0 amps at 12V. It comes with a cooling fan for heavy-duty use. The pressure switch cuts on at 30 psi and cuts off at 45 psi. The loss of flow to the pumping chamber will not damage the pump, as quad pumps can run dry indefinitely. This pump is excellent for home water pressurization and situations where a maximum of flow is required. Duty cycle maximum fifteen minutes. One year warranty.

| 41-241 | Flojet 4300-142 | $119 |

Centrifugal Jet Pumps

Bronze Centrifugal Pump - 12V

This bronze bodied and bronze impelled pump is designed for applications where you have a lot of water to move at a minimum lift. The pump is not self-priming and must be mounted in a dry location where the motor is protected from dampness. Intake and outlet ports have 3/4" internal pipe threads suitable for either brass or plastic pipe fittings. One year warranty.

Performance gpm at total feet of head

2 ft	6 ft	10 ft	12.5 ft
20 gpm	14.5 gpm	7.5 gpm	Shut-off
8.0A	6.7A	5.4A	4.5A

41-301 Bronze Centrifugal Pump - 12V $120

LVM-160 Centrifugal Pump

This 12-volt centrifugal pump is less than 5" long and a little over 2" in diameter. It uses 6.5 amps maximum to pump a maximum of 19 psi (44 ft of head) at a maximum flow rate of 4 gpm. It comes with 3.3' of cable. 3 month warranty.

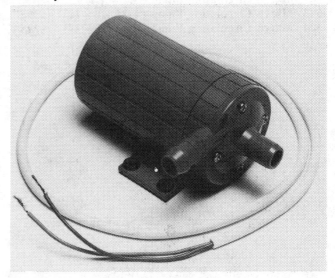

41-321 LVM-160 Centrifugal Pump $59

March 12V & 24V Circulating Pumps

March circulating pumps are used primarily with hot tub, spa, solar thermal, and woodstove hot water systems. They use very little power at either 12 or 24 volts and keep needed water circulation happening.

Models are available with capacities up to 22 gpm. All pumps have a magnetic drive that eliminates the old-fashioned shaft seal. They are easy to service only requiring a screwdriver, and the entire motor assembly can be replaced without draining the system. The 1/100 HP pump runs at 1,950 rpm and the 1/25 HP pump runs at 3,600 rpm. 6-month warranty. Will withstand temperatures to 200 degrees F.

 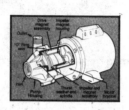

Specifications	1/100 HP	1/25 HP
Flow in gpm	5.5	7.5
Max head	7.1	15.5
Volts	12	12
Amps	1.5	3.8
Weight	7#	7#

41-501 March Circulating Pump - 1/100 HP - 12V $195
41-503 March Circulating Pump - 1/25 HP - 12V $295
41-504 March Circulating Pump - 1/25 HP - 24V $295

Hartell Hot Water Pumps

The Hartell magnetic drive circulator pump series features a DC motor designed to circulate hot water using PV power or low-voltage battery systems. They are ideal for solar water heating or other closed-loop, low-flow pumping applications. Each of the two models available will operate directly off of a solar panel - the brighter the sun, the faster the water is pumped! Both models have a 30,000 hour life expectancy.

Hartell Circulating Pump

This model is designed to be run directly off of a solar panel between 18 and 22 watts. It will pump to a maximum head of 10 feet with a maximum flow of 5 gpm. Will withstand temperatures to 200 degrees F. 6-month warranty.

41-522 Hartell Circulating Pump (CP-10B-12HE) $259

LVM-105 Submersible Pump

This 12-volt submersible pump measures only 1.5" in diameter and 6.5" long, and will pump a maximum of 4 gallons per minute. It will pump a maximum vertical head of 42 feet or 18 psi. At 25' of head the LVM-105 will deliver about 5/8 of a gallon per minute. A strainer is fitted over the input end to prevent large particles from entering the pump. The inlet and outlet sizes are 1/2", it comes with 13 feet of cable with battery clips on the end, and it weighs only 1.2 pounds. Designed for intermittent use only. 3 month warranty

41-671 LVM-105 Submersible Pump $79

Our Flojet pumps can function as booster pumps as well as lifting pumps. They are basically RV pumps, even the fan-cooled heavier duty models. For the serious fulltime booster pump user, the Flowlight Booster Pump can't be beat.

Flowlight Booster Pump

The Flowlight Booster Pump provides city water pressure for homes with 12 and 24-volt power systems. It represents a step up from "RV" pumps (like the Flojet, Shurflo, and Jabsco). The Flowlight will far outlast the RV pumps. It uses the same forged brass, carbon graphite pump head as the Solar Slowpump providing efficient, quiet operation. It will use half the energy of an inverter-driven AC jet pump with 1/5 the starting surge! *A Booster pump pressurizing system is far cheaper than an elevated tank.*

It is typically set up above ground with a conventional pressure tank and switch, but may be suspended above the water in a shallow well (min. 6" casing) and used as both the well and pressurizing pump (specify in well use when ordering.) Because it is non-submersible the well must have a stable water level.

The Flowlight Booster comes in the 5.5 gpm, 50 psi standard model or in the 3 gpm, 65 psi low flow model. If suction lift is greater than 10 vertical feet or frequent filter clogging is expected the low flow model is recommended. Maximum suction lift is 20 feet at sea level (subtract 1 ft. for every 1,000 ft. of elevation). Each model comes in 12 or 24 VDC. Other voltages and AC motors are available upon special request. Filtration is required! One year warranty.

Item #	Description	Price
41-141	Booster Pump Standard (12V)	$545
41-142	Booster Pump Standard (24V)	$545
41-145	Booster Pump Low Flow (12V)	$545
41-146	Booster Pump Low Flow (24V)	$545

Easy Installation Kit:
All the small parts that you need to quickly install your booster pump system: accessory tee, pressure switch and gauge, check valve, drain valve, shut-off valve, pipe nipples, and two 18" flexible pipes with unions. All brass fittings. Often it's very hard to find these parts in out-of-the-way hamlets!

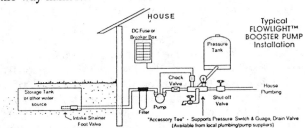

41-143 Easy Installation Kit $105

Refer to Additional Options in SlowPump Section Above - Particularly the inline filter and Dry Run Switch!

Pressure Tanks

Unless you are lucky enough to have a gravity-fed water system, you probably need a pressure tank. Most people are dissatisfied with water pressure less than 15 psi: showers seem to dribble and it seems to take forever to fill the bathtub. Both Paloma and Aquastar instantaneous water heaters require at least 15 psi to work properly. Remember that city water pressure is 40 psi. So, if your water storage is not at least 35 vertical feet above your house, you'll probably be happier with a pressurized water system.

We design many of our Flojet water pump systems with the 2-gallon pressure tank. It prolongs the life of the pump by easing the demand on the pressure switch. It allows the home with minimal pressure (less than 35 feet of head) to use an efficient energy saving demand water heater (see Tankless Demand Water Heater section) for very little expense.

Teel precharged pressure tanks have a permanent, factory pressurized air charge which is totally isolated from the water and can be absorbed as with standard style tanks. The interior of the tanks are epoxy coated for corrosion resistance, and the exterior is a baked green enamel. The 2-gallon tank has a 3/4" MPT connection. The 2-gallon tank is shipped precharged at 20 psi and is easily adjustable. Maximum working pressure is 100 psi, maximum temperature is 120°F.

Item #	Gallon capacity	Style	Maximum drawdown	Diameter	Price
41-401	2	Vertical	0.8 gal	8-3/8"	$55.00
41-402	6	Horizontal	3 gal	10"	$135.00
41-403	12	Horizontal	6 gal	12"	$189.00
41-404	20	Vertical	9.5 gal	16"	$195.00
41-405	36	Vertical	18 gal	20"	$275.00

Deep Well Pumping

Here is another of Jon Vara's installments - this time on deep well water pumping.

In the first installment of this article, I discussed pumping water from driven or dug wells, springs, cisterns, and lakes or streams. Such shallow water sources are vulnerable to contamination from surface runoff, and although it's possible to minimize the danger by careful site selection and proper development, the purity of shallow-source water cannot always be guaranteed. Moreover, in many areas there is simply no shallow water to be had, either because the local water table is uniformly deep, or because shallow sources that do exist are known to be contaminated.

In most cases, the only practical option is a drilled well. An experienced well driller will usually have a good idea of the depth to water and the probable yield, but there are no guarantees; and at a cost of $15 a foot or so, drilling a well can be an expensive, nerve-racking crap shoot.

If you are a lucky crap shooter, however, you may end up with a deep well that can be pumped in exactly the same way as a shallow well. Even where the source of the water lies hundreds of feet below the surface, a quirk of geology known as artesian pressure may force the water upward of its own accord. (Rarely will this come as a complete surprise, however; if artesian pressure commonly - or at least occasionally - occurs in your area, your well driller will certainly know about it.)

In very, very rare instances, artesian pressure will force water from the ground in a steady stream - a fresh-water gusher. More often, though, it will simply rise in the casing and stop somewhere below the surface of the ground. If it happens to rise to within 10 or 15 feet of the surface, and the supply is large enough that withdrawing a few gallons per minute won't cause the level to recede, you can simply drop a suction line into the casing, connect it to one of the shallow-well type pumps mentioned earlier, and you're in business.

In most cases, though, the water level in a drilled well will lie far below the suction limit of any pump, making it necessary to move the pump down, into the well itself. One approach is to lower a small non-submersible pump - such as one of the Flowlight Slowpumps - into the well casing, until the pump itself is within 10 or 15 feet of the water level, and the intake pipe is submerged. Provided that the suction limit is not exceeded, a correctly chosen Slowpump will <u>push</u> water to heads of over 400 feet.

For that to work, however, two important conditions must be met. First, the pump must not draw water from the well more rapidly than the aquifer supplies it, or the level will drop as the pump runs, until the inlet begins to suck air rather than water. A Slowpump that is run dry will soon overheat and destroy itself, and although a dry run switch, to turn off an overheating pump is available, it's better to avoid the problem to begin with.

Secondly, the resting water level in the well casing - what well drillers call the static level - must remain constant, or nearly so, throughout the year. A water level that rises substantially in the spring of the year may flood the pump and short it out; receding water in late summer may leave the intake pipe high and dry.

Finally, bear in mind that when a Slowpump is lowered into a drilled well, out of sight doesn't mean out of mind. It will be necessary to pull the pump from time to time in order to change the foot valve/filter assembly, which protects the pump's inner workings from dirt and grit. That's not necessarily a difficult job, but it's one that must be attended to faithfully. If you're the sort of person who never changes the oil in your car, that may not be easy for you to remember.

Submersible pumps, by contrast, can be lowered directly into the water, which simplifies the problem of a fluctuating static level. One of the simplest ways to pump

At a cost of $15 a foot drilling a well can be an expensive, nerve-racking crap shoot.

water from a deep well is to install a conventional 110-volt AC submersible - the standard deep-well residential pump - and power it through a large inverter. That requires only readily available, well-proven hardware of the sort that all plumbers, well drillers, and other service people are used to dealing with - no small advantage if you are not mechanically inclined yourself. If you already own an inverter with the capacity to start a 1/3 horsepower motor, it's a reasonable inexpensive way to go.

Unfortunately, it's also an energy-inefficient way to go. That's because conventional submersible well pumps, like so many electrically powered devices, are simply not engineered with the electric bill in mind.

Conventional deep well pumps operate on the centrifugal principle. In effect, the motor spins the water in the pump so rapidly that it is pushed against the inside of the housing - as coffee swirls against the sides of the cup when stirred with a spoon - and out and up the discharge pipe.

All truly efficient low powered pumps, by contrast, rely on the general principle of positive displacement. In a positive displacement pump, a unit of water is drawn into a closed chamber and forced out by a rotating vane, oscillating diaphragm, or sliding piston. The Flowlight Slowpumps and Booster Pumps, for example, are vane pumps; the inexpensive RV-type Flojet and the Econosub are diaphragm pumps.

Piston pumps designed for deep well use are generally of the jack pump configuration, in which the pump motor sits on the surface atop the well casing, while the piston and cylinder may be located hundreds of feet below, beneath the static water level. A long "sucker rod" - assembled in sections on the site, as you would screw together a chimney-cleaning rod - transmits the power from motor to piston. (Those on a tight energy budget

Oxen hauling water from deep well in India. Photo by Sean Sprague.

may wish to consider a deep-well hand pump, which - except for being powered by muscles instead of a motor - is similar to a jack pump.)

Jack pumps are simple, rugged, and capable of delivering large volumes of water. They are often used for large-scale irrigation and for similar demanding uses, but because they cost a minimum of several thousand dollars - not including batteries, PV panels, or the cost of the well itself - they are rarely used residentially, unless high water demand and great depth to water make less costly alternatives impractical. **[comment from Windy Dankoff from Flowlight:** *"Jack pumps are rarely used to deliver "large" volumes (certainly not for large-scale irrigation!) Rather, they are used for high lifts (200 to over 1000 feet.) at low to medium volumes."*]

One such alternative that may make sense in your application is a submersible diaphragm pump. Several models are manufactured by Solarjack, which also manufactures a line of high-quality jack pumping equipment. Submersible diaphragm pumps are comparable to the vane-type Flowlight pumps in terms of performance and overall efficiency, although they are substantially higher in cost. Being true submersible pumps, however, they are simpler to install - there is no

danger of the water level rising and flooding the pump - and their diaphragm design enables them to pass particles of grit without damage, eliminating the need for troublesome filtration.

So far, so good. There is, however, one vital aspect of deep-well pumping that I haven't addressed yet, and it's a far-reaching one. In discussing shallow-well pumping, we considered suction limits; now it's time to confront pressure limits, or the maximum head to which any given pump will force water.

Ultimately, suction is regulated by atmospheric pressure and the laws of physics. No matter how powerful a motor is harnessed to a pump, it's simply impossible to suck water more than about 35 feet uphill, and as we've seen, the practical limit is substantially less than that. Pressure, on the other hand, is an engineering problem. With the right pump, a sufficiently powerful motor, and strong enough piping, water can be pressurized to a head of thousands of feet. Theoretically, the sky is the limit.

The practical limit, however, has to do with the cost of a large motor, its physical size, and the amount of power it will require to run. Most efficient, low-voltage pumps will not force water to a total head of greater than three or four hundred feet, and often less.

Furthermore, just as a car with a small engine will have to slow down when climbing a steep hill, a pump that is forcing water to its maximum head will slow down, too. A pump that delivers several gallons per minute at a total head of 20 feet or so may deliver only a fraction of a gallon per minute at a head of 200 feet, and consume twice as much power to do it.

Still, a pump that delivers only 1/4 gallon per minute will deliver 180 gallons in the course of 12 hours, an amount that should easily meet the water need of an ordinary household. The only difficulty is that few households *use* water at the rate of a fourth of a gallon per minute.

Imagine the water use in your own home. For hours on end - during the nighttime hours, for example - no water is used at all. Then, in a great surge of morning showering, afternoon clothes washing, or evening bathing, 40 or 50 gallons may be demanded in the space of half an hour or so - far more than the deep-well pump can deliver.

The solution here is often a two-stage system of some sort, in which a deep-well pump brings up water at a slow, steady rate, and deposits it in a non-pressurized tank buried in the ground or safely stored in the cellar. A second, independent pump is used to pressurize the home water system from that convenient reservoir. That second stage, of course, is identical to the typical shallow-source system described in the first part of this article.

In areas with reliable sun, it is often possible to run the deep well pump directly from an independent array of PV panels. That will reduce the need for long wiring runs and eliminate battery losses. However, it will require a holding tank large enough to carry the household through cloudy spells, when little or no water is pumped from the deep well. Where prolonged cloudy spells are possible, however, it may be impractical to arrange that much storage capacity. In that case, the deep well pump - like the secondary pressuring pump - should be powered directly by the household battery bank.

- **Jon Vara**

There are a number of solar deep-well pumps on the market and we can't possible list them all here. We're providing several to choose from that are easy to figure out and order by mail. If your well is 300' deep or less and you only need around 500 gallons per day, the SolarJack SDS Submersible pump is usually right for the job. If your water level in the well is stable, frequently the SlowPump will work famously with tight elbows, even though it is not submersible. If you need to go deeper, we would prefer to consult with you directly to spec out the proper pump for your application.

SDS-D-224 Deep Well Submersible

The SDS-D-224 deep well submersible pump (formerly called the Econo-Sub) is a low power deep well submersible pump, capable of lifts up to 200 feet using less than 100 watts of 12 or 24-volt DC power, and able to deliver a flow rate from 30 to 100 gallons per hour. It installs easily by hand and is a real breakthrough in deep-well solar pumping. It is made of stainless steel and solid brass parts. The diaphragm and the check valve flaps are EPDM which is a synthetic elastomer that is extremely durable. The check valve housing and relief valves are nylon. This pump is strong! During testing water was brought up from 100' in the well with a 20 watt solar module! The Installation kit (below) is highly recommended. Warranty is six months unless you buy the controller, then it is a full two years.

41-601   SDS-D-224 Submersible Pump $795

SDS SUBMERSIBLE PUMP DATA
SDS-D-224 24 VOLT

PSI	LIFT	GPM	AMPS
0	0	2.75	2.3
10	23.1	2.55	2.8
20	46.2	2.4	3.2
30	69.3	2.25	3.6
40	92.4	2.13	4
50	115.5	2	4.3
60	138.6	1.9	4.5
70	161.7	1.8	4.8
80	184.8	1.7	5
90	207.9	1.6	5.3
100	231	1.5	5.6

SolarJack SDS
Deep Well Installation Kits

SolarJack's new submersible pump kits have been developed to supply all the necessary parts and products for a complete installation of the SDS series deep-well pumps. The only additional component you will need is a support structure for your PV modules. (See Trackers/Mounting Structure section in this Sourcebook - we suggest you use either a tracking device for maximum water yield, or a fixed pole mount.) The kits are prewired, allowing true ease of assembly. All hardware necessary to complete the job is included in each kit. The owners/installation manual included with the pump contains complete, step-by-step instructions written for the non-technical person. Buying the installation kit with the pump extends the warranty on the pump to a full two years. If you choose to buy your installation components locally, you can still obtain the two year warranty by purchasing the SolarJack controller from us.

Each kit includes:

- SolarJack's PC Series Pump Controller
- Submersible Drop Wire
- One Waterproof Wire-Splice Kit
- Pump Drop Pipe
- Pump Safety Rope
- All Clamps and Fittings
- One 4 or 6 inch Sanitary Well Seal
- A 2-Year Limited Warranty on the Pump
- Many Optional Parts and Accessories

PRICING ON KITS FOR VARIOUS DEPTHS

■41-621	50' Installation Kit	$435
■41-622	100' Installation Kit	$475
■41-623	150' Installation Kit	$585
■41-624	200' Installation Kit	$645
■41-625	250' Installation Kit	$695

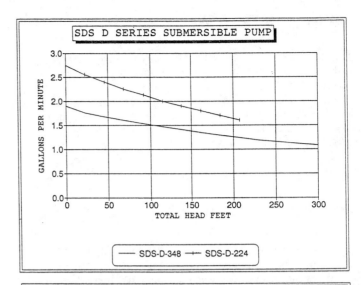

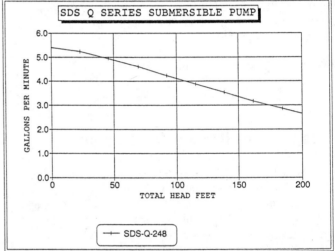

SDS-Q-224 Deep Well Submersible

This new high volume quad pump will deliver 5 gpm at 50-100 feet. It operates at 24VDC with a current of 3 amps, and measures 4.5" in diameter. It can be used in either PV-direct or battery applications. This pump must use a controller for optimum operation - order the **41-631** or the **41-632**. Six month warranty w/o controller; two years with.

■**41-605**   SDS-Q-224 Pump   **$995**

Pump Controller For 24V Pumps

Every 24V deep well submersible pump must have a controller to allow the motor to start only under optimum conditions. This will greatly increase the life of the motor and extend your pump warranty to a full two years. The controller comes equipped with the capability for remote turn-off. The controller can be used for remote water level detection - both high and low level by using three sensors that must be ordered separately.

41-632	**24V Pump Controller**	**$149**
41-636	**Sensors (3) for Level Detection**	**$ 15**

BUILD a nuke. Three Mile Island and Chernobyl never happened. If you want to be remembered for hundreds of generations start your own reactor today! The federal government will give you all the help you need.

SDS-D-348 Deep Well Submersible

This new submersible goes all the way down to 300 feet. It will pump approximately 1 gpm. It operates at 48VDC with a current of 2.5 amps, and measures 3.25" in diameter. To use with solar-direct applications, you must hook up four standard PV modules in series to produce the required voltage. Six month warranty without pump controller, a full two years with the controller.

■41-603   SDS-D-348 Pump $895

SDS-Q-248 Deep Well Submersible

Here is a submersible for which we've had a demand but no product for quite some time. It's a real work horse. It is a quad pump that can reach down to 200 feet where it can deliver 2.5 gpm at 48VDC with a current of 3.5 amps. It measures 4.5" in diameter. Six month warranty w/o controller; two years with.

■41-607   SDS-Q-248 Pump $1,095

Pump Controller for 48V Pumps

Every 48V Submersible Pump must have a controller to allow the motor to start only under optimum conditions. This will greatly increase the life of the motor and extend your pump warranty to a full two years. The controller comes equipped with the capability for remote turn-off. The controller can be used for remote water level detection - both high and low level by using three sensors that must be ordered separately.

■41-634   **Pump Controller - 48V**   **$190**
■41-636   **Sensors (3) for Level Detection**   **$ 35**

Example: 160 foot well, 400 to 800 gallons per day. This system can be installed by hand in a few hours. The system would include:
   SDS-D-224 24V pump
   Pump Controller
   (2) 48 Watt PV Modules
   Solar Tracker / Mounting Rack
   1/2" Polyethylene pipe, fittings, safety rope
   #10-2 Submersible pump cable, splice kit
   Level Sensor Control
   Well Casing Requirement: 4"
   Tank Capacity Required: 2,500 gallon min.

SYSTEM PRICE ESTIMATE: $2,150

Solarjack manufactures a number of other deepwell submersible pumps as well as centrifugal pumps where large volumes of water are desired. In order to size your proposed system properly, please fill out our *Water System Questionnaire* **and call us to aid you in designing your system.**

SDS SUBSERSIBLE PUMP DATA
SDS-D-348 48 VOLT

PSI	TDH	GPM	AMPS
0	0	1.95	0.90
10	23.1	1.75	1.00
20	46.2	1.65	1.15
30	69.3	1.60	1.25
40	92.4	1.55	1.40
50	115.5	1.50	1.50
60	138.6	1.45	1.65
70	161.7	1.35	1.80
80	184.8	1.30	2.00
90	207.9	1.25	2.10
100	231	1.20	2.20
110	254.1	1.15	2.30
120	277.2	1.10	2.40
130	300.3	1.00	2.50

PSI = POUNDS PER SQUARE INCH
TDH = TOTAL DYNAMIC HEAD
GPM = GALLONS PER MINUTE

SDS SUBMERSIBLE PUMP DATA
SDS-Q-248 48 VOLT

PSI	TDH	GPM	AMPS
0	0	5.40	1.88
10	23.1	5.23	1.92
20	46.2	4.94	2.22
30	69.3	4.60	2.38
40	92.4	4.22	2.57
50	115.5	3.86	2.82
60	138.6	3.53	3.00
70	161.7	3.16	3.14
80	184.8	2.85	3.37
90	207.9	2.54	3.50

PSI = POUNDS PER SQUARE INCH
TDH = TOTAL DYNAMIC HEAD
GPM = GALLONS PER MINUTE

Real Goods provides a variety of specialized pumps for use with solar electric power. Many of our customers are confused about total system costs for pumping systems. Throughout the pumping section, we will provide you with typical examples of the systems that we provide to give you an accurate picture of final costs. Our SDS series submersible is a low volume deep well pump that can lift up to 300 feet and produce up to 55 gallons per hour, or 150 gallons per hour at low lifts. It requires only one to four 40-60 watt photovoltaic panels and fits inside a 4" well casing.

Solarjack Jack Pump

For your deep well pumping needs (well depths to 1,000 feet and from 1/2 gallon to 28 gallons per minute) we offer the Solarjack Jack Pump water pumping systems; one of the most cost-effective and reliable photovoltaic solar pumping units available.

Solarjack has two Pump Jack models to choose from: the SJA (3/4 HP) and the SJB (1 HP). Each system includes the Pump Jack, motor, controller, bridle assembly, and anchor bolts.

The SJA model is limited to 500' of total head and a maximum of 600 peak solar module watts while the SJB is capable of lifting water from 1000' and limited to 1080 peak module watts.

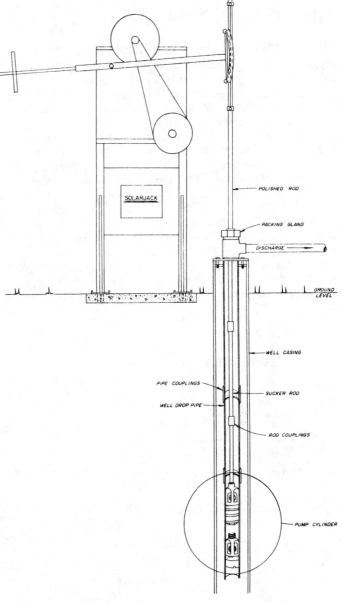

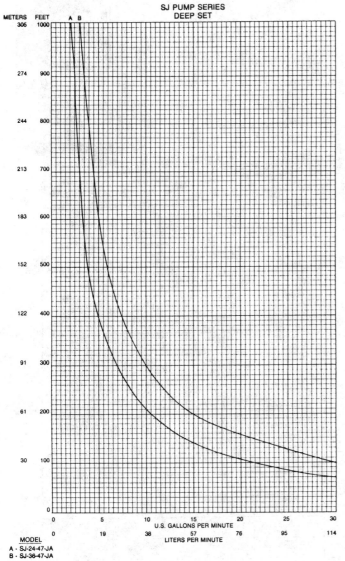

Installation

Each Pump Jack is mounted on a steel base which is then anchored to a concrete floor. (Anchors and bolts are included) The solar modules are mounted nearby on a frame and anchored to a foundation. Drop pipe, sucker rod, polished rod, packing gland, and pump cylinder are necessary to complete the installation.

Maintenance

Very little supervision and maintenance is required. An occasional check of the oil level (SJA Jack) and lubrication of grease fittings is recommended, and it should operate for 20 years or more, according to the manufacturer, if maintained properly.

Pumps start at $3,000 for the basic unit. After adding the additional accessories, you're looking at a minimum of $5,000. Please write or call one of our technicians for an exact price quotation.

A.Y. McDonald Solar Sub

This new 4" submersible pumping system from A.Y. McDonald is designed for the larger-scale or commercial user. Possible application would include livestock watering, nurseries, irrigation, or village water supplies. Depending on the model and the available voltage, it can pump anywhere from 90 to 4,000 gph.

This pump is designed to be run directly off of solar panels, without an inverter or control system. It can also be run off of a battery bank. The voltage and pumping capacity can be increased by adding more solar modules.

The submersible pump/motor combination fits 4" diameter wells or larger. Its construction of stainless steel, brass, and engineered plastic is corrosion resistant. A built-in check valve prevents backspin on the NEMA standard motor, and the pump connects directly for positive torque without slippage. No priming is required for this pump and it has built-in lightning protection. Its operating voltage range is from 45 volts or less up to 90 volts.

Specification data

For well diameters 4" ID (100 MM) and larger

Model	Nominal Discharge voltage	watts	Size	Length	Weight
180809DM	84 VDC	1,100	1-1/4" NPT	55"	68#
180810DP	84 VDC	1,100	1-1/4" NPT	55"	68#
180809DM	84 VDC	1,100	1-1/4" NPT	55"	68#
180810DP	84 VDC	1,100	1-1/4" NPT	55"	68#
180815DK	84 VDC	1,100	1-1/4" NPT	55"	68#
180825DJ	84 VDC	1,100	1" NPT	70"	74#

Please contact us for pricing information.

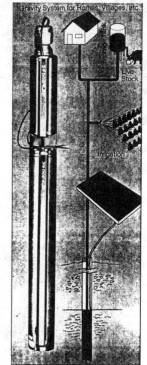
Gravity System for Homes, Villages, etc.

HOW TO USE THE CHART:
1. When the Depth To Water and Total Lift (Head) is known, select the column closest to the lift required.
2. Select the Gallons Per Day closest to the amount needed by moving down the selected lift column to find the best pump.
3. Move horizontally to the left to find the quantity of 53-Watt Modules required.
NOTE: G.P.D. are based on a good Solar Day of 6 KWM/1 Sq. Meter (Sun Intensity). Your pumping output will vary depending on the actual modules used, tilt angle and seasonal weather conditions as found in your geographical area.

CONVERSION FACTORS:

Meters to Feet — Multiply Meters by 3.28

U.S. G.P.D. to Cubic Meters Per Day — Divide G.P.D. by 264.2

Feet of Lift to PSIG — Divide Feet by 2.31

A.Y. McDonald's "Solar Sub" Performance in U.S. Gallons Per Day (GPD) and Maximum Gallons Per Minute (MGPM) Based on Standard Solar Day of 6 KWH/Sq.M.

QUANTITY OF 53 WATT PV MODULES	VOLTS	AMPS	15m (49') PUMP MODEL NO. GPD / MGPM	20m (66') PUMP MODEL NO. GPD / MGPM	25m (82') PUMP MODEL NO. GPD / MGPM	30m (98') PUMP MODEL NO. GPD / MGPM	40m (131') PUMP MODEL NO. GPD / MGPM	50m (164') PUMP MODEL NO. GPD / MGPM	60m (197') PUMP MODEL NO. GPD / MGPM
6 3Series × 2Parallel	45.1	5.9	180810DP 4224 / 10.9	180815DK 2531 / 5.9	180815DK 2075 / 5.0	180815DK 1599 / 4.1	180825DJ 756 / 2.2	180825DJ 417 / 1.3	180825DJ 189 / 0.4
9 3Series × 3Parallel	49.2	6.7	180810DP 6362 / 13.1	180810DP 4464 / 9.5	180815DK 3133 / 6.5	180815DK 2643 / 5.5	180815DK 1494 / 3.3	180825DJ 887 / 2.0	180825DJ 440 / 1.2
12 4Series × 3Parallel	54.9	8.7	180810DP 8474 / 18.6	180810DP 7090 / 16.3	180815DK 3966 / 8.8	180815DK 3545 / 8.0	180815DK 2651 / 6.5	180815DK 1731 / 4.6	180825DJ 1167 / 3.1
15 5Series × 3Parallel	68.8	10.0	180809DM 9448 / 23.4	180810DP 8002 / 19.6	180810DP 6722 / 17.4	180810DP 5460 / 14.9	180815DK 3097 / 8.2	180815DK 2329 / 6.9	180815DK 1646 / 5.3
16 4Series × 4Parallel	64.8	9.4	180809DM 10587 / 21.8	180809DM 8783 / 18.4	180810DP 7429 / 15.5	180810DP 5786 / 12.3	180815DK 3523 / 7.5	180815DK 2632 / 5.9	180825DJ 1728 / 3.8
20 5Series × 4Parallel	77.0	11.2	180809DM 12463 / 26.9	180809DM 10963 / 24.4	180810DP 9304 / 20.5	180810DP 8171 / 18.5	180810DP 5623 / 13.6	180815DK 3678 / 8.6	180815DK 2977 / 7.4
25 5Series × 5Parallel	81.0	11.9	180809DM 14373 / 28.5	180809DM 12692 / 25.9	180809DM 11177 / 23.3	180810DP 9681 / 20.1	180810DP 7177 / 15.4	180815DK 4481 / 9.5	180815DK 3788 / 8.3
30 5Series × 6Parallel	83.6	12.2	180809DM 15734 / 29.5	180809DM 14116 / 27.0	180809DM 12437 / 24.4	180809DM 10732 / 21.5	180810DP 8230 / 16.7	180810DP 5790 / 11.9	180815DK 4360 / 8.8

Continued Except for 6 and 9 Module Systems. N/A Above 200' Lift.

QUANTITY OF 53 WATT PV MODULES	VOLTS	AMPS	70m (230') PUMP MODEL NO. GPD / MGPM	80m (262') PUMP MODEL NO. GPD / MGPM	90m (295') PUMP MODEL NO. GPD / MGPM	100m (328') PUMP MODEL NO. GPD / MGPM	120m (394') PUMP MODEL NO. GPD / MGPM	130m (426') PUMP MODEL NO. GPD / MGPM	150m (450') PUMP MODEL NO. GPD / MGPM
12 4Series × 3Parallel	54.9	8.7	180825DJ 910 / 2.5	180825DJ 645 / 1.9	180825DJ 366 / 1.1	180825DJ 106 / 0.4			
15 5Series × 3Parallel	68.8	10.0	180825DJ 1179 / 3.6	180825DJ 945 / 3.0	180825DJ 717 / 2.4	180825DJ 492 / 1.8			
16 4Series × 3Parallel	64.8	9.4	180825DJ 1392 / 3.1	180825DJ 1052 / 2.5	180825DJ 687 / 1.7				
20 5Series × 4Parallel	77.0	11.2	180825DK 2232 / 6.0	180815DK 1711 / 4.6	180825DJ 1343 / 3.5	180825DJ 1062 / 2.9	180825DJ 625 / 1.8	180825DJ 395 / 1.2	180825DJ 223 / 0.8
25 5Series × 5Parallel	81.0	11.9	180815DK 3078 / 6.9	180815DK 2413 / 5.6	180825DJ 1801 / 4.0	180825DJ 1508 / 3.5	180825DJ 934 / 2.3	180825DJ 650 / 1.7	180825DJ 468 / 1.3
30 5Series × 6Parallel	83.6	12.2	180815DK 3640 / 7.5	180815DK 2963 / 6.2	180815DK 2234 / 4.8	180825DJ 1818 / 3.8	180825DJ 1226 / 2.7	180825DJ 909 / 2.1	180825DJ 681 / 1.6

Bowjon Wind-Powered Pumps

We had limited experience with Bowjon Pumps back in the late seventies and early eighties; a few raves and a few problems, then they disappeared from the market for several years. Now Bowjon is back promising a perfected product of rigid integrity with two years in redesign and development. So far, we're pleased with what we see.

Bowjon utilizes a unique "air injection" system that is ideal in areas with high wind velocities. You can place your Bowjon where the wind blows best, as far as 1/4 mile away from your well. It uses five high-torque, heavy-duty aluminum blades. It is quickly installed on a single post tower. One person can assemble it without special rigs, tools, or equipment. Its minimal maintenance only requires checking the compressor oil every six months. There are no moving parts, leathers to fight, and there are no cylinders or plunger assembly - allowing the Bowjon to run dry without harm.

The Bowjon is available with an eight foot diameter or a six foot diameter single or twin in-line compressor with Timken Bearings. The Bowjon's pumping system is capable of effectively lifting water up to 7 gallons per minute. More than one windmill can service the same well. The hub is a massive 17 lbs. of tempered aluminum that will withstand 3,200 psi. Bowjon's five blades are made of 6061-T6 tempered aluminum, each consisting of three layers for the maximum strength that's required in extremely high winds.

Bowjon Specifications:

Propeller: 8'8" or 6'8" diameter tempered triple layer reinforced blades to minimize flex in high winds. Blades have a unique varied pitch to allow low wind, high torque start-up. They have high RPM ability.

Compressor: Industrial strength twin or single cylinder depending on air pressure needs for required water demand. Compressor is capable of five cubic feet of air volume per minute.

Start-Up: 5-8 MPH

Pump: Air injection. No moving parts, no cylinders, valves, rods, or leathers to wear out. Will run dry, accepting silt, sludge, and sand without harm. Length is five feet. Minimum well casing diameter is two inches.

Air Line: 3/8" Polyethylene Tubing (200psi rated).

Water Line: 0-50' Lift: 1" tubing or PVC Schedule 40 plastic pipe. 50'-100' Lift: 3/4" tubing or PVC Schedule 40 plastic pipe. 100' and over Lift: 1/2" tubing or PVC Schedule 40 plastic pipe.

Submersion: (Minimum) equal to 30% of the vertical lift distance from water level in the well to the highest point of delivery or storage. Where minimum recommended submersion is not possible, use of a collector tank and a regulator in the air line will prevent excessive air pressure and volume. Submersions less than 30' restrict vertical lift; but where water is only to be lifted a few feet, submersions of 5' or 10' with air tank and regulator are practical.

Lift/Submergence Ratios:

Vertical Lift	Submergence Below Water	Optimum Total Well Depth
50'	35'	90'
80'	56'	140'
120'	85'	210'
200'	100'(50% of lift)	350'

(If the submergence is too low for the amount of lift, the air will separate from the water. If the submergence is too high, the air will not lift the water. For these reasons, the submergence of the pump must be calculated carefully.)

Homesteader Model Bowjon Pump

The Homesteader model comes with 6'8" diameter blades with a single cylinder compressor. It has a 250' x 3/8" sir line and AL2 air injection pump. One-year warranty.

■**41-701   Homesteader Bowjon Pump $1295**
Shipped freight collect from So. California

Rancher Model Bowjon Pump

The Rancher model comes with 8'8" diameter blades with a twin cylinder compressor. It has a 250' x 3/8" air line and an AL3 air injection pump. One-year warranty.

■**41-711   Rancher Model Bowjon Pump $1595**

Ram Pumps

The ram pump works on a hydraulic principle using the liquid itself as a power source. The only moving parts are two valves. In operation, the liquid (usually water) flows from a source down a "drive" pipe to the ram. Once each cycle, a valve closes causing the liquid in the drive pipe to suddenly stop. This causes a water hammer effect and high pressure in a one-way valve leading to the "delivery" pipe of the ram, thus forcing a small amount of water up the pipe and into a holding tank. In effect the ram uses the energy of a large amount of water falling a short distance to lift a small amount of water a greater distance. The ram itself is a highly efficient device; however, only 6% to 10% of the liquid is recoverable, so a large amount of water is required. The maximum head or vertical lift of a ram is about 500 ft.

Selecting a Ram

Estimate Amount of Water Available to Operate the Ram

This can be determined by the rate the source will fill a container. Avoid selecting a ram that uses more water than available.

Estimate Amount of Fall Available

The fall is the vertical distance between the surface of the water source and the selected ram site. Be sure the ram site has suitable drainage for the tailing water. Often a small stream can be dammed to provide the 1.5 feet or more head required to operate the ram.

Estimate Amount of Lift Required

This is the vertical distance between the ram and the water storage tank or use point. The storage tank can be located on a hill or stand above the use point to provide pressurized water. Twenty or thirty feet water head will provide sufficient pressure for household or garden use.

Estimate Amount of Water Required at the Storage Tank

This is the water needed for your use in gallons per day. As examples, a normal household uses 100 to 300 gallons per day, much less with conservation. A 20 by 100 foot garden uses about 50 gallons per day. When supplying potable water, purity of the source must be considered.

Using these estimates, the ram can be selected from the following performance charts. The ram installation will also require a drive pipe five to ten times as long as the fall, an inlet strainer, and a delivery pipe to the storage tank or use point. These can be obtained from your local hardware or plumbing supply house. Further questions regarding suitability and selection of a ram for your application will be promptly answered by our engineering staff.

Aqua Environment Rams

We've sold these fine rams by Aqua Environment for over 10 years now with virtually no problems. Careful attention to design has resulted in extremely reliable rams with the best efficiencies and lift to fall ratio available.

Construction is of all bronze with O ring seal valves. The outlet gage and valve permit easy start up.

Each unit comes with complete installation and operating instructions.

Typical Performance and Specifications					
Vertical Fall (feet)	Vertical Lift (feet)	Pump Rate (Gallons/Day)			
		3/4" Ram	1" Ram	1¼" Ram	1½" Ram
20	50	650	1350	2250	3200
20	100	325	670	1120	1600
20	200	150	320	530	750
10	50	300	650	1100	1600
10	100	150	320	530	750
10	150	100	220	340	460
5	30	200	430	690	960
5	50	100	220	340	460
5	100	40	90	150	210
1.5	30	40	80	130	190
1.5	50	20	40	70	100
1.5	100	6	12	18	25

Water Required to Operate Ram

3/4" Ram - 2 gallons/minute   Maximum Fall - 25 feet
1" Ram - 4 gallons/minute   Minimum Fall - 1.5 feet
1¼" Ram - 6 gallons/minute   Maximum Lift - 250 feet
1½" Ram - 8 gallons/minute

■41-811	3/4" Ram (#201)	$195
■41-812	1" Ram (#202)	$195
■41-813	1-1/4" Ram (#203)	$235
■41-814	1-1/2" Ram (#204)	$235

MAKE a noise. Liven things up a little. The great out of doors can get awfully quiet unless you do something about it; loud music and open spaces go hand in hand.

High Lifter Pressure Intensifier Pump

A new development in water pumping technology, the High Lifter Water Pump offers unique advantages for the rural user. Developed in Mendocino County, California expressly for mountainous terrain and low summertime water flow, this water powered pump is capable of extremely high lifts in ratio to fall (over 1,000 feet) with just a trickle of inlet water. For example, if you have a flow of 2 gallons per minute and a fall of 40-feet from a water source (spring, pond, creek, etc.) to the High Lifter pump with a 200-foot rise (head) from the pump up to a holding tank you would have a total head of 240 feet. The lift fall ratio between the 40 foot fall and the 200 foot lift would be 6:1. The High Lifter pump would deliver 300 gallons of water per day from a 6:1 working ratio.

The High Lifter pump has many advantages over a ram (the only other water powered pump). Instead of using a "water hammer" effect to lift water as a ram does, the High Lifter is a positive displacement pump that uses pistons to create a kind of hydraulic lever that converts a larger volume of low-pressure water into a smaller volume of high-pressure water. This means that the pump can operate over a broad range of flows and pressures with great mechanical efficiency. This efficiency means more recovered water. While water recovery with a ram is normally about 6%, up to 20% recovery may be achieved with the High Lifter Pump.

In addition, unlike the ram pump, no "start up tuning" or special drive lines are necessary. This pump is also quiet.

The High Lifter pressure intensifier pump is economical compared to gas and electric pumps, because no fuel is used and no extensive water source development is necessary.

There are two model High Lifter pumps available with a 4.5:1 and 9.1:1 working ratio. A kit to change the working ratio of either pump after purchase is available. A maintenance kit is available too. Choose your model High Lifter pump from the specifications and High Lifter performance curves. One year parts & labor warranty.

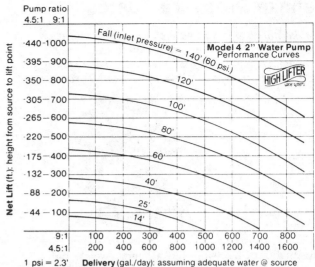

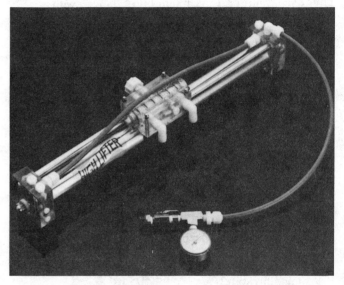

■ 41-801   High Lifter Pump - 4.5:1 Ratio $895
■ 41-802   High Lifter Pump - 9:1 Ratio $895
■ 41-803   Conversion kit (to other ratio) $ 97

TYPICAL APPLICATION

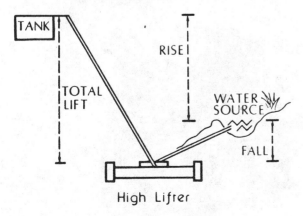

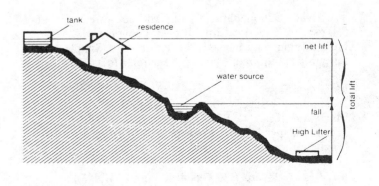

Tanaka Pressure Pumps

Tanaka makes a great two-cycle gas pump that is powerful, compact, easy to transport, and easy to use. Their only disadvantage is that they are very loud, like a chain saw. If noise isn't a problem they're hard to beat! 2 year warranty.

QCP-121

The smaller of the two Tanaka pumps, weighs only 13 pounds and will deliver a maximum of 1,900 gallons per hour (31.7 gpm) or a maximum lift of 98 feet, with a maximum suction of 23 feet. Fittings are 1" inlet and outlet and the pump comes with 10' of suction hose.

41-901   Tanaka Pump - QCP-121 $359

High-Pressure Gasoline Diaphragm Pumps

Hypro D19

The D19 comes with a 3-1/2 hp Honda engine with gear reduction and a control unit. This pump will deliver 5.3 gpm at 275 psi (635 feet of lift vertically). Simple assembly is required. Maximum suction lift is 10'.

Maximum flow:	6 gpm
Maximum pressure:	275 psi
Maximum lift:	635 feet
Maximum speed:	650 rpm
Maximum temp.:	140°F

■41-911   Hypro D19 Gasoline Pump $1,095
pump & engine shipped in separate boxes

Hypro D30

The Hypro D30 diaphragm pump will deliver 8.5 gpm at 550 psi (1,270' vertical lift). It comes coupled with a 5 hp Honda engine and includes the control unit. Simple assembly is required. Maximum suction lift is 10'.

Maximum flow:	9.5 gpm
Maximum pressure:	550 psi
Maximum lift:	1,270 feet
Maximum speed:	540 rpm
Maximum temp.:	140°F

■41-912   Hypro D30 Gasoline Pump $1,495
pump and engine shipped in separate boxes
Larger hypro pumps are available on request.

TCP-381

This Tanaka pump weighs only 21 pounds and will pump a maximum of 4,750 gallons per hour (79 gpm) or a maximum lift of 131 feet. The inlet and outlet are 1-1/2". At 50' of head this pump will deliver 75 gpm, at 75' it will deliver 50 gpm, at 100' it will deliver 40 gpm, and at 120' it will deliver 20 gpm.

41-902   Tanaka Pump - TCP-381 $399

For high-pressure situations where a large vertical lift must be achieved, diaphragm pumps are the most reliable and most durable pumps available. These pump/motor combinations consist of the ultraquiet Honda engines coupled with the highly efficient and durable Hypro pumps. They can be run dry without harm, and have built-in pulsation dampeners that act as shock absorbers. Maintenance on the pumps is simple and diaphragms are very easy to replace.

Pump Accessories and Controls

Check Valves

A check valve must be installed in pressurized systems. Check valves seal the water system between the pump or pressure tank and the faucets in the home. It prevents reverse flow of water back into the pump and maintains pressure switch settings. All of our check valves come with accessory holes for a pressure gauge and pressure switch.

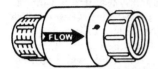

41-410	Check Valve - Bronze - 3/4"	$18
41-411	Check Valve - Bronze - 1"	$19
41-412	Check Valve - Bronze - 1-1/4"	$23
41-413	Check Valve - Bronze - 1-1/2"	$28
41-414	Check Valve - Bronze - 2"	$39

Foot Valves

Priming a pump is necessary to ensure that it isn't run dry. A foot valve prevents the need to do this every time that you start the pump. It is a one-way valve that lets water into the inlet line without letting it drain back into your water supply.

41-421	Foot Valve - Bronze - 1"	$17
41-422	Foot Valve - Bronze - 1-1/4"	$19
41-423	Foot Valve - Bronze - 1-1/2"	$35
41-424	Foot Valve - Bronze - 2"	$45

Gate Valves

Gate valves should be placed strategically in a system so that water may be shut off and equipment may be replaced or repaired without the user having to empty a holding tank or try to effect repairs while being sprayed with water.

41-431	Gate Valve - Bronze - 3/4"	$ 8
41-432	Gate Valve - Bronze - 1"	$10
41-433	Gate Valve - Bronze - 1-1/4"	$16
41-434	Gate Valve - Bronze - 1-1/2"	$19
41-435	Gate Valve - Bronze - 2"	$28

Line Strainers

Roller pumps and most high-head pumps are particularly vulnerable to dirty water. A line strainer of at least 40 mesh should be used. We recommend 80 mesh for additional safety. These line strainers have clear bottoms so that you can see if they need to be cleaned.

42-508	Line Strainer - 80 Mesh - 3/4"	$35
42-509	Line Strainer - 80 Mesh - 1"	$65

Arkal Filters

The Arkal filter is probably the best designed filter for drip irrigation systems. It also works famously for all filtration systems. Instead of using the traditional cartridge or fine mesh screen for filtering, the Arkal uses an assembly of thin rings. A spring holds the rings tightly together when the filter cover is on and lets them separate when the cover is removed. This makes cleaning the filter very quick and easy, as well as avoiding the expense of replacing cartridges.

Another advantage of this system is that it can maintain a higher gph and psi than with a cartridge or mesh system.

Filtrap 3/4"

The 3/4" filter has a built-in valve that can be used both to turn your water flow off and to regulate pressure up to 50 psi.

Mesh: 140
Inlet & outlet diameter: 3/4"
Max. operating pressure: 140 psi
Max. capacity: 18 gpm
Filtering volume: 37"
Filtering area: 63"

43-201	Arkal Filter with shutoff (3/4")	$29.00
43-202	3/4" Arkal w/o shutoff	$22.00

Water Storage

Polyethylene Storage Tanks

These poly tanks are also approved for drinking water. They are slightly more durable than fiberglass and probably last about as long - some claim up to 30 years for either tank. The 1,400-gallon tank will nicely fit on any pickup truck. Tanks come with two 1-1/4" fittings at the top and one 1-1/4" fitting at the bottom. *All tanks are shipped freight collect from the manufacturer (Quadel) in Oregon.* The widths below are for the diameter on top. On the 1400 gallon tank the diameter on the bottom is 62".

Item #	Gallons	Width	Depth	Price
■47-201	1,400	72"	97"	$795
■47-202	500	48"	76"	$555

Kolaps-a-Tank

These handy and durable nylon tanks fold into a small package or expand into a very large storage tank. They are approved for drinking water and withstand temperatures to 140°F, and fit into the beds of several truck sizes. They will hold up under very rugged conditions, are self-supported, and can be tied down with D-rings. Our most popular size is the FDA98MT (525 gallons) which fits into a standard long-wide bed (5' x 8') pickup.

Item #	Model no.	Size	Gallons	Price
■47-401	FDA50MT	40"x50"x12"	73	$335
■47-402	FDA73MT	80"x73"x16"	275	$395
■47-403	FDA98MT	65"x98"x18"	525	$495
■47-404	FDA610MT	6'x10'x2'	800	$575
■47-405	FDA712MT	7'x12'x2'	1140	$825

Doughboy Swimming Pools/Storage Tanks

Over 45 years ago Doughboy introduced the first storable, above-ground swimming pool. These pools also serve beautifully as economical water storage devices (around $0.15/gallon for the 7,000 Timberline to $0.18/gallon for the 16,300 gallon Cedar Ridge). Doughboy produces its own virgin vinyl liner material in a variety of thicknesses from 15 mil (10 yr warranty) to 25 mil (20 yr warranty). To support the inner liner, Doughboy features hot-dipped galvanized steel walls with seven protective coatings, with a slide joint connector system for structural integrity and easy assembly. All of the tanks listed below are 20 mil thick with a 15-year prorated warranty. For water storage the Timberline series has a 4" top seat and is more economical. For swimming pools we recommend the Cedar Ridge series with a 6" top seat. All pools are 4 feet deep, but can be stretched to allow for a "deep end" by digging into the ground.

Select the standard skimmer and Sequel I filter with a 3/4 hp pump for the Timberline series. Select the deluxe skimmer and the Sequel II filter with 1 hp pump for the Cedar Ridge series.

Timberline Series

Item #	Size	Gallons	Price
■47-301	16' round	5,500	$ 775
■47-302	18' round	6,700	$ 895
■47-303	12'x24'	7,000$	1,395

Cedar Ridge Series

Item #	Size	Gallons	Price
■47-321	24' round	12,160	$1,475
■47-322	16'x32'	12,250	$2,150
■47-323	18'x38'	16,200	$2,995

Pump & Filtering Equipment

■47-311	Standard Timberline Skimmer	$ 50
■47-312	Sequel I filter w/3/4 HP pump	$445
■47-341	Deluxe Cedar Ridge Skimmer	$ 89
■47-342	Sequel II filter w/1 HP pump	$575

Water Heating

Instantaneous Demand Water Heating

We can't recommend instantaneous water heaters highly enough. Standard tank-type water heaters account for about 20% of all the energy we use in our homes. Many people keep their water heaters at 140°, wasting energy and shortening the tank's life. These energy-saving tankless heaters have been in use almost exclusively for years in Europe and Japan. In fact, America is one of the few civilized countries in the world backward enough to still use the archaic technology of storage tanks. These new tankless gas water heaters will save many utility dollars because they rid one of the need for a water storage tank. They provide instant hot water when you need it, eliminating the need to heat 30 or more gallons in anticipation of your hot water demands and cut the need for a two-tank solar hot water system to a one-tank storage system. They are also ideal for heating hot tubs and spas.

Our favorite analogy that illustrates the stupidity of tank-type water heaters is the one about the car. Keeping 40 gallons of water hot at all times *just in case* you might need it is the same as leaving your car running in your garage 24 hours per day, seven days a week, *just in case* you decide you need to go for a drive! **Doesn't it make more sense to only heat the water when you need it?**

These heaters are always ready to ignite using piezo starters (Paloma only); they use limited space because they mount on the wall, and are very easy to control.

We offer two brands of tankless water heaters, the Paloma and the Aquastar. We have numerous installations using both, and virtually all our customers are happy with them, encountering few problems. Once in a great while the gas jet may clog requiring cleaning, but that is about all the maintenance needed.

Both units will operate from pressures as low as 15 psi, and can be used with small low-voltage pumps and in most gravity-fed situations.

Paloma and Aquastar have proved to be very reliable heaters for us. The Paloma from Japan and the Aquastar from France are both made with copper, brass, and stainless steel parts. The Paloma PH-6 is by far our best seller providing 1.4 gpm at a 50° F temperature rise and is adequate for one tap at a time use as is the Aquastar 80VP providing 2.4 gpm @ a 50° F temperature rise. Most customers are satisfied with the shower provided by the PH-6. The PH-5 is our most popular heater among RV users as the complete unit measures less than 16" high for easy installation into shower stalls or under counters. The PH-5 produces nearly the volume of the PH-6 providing 1.2 gpm at a 50° temperature rise. Where larger volumes of hot water are needed we recommend either the Paloma PH-12 (2.9 gpm @ 50° rise) or the AquaStar 125LP (3.8 gpm @ 50° rise).

Sizing Your Tankless Water Heater

The most important charts for you will be the water flow rate (gpm, gallons per minute) for determining the size heater you need. As an example, the Paloma chart's top line indicates the flow of water in gpm. The left hand column indicates the model number. Note in the third column that at 1.5 gpm, the PH-6 heater will raise the water temperature 47°. If the water from the delivery system is 50° (the usual well water temperature), the hot water delivered at 1.5 gpm will be only 97°. The PH-12, however, would deliver water at 185° under the same circumstances. Be sure to take these facts into consideration before ordering:

- How much water will your system be pumping?
- How high a water temperature do you desire?
- How long are your pipe runs?

You can lose up to 50% of your heat through uninsulated pipes, and the longer the run, the more heat loss involved. Insulate the pipes and make the run as short as possible when planning your system.

Gas consumption is given in Btu/hr., and you might keep in mind that there are 91,500 Btus in one gallon of propane and 1,000 Btus per cubic foot of natural gas. A little figuring will provide you with the cost savings these units will provide when comparing the operating time and Btu/hr. consumed with these tankless heaters to the number of times a traditional heater comes on to maintain the temperature of standing water. Some other useful conversion numbers:

1 Btu = .293 Wh (watthours),
1 Btu/hr = .293 watts,
1 Btu = 1.06 kJ = 1.06 kVAsec.

Comparing Demand Water Heaters

We've recently discovered the Rodale Product Testing Report on Instantaneous Water Heaters, and have received added feedback from one of the largest distributors of demand water heaters in the country, which has sold and serviced both Paloma and Aquastar for years. These two sources have led us to conclude overall that Paloma has the edge, but that Aquastar is still a fine unit and slightly more cost-effective in dollars per BTU. Here is the comparison:

Efficiency: (source: Rodale Press)
Paloma: average 77.5%
Aquastar: average 74.8%
Construction type:
Paloma: Fully Enameled housing heavy gauge steel; two pieces
Aquastar: Partially enameled light gauge steel; four pieces
Minimum start-up flow rate:
Paloma: PH12, PH16: 0.48gpm, PH24 0.52gpm
Aquastar: 0.75gpm
Heavy dutiness:
PH16: 121,500 btu, 60 lb.
PH12: 89,300 btu, 36 lb.
Aquastar 125: 125,000 btu, 37 lb.
Aquastar 80: 77,500 btu, 31 lb.
Ease of installation:
Paloma: Fully assembled, ready to mount on wall with one piece mounting bracket. Includes stainless steel flexible water connections. Includes three drain plugs.
Aquastar: Partially assembled, no flexible water connections, no drain plugs.

Customers frequently ask us about clearances for installing our Aquastars and Palomas. We've developed the following chart to help you plan your installation before you buy.

Water Heater Clearance Chart
(clearance from heater to combustibles)

Heater	Top	Bottom	Side	Front	Wall to Flue C/L
Aquastar 80	12"	12"	1"	6"	5-1/8"
Aquastar 125	12"	12"	6"	36"	5-1/8"
Aquastar 170	12"	12"	6"	open only	7-3/4"
Paloma PH-5	16"	6"	6"	6"	5"
Paloma PH-6	16"	6"	6"	6"	5-1/8"
Paloma PH-12	16"	6"	6"	6"	6-1/2"
Paloma PH-16	16"	6"	6"	6"	7"
Paloma PH-24	16"	6"	6"	6"	7-1/8"

Aquastar Tankless Heaters

The French-made Aquastar tankless water heater performs much the same way as the Paloma except that it is thermostatically controlled making it an excellent unit for a solar hot water or wood stove hot water system's backup. Aquastar was rated #1 by Consumer Reports Magazine. The Aquastar features a safety thermocouple at the burner and pilot, an overheat fuse, a manual burner control adjustment for finer temperature control, and built-in gas shut-off valves.

The Aquastar 80 is designed for use with one tap at a time and will produce 1.8 gallons per minute at a 60° temperature rise. The Aquastar 125 will produce 3.25 gpm at a 60° temperature rise. The Aquastar is really the only instantaneous water heater that should be used with preheated water systems. The "S" series is designed for use with solar or woodstove preheated water and is "0 modulated." What this means is that if the incoming water is hot enough the Aquastar's burner will not come on at all. The Paloma, on the other hand, can be turned down to 20,000 Btu/hr., which means if the water is preheated the burners will fire at least 20,000 Btu/hr.

All Aquastars have a 10-year warranty on the heat exchanger and a 2-year warranty on all other parts. *With recent increases in the Japanese Yen, the AquaStar is a much better buy than Paloma in terms of dollars per Btu/hr.* You must order either in Propane (LP) or Natural Gas (NG).

Item #	Model	Btu/hr	SALE
45-102-P	Aquastar 80LP	77,500	$495
45-102-N	Aquastar 80NG	77,500	$495
45-105-P	Aquastar 125LP	125,000	$595
45-105-N	Aquastar 125NG	125,000	$595
45-106-P	Aquastar 80LPS	77,500	$550
45-106-N	Aquastar 80NGS	77,500	$550
45-107-P	Aquastar 125LPS	125,000	$595
45-107-N	Aquastar 125NGS	125,000	$595
■45-104-P	Aquastar 170LP	165,000	$895
■45-104-N	Aquastar 170NG	165,000	$895

Model 170 only shipped freight collect from Vermont or L.A.

Be sure to use correct item number to indicate PROPANE (LP) or NATURAL GAS (NG).

Specifications	Model 38	Model 80	Model 125	Model 170
Power (BTU input)	38,700	77,500	125,000	165,000
Recovery time (gallons/hour)	45	78	126	174
Gallons/Minute				
60° rise to 115°	1.1	1.8	3.3	4.5
70° rise to 125°	.9	1.7	2.8	3.8
90° rise to 145°	.7	1.3	2.1	2.9
Minimum flow to activate burners (gallons/minute)	.75	.75	.75	1.0
Vent Size	3"	4"	5"	6"
connections				
Water		.5" Cu sweat	.5" Cu sweat	.75" NPT male
Gas		.75" NPT male	.75" NPT male	.75" NPT male
water pressure (psi)				
Minimum	15	15	15	15
Maximum	150	150	150	150
Height x Width x Depth	18.5x8.38x7	27.5x12x9.5	27.5x17x9.5	not given

The Paloma PH Series

The Paloma is regulated by restricting the water flow to raise water temperature. This is a drawback for use as a solar backup which really requires thermostatic gas control for automatic temperature regulation. The Paloma PH-6 is our best seller providing 1.4 gpm at a 50° F temperature rise and is adequate for one tap at a time. Most customers are satisfied with the shower provided by the PH-6. The PH-5 is our most popular heater among RV users as the complete unit measures less than 16" high for easy installation into shower stalls or under counters. The PH-5 produces nearly the volume of the PH-6 providing 1.2 gpm at a 50° temperature rise. The Paloma PH-12 will produce 2.9 gpm @ 50° rise). The Paloma carries a limited 5-year warranty on the heat exchanger and a 3-year warranty on all other parts and is available in white only. *Be sure to order either Natural Gas (NG) or Propane (LP).*

Important note: All Palomas are available in either propane or natural gas models. Since we sell propane (LP) powered units by 30 to 1 over natural gas (NG) powered units, we will ship you a propane unit unless otherwise specified and *you will be responsible for the return freight if you order the wrong one.*

Item #	Model no.	Btu/hr	Price
45-201-P	PH5-LP	38,100	$245
■45-201-N	PH5-NG	38,100	$245
45-202-P	PH6-LP	43,800	$375
■45-202-N	PH6-NG	43,800	$375
■45-203-P	PH12-LP	89,300	$745
■45-203-N	PH12-NG	89,300	$745
■45-204-P	PH16-LP	121,500	$995
■45-204-N	PH16-NG	121,500	$995

All units shipped with draft hood except PH5

45-231	Draft hood for PH5	$ 55

	PH-5-3F	PH-6D	PH-12MD	PH-16MD	PH-24MD
BTU/hr Gas input	38,100	43,800	30,000 (Min.) 89,300 (Max.)	30,000 (Min.) 121,500 (Max.)	37,700 (Min.) 178,500 (Max.)
Net weight in lbs.	13.7	20	36	66	76
Hot water output g/m g/h	g/m g/h	g/m g/h	g/m g/h	g/m g/h	g/m g/h
50°F rise	1.2 72	1.4 84	(2.85) (171)	(3.8) (228)	(5.7) (342)
100°F rise	.6 36	.7 42	1.43 86	1.9 114	2.85 171
DIMENSIONS:					
Height in.	15-13/16	29-5/8	35-3/4	40-3/4	40-3/4
Width in.	11-17/32	10-1/4	13-3/4	18-3/4	18-3/4
Depth in.	9	9-7/8	11-3/4	13	14-7/8
Vent size in.	4	4	5	7	7
Gas connection in.	1/2 Male	1/2 Female	1/2 Female	1/2 Female	3/4 Female
Water connection in.	1/2 Male	1/2 Male	3/4 Male	3/4 Male	3/4 Male
Min water pressure	4.3 psi	4.3 psi	2.1 psi	2.1 psi	2.4 psi

TEMPERATURE RISE (°F)

( ) Mixing with cold water

	WATER FLOW RATE (Gallons per Minute)								
Model	0.5	1.0	1.5	2.0	2.5	3.0	3.5	4.0	4.5
PH-5-3F	119	59	40						
PH-6D	110/.6	71	48						
PH-12MD									
Hot	100/.48	100	100/1.43	90/1.59					
Warm	60/.79	60	60	60/1.59	(57)				
PH-16MD									
Hot	100/.48	100	100	100/1.9	90/2.11				
Warm	60/.79	60	60	60	60/2.11	(63)			
PH-24MD									
Hot	100/.52	100	100	100	100	100/2.85	90/3.17		
Warm	60/.87	60	60	60	60	60/3.17	60/3.17	(71)	(63)

WATER FLOW RATE AND TEMPERATURE RISE

( ) Mixing with cold water

		Temperature Adjusting Knob						
Model		Warm	1	2	3	4	5	Hot
PH-5-3F	Water flow (gpm)	1.5	1.5	1.5	1.3	1.1	0.8	0.6
	Max. Temp. Rise (°F)	40	40	40	46	54	75	100
PH-6D	Water flow (gpm)	2.0	1.8	1.5	1.0	0.8	0.7	0.6
	Max. Temp. Rise (°F)	35	39	47	70	88	102	117

		Warm		• • •		Hot		
PH-12MD	Water flow (gpm)	.79 – 1.59				.48 – 1.43 – 1.59		
	Temp. Rise (°F)	60		60 – 100		100 90		
PH-16MD	Water flow (gpm)	.79 – 2.11				.48 – 1.9 – 2.11		
	Temp. Rise (°F)	60		60 – 100		100 90		
PH-24MD	Water flow (gpm)	.87 – 3.17				.52 – 2.85 – 3.17		
	Temp. Rise (°F)	60		60 – 100		100 90		

Tankless Water Heater Booster Kit

Both the AquaStar and Paloma water heaters require at least 15 psi to operate. Many of our rural customers have low water pressure and are unable to operate these heaters. For them we've developed our booster kit to allow a full pressure hot shower for even the most remote dwellers. The kit consists of three basic components: a 4100-143 Flojet 12V pump (the highest flow model), a twenty gallon bladder-type pressure tank to even out the pulses, and a pressure switch to control the system. The standard pipe fittings and 12V battery are not included.

45-235   Booster Kit $350

Understanding Btus

A Btu, or British Thermal Unit is a unit of heat. It is the quantity of heat required to raise the temperature of one pound of water one degree Fahrenheit. You may find these conversions useful: 1 Btu = .293 watthours, or 1 Btu/hour = .293 watts, or 1 Btu = 1.06 kJ = 1.06 kVAseconds. Unfortunately it is the custom in the gas industry to express power input and output incorrectly in Btu when in fact they mean Btu/hour.

WASH a car.
There's little else to do out of doors, so fight boredom by bathing your limo every day.

Un-Clog-It Descaling Kit

The Un-Clog-It works great on tankless water heaters, as well as spa/hot-tub heaters, humidifiers, ice machines, water coolers, air conditioners, and cooling jackets. It will dissolve harmful mineral buildup from water lines. Minerals occurring in hard water will adhere to the sides of copper pipe, gradually choking the water path and eventually interfering with the normal operation of your water heating and cooling appliances. This kit dissolves the mineral buildup by circulating a safe hot acid solution. It The kit contains: a submersible pump, 1/2" clear plastic hoses, heavy duty brass swivel fittings, and the descaling agent consisting of 2 lb of sulfamic acid which will yield four one-gallon treatments. The kit is available either with 3/8", 1/2", or 3/4" fittings.

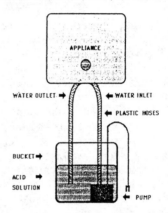

■**45-490   Un-Clog-It Descaling Kit (3/8") $115**
■**45-491   Un-Clog-It Descaling Kit (1/2") $115**
■**45-492   Un-Clog-It Descaling Kit (3/4") $115**

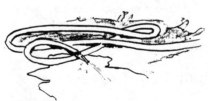

RUN a hose. Tired of all those water restrictions? Well then, show your independence. Open the spigot; let 'er rip. An all-night soaking costs only pennies, and no one can prove you did it.

Copper Cricket Solar Water Heater

"The most exciting, cost-effective solar hot water system to be developed in the last decade." - Worldwatch Institute.

"It is what solar always should have been...it ought to replace pumped and controlled active systems...Personally I'd recommend it as the best system on the market." - Amory Lovins

After months of research we have found the solar hot water heating system that meets our standards - the Copper Cricket. It has the advantage of active systems in that there is a remote flat plate collector that can be as much as 36 feet above the storage tank. But, it is a passive system as well with *ABSOLUTELY NO controls, pumps, or freeze protection required!*

The Copper Cricket will provide 52 gallons per day of hot water heated to 160° or 80 gallons per day heated to 120° under ideal conditions. This will meet the needs of 3.4 people in a standard situation. A good handyperson can install the system in a weekend. The Copper Cricket uses the "geyser effect" to pump the water up the collector much like a coffee percolator and as reliable as sunshine. The fluid is a water and 15% methyl alcohol mix and the system's thermal rating is 22,000 Btu per day. The Copper Cricket makes use of your existing electric water heater and not an expensive "solar" storage tank. The solar pad heat exchanger sits underneath your tank and heats up the water. The complete system includes the solar pad heat exchanger, valve pak with mounting and connecting hardware.

System life expectancy is over 30 years. Popular Science picked this as one of the most significant scientific breakthroughs of the year in its December '89 issue. The installation kit includes drain valves, hand vacuum pump with access valves, miscellaneous parts & pieces, etc.

The copper cricket package includes the solar collector, "Solar Pad" heat exchanger and all parts necessary for a typical installation: the roof-jack, anchor

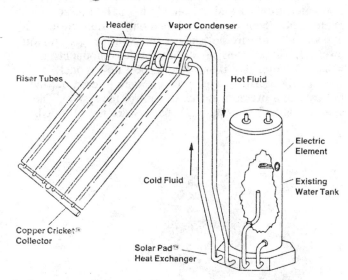

bolts, copper fittings, (except those required for plumbing between the collector and the heat exchanger), and a couple of extra nuts and washers just in case some roll off the roof.

The **Installation Kit** includes drain valves, hand vacuum pump with access valves, miscellaneous parts and pieces, etc. Note: $125 less the cost of any materials used will be refunded upon return of the installation kit. *You will need to buy the Installation Kit to properly install your Copper Cricket!*

■45-405	Copper Cricket	$2,180
■05-230	Cricket Crating Charge	$ 50
■45-415	Installation Kit	$ 150

Shipped Freight Collect from Oregon
Allow 4 weeks for delivery

Heliocol Solar Pool Heaters

Solar swimming pool heating is the most cost effective and technically proven form of thermal solar heating. A solar pool heater will extend your pool heating season and maintain a pleasantly warm pool during the summer. Our supplier, Heliocol, produces more square footage of solar thermal collectors than any other manufacturer in the world. These collectors are made of polypropylene and carry a ten year full warranty, parts and labor included.

Typically a Heliocol solar pool heater will extend your pool swimming season by two or more months and give you a toasty pool in the summer. There are too many variables to accurately quote you a price in the space we have here. To give you an idea on pricing: A typical 16' x 32' pool will require around seven 4' x 10' solar panels at a total system cost of less than $2,000.

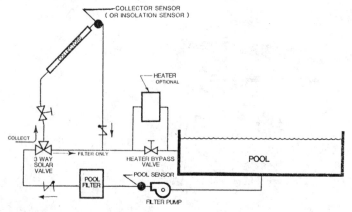

45-406 Installation Video $20
(Refundable with purchase of system)

Wood Burning Water Heaters

Chofu Wood-Fired Water Heater

The Chofu is a precision-built, high-efficiency, wood-fired water heater. It circulates water without need for a pump by thermosyphon. The Chofu can be used for hot tubs, domestic hot water systems, thermal-mass heating, or solar collector back-up.

The stove body of the Chofu is constructed of durable 22 gauge stainless steel with a 1⅛" thick water jacket surrounding the firebox. The Chofu is 23"L x 16" in diameter. Heat transfers very rapidly through 9 sq ft of heat exchange surface. The Chofu will heat 200 gallons of water from 55 to 105° F in 2-1/2 hours. It takes up to 17" wood and uses about two 5-gallon buckets of wood to heat a 200-gallon hot tub.

The Chofu is manufactured in Japan to a very high standard of quality. It was developed to maximize the heat output from the limited supply of wood available in Japan. (The Japanese have a passion for soaking in the "ofuro" or hot bath and use it daily.) The front of the stove is constructed of ⅛" steel. The circulation pipes are 1-1/2". The chimney stack is 4" and a drain plug provides freeze protection.

Standard accessories include: through-wall tub ports, connecting hoses, ash rake, drill bit, and screws. Many customers use galvanized "stock tanks" for hot tubs because of their low cost, wide availability, and ideal 24" depth. We have an enclosure kit available for converting a stock tank into a hot tub including foam insulation, cedar tongue & groove paneling (that makes the unit very attractive), stainless straps, and a drain system. A variety of accessories are available including vinyl covers, filters, stainless stove pipe, and more.

The photograph shows a 5-foot-diameter stock tank with the cedar enclosure kit.

"Since my introduction to the Japanese bath many years ago I use the Chofu to heat my small hot tub almost every day. The Chofu has served me as a faithful friend for 4 years. The grates, however are now beginning to break and I will need to get some new ones. Thank you for the best purchase I have ever made!" **Eric Nelson, Vashon, WA**

■**45-402**   Chofu heater $575
■**45-403**   Cedar enclosure kit (Chofu) $425
Specify whether your tub is more or less than 1/2"
Shipped freight collect from Washington

Send SASE for brochure on Chofu or conversion kit

The Snorkel Hot Tub Heater

The Snorkel Stove is a wood-burning hot tub heater that brings the soothing, therapeutic benefits of the hot tub experience into the price range of the average individual. Besides being inexpensive, simple to install, and easy to use, it is extremely efficient and heats water quite rapidly. The stove may be used with or without conventional pumps, filters, and chemicals. The average tub with a 450-gallon capacity heats up at the rate of 30° F or more per hour. Once the tub reaches the 100° range, a small fire will maintain a steaming, luxurious hot bath.

Stoves are made of heavy-duty, marine-grade plate aluminum. This material is very light, corrosion resistant, and very strong. Aluminum is also a great conductor of heat, three times faster than steel. The Snorkel hot tub stove will heat up a hot tub 50% faster than an 85,000-Btu/hr. gas heater and 3 times as fast as a 12-kilowatt electric heater!

"Thank you very much for the best buy we've ever made! We had no problems with the stove and we were amazed at how well it worked." *Don Evans, Zenia, CA*

■**45-401**   Snorkel stove $595
Shipped freight collect from Washington State

Wood-Flame Water Heater

The Wood-Flame water heater allows you to heat your water when you need it quickly and very efficiently. It works well in conjunction with solar collectors as a back-up system or as a stand-alone system. It's efficient and economical as a spa or hot tub heater as well. In conjunction with a small circulating pump it can heat vast quantities of water.

The Wood-Flame water heater will provide very hot water in 5-10 minutes after it is lighted. Although the capacity is only slightly over 6 gallons, when mixed with cold water its functional capacity is much greater. The heater will heat 200 gallons of water from 73° F to 108° F in two hours with firings at regular intervals. A completed two-hour firing should consume about 2/3 of a garbage can full of kindling. Properly insulated, one or two small firings the next day should compensate for heat loss.

The heater is manufactured in Mexico where it has been used for many years providing virtually free hot water. The Wood-Flame water heater is constructed of galvanized steel in the tank and firebox. The firebox door and cover are constructed of cast iron. The internal grate is cast iron, the base and ash pan are steel. The piping is 1/2" galvanized, chimney stack is 4". Each water heater is tested to 80 psi. The heater will burn almost anything including kindling, corn cobs, paper, leaves, and straw.

I have used a Wood-Flame outside for 5 years. The bottom is just starting to rust but nowhere near falling apart yet. I love it. They are a lot of fun to fire up for the hot tub - flames shoot up 18" out of the stack. For protection from rain I put a coffee can over it. The only problem with them is that they have no drain, so if they are to be used outdoors they have to be protected against freezing; I use quick release connections so the heater can be turned upside down to drain. **D. Carpenter, TX**

■45-404 Wood Burning Water Heater $229
Shipped freight collect from Texas

Note: The availability of the wood burning water heaters is subject to our source of supply in Mexico. We recommend you call first before ordering to ensure availability.

Send SASE form more literature.

Real Goods Analysis of Wood-Burning Water Heaters

Now that we're featuring three different wood-burning water heaters we thought it might be time for some objective analysis. Below we list the weak and strong points of each heater, with comparisons of each heaters' specifications:

Chofu heater
Heat exchange surface area	9 sq ft
Approx. output	32,000 Btu/hr.
Firebox dimensions	18"L x 14"W x 8.5"H
Wood length used	17"
Time between refills at fast burn	45 minutes

Snorkel Stove
Heat exchange surface area	22 sq ft
Approx. output	120,000 Btu/hr.
Firebox dimensions	27"L x 12"W x 19"H
Wood length used	26"
Time between refills at fast burn	1 hour

Corona Heater
Heat exchange surface area	4 sq ft
Approx. output	17,000 Btu/hr.
Firebox dimensions	10" dia. x 19"H
Wood length used	6"-10"
Time between refills at fast burn	15 minutes

Here's a letter we received from Frazier Mann in Clinton, Washington:

"The Chofu heater and Snorkel stove are quite comparable in effectiveness as hot tub heaters but differ greatly in the type and size of tub they are best suited for. The Chofu cannot practically heat over 350 gallons and therefore is best for heating stock tanks or other low volume tubs. The Snorkel requires a larger tub to accomodate it inside and typically is used with a wooden tub of 450-900 gallons. From a survey of hot tub users we found that a four-person stock tank heated by the Chofu has a similar heating time as a four-person wooden tub heated by the Snorkel. A complete system costs a little less with the Chofu. The choice is a matter of style, cost, and water and wood consumption."

Gas Heaters & Appliances

Space heating, like water heating, is one of the homestead's largest power consumers. Wood has traditionally been the most sensible fuel for heating, especially for those with renewable wood lots on the "back 40." But for those of you who have hung up your splitting mauls or who just don't have access to a reliable wood supply, there is always propane, which is still fairly cheap in most parts on the country and quite an efficient fuel as a heating source.

Valor Unvented Heaters

English made Valor unvented heaters are 99.9% efficient and will run on either propane or natural gas. Heating up to 1,100 square feet, these attractive units are safe, too: an oxygen depletion sensor automatically shuts down when there is insufficient oxygen in the room. The Valor comes in two sizes - 15,000 Btu/hr. and 25,000 Btu/hr. (propane); and 18,000 BTU and 30,000 (natural gas). The ignition is piezo-electric and there is a three-position heat control knob for adjustable output. If the pilot light goes out, the gas flow is immediately stopped by the safety thermocouple.

The Valor comes in a woodgrain and brass cabinet which is either wall or floor mountable. The 15,000 Btu/hr. propane unit (VU-184P) measures 18.5"H x 6-7/8"D x 19-3/8"W, and the 25,000 Btu/hr. propane unit measures 18.5"H x 6-7/8"D x 26-1/2"W. Order the "P" model for propane and the "NG" model for natural gas.

65-101	Valor VU-184-P (15,000 Btu/h)	$295
65-102	Valor VU-185-P (25,000 Btu/h)	$399
■65-103	Valor VU-184-NG (18,000 Btu/h)	$325
■65-104	Valor VU-185-NG (30,000 Btu/h)	$435

Unvented heaters are not approved by building officials in some areas. Check with your local authorities before purchasing.

Catalytic Propane Heaters

The "CAT" is made in Washington State and is the only vented propane catalytic heater on the market approved by the AGA (American Gas Association) for use in a home. It is a direct vented heater that allows no combustion inside the room. It has a thermostat and has passed NFPA-501-C making it fully legal in California for either homes or RV's. A 6P-12A (the 12V unit) keeps our office toasty and a 1500-XL (off of our Trace 2012 inverter) heats one of our showrooms. They're quiet, easy to maintain, thermostatically controlled, and very economical with propane! Catalytic heaters work like the radiant heat of the sun; they heat people and not objects like furniture.

CAT 6P-12A

The 6P-12A is a 6,000 Btu/hr. heater with 12V automatic ignition, using less than 1/2 amp. Its propane consumption is only 1/4 lb. per hour. Its penetrating radiant heat warms people first, not objects. It will warm an area up to 500 square feet.

65-201   6P-12A 12V Catalytic Heater $445

CAT 1500-XL

The 1500-XL is a propane catalytic heater designed for use in homes. It has 110V automatic ignition. It is vented, thermostatically controlled, and fully approved by the AGA. It too will warm an area up to 500 square feet.

■65-202   1500-XL 110V Catalytic Heater $475

Cozy Vented Heaters

Because the Valor unvented propane heaters aren't officially legal in California and several other states, we're introducing a high quality vented heater. Cozy is made by a very reliable American manufacturer. All units feature the low-Btu/hr. safety pilot and are A.G.A. approved for both natural gas and propane.

Cozy Vented Console Heaters

The Cozy closed front vented console heaters are extremely efficient and reliable. The 20,000 Btu/hr. unit will heat 500 square feet comfortably and more to just take the chill off. The 35,000 Btu/hr. unit will heat around 700 square feet. The heaters come in a textured brown baked-enamel finish, and will operate on either natural gas or propane. Safety is assured with the "vent spill switch", which automatically shuts the unit off in the event of flue blockage or improper installation. The 20,000 Btu/hr. heater is available without a thermostat (C420A) and with a thermostat (C220A). The 35,000 Btu/hr. heater (C235A) comes standard with a thermostat. The cast iron slotted-port burner is designed to provide economic operation and quiet ignition and extinction and is guaranteed for the life of the unit.

■65-342	20,000 Btu/h w/o Thermostat	$395
■65-320	20,000 Btu/h w/Thermostat	$495
■65-335	35,000 Btu/h w/Thermostat	$545

The 35,000 Btu/hr. unit is shipped freight collect

We also have available Cozy direct-vented heaters. Call one of our technicians for details.

Hardwick Gas Ranges

Customers have been asking us for years to come up with a good basic gas cook stove and we've finally found one that we're real happy with. Hardwick ranges are manufactured in Cleveland, Tenessee. They come with a universal orifice so they'll accept either propane or natural gas, and we've culled out only the models without pilotless ignition so you don't need 110V AC to light them. We're introducing them at very attractive prices. We checked with our local gas company and set our prices 15%-20% below them.

30" Standard Range

The Hardwick C9616-79R is a standard 30" range that will operate on either propane or natural gas. It comes with a standing pilot and is not "pilotless ignition." It measures 40" High x 30" Wide x 25-5/8" Deep. It can be ordered in either almond or white. It comes with a low profile 4" backguard. This unit normally retails for $495.

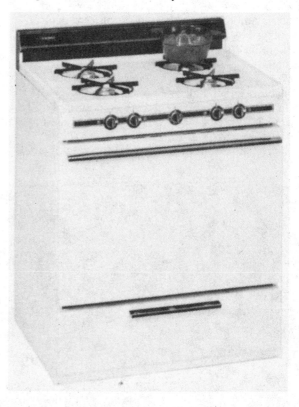

■61-102　30" Standard Range　$395
Shipped freight collect

To determine the number of Btus/hour required to heat your space with a propane heater, first calculate the cubic feet in the space to be heated. Multiply this number by 6 for sunbelt states, CA, OR, WA; by 8 for the midwest and east, and by 10 for Alaska. This total will give you the number of Btus/hour required to heat the space.

Refrigeration

We often speak of the need to reduce dependence on traditional electrical usage to design an efficient alternative energy system. Up until early 1983, refrigeration rivaled air conditioning and electric heating for inefficiency. A standard size 17-cubic-foot AC unit, even with energy-saving design, uses from 3 kWh to 5 kWh per day to operate. The standard AC refrigerator is one of the most poorly designed and inefficient appliances known to humankind. The refrigerators produced back in the 1940's are more efficient than 90% of the refrigerators designed today!

Finally, solutions to the traditional refrigerator inefficiencies are coming to bear. The first is propane or LPG gas-fired refrigeration and the second is the standard and new low-voltage ice boxes and chests using advanced compressor design. Some of the newest designs incorporate eutectic plate technology, although a good percentage of the efficiency gain is achieved by extra insulation and by simply placing the heat-generating electronics and mechanics, like the compressor, on top of the unit rather than underneath, where the heat can penetrate the interior. You'd think the appliance giants would have thought of that one, wouldn't you - almost makes you think they're in collusion with the power companies! The chest-type design is a factor also, because it helps keep cold air in when opening the door.

The new low-voltage refrigerator compressors require very little maintenance, and some are brushless. Most carry a five-year warranty. If you live in an area where temperatures are high, more power is required the higher the temperature. Be sure you plan for this variation when considering photovoltaics or wind generation.

Both propane gas and all except one model of low-voltage refrigerator suffer from a common problem of size. The largest propane unit available is 8 cu ft and the same is true for most low-voltage refrigerators with one notable exception - the custom-built Sunfrost. It can be ordered in an impressive 19 cu ft unit and uses only about 550 watthours per day at an ambient temperature of 72 degrees. That is a real efficiency breakthrough.

Propane gas will be around for a long time as it is a byproduct of all fossil fuel production. Currently the cost of operating a gas refrigerator is about $6 per month (at $1 per gallon of propane) while a conventional electric refrigerator using 3kWh per day costs about $11 per month (at $0.12 per kWh). However, there are a few drawbacks to propane refrigerators. In very hot climates they become less efficient, and in 100-degree heat it's hard to make ice. They *should* be vented, especially when used in today's almost airtight homes, because they combust oxygen and can release toxic fumes. In reality, almost none of our customers has ever vented a Sibir or new Servel propane refrigerator, and we have not heard of a problem to date. The advantages of propane refrigerators are no moving parts, little maintenance, and longevity. Gas refrigerators are known to operate for 40 years or more.

Many of you are familiar with the old Servel propane refrigerators, the mainstay of refrigeration in the 1920's through the 1950's. While you still see a few of these kicking around they were (are) gas hogs using up to 5 gallons of propane per week. (Not to be confused with the new highly efficient Servels which use only 1.5 gallons per week.)

Sunfrost Refrigerator/Freezers

The Sunfrost is a design breakthrough: for the first time it is practical to power a refrigerator on 12 volt DC. Refrigeration is typically the largest consumer of electrical energy in an energy-efficient home. Conventional AC refrigerators draw 3,000 watthours (3 kWh) per day but the Sunfrost uses only 350 watthours (1/3 kWh) per day, making it by far the world's most efficient electric refrigerator. The unit is cooled by two highly efficient top-mounted hermetically sealed compressors. The compressors and condenser are top mounted so they cool efficiently without having the heat re-enter the cabinet.

The walls of the refrigerator contain 3" of insulation and the freezer section up to 6" of polyurethane foam. Frost buildup in the freezer is very slow because there is no air circulating. The entire cooling system contains only one moving part, powered by a brushless DC motor. The Sunfrost is quiet without the loud hum of AC refrigerators. All models are available in 12V or 24VDC, or 120VAC.

The manufacturer claims that the most popular unit, the RF-12 12-cubic-foot refrigerator-freezer, draws 28 amphours per day with a 70° F ambient temperature. Tests have shown the unit to exceed these specs. These figures mean that the entire refrigeration system can be powered from only two 48-watt PV modules! Although the initial price seems high, the Sunfrost is actually the cheapest source for electric refrigeration when amortized over a ten-year period.

All Sunfrosts are guaranteed for one year. Besides the 12 cubic foot refrigerator-freezer (the most popular) Sunfrosts are also available from 4 to 19 cubic feet.

In the chart below, R stands for refrigerator, F for freezer; the amphours per day energy consumption figures assume a 70° F ambient temperature and are given for 12VDC. Increase by about 50% for a 90° F ambient temperature. There is a $100 charge for any color other than white.

Item #	Model	Watthours per day	Height	Width	Depth	Price
■62-134	RF-16	540	62.5"	34.5"	27.5"	$2,595
■62-135	RF-19	744	64.0"	34.5"	27.5"	2,695
■62-125	R-19	360	64.0"	34.5"	27.5"	2,450
■62-115	F-19	1200	64.0"	34.5"	27.5"	2,795
■62-133	RF-12	336	49.5"	34.5"	27.5"	1,750
■62-122	R-10	180	43.5"	34.5"	27.5"	1,550
■62-112	F-10	660	43.5"	34.5"	27.5"	1,650
■62-131	RF-4	156	31.5"	34.5"	27.5"	1,450
■62-121	R-4	108	31.5"	34.5"	27.5"	1,450
■62-111	F-4	336	31.5"	34.5"	27.5"	1,395
■62-139	RF-16 - 110VAC		62.5	34.5	27.5	$2,595

05-200 Crating Charge on Sunfrost (all models)
Shipped Freight Collect from Northern California $ 60

All Sunfrosts are made to order - allow 4-6 weeks for delivery. Be sure to specify if you want the hinge on the right or the left and the voltage!

Here's an interesting aside to Sunfrost refrigerators that we calculated recently when *Bay Area Backroads*, a San Francisco TV station, came to do a feature on us. **Consider this:** A Sunfrost RF-16 uses approximately 180 kWh of electricity per year compared to approximately 1080 kWh per year for a standard "energy saving" refrigerator. **This means the Sunfrost saves 900 kWh per year.** If every American home (of which there are 100 million) had a Sunfrost, our country would save 90 billion kWh every year. A large nuclear power plant generates approximately 5 billion kWh per year. This means that if **every American home had a Sunfrost, we could immediately shut down 18 large nuclear power plants!** Now, just think... if George Bush were to put his money into Sunfrosts instead of into marine power in the Saudi Arabian desert, maybe we wouldn't even need that oil...

Ecologue, a new consumer's guide to environmentally safe products, lists SunFrost frrigerators as one of its ten best. Also included in the top ten list is the Honda Civic. Sunfrost refrigerators have been shipped to over 50 countries world-wide.

P-16 Refrigerator

This is a brand new 16 cubic foot refrigerator with a large 4-cubic foot freezer has just come on the market in direct competition with Sunfrost. It's available in either 12 or 24 Volt DC or 115 Volt AC. In our preliminary tests in our showroom between the Sunfrost RF-16 and the new P-16, we find that the P-16 uses approximately 30% to 50% more electricity to run with freezer being at 18 degrees F and refrigerator at 37 degrees F. (The Sunfrost is typically 0 degrees F for the freezer and 34 degrees F for the refrigerator.) We have yet to test the P-16 at a 90° ambient temperature (where traditionally one-compressor refrigerators suffer) but we suspect that it won't meet the muster of the Sunfrost. To its credit it features extra-quiet operation (like the Sunfrost) without the typical HUM of standard refrigerators. It includes an inside light, a *"low-voltage warning"* indicator light and deep adjustable shelves. The P-16 is an off-white color, measures 32" Wide x 28" Deep x 71" High and weighs 185 lbs. (200 lbs. crated). Two doors are included for maximum efficiency. The unit typically uses 800 watthours per day depending on climate and operating conditions which means it can be operated with 3-5 48-watt solar modules. The P-16 utilizes one compressor compared to Sunfrost's two compressors. We're introducing the brand new P-16 at a very affordable price. **Be sure to specify 12V, 24V, or 110V and allow up to four weeks for delivery.** The manufacturer has just developed a "Turbo Cooling Fan" that will increase efficiency in very hot climates. We strongly recommend this option if your ambient summer temperatures are greater than 90 degrees F.

■62-145	P-16 Refrigerator	$1,995
■05-215	Crating Charge for P-16	$ 60
■62-146	Turbo Coooling Fan	$ 95

Shipped Freight Collect from Northeastern California

Servel Propane Refrigerators

Servel has been a household word in gas refrigeration for over 50 years, when it was first marketed by the Swedish company Electrolux in 1925. In 1956, the rights to Servel were acquired by Whirlpool, which was unsuccessful with the unit, and it disappeared for 30 years. Now Servel is back with the state-of-the-art household propane refrigerator.

We've sold well over 400 of the new Servels since their reintroduction in April of 1989 (aided by the fact that Sibir went out of business in June, 1990), and have one hooked up in our showroom. All continue to perform flawlessly and appear to be extremely well built. They draw less than 1-1/2 gallons of propane per week! The body is all-white and the door is hinged on the right. The unit comes with four refrigerator and two freezer rust-proof racks. The spacious interior has two vegetable bins, egg and dairy racks, frozen juice rack, ice cube trays, and an ice bucket. An optional 2-year warranty can be purchased for $39.95. The total volume is 8 cubic feet (6.3 for the refrigerator, 1.7 for the freezer). Total shelf space is 12.4 square feet. The overall dimensions are 57-3/4" high by 24-3/4" wide by 24-3/4" deep. The net weight is 181 lb. It is operable on either propane or 110VAC (where it draws 250 watts of power).

■62-300	Servel Propane Refrigerator	$1,395

Shipped freight collect

Norcold Refrigerators

Norcold manufactures some very efficient refrigerators that run on 12VDC, and a few on propane. The core of their 12V refrigerators is the 40-watt compressor unit manufactured by Fuji for Norcold for the last 25 years. While not as efficient as Sunfrost 12V refrigerators, they are more versatile and less expensive.

DE-251 Norcold

The 2.0-cubic-foot DE-251 is the best selling refrigerator in the entire Norcold line. It has one hermetically sealed 40-watt 12V compressor. It's designed to be built in so the sides and top have been left unfinished, allowing you to slip it under a counter, into a closet, or into a wall. It draws 3.75 amps for 12 hours per day (540 watthours per day) and measures 20-1/4"H x 17-3/4"W x 20-3/8"D and weighs 56 lb. It's available as an AC/DC unit as the DE-251 or can be purchased as 12V only as the DC-254 (for cheaper!). one-year warranty.

■62-431   Norcold DE-251 (AC/DC) $485
■62-432   Norcold DC-254 (12VDC only) $425
Shipped freight collect from Ohio

DE-704 Norcold

The DE-704 is an AC/DC 3-cubic-foot refrigerator that draws 3.75 amps approximately 12 hours per day (540 watthours per day). It has a cross-top freezing compartment. The cabinet is unfinished for built-ins. The unit measures 32-7/8"H x 22"W x 21-7/8"D and weighs 85 lb. one-year warranty.

■62-420   Norcold DE-704 $665
Shipped freight collect from Ohio

DE-828 Norcold

The DE-828 is an AC/DC 6.2-cubic-foot, double door refrigerator/freezer. It has two 40-watt compressors and draws 7 amps for 12 hours per day (1008 watthours per day). The spacious refrigerator has three removable shelves, twin crisper, and in-door storage. The big 40-lb freezer also has in-door storage and two ice trays. The unit measures 51-7/8"H x 23-15/16"W x 23"D and weighs 132 lb. one-year warranty.

■62-410   Norcold DE-828 $925
Shipped freight collect from Ohio

Mariner 12V Ice Cube Maker

Mariner 12V ice cube makers assure ice at all times. They are fully automatic with no special startup required. The model SFI-55 makes up to 13 lb per day (435 ice cubes) and stores them. The unit measures 24.5" high x 14" wide x 16.5" deep. It will run on either AC or 12VDC and uses 6 amps to harvest the ice and only 1 amp to maintain the ice cubes. It will harvest around one pound of ice per hour.

■62-531 Mariner SFI-55 (AC/DC) ice maker $1,150
Shipped freight collect from Los Angeles

MRFT Series Norcold 12V Refrigerator or Freezer Combinations

The MRFT series can be switched back and forth between refrigerator and freezer mode. They are chest type units in charcoal grey color. The MRFT-630 is a 1.06-cubic-foot unit measuring 14-3/4"H x 25"W x 14"D, with a 50 lb capacity, and will hold two cases of beverages. The MRFT-640 is a 1.5-cubic-foot unit measuring 18-7/8"H x 24-13/16"W x 14-1/8"D, with a 75 lb capacity, and will hold three cases of beverages. Both the 630 and the 640 use a single 40-watt compressor and draw 540 watthours per day. The MRFT-660 is a 2.15-cubic-foot unit measuring 17-3/8"H x 31-1/8"W x 19-1/4"D, with a 100 lb storage capacity. It has a larger 60-watt compressor and draws 720 watthours per day. All units can be switched back and forth from refrigerator to freezer mode with the switch of a dial. one-year warranty.

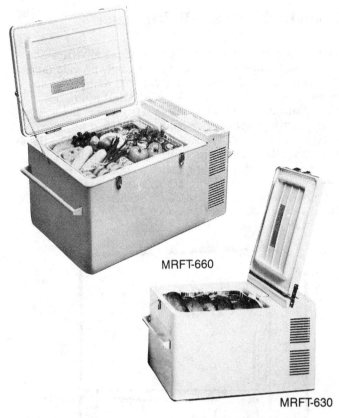

MRFT-660

MRFT-630

■62-441	Norcold MRFT-630	$595
■62-442	Norcold MRFT-640	$695
■62-443	Norcold MRFT-660	$895

Shipped freight collect from Ohio

Koolatron 12V Refrigerators

Koolatron (of Canada) is the world's largest manufacturer of 12-volt coolers. The secret of their cooler is a miniature thermoelectric module that effectively replaces bulky piping coils, compressors, and loud motors used in conventional refrigeration units. For cooling, voltage passing through the metal and silicon module draws heat from the cooler's interior and forces it to flow to the exterior, where it is fanned through the outer grill. The Koolatron has a relatively low amp draw for a refrigerator - averaging 2.5 amps (max. amp draw is 4.0 amps).

The unit will maintain approximately 40-50° F below outside temperature. It has an interior temperature indicator as well as a thermostat and can be used on 12VDC or adapted to 110VAC (*by ordering the optional adaptor*). The unit is constructed of high impact plastic and super urethane foam insulation, and comes with a 10-foot detachable 12V cord. The Scotty cools to 40 degrees below outside temperature and the Super Scotty to 50 degrees below. All models are warranted for at least one year. In addition the Scotty II has a five year warranty on the cooling module, and the Super Scotty and Caddy II have 10-year warranties on their cooling modules.

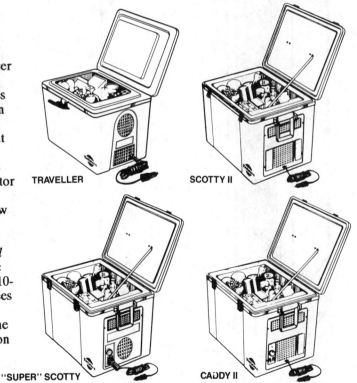

TRAVELLER SCOTTY II

"SUPER" SCOTTY CADDY II

	Model	Quarts	Cu.ft.	Dimensions	Price
■62-521	Traveller	7.5	0.25	10"W x 12.5"H x 15"L	$135
■62-522	Scotty II	27	0.9	16.5"W x 16.5"H x 20"L	$195
■62-523	Super Scotty	27	0.9	16.5"W x 16.5"H x 20"L	$250
■62-524	Caddy II	36	0.2	16.5"W x 16.5"H x 22"L	$325
■62-525	AC Adaptor			(works on all models)	$ 59

Dometic Kerosene Refrigerator

Kerosene refrigerators are in extensive use in Africa and lots of third world countries. We frequently provide them for our Amish and Mennonite customers who prefer not to use propane. The RAK-1302 is identical in interior dimensions to our Servel propane refrigerator. It has a 3.7 gallon kerosene tank which holds enough fuel for approximately three weeks of normal operation. The flame of a simple kerosene burner operates the ingenious Dometic absorption cooling unit. The operation is completely silent, and there are no moving parts to wear out. Outside dimensions are 24.75" Wide x 68.25" High x 24.75" Deep. The freezer measures 2.74 square feet. The refrigerator weighs 225 lbs.

Dometic RC-65 Propane Deep Freezer

The Dometic RC-65 is a very dependable deep freezer that gives efficient, trouble-free performance. Absolutely no electricity is required in this propane-powered unit. It comes with thermostatic control, a convenient defrosting drain system, and a childproof lock. It also has a built-in spriit level to make sure the freezer is properly leveled, a thermo-electric safety valve for gas supply, a built-in easy to use flame-igniter, and a stainless steel wear strip for scratch-free loading and unloading. The RC-65 may also be converted, in less than 10 seconds, to a beverage cooler. Capacity is 5.5 cu. ft. (155 liters). It measures 40.9"H x 37.4"W x 30.1"D.

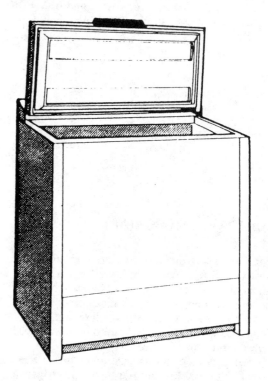

■62-512   Dometic RC-65 Propane Freezer $1,895
Shipped freight collect from Indiana

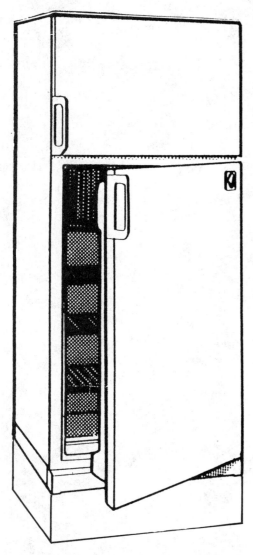

■62-302   Dometic Kerosene Refrigerator $1,650
Shipped freight collect from Ohio

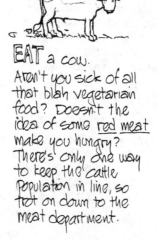

EAT a cow.
Aren't you sick of all that blah vegetarian food? Doesn't the idea of some red meat make you hungry? There's only one way to keep the cattle population in line, so trot on down to the meat department.

Solar Cooling

A wide range of DC air-moving systems and equipment is available to operate directly from the sun (sun synchronous) or from battery power. Everything, from cooling a home via evaporation to grain drying and greenhouse ventilating, may be accomplished. Browse through these pages and use your imagination.

Air-moving Terminology

Here are a few terms and formulas that will be helpful in choosing the size fans you need and understanding some of the charts that appear with some equipment.

cfm - Cubic feet per minute.

Free Air Delivery - No effective resistance or restriction of air flow (no static pressure) at the intake or outlet of an air-moving device.

rpm - Revolutions per minute.

SP - Static pressure - a measure of resistance to movement of forced air through a system of ductwork, louvers, screens, etc. Measured in inches of water gage (w.g.); the height, in inches, to which the pressure will lift a column of water.

To choose the right fan; figure the volume of space to be ventilated by multiplying the building or room's length by its width and by its height in feet. Exclude basements, closets, pantries, storage areas, and attics. Add room totals to determine the total cubic feet of space to be ventilated. A fan capacity is given as cfm (cubic feet per minute). Experts recommend that a hole supplying at least one square foot of net free open area be provided for each 750 CFM of fan or blower capacity at 0.10" SP (static pressure). If not, add more fans or select a larger size fan blade or both. Louvers with insect or rodent screens may only be equal to one half of the free open area required depending on the hole size of the screening, so be sure to take this into account in your planning. A final note: *you can't blow air into a closed container.* This will cause high static pressures and low CFM.

12V - 16" Intake/Exhaust Fan

For moving large volumes of air, this 16" 12V fan can't be beat. It moves a full 1,000 cfm and comes with a reversible permanent magnet motor, a 16" three-wing aluminum blade, and a 20" square steel venturi. An 18" hole is required for mounting. It draws a meager 1.1 amps at 12VDC.

64-221   16" Intake/Exhaust Fan $125

12V - 12" Intake/Exhaust Fan

This is a smaller version of the 16" fan. It delivers 550 cfm, comes with the same reversible PM motor, has a 12" three-wing aluminum blade, and a 16" square steel venturi. It requires a 14" hole for mounting. Uses 1.1 amps at 12VDC.

64-222   12" Intake/Exhaust Fan $119

Shaded Pole Blower

This 12VDC air delivery system can be used to deliver heat to other rooms in the house through ducting or to move heat up into a loft. It moves up to 137 cubic feet of air a minute and uses only 5.1 amps at 12V. UL approved.

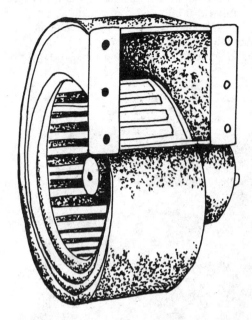

64-200   Shaded Pole Blower (12V) $65

12V Fans

These 12VDC axial fans are ideal for moving woodstove heat throughout the house. The brushless motor design minimizes electromagnetic interference and radio frequency (rf) interference. The fans have PBT plastic housings with permanently lubricated ball bearings. All motors are polarity protected. The voltage range for the nominal 12VDC fan is 6 to 16 VDC. The two smaller fans are 1" in depth and the larger fan is 1.5" deep.

Item #	Model	CFM	Amps	H	W	Price
64-211	4C909	15	0.24	2-3/8"	2-3/8"	$42
64-212	4C911	32	0.25	3-1/8"	3-1/8"	$42
64-213	4C918	105	0.55	4-11/16	4-11/16	$45

Sunvent

The Sunvent is a small but powerful solar-powered ventilating fan for any application where humid or stale air collects. It is great for boats, motorhomes, travel trailers, and greenhouses. It's simple to install, requires no electric hookup, is solar powered, and is completely weatherproof. It extracts 680 cubic feet of air per hour under normal working conditions. The Sunvent's simplicity is its greatest virtue, as it works whenever the sun is shining, pulling out hot air the fastest when the sun is at its highest. This is the perfect sun-direct application! It requires a 6" round mounting hole.

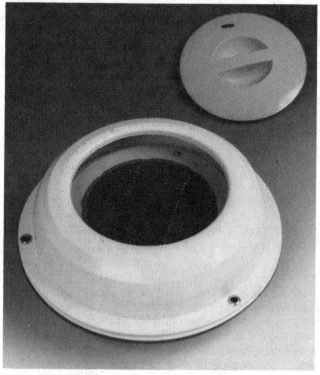

64-242   Sunvent Fan $49

12-Volt Fan Motors

This 1/35 hp 2,350 rpm motor is excellent for powering fans to cool greenhouses, exhaust kitchen range hoods, or move air for cooling or heating from one room to another. The possibilities are limitless. The motor uses 4 amps at 12V (48 watts).

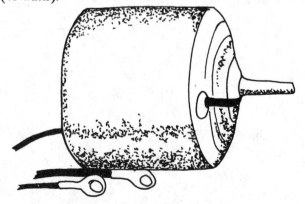

64-401   Fan Motor (12V) $27.50

Thermofor Solar Ventilator Opener

The Thermofor is a compact, solar-powered device which regulates window, skylight, or greenhouse ventilation according to the temperature using no electricity and requiring no wiring. You can set the temperature at which the window starts to open between 55 and 85° F. As the temperature rises 20 degrees, the window will open up to 12 inches. The 15-lb thrust will open a 30-lb hinged window, and it can be fitted in multiple on long or heavy vents. These units are ideal for greenhouses, animal houses, solar collectors, cold frames, and skylights.

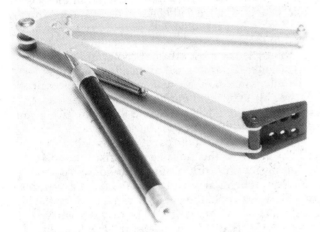

64-302 Thermofor Solar Vent Opener $59

12V Brass Ceiling Fan

This is our old standby ceiling fan that comes equipped with three 44" (measured from blade tip to tip) mahogany finished wood blades and has a tiny draw of 1/2 to 3/4 of an amp (6-9 watts!). The 12" long base fastens to a ceiling with regular butterfly fasteners or may be screwed to wood beams. Base size is 10" x 8 1/2". Fan comes with an attractive brass base. The fan will run solar direct with a 20 watt PV panel. The 4 amp variable speed control switch (Conserve) also has a forward/reverse feature for summer/winter air direction.

| 64-232 | Brass base 12V Ceiling Fan | $225 |
| 24-102 | Variable speed control switch | $ 28 |

Two-Speed Sanno Table Fan

This 8" two speed oscillating fan is quiet and unobtrusive. It can be operated on either 12 or 24 volts. It runs at 2,500 rpm and draws 1.25 amps at low setting, and 2 amps at high setting (when on 12V). It can be table or wall mounted.

64-241 Table Fan (12V or 24V) $35

12V Ultra-Efficient Ceiling Fan

We've just located this new 12V ceiling fan that only draw a scant 0.4A (we measured at 0.3A) at 12VDC. This fan is made of a one-piece black ABS vacuformed body. It is available in two different sizes both wih reversible oak blades. The smaller model is a three-blade 44" diameter fan. The larger model moves 33% more air and is a four-blade 46" diameter fan with the same low 0.4 amp draw. Both fans come with a permanent magnet, heavy-duty, high-torque, reversible, UL-approved motor. In two years of production, the manufacturer has had no returns! Being reversible the fan can do double duty cooling in the summer, and efficiently moving warm air in the winter. (We recommend also ordering the variable speed control switch with forward/reverse feature - 24-102)

| 64-245 | 12V Ultra Ceiling Fan - 44" | $175 |
| 64-246 | 12V Ultra Ceiling Fan - 46" | $210 |

Efficient solar or battery powered air conditioning is still a ways in the future. In the meantime, evaporative cooling can be very effective, especially in areas where humidity is low. The higher the humidity, the less effective evaporative cooling is.

Everyone has experienced natural evaporative cooling. When you step, dripping wet from the shower or swimming pool, evaporative cooling makes you feel cooler, even though the air may be warm. At lakeside or seashore it is evaporation from the water surface that makes the breeze refreshingly cool. In a forest or green field, the air is freshened and cooled by evaporation of moisture from plant and tree leaves, and the human body itself is cooled primarily by the evaporation of perspiration.

Evaporative cooling is undoubtedly our oldest method of finding comfort in hot climates. Ancient records reveal that wetted grass mats and porous jars were used to cool air or water by the evaporative process.

Whenever water is evaporated, heat is absorbed. Wet the back of your hand; then blow on it. The skin surface is immediately cooled. You have just demonstrated the basic principle of evaporative cooling.

The fresh, cool air provided by an evaporative cooler is just like the fresh, cool breeze people enjoy at the seashore. In each case, the two essential requirements are a wetted surface and a source of moving air. The filter pad (aspen fibers) provides the wetted surface. A blower, as in the case of the solar cooler, moves the air. With evaporative cooling, a complete air change occurs every one to three minutes (unlike refrigerated air conditioning, which is a complicated "closed" system that recirculates the same stale dry air over and over).

Relative humidity plays an important part in the effectiveness of evaporative cooling. Here is a performance table of air discharge temperature relative to outside dry bulb temperature and percent of humidity.

Solar Evaporative Air Conditioner

The Recair 18A2 (formerly called the New Breeze or Cool Breeze - this product is marketed by Photocomm still as the "New Breeze") turns hot dry summer air into a refreshing indoor climate, while filtering out dust and pollutants. It can cool up to 400 square feet, using only one 48-watt solar panel and 2-4 gallons of water per day to cool and clean the air. Unlike cellulose cooler filters, it will not promote bacterial growth or foul odors. It will move 700 cfm drawing 3 amps @ 12VDC off of one 48-watt solar panel, or move 850 cfm and draw 4 amps from a 12V battery. It is installed with simple hand tools and connected to a garden hose.

Recair is the most efficient cooler of its type on the market today. Conventional air conditioners recirculate the same stale air within your house. Recair uses outside air and filters out any impurities such as pollen or dust, while adding moisture to the air giving you a more healthful, cool indoor air environment.

Installed with a solar panel, Recair will cool when it is needed most. As the morning sun strikes the solar panel, Recair goes to work cooling your house. As the dry air is forced through the wet filter by the fan, the water is evaporated, which cools the outgoing air. The drier the air, the greater the temperature drop. The only drawback is that it will not work efficiently where relative humidity exceeds 40%. Dimensions: 17"W X 18.5"H X 14.25"D

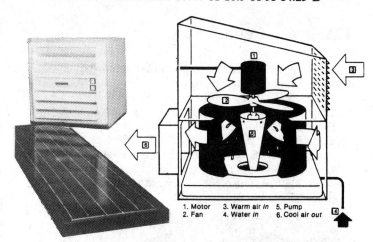

1. Motor 3. Warm air *in* 5. Pump
2. Fan   4. Water *in*   6. Cool air *out*

■ **64-201 Solar Evaporative Cooler $495**
■ **64-202 3-gallon hand pressure tank $ 65**

Table For Evaporative Cooler Discharge Temperatures

Outside Dry Bulb Temperature	Outside Relative Humidity							
	5%	10%	15%	20%	30%	40%	50%	60%
80°F	56°F	58°F	60°F	61°F	64°F	67°F	69°F	72°F
90	62	64	66	68	72	75	78	81
100	68	70	73	75	79	83	87	90
110	74	77	80	82	87	92	95	—
120	79	83	86	89	95	100	—	—

Chapter 5

Sanitation & Conservation

Waterless Composting Toilets

Since the beginning of time, humankind has struggled with the dilemma of how to properly dispose of human waste. How many of us nearly-twenty-first century, enlightened conscious beings really know what happens to our waste after we hit the flush handle on our toilets? In "wet," or waterborne, sanitation systems, excrement is flushed away through a network of pipes along with other household wastes and effluent from business and industry. Raw sewage is typically simply discharged into nearby rivers, lakes, or coastal waters. Most industrial cities deliver household sewage to central treatment plants, where various combinations of physical, biological, and chemical actions eventually yield a purified wastewater effluent and the nutrient-rich substance called sludge.

Affordable options for disposing of sludge are becoming increasingly rare. Landfilling is banned in many places, and where it is allowed it can cost up to $150 a ton to dispose of it. Conventional sludge disposal is becoming so expensive that now even countries that pioneered the

most elaborate sanitation methods are shifting to recycling. Sludge can be applied to the top of soil or plowed into the upper surface for effective nutrification of the soil. The city of Seattle recycles every ounce of the 100,000 tons of sludge its 1.1 million residents produce each year.

But what about the rest of this nutrient-rich and valuable fertilizer? City dwellers worldwide generate between 100 and 150 million tons of nutrient-rich human waste every day, most of it passed along to rivers, streams, and oceans without a thought toward reuse. The same basic nutrients, nitrogen, phosphorus, and potassium, that farmers purchase in chemical fertilizers are treated like garbage. The Environmental Protection Agency estimates that American sludge has a total nutrient content equivalent of 10% of the chemical fertilizers farmers purchase - or a value of over $1 billion per year!

What a contrast from many third world countries where soil fertility probably could not have been

Bio-gas generator, Pokhara, Nepal.
Photo by Sean Sprague

maintained over the last 4,000 years without human waste
nutrient recycling. Chinese fish farmers continue to
fertilize their ponds with nightsoil as they have for
thousands of years. When the fish farmer adds nightsoil to
an aquaculture pond, plants and microorganisms feed on
it, and, in turn, become the fish food.

In a world of finite resources, closing nutrient cycles
is one of the keys to building a sustainable society.
Composting toilets make sense from numerous
perspectives. They are environmentally and ecologically
correct, and are economically efficient costing far less than
standard septic systems. Over 200,000 composting toilets
have now been sold worldwide. Composting toilets are
environmentally safe: they require no septic system, no
holding tank, use no chemicals, and produce no pollutants,
while facilitating the work of nature.

Dear Real Goods,

*Enclosed is a check for yet another Sun-Mar Non-Electric
toilet. My neighbors love mine so much, they want one of
their own. All of my original apprehensions have been
completely answered. We've had our Bowli installed 4
months here in Hawaii and there is no odor whatsoever.
When we dump the compost it's a fine black humus that
we can put right on our vegetable garden. How we could
have ever contemplated spending $3,000 for a septic system
is beyond me. Thanks for saving us so much money, and
for making our lives simpler.*

Stan Dzura, Kealekekua, HI.

Sun-Mar Composting Toilets

We have now supplied well over 500 Sun-Mar toilets (formerly Bowli's) to our customers, and we are pleased to report that these toilets have exceeded our expectations and the manufacturer's specifications. We're convinced that this is the best small composting toilet system on the market. The old problem of resistance from doubting building inspectors has virtually been eliminated with the recent NSF approval (on the XL).

The heart of the Sun-Mar system lies in the revolutionary "Bio-Drum" composting process. The toilet's inventors are the same people who were involved in the original Swedish composting systems 27 years ago. The inventor was the recipient of the Gold Medal for the best invention at the International Environmental Exhibition in Geneva, Switzerland.

Sun-Mar composting toilets work like a compost pile. Human waste, peat moss, and kitchen scraps are introduced into the toilet; heat and oxygen transform this mixture into good fertilizing soil. The BIO-DRUM ensures effective aeration and sterilization, killing anaerobic microbes and mixing the compost well. Turning the drum periodically maintains the aerobic composting process. Oxygen is provided by the ventilation system. The material entering the toilet is approximately 90% water, which is evaporated into water vapor and carried outside through the venting system. The remaining waste is transformed into an inoffensive earth-like substance.

Freezing temperatures WILL NOT damage the toilet or the compost; however, in temperatures below 50° F the composting action decreases. Toilet paper is composted along with the rest of the material. Composted material is removed one to four times per year, depending on use. Residential use may require removal slightly more often. This residual compost is the best garden fertilizer you can get.

How can the compost be dumped onto the garden when fresh waste is present? The device's bottom drawer serves the purpose: when the drum is approximately half to two-thirds full, some of the compost is cranked into the bottom drawer where final composting action occurs prior to use in the garden.

We have only very rarely had a Sun-Mar customer complain about odor. The air flow provides a negative pressure which ensures no back draft. The air is admitted through ventilation holes in the front. The rotation and aeration by the Bio-Drum or shaft-mixers along with the addition of organic material ensures a fast, odorless, aerobic breakdown of the compost.

All Sun-Mar toilets need a minimum of maintenance. All that needs to be added is a cup of peat moss per person per day, plus if available some other organic material such as vegetable cuttings, greens, and old bread. If the toilet is used continuously, once every third day the compost needs to be aerated and mixed. This is simply done by giving the handle a few turns.

All Sun-Mar toilets are fully guaranteed for 2 years, and come with vent pipe and everything necessary for a do-it-yourself installation. All units are fully approved by the Canadian Ministry of the Environment and the CSA. As of mid-July, 1989 the XL is fully N.S.F. (National Sanitary Foundation) approved.

Sunmar N.E. (Non-Electric) Composting Toilet

The Non-Electric is our most popular selling composting toilet. It's perfect for many of our customers living off the power line and not wanting to be dependent on their inverters. The tremendous aeration and mixing action of the Bio-Drum, coupled with the help of a 4" vent pipe and the heat from the compost creates a "chimney" effect which draws air through the system similar to a wood stove. The N.E. is designed for one to three people in residential use and four to six people in cottage use (Sun-Mar rates usage **very** conservatively). It is 23" wide, 28" high, and 32" long. Because the N.E. lacks a heater like the larger models, it has a small evaporation capacity. It therefore requires a small drain connected to a small 1' x 1' drain pit for occasional overloads: 8 ft of 3/4" pipe is included for this purpose.

44-101 N.E. Composting Toilet $1,095
44-001 Add for locations west of the Rockies $ 40

Shipped Freight Collect from Buffalo, NY to locations east of the Rockies, and from our warehouse to locations west of the Rockies.

Send an SASE for a color brochure on the entire Sun-Mar line.

The X.L.

The Sun-Mar XL has received full N.S.F. (National Sanitation Federation) approval after being put through rigorous testing for over a year. For those with access to 110V electricity, the XL is a high-capacity unit ideal for year round or seasonal use, using a maximum of 280 watts for the thermostatically controlled heater and fan. It is designed for five to seven people in cottage use and two to four in residential use. For short periods the XL will safely handle double these numbers.

The Bio-Drum gives the XL (and other Sun-Mar models) an incredible ease of maintenance. You simply add peat moss and turn the handle every third day if the toilet is used continuously or once before leaving if used on weekends only.

Dimensions are 22.5" wide, 29.5" high, and 33" long.

■44-102   X.L. Composting Toilet $1,295
Shipped Freight Collect from Buffalo, NY

The WCM

WCM stands for "Water Closet Multrum." The WCM is a high-capacity remote composter which operates in conjunction with a one-pint flush toilet. The toilet (or toilets) is connected to, and flushes into, the remote composter via a gravity fed 3" ABS pipe. The WCM composter can be located at some distance from the toilet, and placed either outside, or in a basement. For outside installation, a minimum of 31" (and preferably 3-4 feet) is required under the dwelling to permit installation.

The WCM has the Bio-Drum contained within and a thermostatically controlled heating element and electric fan. This unit is ideal for a larger family or people who don't want the compost in the same room with the toilet. It is large enough for a family of three to five in residential use or six to eight in cottage use. (For a short period these numbers can be doubled.) The WCM's fiberglass and stainless steel construction makes it resistant to exposure to the elements. The WCM is the only flush toilet system that does not require a septic field or holding tank for the handling of human waste.

We offer the "Aqua Magic" low-flush toilet, made of plastic, and the Sealand (for additional information on the Sealand see the low-flush toilet section of the Sourcebook) low-flush vitreous china toilet with the WCM - both use less than one pint of water to flush. The compact low-flush toilet is mounted on a standard 3" floor fitting and has 3/8" water connections which require a minimum of water pressure.

As an option the WCM is also available in a Non-Electric version. However, it is supplied with a 4" vent pipe which has to be installed vertically (no elbows). In the example, the WCM-NE is installed only partially under the dwelling so the vent pipe can run on the outside wall of the cottage. The WCM-NE capacity is rated at five to seven people in cottage use and two to four people in residential use. For residential applications where no electricity is available we strongly recommend using the WCM-NE

instead of the standard NE. We feel so strongly about this that we have just lowered our price (below our normal profit margin) on the WCM-NE to encourage you to purchase it instead on the NE. Most people prefer to have the composting action occurring in a location other than their own bathroom!

■44-201   WCM Composting Toilet $1,295
■44-202   WCM-NE Composting Toilet $1,095
Shipped Freight Collect from Buffalo

■44-203   Aqua Magic plastic toilet $ 139
■44-405   Sealand China toilet $ 189
Shipped Freight Collect when ordered with Sunmar toilets; shipped UPS when ordered alone.

Composting toilets could help save harbor

By BETH MARLIN
The Spectator

HAMILTON HARBOR would be relieved of hundreds of tonnes of pollutants each year if one in ten regional households would just stop flushing their toilets into the sewer system, say Environment Canada researchers.

According to a recent study by the Canada Centre for Inland Waters in Burlington, composting toilets could reduce total nitrogen, phosphorus and potassium discharges into the bay by 430 tonnes (443 tons) a year if used by 10 per cent of local households.

And using the non-flush toilets, which decompose waste organically through electric heating coils, oxygen and a bulking agent in a large holding tank below, would conserve almost 2 billion litres (400 million gallons) of water a year, says the study.

The compost, which is produced over three or four months, resembles soil and can be added to gardens as a soil conditioner or fertilizer.

According to the study, many local residents seem open to the idea of using composting toilets as one potential solution to the degradation of Hamilton Harbor.

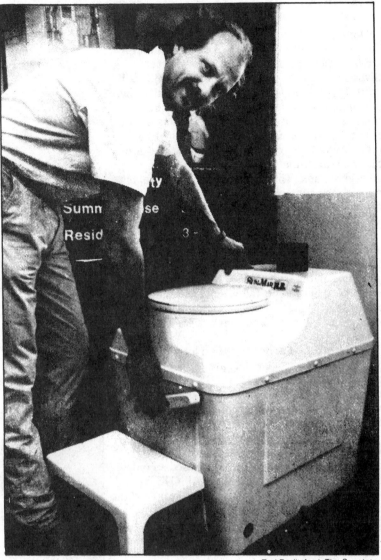

Ted Brellisford, The Spectator

Henric Sundberg of Sun-Mar in Burlington demonstrates toilet.

12V Toilet Exhaust Fan

At last a powerful 12V fan designed to fit easily over a standard 4" ABS vent pipe!. When used as a standard bathroom ventilation fan it can be wired in to your 12 volt lighting system via a switch.

It can also be used on a continuous daytime basis by connecting an MSX10 solar module. This combination works great for ventilating composting toilets such as the Sun-Mar N.E. It is also a good system for greenhouse ventilation.

The Bathroom Fan uses only 9 watts (3/4 amp at 12 volts) and the fan motor is encapsulated to protect it from corrosion and locally fused to protect against motor burnout.

This exhaust fan is highly recommended for residential use and in locations where down draft can occur, such as areas surrounded by mountains or high trees. It will greatly improve airflow and capacity.

44-802 12V Toilet Exhaust Fan $69

CTS 410 Composting Toilet

While our Sunmar toilets work great for small capacity systems, *Composting Toilet Systems* makes a great unit for larger capacities. This unit is nearly identical to the original Clivus Multrum. It is constructed of double wall fiberglass with 1" insulation (R-8) between the walls. The CTS uses absolutely no water, no chemicals, and has no adverse impact upon the environment. CTS toilets are in use by the Army Corps of Engineers, the National Park Service in the Grand Canyon, and lots of residential users.

Aerobic decomposition takes place in the fully insulated fiberglass tank. This digester tank has a sloping floor upon which the composting waste pile is built. Organic materials such as waste, tissues, toilet paper, and wood fibers are accumulated over a period of time. Baffle walls and air channels are part of the digester tank, creating an atmosphere rich with oxygen, which is the ideal environment for micro organisms to digest the organic waste. The natural air flow in the stack is assisted with either an AC or DC solar fan which creates a vacuum insider the digester tank. The end product which is a fertile organic humus, has no odor and is easily removed.

The CTS 410 is designed for full-time long-term use for up to four people. Larger models are available for public facilities and large groups. The CTS can accommodate up to 40,000 uses per year. While the Sunmar is generally dumped 2-12 times per year, the CTS 410 is dumped only once a year after the second year of use. Only fully digested humus is removed, about two bushels worth annually. The small fan (specify 12V or 110V) is used only when the toilet is being used so only very minimal amounts of power are used. All CTS products are warranted for 3 years, except one year for the fan. **Write for free brochure.**

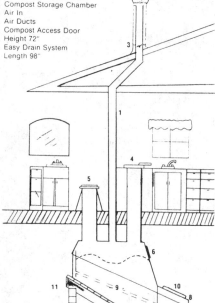

TANK DIAGRAM LETTERING LIST
1. Vent Pipe
2. Turbine Fan
3. Electric Fan
4. Kitchen Garbage Chute
5. Toilet
6. Emergency Access Door
7. Compost Storage Chamber
8. Air In
9. Air Ducts
10. Compost Access Door
11. Height 72"
12. Easy Drain System
13. Length 98"

This is the basic installation, two toilets, either Back to Back or Upstairs & Down can be connected to one CTS tank.

BioActivator Compost Helper

Our BioActivator greatly aids in the composting process. It greatly speeds the rotting of residue and works wonders when all else fails. This has been the perfect solution for many of our customers who have a difficult time maintaining an active composting environment. You can bring compost to life by using only one tablespoon per cubic foot of material or 1/2 lb. per ton. Each 2 oz. packet contains over 50 billion beneficial bacteria in a nutrient-rich mix of kelp and bran that will turn 500 lbs (10 cubic feet) of organic waste into valuable compost.

44-250 BioActivator (2 oz.) $2.50

■44-601   CTS 410 Composting Toilet $3,650
Shipped Freight Collect from Washington State

Water Conservation

"By installing a few simple water-saving devices, costing less than $50, the average household can save more than 30,000 gallons of water and over $60 in water and energy costs each year. If every American made this investment as an Earth Day project, together we would save enough water to cover a football field 1,500 miles high, energy equivalent to 7 huge power plants, and over $1.3 billion per year." - **Amory Lovins**, Director of Research, Rocky Mountain Institute.

Conserving water saves energy and money and preserves fresh water habitat. By conserving water it is possible to prevent some of the pollution caused by excessive energy use, such as global warming and acid rain. Increasing amounts of water are being diverted from major rivers to wasteful city dwellers. This diversion often leads to the destruction of wildlife. When rivers shrink, fish can no longer follow their normal paths of migration to spawn and may fail to reproduce.

Much of the water we consume comes from underground reserves. If this water is used faster than it is replenished, it can cause land to sink, a process called subsidence. Once subsidence occurs, the underground aquifers where water was stored cannot be reformed. According to the U.S. Geological Survey, 35 states are pumping groundwater faster than it is being replenished.

The United States uses two to four times as much water per person as the countries of Europe. Because there are so many inexpensive and simple ways to conserve water, this fact is a national disgrace. Since 70% of all indoor hourshold water is used in the bathroom, we'll begin with ways of conserving water there.

Water Conservation Kit

An average family of four will save 30,000 gallons of water per year by simply installing one low-flow showerhead, two faucet aerators, and a set of toilet dams. We've packaged all these water-savers together with a toilet leak detection kit, an instruction card, and a 26-page booklet on saving water. This is the ideal drought-resistance kit for California, now in its fifth of seven drought years. This is a $45 retail value but we're offering this kit at a price that will make saving water irresistable.

46-109   Water Conservation Kit $19

Toilet Dams

Each time your toilet is flushed, it uses 5 to 7 gallons of water. Toilet dams are plastic barriers that isolate part of your toilet's tank so that water in this section does not run out with the flush. By installing two toilet dams you can save two to three gallons per flush. For a family of four that amounts to a savings of up to 12,000 gallons per year! Our toilet dams install easily in seconds without tools into all standard toilets. Pressure is maintained for a forceful flush. Dams are constructed of stainless steel and thermoplastic and do not interfere with a normal flush. If only 10,000 people installed our toilet dams, 120 million gallons of water per year could be saved.

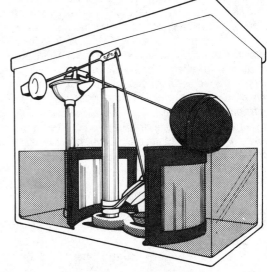

46-201   Set of 2 Toilet Dams $11.50

Ultra-Low Flush Toilets

An average family of four wastes more than 70 gallons of water each day with the use of a standard toilet. This represents about 40% of a family's daily water consumption. These wasted gallons aggravate water shortages. They also overload the sewer system and cause a real problem for sewage treatment facilities and owners of septic tanks.

Aside from the obvious dollar and water savings afforded the user of a low-flush toilet, there are other benefits as well. Many homesteaders are faced with the dilemma of poor water percolation from clay soils and very expensive leach field systems and septic tanks. Most building inspectors around the country will allow a 25-30% reduction in leach field sizing with the use of a low-

flush toilet. The cost savings are dramatic. Typical leach lines cost approximately $10 per foot. A building inspector who requires 400 feet of leach lines for a normal toilet will require only 300 or less for a low flush toilet. That is a savings of $1,000, which will pay for quite a few toilets!

We're constantly researching new toilets on the market and endeavoring to provide you with the latest possible recommendations. Refer to our toilet comparison chart to see what toilets are on the market and how well they fare against competition. If you're considering a no-flush or composting toilet, we urge you to check out that section in this AE Sourcebook. All of our low flush toilets use a standard 12" rough-in.

Aqualine (Eljer) Ultra Low Flush Toilets

The Aqualine's (now owned by Eljer) compact tank and pedestal base have the smooth lines of the "European look." They are manufactured in the USA of vitreous china and available in a variety of colors - white, natural, desert sand, silver mist, and aqua blue. The flush is a scant 1-1/2 gallons.

Tank height is 14" for a good, high head of water. The completely slotted rim feeds water through in vortex to achieve a total, thorough wash in the bowl. *Toilets do not come with seats although an oak seat is shown in the illustration.*

Dear RG,

The Aqualine toilet is the solution I have been looking for a long time. We have a 70-foot-deep hand-pump well so we're highly conscious of water usage. In the five months we've had this toilet, I've been astounded how little water the thing takes, and the fact that it has never clogged. We have a number of conventional friends who visit our earth-sheltered solar home, so we're also conscious of style, ease of cleaning and use, and overall approximation to the traditional toilet. The Aqualine is the answer in all areas. In fact, it is much less noisy than a conventional toilet, and more comfortable to sit on. I think composting toilets are a good idea, but my hat is off to whoever designed the Aqualine!

D. Geery, Shelley, ID

■44-501	Aqualine toilets (White)	$199
■44-502	Aqualine toilets (Colors)	$275
■05-200	Pallet charge	$ 20
■05-210	Air Freight crating charge	$ 50

Add one (only) $20 pallet charge (or air freight charge for HI/AK) whether you order one or many toilets. All toilets shipped freight collect from Northeastern California.

Toto LF-16 Toilet

The Toto LF-16 is a 1.6 gallon per flush (gpf) toilet, which has the highest performance results of any 1.6 gpf toilet on the market according to Uniform Plumbing Code requirements. Manufactured in Japan by the largest plumbing manufacturer in the world and imported by Microphor, the Toto is easy to install and uses standard U.S. flush mechanisms. Constructed of high quality vitreous china, it meets all current code standard requirements (IAPMO, BOCA, SBCCI.) The water surface area is 8" x 7.5" or 60 square inches. *Seat not included.*

■44-701-W	Toto LF-16 (white)	$190
■44-701-B	Toto LF-16 (bone)	$215

Shipped UPS from Maryland.

Ifo Cascade Toilets

Many of our veteran readers will remember the beautiful Ifo toilets that we featured several years ago. We discontinued them because the price went over $400 per toilet with exchange rate fluctuations. Now Ifo has liscensed a USA factory to manufacture their fine toilets and we're proud to re-introduce them to you. Ifo has always stood out because of its beautiful European styled tank shape and its fine working mechanisms. Ifo has nearly 50 years of worldwide experience with flush toilets.

Ifo 3180

The 3180 is Ifo's most popular toilet. It has an adjustable flush from one to one and a half gallons. It comes with a standard 12" rough-in and has a water surface of 5-1/2" x 5-3/4". It accepts a standard round seat (not included).

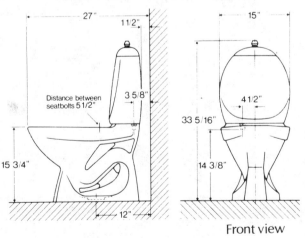

Front view

■44-511	Ifo 3180 - White	$269
■44-512	Ifo 3180 - Bone	$339
■05-200	Pallet charge	$ 20
■05-210	Air Freight crating charge	$ 50

Add one (only) $20 pallet charge (or air freight crating charge for AK/HI) whether you order one or many toilets. All toilets shipped freight collect from Northeastern California

Ifo 3190

The 3190 was designed for the American market and is IAPMO listed. It is a 1.5 gallon flush toilet that comes with a standard 12" rough-in. The water surface is much larger than the 3180 - 8.5" x 9" allowing for easier cleaning. It accepts a standard elongated seat (not included).

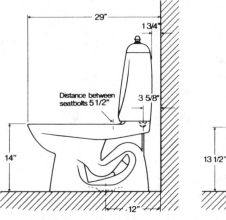

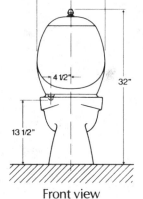

Front view

■44-516	Ifo 3190 - White	$329
■44-517	Ifo 3190 - Bone	$389
■05-200	Pallet charge	$ 20
■05-210	Air Crating charge	$ 50

Add one (only) $20 pallet charge (or Air Crating charge for AK/HI) whether you order one or many toilets. All toilets shipped freight collect from Northeastern California

Sealand China Toilet

The Sealand vitreous china toilet uses less than one pint per flush and is the toilet preferred by most of our customers purchasing the Sun-Mar Water Closet Multrum (WCM) composting toilet system. The Sealand has a very effective flushing mechanism with a 360-degree rim flush. Installation is fast and easy with four easy-to-reach closet bolts that secure to the floor and the base cover snaps in place. There is usually no need to alter plumbing fittings or connections, with the water valve on the side where service and connection are easy. All Sealand toilets are covered by a two-year warranty and are built in the USA. Recommended for WCM or R.V. use only. These toilets cannot be connected to a sewer system. The 910 sits 19-1/4" away from the wall and is 15-1/2" wide and designed for areas where space is limited.

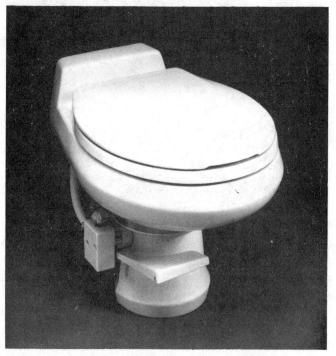

■44-405	Sealand 910 (White)	$189
■05-200	Pallet Charge	$ 20

Shipped Freight Collect when ordered with Sunmar toilets; shipped UPS when ordered alone.

Dear Real Goods,
   Thanks for your great catalog! What a source of ideas! I notice that you list the Sealand low-flush toilet only as a Bowli accessory. You should promote these separately! They're fantastic. Much less flush water (1 quart) than even the Aqua-Line (1.5 gallons - that's not ultra-low... low maybe!)
   I've lived with three in my home for five years, and I'm very pleased. The china bowl is easy to keep clean and is very durable, and they're solidly built, providing a steady seat!

James Fackert, Brighton, MI

LOW CONSUMPTION TOILET PERFORMANCE TEST RESULTS

Year of Approval	Manufacturer Model	Test Lab	Ball Rough Test 100	Granules Remaining 0	Cylinders Passed 20	Inkline Total Remaining 0	Dye Test Dilution Ratio 1000	P=Pressure F=Flushometer Assist G=Gravity	Water Surface LxW
1989	Microphor/ Toto LF-16 Round	PSI	12	100	0.4	20	0	1000	G	8x7
1989	Gerber 21-362 Elongated	UST	12	100	10.4	20	0	1000	P	12x10
1989	Peerless Hydromiser 5160 Round	UST	12	99	1	20	0	1000	G	8x9
1989	Gerber 21-352 Round	UST	12	100	13.2	20	0	500	P	12x10
1989	Briggs Turboflush 4700	PSI	12	100	6.4	20	0	500	P	15x11.5
1988	Kohler Wellworth Life K3421	UST	12	100	2	20	0	500	G	9x8
1988	Crane Economizer 3.80 Round	UST	12	100	0.6	20	3/8"	833	P	11x10
1989	Peerless Hydromiser 5164 14"	UST	14	97.2	14.2	20	5/8"	1000	G	10x9
1989	Gerber Flushometer bowl 25-830	UST	12	100	18.2	20	1/2"	666	F	13x10
1988	Flor Ultra 1G 091-4805 Elongated	PSI	12	98.4	12.6	20	0	600	G	4.6x5
1989	Mansfield Allegro 130-16	UST	12	98	8	20	1 1/8"	1000	G	6.5x5
1989	Peerless Hydromiser 5660	UST	12	96	22.8	20	0	833	G	9.5x9
1989	Peerless Hydromiser 5161	UST	10	96	35.6	20	9/18"	833	G	10x9
1988	Crane Economizer 360 Elongated	UST	12	100	18.2	20	0	300	P	13x9
1989	Fowler Soveriegn ULF-1000	UST	12	91	5.2	20	0	200	G	8.5x7
1988	Porcher Venelo 971		12	83	38.8	20	1/2"	200	G	5.5x5
1989	American Standard Cadet Aquameter Round	UST	12	100	19	20	5/8"	100	P	12x11
1989	American Standard Cadet Aquameter Elongated	UST	12	98	63	20	3/8"	200	P	12x11
1988	Eler Preserver 019 4500 Round	PSI	12	83	71.4	20	0	500	G	4x5.5
1989	Mansfield Cascade 3190 Elongated		12	79	76	20	0	200	G	9x9
1985	Eler Ultra 1G Round	STE	12	97	13	17	0	183	G	4x5
1985	Eler Ultra 1G Elongated	STE	12	82	13	17	0	175	G	4x5
1989	Crane Cranemiser 3-660 Round	UST	10	95	19.4	18	1/4"	100	G	9x8.5
1989	Crane Cranemiser 3-662 Round	UST	12	98	22	17	1/4"	200	G	9x8.5
1989	Crane Cranemiser 3-664 Round	UST	14	94	27.4	17	0	200	G	9x8.5

Water-Saving Shower Heads

If a family of four takes 5-minute showers each day, they will use more than 700 gallons of water every week - the equivalent of a three-year supply of drinking water for one person. -50 Simple Things You Can Do To Save The Earth.

Showers typically account for 32% of home water use. A standard shower head uses about five to seven gallons of water per minute, so even a five-minute shower can consume 35 gallons. According to the Department of Energy, heating water is "the second largest residential energy user." With a low-flow shower head, energy use and costs for heating hot water for showers may drop as much as 50%. Add one of our instantaneous water heaters for even greater savings. Our low-flow shower heads can easily cut shower water usage by 50%. A family of four which normally takes 5-minute showers saves at least 14,000 gallons of water per year. A recent study showed that changing to a low-flow showerhead saved $0.27 of water per day and $0.51 of electricity for a family of four. So, besides from being good for the Earth, a low-flow shower head will pay for itself in about two months!

The three showerheads listed immediately below all work incredibly well. In our showroom we have a showerhead testing module which holds six heads and measures both pressure (psi) and flow (gallons-per-minute). We can unequivocally recommend all three of these showerheads for both performance and styling! We invite you to visit our showroom and judge for yourself.

A close-up from our working showerhead display in our showroom. On the right is the "hardware store low-flow" which actually uses close to 3 gpm. On the left is the Spradius Aerator. Next from the left is the Lowest Flow (46-104), then the Plusational (46-107), the Deluxe (46-106), and second from the right is the Spa 2000 (46-101).

SPA 2000

A good low-flow, high-performance showerhead reduces water consumption and heating costs significantly while still getting the deed done admirably. The SPA 2000 showerhead is one of the best available. Its uniquely engineered "venturi" design delivers a vigorous, pulsating spray using just 2.5 gpm and its handy built-in pressure-reducing lever can be used to save even more water in high-pressure areas or simply to reduce flow when you are soaping up. Durably constructed, its head is made from white Celcon plastic, and the showerarm coupling is chrome-plated brass. Fits standard, threaded showerarms.

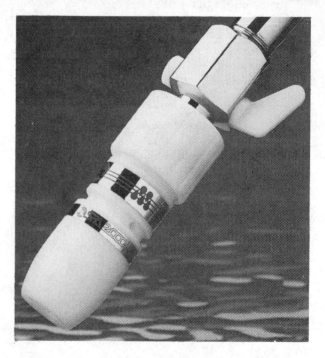

46-101 Spa 2000 Showerhead $24

Lowest Flow Showerhead

This is by far our best selling and finest designed showerhead. It can save up to $250 for a family of four by cutting hot & cold water use by up to 70%. At 40 psi it delivers 1.81 gpm. Manufactured in the USA of solid brass, chrome plated, it exceeds California Energy Commission Standards. It comes standard with a built-in On/Off button for soaping up and standard 1/2" threads so that a wrench is all that's necessary for installation. Fully guaranteed for 20 years. Specified by the city of Los Angeles.

46-104　　　Lowest Flow Shower Head　　$12

Plusational Showerhead

This brand new technology showerhead gives an excellent water stream. It features a new minimum/maximum water saving feature. You can adjust the shower to give 1.0 gpm, 1.4 gpm, or 1.7 gpm at 40 psi. The stream this head puts out is very close to a pulsating showerhead. It means much greater rinsing efficiency at flows of 4 quarts per minute and less. It is made in the USA of solid brass, and is chrome-plated with a 20 year guarantee. It also includes an on/off valve. Specified by the city of Santa Monica.

46-107　Plusational Showerhead　　　$17

Deluxe Low Flow Showerhead

Made from solid brass and with real 14 kt. gold plating, this Deluxe shower head is identical to our Lowest Flow Shower Head but much more elegant looking (if you like gold!). This showerhead is engineered for low-flow and does not use flow restrictors or pressure compensators to reduce flows. This head gives a superior spray at all pressures and is fully guaranteed for 20 years. Also equipped with an on/off valve. The Deluxe looks like the Plusational showerhead but has a superior full conical spray.

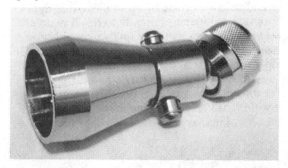

46-106　Deluxe Low Flow Showerhead　　$18

Five Gallon Solar Shower

This incredible low-tech invention uses solar energy to heat water for all your washing needs. The large 5 gallon capacity provides ample hot water for at least four hot showers. On a 70° day the Solar Shower will heat 60° water to 108° in only 3 hours for a tingling-hot shower. This unit is built of 4-ply construction for greatest durability and efficiency. We've seen this shower selling for as much as $20!

90-416　Solar Shower　$12

Low-Flow Faucet Aerators

According to Home Energy magazine, we would save over 250 million gallons of water every day if every American home installed faucet aerators. Installing aerators on kitchen and bathroom sink faucets will cut water use by as much as 280 gallons per month for a typical family of 4.

Deluxe Faucet Aerator

This low-flow faucet aerator with finger tip On/Off lever will cut hot and cold water use by up to 60%. It is dual threaded both internally and externally to fit all male and female faucets. It limits faucet flow to 2.75 gpm at any water pressure. It's made of solid brass, chrome plated. It installs simply in seconds and can save thousands of gallons of wasted water every year. The on/off finger-tip control lever allows the user to temporarily discontinue the flow of water without readjusting at the hot and cold controls. It's ideal for shaving, brushing, & washing dishes.

46-103   Deluxe Faucet Aerator $7.50

Bathroom Faucet Aerator

This is one of the aerators included in our Water Conservation Kit. Installed in the bathroom you'll (a typical family of four) save up to 100 gallons every day. Very simple to install and constructed of solid brass, chrome plated. Sold in sets of two. Uses 1.5 gpm at 30 psi. Fits both male & female faucets.

46-108   Bathroom Aerator (set of 2) $4.50

Kitchen Faucet Aerator

This is one of the aerators included in our Water Conservation Kit. Installed in the kitchen sink, you'll (a typical family of four) save up to 50 gallons every day. Very simple to install and constructed of solid brass, chrome plated. Sold in sets of two. Uses 2.5 gpm at 30 psi. Fits both male & female faucets.

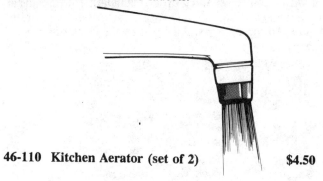

46-110   Kitchen Aerator (set of 2) $4.50

Spradius Kitchen Aerator

Here is a new 2-position faucet aerator that saves lots of water. It's rated at 2.5 gpm and it swivels 360 degrees to direct spray or stream to every part of your sink. When pushed up it functions as an aerator keeping water from splashing out of the sink by injecting air into the water to soften it. When pulled down for spray it boosts pressure and saves water with a great spray pattern that's handy for cleaning or rinsing. It installs quickly and without tools. The double swivel design allows you to rinse the entire sink all the way up to the edges, and eliminates the need for an expensive sprayer hose.

46-131   Spradius Kitchen Aerator $11

Recycled Products & Recycling Aids

Personal Recycling File

If you're like us, you end up shoving rejected xerox copies and potential scratch paper under a desk to await spring cleaning. This new elegant and simple component sorter allows you to separate the paper you will be recycling. It's constructed of strong, recycled corrugated fiberboard and has two trays for easy sorting. Each desk needs one and don't forget the copy machine!

51-204   Personal Recycling File $8

Paper Boy Newspaper Recycling Aid

Here is an inexpensive, easy to assemble newspaper holder and bundler. It is a durable product made from recycled, biodegradable cardboard material and makes recycling newspaper easy. The Paper Boy has a storage compartment for scissors and twine. Its unique design makes bundling easy. It suspends the stack of newspapers so the bundler can reach under it from four directions to place twine. Then, tie the knots, remove the bundle and PaperBoy is ready for the next stack of newspapers. Does not include twine or scissors.

51-205   Paper Boy $6

Recycling Bins

Our reinforced space-saving recycling bins are ideal for home recycling. They come with easy-grip handles and the legs lock securely for safe stacking. Each bin comes with a drop front to allow access when stacked. Bins are made from a tough polyethylene that won't rust corrode or mildew. Get one for aluminum, one for newspaper, and one for glass or plastics. Dimensions: 20-7/8" Long x 15-1/8" Wide x 13-3/8" High.

51-201	Recycling Bins (each)	$14
51-203	Recycling Bins (Set of 3)	$39

Reusable Lunchbags

Our new nylon lunchbags eliminate the need for paper bags. They're washable and include an easy to seal velcro closure. Each bag is decorated with an attractive earth logo. 4% of all sales proceeds are donated to *Friends of the Earth.* Cheaper and easier to use than lunch boxes! Colors are green & turquoise background with a black border around the earth. 12"H x 7"W x 5"D.

51-221   Reusable Lunchbag $7

Aluminum Can Crusher

Our new can crusher allows more than five times as many flattened cans to fit in the same storage space as regular uncrushed cans. Constructed of heavy gauge steel with a baked enamel finish this unit is far stronger than the nylon model we used to sell. It also crushes cans to half the size! It mounts very easily in your kitchen or next to your soda machine at work. It measures 14"H x 15"W x 4-1/2" diameter. Th large compaction chamber accepts 12 oz. or 16 oz. cans without adjustment. Mounting screws not included.

51-209   Can Crusher $19

Recycled Rags

We have located an environmentally conscious company that recycles cotton clothing for use as rags. They have been committed to recycling for 15 years and are responsible for keeping six to seven million pounds of compressed clothing (266,000 sq. yds.) out of the landfill annually. If you're like us, it's not always easy to generate your own rags if you don't wear out clothes quickly! This is a great way to conserve on paper. Each year over 50 million trees are cut for paper products. A portion of each sale is donated to environmental organizations. Each 2.5 lb. box of rags is approximately 90% cotton and comes in a box 7.5" x 7.5" x 7.5". *Originally sold for $5, we've just lowered the price!*

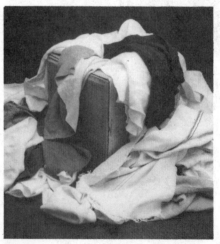

51-207   Recycled Rags (2.5 lbs.) $4

Cotton String Shopping Bags

Made from 100% cotton and manufactured entirely in the USA our string bags are the finest on the market, made of a heavier construction than just about all that you've seen. They will easily hold two to three times the weight of a common grocery store plastic bag and a little more volume. Under stress (dropping, etc.) they will greatly outperform either plastic or paper. The soft cotton handle is comfortable even when fully loaded. They're also ideal for picnics, trips to the beach, and even for storing food in your kitchen. When not in use they fold up easily and can be carried in your pocket.

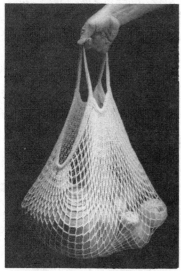

51-901   String Bags $5

Canvas Shopping Bags

It takes one tree to provide 700 paper shopping bags and 500 paper bags are used every hour in one supermarket. This means that 8 to 9 trees per store are destroyed every day in the name of convenience. Any item purchased and sacked can go into our canvas bags. We've been deluged since Earth Day with samples of canvas bags sprouting up from new cottage industries across the country, so we were able to pick the best. Our bags are made of 100% cotton canvas in 10 oz. weight and natural color to prevent chemical dies. The bags measure 8" x 13" x 17". The top hem is double sewn to provide the handles with maximum strength for heavy loads. They are designed with a flat bottom which allows them to stand up for easy filling, just like paper.

51-902   Canvas Shopping Bag $9

Multi-Bag System For Eco-Shopping

Eco Tote is a complete shopping bag system - Not *one* string bag or *one* canvas bag that will only carry limited groceries, but *four* bags made of strudy 12 oz. cotton to take shopping! Three of the bags are carried in the 4th bag and removed for use as needed. Equivalent to six to seven paper or plastic bags, the extra bags fit into pockets in the main bag and after they're removed, the main bag is free for more merchandise. The pockets make great holders for bread or wine bottles. You'll save on resources and money too. If sold separately each bag could retail for $9 or more.

51-903   Eco-Tote Multi-Bag $25

Recycled Paper Products

It takes an entire forest - over 500,000 trees - to supply Americans with their Sunday newspapers every week. -50 Simple Things You Can Do To Save The Earth.

Consider a few telling facts: Each ton of recycled paper produced saves approximately 17 trees. Each ton of recycled paper produced saves approximately 4,102 kWh of energy compared to virgin paper production which is enough to power the average home for six months. The manufacture of recycled paper requires 7,000 gallons less of water per ton compared to virgin paper. The manufacturing of recycled paper reduces overall emission of air pollution by 60 pounds per one ton of paper produced. For every ton of recycled paper produced, land fill is reduced by three cubic yards.

Americans use 50 million tons of paper annually - which means we consume more than 850 million trees, or 580 pounds of paper for each American. We know that waste paper and paper products constitute nearly half of all municipal solid waste. Making new paper from old paper uses 30% to 55% less energy than making paper from trees - and it reduces related air pollution by 95%.

We've researched the market for a manufacturer who uses a maximum of post-consumer waste in making recycled paper products and are happy with our choice of the "Envision" line of toilet paper, paper towels, and facial tissue. These products are made from virtually 100% recovered materials!

Soft Facial Tissues

Made from 100% recycled paper, and meeting the EPA's guidelines for post-consumer waste, these 2-ply facial tissues are extremely soft and gentle. Undistinguishable from your usual brand, these tissues come 100 sheets to the box.

51-106	Facial Tissue Full Case (30 boxes)	$26
51-105	Facial Tissue Half Case (15 boxes)	$14
51-109	Facial Tissue Sampler (6 boxes)	$ 6

Toilet Paper

Made from 100% recycled paper, and made from 100% post-consumer waste, this 2-ply toilet paper is unbleached and Dioxin free. It's identical to the best you've been used to. 500 sheets to the roll (compared to 300 for standard virgin toilet paper), it's very soft and you'll never know the difference...Because of excessive weight, we must charge extra for Alaska & Hawaii.

51-102	Toilet Paper Full Case (96 rolls)	$55
51-101	Toilet Paper Half Case (45 rolls)	$29
51-107	Toilet Paper Sampler (12 rolls)	$ 8
51-195	Add'l freight on full case to HI & AK	$15
51-196	Add'l freight on half case to HI & AK	$ 8

Strong Paper Towels

Made from 100% recycled paper, and meeting the EPA's guidelines for post-consumer waste content, these 2-ply paper towels are very strong and extremely absorbent. In fact, they're identical to the best non-recycled paper towels you've ever used! They come 100 sheets to the roll; each one 11" x 9".

51-104	Paper Towels Full Case (30 rolls)	$34
51-103	Paper Towels Half Case (15 rolls)	$18
51-108	Paper Towel Sampler (6 rolls)	$ 8
51-197	Add'l freight on full case to HI & AK	$10
51-198	Add'l freight on half case to HI & AK	$ 6

Americans produce 154 million tons of garbage every year - enough to fill the New Orleans Superdome from top to bottom, twice a day, every day. 50% of this trash is recyclable! - 50 Simple Things You Can Do To Save The Earth.

Recycling saves energy, thus reducing acid rain, global warming and air pollution. Recycling paper uses 60% less energy than manufacturing paper from virgin timber. Recycling a glass jar saves enough energy to light a 100-watt light bulb for four hours. 75,000 trees are used for the Sunday edition of the New York Times each week, yet only 30% of newspapers are recycled in the United States. If we all recycled our Sunday papers, we could save over 500,000 trees every week.

Recycling cuts down on landfill. The average American throws away four pounds of garbage per day. By 1994, half the cities in the U.S. will run out of landfill space. By recycling it is possible to cut our waste stream by 80%.

If you need information on recycling programs in your area call the Environmental Defense Fund @ 1-800-CALL-EDF.

Reusable Coffee Filters

Our dioxin free, 100% cotton muslin reusable coffee filters contain absolutely no chlorine. They replace standard throwaway filters and will last approximately two years with daily use. They are easy to wash in hot water, they won't tear, and stains can be easily removed by an occasional soaking in boiling water with baking soda. Sold in packages of 2. The basket filter is for automatic coffee makers. The #2, #4, & #6 are Melita sizes.

51-211	#2 Filter (2 ea.)	**$5.50**
51-212	#4 Filter (2 ea.)	**$6.50**
51-213	#6 Filter (2 ea.)	**$7.50**
51-214	Basket Filter (2)	**$8.75**

Don't Buy These Plastic Trash Bags!!!

We don't like using plastic trash bags any more than you do. They fill up our landfills and make the garbage within take all that much longer to decompose. So we heartily recommend that you don't use any plastic products.

However, we also understand these bags have become a necessary evil to many peoples' lifestyles and thought it would be prudent to find the absolute least offensive variety. We've come up with these new 33-gallon trash bags that are made of 80% recycled plastic. Our bags may eventually degrade but this will take many many years. These bags have received the *Green Cross Certification* confirming that they meet the manufacturer's claims. For those of you who absolutely *must* use trash bags, *please use these.* Each pack contains 30 each 33-gallon bags.

51-150 Trash Bags from Recycled Plastic $9.50

Non-Toxic Household Products

Citra-Solv Natural Citrus Solvent

Citra-Solv is one of our favorite new products - it will handle nearly all your cleaning needs. It dissolves grease, oil, tar, ink, gum, blood, fresh paint, and stains to name just a few. It replaces carcinogenic solvents such as lacquer and paint thinner. It replaces toxic drain cleaners and caustic oven and grill cleaners - *you won't believe how well it cleans your oven!* It also replaces certain unsafe room deoderizers - you can dilute Citra-Solv in water, pour into a trigger sprayer, and you have a fragrant orange smelling deodorizer with no aerosol propellants. It replaces soap scum removers which use bleaching agents. Citra-Solv is composed of natural citrus extracts derived from the peels and pulp stemming from the making of orange juice. It is highly concentrated and can be used on almost any fiber or surface in the home or workplace *except plastic*. Citra-Solv is 100% biodegradable and the packaging for Citra-Solv is 100% recyclable.

54-140 Citra-Solv (16 oz. bottle)   **$8**

Freshen Your Air With an Orange Scent

This fragrant spray will purify your air while you freshen it. The orange aerosol functions similary to an ionizer machine in that each droplet conatains millions of active electrical charges (ions) that attract and neutralize offensive orders. It contains no unnecessary artifical ingredients or additives - only essential oils distilled from citrus fruits and exotic herbs. The spraying mechanism is a spray pump which uses no fluorocarbons. This is very effective on all household odors. No animal testing is done on the product. This size (16 oz.) will last for at least two months under normal usuage - making it very economical.

54-143 Orange Air Therapy $12

Dear John,

I enjoy the Real Goods catalog very much. I was reading your blurb on **Citra-Solv.** I was horrified to see that it is being advocated as a safe air deodorizer when used diluted in trigger sprayer, and that people are bathing their pets in it! I have been using this product for some time - it is a terrific cleaner and degreaser. However, it is not <u>non-toxic</u>. Many people make the mistake of confusing biodegradable with non-toxic. (For example, carbon monoxide is biodegradable and toxic.) I like to be an educated consumer, so I often write to manufacturers for the Material Safety Data Sheet on their products, so I can find out just what's in it. I'm sending along a copy of the MSDS on Citra-Solv that I have. I use this solvent because: 1) it is biodegradable, and 2) it is not derived from petroleum, a process which releases hydrocarbons during refinement. I always wear gloves when I use it, and I open the window in the room and place a fan in the window blowing the air <u>out</u>. Then I rinse the surface with plain water. I wash any surface that will touch food with mild soap and water after. People should avoid breathing and skin contact with Citra-Solv. Never bathe your dog in it! Just because the odor is pleasant doesn't mean that it isn't harmful... Do continue to carry this great product! But please amend your suggested uses, and tell people to stop bathing their pets in it. - **Lisa Shaftel, New York, NY**

Smart Mouse Trap

Often, the first interaction that a child has with "wild" amimals is a mouse caught and deformed in a standard mouse trap. The Smart Mouse Trap teaches a child the idea of working in harmony with animals and the ecosystem. This is a humane and effective trap - a forgiving answer to unwanted visitors. Why kill? For a little more effort, you can take the mouse to the nearest brush or wooded area and release it to live out its tiny life. It is a joy to feel the shared compassion with a child watching a trapped mouse escape to freedom. This trap is simple to set - the bait is a soda cracker and it will catch mice that standard traps can't. The trap is make of Kodar plastic with two stainless steel springs. It measures 2" x 3" x 7".

54-202   Smart Mouse Trap $10

RCI Environmental Test Kit

This kit contains everything you need to collect samples of possible cancer causing materials commonly found around the home. You collect the samples and send mail them to the lab in the containers provided. The accredited laboratory uses procedures and protocols established by EPA, OSHA & ASTM. The kit contains an informative instruction manual, a sample collection work sheet, a laboratory screening sheet, sampling tools, sample containers, and safety equipment.

Two sample tests (a <u>radon</u> test and a test for <u>lead in drinking water</u>) are included in the price of the kit. Additional tests are paid for on a test by test basis when you send samples to the lab. *The additional tests available are as follows: FOR BUILDING MATERIALS: Lead, Formaldehyde, Asbestos; WATER: Radon only, Inorganic, Organic; SOIL: Petroleums/Solvents, Insecticides-PCB.*

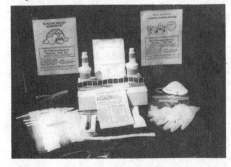

57-001   RCI Environmental Test Kit $59

Sorry, no returns on this item.

Yellow Jacket Inn

Since so many of our customers live in wilderness situations, we thought it appropriate to offer some relief from the yellow jacket menace. These re-usable traps capture incredibly large numbers of the stinging pests - enough to make outdoor eating a pleasure again. They use natural bait such as tuna (only the dolphin-safe variety!) or chicken (or chicken-scented tofu). They're ideal for taking camping or fishing as they weigh only 6 oz. and can be re-used over and over. These traps are made by Seabright Laboratories, a company dedicated to non-toxic devices and humane ways of dealing with pests. Seabright is very environmentally conscientious and even ships with popcorn instead of styrofoam!

54-201   Yellow Jacket Inn $6

Cedar Blocks Moth-Proof Clothes Naturally

Our cedar products are perfect for keeping moths out of drawers, suitcases and garment bags - and these blocks also have a wonderfully pungent cedar scent. Each block lasts a lifetime - *light sanding renews the fresh cedar scent.* The smaller moth blockers can be used anywhere that you store clothing. They protect clothes from moths without the nasty odor associated with naphtha (moth balls). Both of these cedar products are non-toxic and safe in homes with kids and pets. We checked into the harvesting processes to be sure there was no adverse effect and found out that they come from cedars grown on private lands in the Ozarks. All of the cedar is used and replanting is encouraged - *and cedar trees are not endangered.*

54-145   12 Cedar Blocks $ 9
54-146   50 Small Blockers $11

Livos Non-Toxic Household Products

Livos is an American company that imports non-toxic, and non-animal tested household products from Germany. Livos is dedicated to good air, clean water, healthy food, and environmental protection. We're proud to introduce their fine line of products here.

Liquid Wax

Made from biodegradable beeswax, our liquid wax is the perfect product for cleaning and refinishing furniture, waxed floors, shelves, wood panels, posts, beams, and other large surface areas. It also does a great and economical job on large surface areas like architectural woodwork, stone, metal, and plastic laminates. It enhances and protects without using petroleum-based paraffin and is completely environmentally safe.

54-102   Liquid Wax (8.5-oz bottle) $8.50

Bee & Resin Ointment

The best thing about working with this bee & resin ointment is the heavenly smell! This natural beeswax paste builds a tight and dense surface providing excellent dirt resistance and preventing electrostatic build-up. It's ideal for furniture, interior window frames, stone, metal and plastic laminates. A little bit goes a long way!

54-103   Bee & Resin Ointment (6.8-oz jar) $11.50

Leather Seal

Leather seal can be used where you've traditionally used mink oil or saddle soap. It's great for boots exposed to rain and snow, saddles, belts, briefcases, purses, and other items of leather. Made of all natural and biodegradable ingredients.

54-135   Leather Seal (3-oz jar) $9

Water Miser Car-Wash

Water Miser car-wash solution cleans your car and saves up to 150 gallons of water on every car wash. We were skeptical and hesitant to carry this product until we tested it and found out it really works. It is non-toxic, non-abrasive, biodegradable, and silicone free. To use the product, you dilute one capful of the soapy solution in 2 quarts of water to wash a small compact, two capfuls in a gallon for larger autos. Apply the cleaner to a small section of the car with a wash mitt or sponge, then dry it with a chamois or soft cloth. Our 8-oz bottle will wash 18 to 30 cars. This is an ideal product for drought-ridden California!

Shoe & Leather Polish

Most people are unaware that most conventional shoe polish contains up to seven highly toxic ingredients including methylene chloride. The inhalation of these vapors can cause carbon monoxide accumulation in the blood stream. Our shoe polish contains only natural ingredients, organic polishing waxes, sustaining plant oils, and plant and earth pigments. There are **no animal ingredients** used. Our polish cleans, maintains, and preserves all leathers and its penetrating waxes are easily buffed to a high shine. Only the pleasant fragrances of beeswax, natural resins, and essential oils are emitted.

54-132-BL   Shoe Polish (black) 1.7-oz $6
54-132-BR   Shoe Polish (brown) 1.7-oz $6
54-132-CL   Shoe Polish (clear) 1.7-oz $6

"I live five miles up a dirt road and firmly believe in washing my car twice a year. So when I was asked to try out this car wash stuff with no rinsing, I was pretty skeptical. Surprise! This stuff did a nice job, got my car clean, and only used a gallon of water. Good job!" **-Doug Pratt, Ukiah, CA**

54-171 Water Miser Car Wash $5

On-Site Greywater Recycling

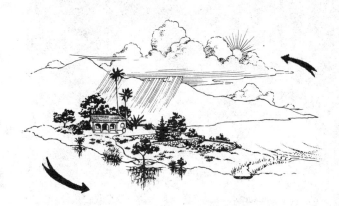

Taking responsibility for the proper handling of one's own household greywater is a way individuals can truly improve their relationship with natural cycles as well as save money and improve the health and vigor of their gardens. When household greywater is used for irrigating the yard, it is naturally purified by biological activity in topsoil. Soil microorganisms break down organic contaminants (including bacteria,. viruses, and biocompatible cleaners) into water soluble plant nutrients. Plant roots take up these nutrients and much of the water and the pure water left over percolates down and recharges the groundwater. Properly utilized natural biological treatment is more effective than municipal treatment. It also reduces both fresh water use and loading on treatment plants or septic tanks. With on-site greywater recycling you can be a crucial link in the global nutrient and water cycles instead of being a drag on them!

Cleaners for Greywater Systems

Most substances found in household greywater biodegrade into plant nutrients; household cleaners can be the exception. Household cleaners vary in their toxicity to plants and soil, as different plants vary in their susceptibility. Few cleaners are so bad that they will immediately kill plants.and insensitive plants may be watered for years with greywater containing non-biocompatible cleaning products. However, toxins from these cleaners can eventually create a problem in the yard. When rainfall flushes toxins from these cleaners out of the topsoil it just introduces the problem into the groundwater. In general:
• Avoid washing more often or using more cleaner than needed.
• Never use cleaners which contain boron (borax), a potent plant toxin.
• Avoid the use of cleaners which contain sodium or chlorine. Liquid cleaners and laundry detergents typically contain less sodium than powders.
• Use biocompatible cleaners. Oasis biocompatible laundry detergent and biocompatible all-purpose cleaner are the only cleaners we know of which have been specifically designed and tested for greywater systems. Oasis is currently testing other biocompatible cleaning products.

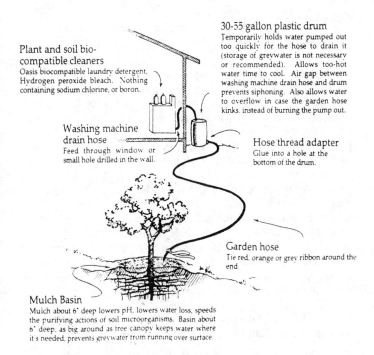

Plant and soil bio-compatible cleaners
Oasis biocompatible laundry detergent, Hydrogen peroxide bleach. Nothing containing sodium chlorine, or boron.

30-55 gallon plastic drum
Temporarily holds water pumped out too quickly for the hose to drain it (storage of greywater is not necessary or recommended). Allows too-hot water time to cool. Air gap between washing machine drain hose and drum prevents siphoning. Also allows water to overflow in case the garden hose kinks, instead of burning the pump out.

Washing machine drain hose
Feed through window or small hole drilled in the wall.

Hose thread adapter
Glue into a hole at the bottom of the drum.

Garden hose
Tie red, orange or grey ribbon around the end.

Mulch Basin
Mulch about 6' deep lowers pH, lowers water loss, speeds the purifying actions of soil microorganisms. Basin about 6' deep, as big around as tree canopy keeps water where it's needed, prevents greywater from running over surface.

Illustration from the Oasis Greywater Booklet showing the simplest way of delivering laundry water (the most easily accessible greywater) to the yard.

Greywater Information Booklet

This is an excellent small booklet on just about anything you ever wanted to know about greywater disposal and usage. It contains all the information you need to construct and use a greywater system. It was written by the Goleta, California Water District for its customers, and helped along with additional information from the folks at Oasis Products.

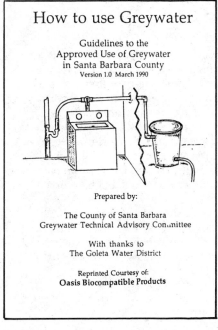

How to use Greywater

Guidelines to the Approved Use of Greywater in Santa Barbara County
Version 1.0 March 1990

Prepared by:

The County of Santa Barbara Greywater Technical Advisory Committee

With thanks to The Goleta Water District

Reprinted Courtesy of:
Oasis Biocompatible Products

80-203 Greywater Information Booklet \$4

Oasis Bio-Compatible
Laundry Detergent

Wash water containing Oasis can be used in the garden without concern for harming plants. It will generally produce better growth than plain water, because it biodegrades into a balanced mix of essential plant nutrients including carbon dioxide, nitrogen, phosphorous, potassium, and other nutrients that promote beneficial plant growth. By running a hose from your washing machine to your garden you can capture up to 1,000 gallons a month headed right down the drain. Your clothes will come out just as clean (we tested it) and your tomatoes will grow stronger with Oasis as the sole source of nutrients! Oasis has been field tested on dozens of plant species and soil types and has proven to be entirely safe and beneficial. Oasis is super concentrated: using 1/5 cup per load of laundry, you can do 80 loads per gallon or 20 loads per quart of detergent.

54-151	Oasis Laundry Detergent (Quart)	$ 9
54-152	Oasis Laundry Detergent (Gallon)	$25

Oasis Biocompatible
All-Purpose Cleaner / Dishwash

Our All-Purpose Cleaner / Dishwash is the same idea as our Oasis Biocompatible laundry detergent, but it's used for dishwashing and general cleaning. It is beneficial for irrigation of all plants. It is non-harmful for septic tank or sewer disposal as well.

54-155	Oasis All-Purpose Cleaner (Quart)	$ 7.25
54-156	Oasis All-Purpose Cleaner (Gallon)	$20

Ossengal Spot Remover

Ossengal Stick is wonderful stuff. It gets rid of spots and stains better than anything we've ever tried - AND it's organic, odorless, chemical-free and easy to use. It's made from the purified gall of oxen by the Dutch, and when you use it you can understand why they're famous for getting things clean. We asked a chemist why Ossengall works so well. He sent a page of words like: "...short chain organic acids, phospholipids, enzymes,..." The gist of it seems to be that an ox needs tough-acting stuff to break down what it eats, and it acts the same way on spots and stains on fabric.

Ossengal is a pure white stick of soaplike substance, packaged in a small pocket-size pushup tube. To use it you just moisten the spot, rub it well with the stick, bursh or knead it in, then rinse. Most stains just disappear.

54-101   Ossengal Stick Spot Remover $6

Radiation Control

The Dangers of Magnetic Radiation

Long suppressed studies have recently come to light regarding the health effects of magnetic radiation. The EPA staff has classified magnetic fields as B1 carcinogens (same as DDT, PCBs, and Dioxins) and has written that there is enough evidence to statistically link them to cancer. This historical warning aired on CBS News with Dan Rather on 28 May 1990.

Over the past ten years reports of miscarriages, leukemia, and cancer caused by exposure to the magnetic fields emitted by utility power lines and computer terminals have often made the evening news and then quietly disappeared from sight.

Three June issues of *The New Yorker* magazine carried a series of articles entitled "The Annals of Radiation"(call the New Yorker @212/840-3800 & ask for June 12,19, & 26, 1989 back issues) which offers the first in depth and well documented look at the health risks of VLF (very low frequency) and ELF (extremely low frequency) magnetic radiation.

· There are currently more than 100 lawsuits concerning property values being lowered by the presence of power lines. Even more interesting is the fact that computer companies pay people to sign "gag" papers to keep this issue out of the news.

We are concerned about the health of our customers and know that all of you, especially those of you living beyond the utility power line grid are very concerned about the quality of your living environment.

So in addition to our magnetically clean DC lights and appliances, we are offering several products to help you find and clean up the invisible Magnetic Radiation in your home or work place. *We've had a lot of customers abuse our return policy with radiation meters by testing their homes, then returning the meter. For this reason we can no longer accept returns on radiation meters unless they truly are defective!*

The Safe Computer Monitor

This is worlds first *radiation free* computer monitor. Most standard monochrome monitors send magnetic fields 10 to 15 feet (in all directions) from the monitor and color monitors can radiate as far as 25 feet. Even laptop computers give off magnetic fields (from backplane drives and screen lighting) which travel 4 to 6 feet. These fields penetrate almost everything, *including all anti-glare screens*, so don't put your computer on the wall outside of your childs bedroom.

The Safe Computer Monitor consists of a monochrome style shielded liquid crystal display with an 8-1/2" by 5-1/2" Screen with 640 X 200 pixel resolution. Onboard electronics are are completely shielded so it emits *zero* magnetic radiation. This monitor will work with the IBM PC, XT, AT, and any of the compatibles

■57-110   Safe Monitor $1,095

How to Reduce Magnetic Fields in Your Home (Video)

The Current Switch is an interesting and informative 68 minute video tape showing how to reduce or eliminate harmful Electro-Magnetic Fields (EMF's)in your home. It describes a number of negative physical, mental, and biological health effects (from cancer and leukemia to sudden death) in which EMF's have been implicated. This video also shows how to map EMF hazards in your home environment, rearrange home appliances, check wiring for proper phasing and grounding, etc. The Current Switch makes an excellent companion to our magnetic field meter. With these two items you can greatly reduce the Electro-Magnetic Radiation exposure of yourself and your family. Even if your home is powered by DC (properly installed DC wiring emits virtually no EMF's) you will still find this tape useful for plotting safe distances from microwave stations and military or civilian radar transmitters. Refer to the book section for additional books on magnetic radiation.

80-112   Current Switch Video $55

Adapter Card (MCGA)

If you don't have a Hercules compatible Monochrome Graphics Adapter Card you will need one of these to connect the Safe Computer Monitor to your PC.

■57-111   Adapter Card $150

Pocket Radiation Meters

Magnetic Field Meter

Our finest Magnetic Field Meter (milliGauss meter) will measure ELF & VLF magnetic radiation. With a large digital readout, it reads magnetic fields from .001 milligauss to 300 milligauss with seven different ranges. (The Denver childhood leukemia study showed increased leukemia rates in magnetic fields as low as 1 to 2 milligauss). This meter will measure magnetic fields from Computer monitors, AC power lines, TV's, electric blankets, fluorescent lights, appliances, microwaves, and transformers. It features a flat frequency response curve which allows it to accurately measure complicated 60 Hz sources. Walking through an AC powered house with one of these meters can be a real eye opener. This is by far the most accurate magnetic field meter that we've been able to find.

57-103   Magnetic Field Meter $259
Sorry, no returns on this item.

Magalert 660 Field Meter

Magalert is a less expensive magnetic field meter designed for ease of use by the householder or concerned citizen. With a single push-button control, it shows field strengths ranging from .1 to 100 milligauss, from the lowest field background levels to those from major power lines. You can check exposure to sources in the home or office and assess the daily changes in background levels which are now affecting property values. You can survey AC magnetic fields in proximity to electric blankets, TVs, video data terminals, and unbalanced household wiring. Determine your own exposure level. Instrument accuracy of plus or minus 10% is more than adequate for general surveys. Magalert 600 contains a high quality 9-volt alkaline battery, assuring long and dependable operating life. *Sorry, no returns on this item.*

57-104   Magalert 660 $159

Power Pet Magnetic Radiation Warning

This is the simple version of our "Magnetic Field Meter." A green light turns red when the device senses that magnetic radiation is above one milliGauss. It's a real eye opener taking it around your home or office. You can see at a glance how far away you need to sit from your TV or video monitor. The box measures 2.5" x 2.5" x 1.25" and easily fits in the palm of your hand. *Sorry, no returns on this item.*

Pocket Geiger Counter

The Monitor 4 is the best low cost Nuclear Radiation Monitor we could find for giving instantaneous readings. Other Nuclear Monitors cost thousands of dollars, are bulky to carry, and use hard to find 22.5-volt batteries. The Monitor 4 is small enough to fit in a shirt pocket and uses standard 9-volt batteries. It's a must have item if you live near a nuclear plant (Plant operators don't usually tell you about spills and accidents until it's too late!) The Monitor 4 reads ranges from .5 to 50 mR/hour. It has a switch selectable Audio/Visual count indicator. *Sorry, no returns on this item.*

57-105   Geiger Counter $319

57-109   Power Pet $79

Water Purification

We've researched good water filtration systems for several years now, because many customers have encouraged us to do so. The drinking water purifier market is crowded and expanding rapidly as people become increasingly aware of the insidious pollution in our water. This boom has spawned *lots* of new companies. We commissioned our staff "Water Guru," Randy Wimer, to do an objective analysis of the different methods of filtration available with a critical eye toward separating the hype from the hip. His entire report is reprinted below.

Water - Who Needs It?

Water is absolutely vital to our bodies. Just like the surface of the earth, our bodies are mostly made up of water. The average adult contains 40 to 50 quarts of water! The water in your body must be renewed every 10-15 days. With the intake of foods such as fruits and vegetables, you are receiving water, but you must drink at least 6 glasses of water daily to enable your body to function properly - water is the base for all bodily functions. *Your body depends upon what you drink!*

The EPA released to the news media on December 14, 1988, information that stated there is some kind of toxic substance in our ground water no matter where we live in the United States. We have all heard some reference to the problems resident in our drinking water in this decade. Even materials added to our drinking water to "protect" us (such as chlorine) are being found to promote cancer and form other toxic compounds which adversely affect us.

The old adage "If you want something done, do it yourself" applies to our drinking water also. The most sensible solution to pollution is a point-of-use water purification device. Point of use refers to the tap, which is the location from which we draw our water. A system here is the end of the road for the water we and our families consume. After the tap, there are no more pipes or conduits which can leach elements into our drinking water that we don't want to be there.

To help make the best choice for a water purification system which will suit your needs, let's summarize the problems we are faced with.

Biological Impurities

Years ago, waterborne diseases - from bacteria, viruses, and parasites - accounted for millions of deaths. Even today in underdeveloped countries, an estimated 25,000 people will die daily from waterborne disease. Effects of waterborne microorganisms can be immediate and devastating. Therefore, microorganisms are the first and most important consideration in making water acceptable for human consumption.

Generally speaking, modern municipal supplies are relatively free from harmful organisms because of routine disinfection with chlorine or chloramines and frequent sampling. *This does not mean municipal water is free of all bacteria.*

Those of us with private wells and small rural water systems have reason to be more concerned about the possibility of microorganism contamination from septic tanks, animal wastes, and other problems. There is a little community in California, for example; where 4 million gallons of urine hits the ground daily from dairy cows!

Authorities say that at least 4,000 cases of waterborne diseases are reported every year in the United States. They also estimate that most of the temporary ills and everyday gastrointestinal disorders that go routinely unreported can be attributed to organisms found in our water supplies.

Inorganic Impurities

Dirt and Sediment or Turbidity

Most waters contain some suspended particles which may consist of fine sand, clay, soil, and precipitated salts. Turbidity is unpleasant to look at, can be a source of food and lodging for bacteria, and can interfere with effective disinfection.

Total Dissolved Solids (TDS)

These substances are dissolved rock and other compounds from the earth. The entire list of them could fill this page. The presence and amount of total dissolved solids in water represents the point of greatest controversy among those who promote water treatment products. Here are some facts about the consequences of higher levels of TDS in water:

1. High TDS results in undesirable tastes that are salty, bitter, or metallic.
2. High-TDS water is less thirst quenching.
3. Some of the individual mineral salts that make up TDS pose a variety of health hazards. The most problematic are nitrates, sodium, sulfates, barium, copper, and fluoride.
4. The EPA Secondary Regulations advise a maximum level of 500 mg/liter (500 parts per million-ppm) for TDS. Numerous water supplies exceed this level. When TDS levels exceed 1,000 mg/L in water it is generally considered unfit for human consumption.
5. High TDS interferes with the taste of foods and beverages, and makes them less desirable to consume.
6. High TDS makes ice cubes cloudy, softer, and faster melting.
7. Minerals exist in water mostly as *inorganic* salts. In contrast, minerals having passed through a living system are known as *organic* minerals. They are combined with proteins and sugars. According to many nutritionists, minerals are much easier to assimilate when they come from foods. Can you imagine going out to your garden for a cup of dirt to eat rather than a nice carrot; or drinking a whole bathtub of water for calcium instead of an 8-ounce glass of milk?
8. Water with higher TDS is considered by health advocates to have a poorer cleansing effect in the body than water with a low level of TDS. This is because water with low dissolved solids has a greater capacity of absorption than water with higher solids.

Toxic Metals or Heavy Metals

Among the greatest threats to health are the presence of high levels of toxic metals in drinking water - arsenic, cadmium, lead, mercury, and silver. Maximum limits for each are established by the EPA Primary Drinking Water Regulations. Other metals such as chromium and selenium, while essential trace elements in our diets, have limits imposed upon them when in water because the form in which they exist may pose a health hazard. Toxic metals are associated with nerve damage, birth defects, mental retardation, certain cancers, and increased susceptibility to disease.

Asbestos

Asbestos exists as microscopic suspended mineral fibers in water. Its primary source is asbestos-cement pipe, which was commonly used after World War II for city water supplies. It has been estimated that some 200,000 miles of this pipe is presently in use to transport our drinking water. Because these pipes are wearing, the deadly substance of asbestos is showing up with increasing frequency in drinking water. It has been linked with gastrointestinal cancer.

Radioactivity

Even though trace amounts of radioactive elements can be found in almost all drinking water, levels that pose serious health hazards are fairly rare - for now. Radioactive wastes leach from mining operations into groundwater supplies. The greatest threat is posed by nuclear accidents, nuclear processing plants, and radioactive waste disposal sites. As containers containing these wastes deteriorate with time, the risk of contaminating our aquifers grows into a toxic time bomb.

Organic Impurities

Tastes and Odors

If your water has a disagreeable taste or odor, chances are it is due to one or more of many organic substances ranging from decaying vegetation to algae; hydrocarbons to phenols. It could also be TDS and a host of other items.

Pesticides and Herbicides

The increasing use of pesticides and herbicides in agriculture shows up in the water we drink. Rain and irrigation carry these deadly chemicals down into the groundwater as well as into surface waters. More than 100-million people in the United States depend upon groundwater for all or part of their drinking water. As our reliance upon groundwater is escalating, so is its contamination. Our own household use of herbicide and pesticide substances also contributes to actual contamination. These chemicals can cause circulatory, respiratory and nerve disorders.

Toxic Organic Chemicals

The most pressing and widespread water contamination problem is a result of the organic chemicals created by industry. The American Chemical Society lists 4,039,907 distinct chemical compounds as of late 1977, which includes only those chemicals reported since 1965. The list has been growing by some 6,000 chemicals per week! Some 70,000 chemicals may still be in production in the United States. As of December 1978, 50 chemicals were being produced in greater quantities than 1.3 billion pounds per year in this country alone. There are 115,000 establishments involved in the production and distribution of chemicals, with the business being worth $113 billion per year.

There is some kind of toxic substance in our ground water no matter where we live in the United States.

According to the EPA, there are 77 billion pounds of hazardous waste being generated each year in the United States. Ninety percent of this is not disposed of properly. This would equal 21,300 pounds of hazardous waste disposed each year on every square mile of land and water surface in the United States including Alaska and Hawaii!!

There are 181,000 manmade lagoons at United States industrial and municipal sites; at least 75% of these are unlined. Even the lined ones will leak, according to the EPA. Some of these are within 1 mile of wells or water supplies. There is still a lack of information on the location of these sites, their condition, and containment. *This is a horror story!*

Little River, Mendocino, CA. Photo by Sean Sprague

Chemicals end up in our drinking water from hundreds of different sources. There are hundreds of publications each year highlighting this problem.

The effects of chronic long-term exposure to toxic organics in contaminated drinking water include recurring headache, rash, or fatigue - all of which are hard to diagnose as being water related. More serious consequences of drinking tainted water are higher cancer rates, birth defects, growth abnormalities, infertility, and nerve and organ damage. Some of these disorders may go unnoticed for decades!

Just how toxic these chemicals are may be illustrated by looking at two examples:

TCE is a widely used chemical which routinely shows up in water supplies. Just two glassfuls of TCE can contaminate 27 million gallons of drinking water!

One pound of the pesticide Endrin can contaminate 5 billion gallons of water.

Chlorine

Trihalomethanes are formed when chlorine, used to disinfect water supplies, interacts with natural organic materials (by-products of decayed vegetation, algae, etc.). This creates toxic organic chemicals such as chloroform and bromo-dichloromethane. A further word about chlorine: Scientists at Columbia University found that women who drank chlorinated water ran a 44% greater risk of dying of cancer of the gastrointestinal or urinary tract than did women who drank non-chlorinated water! Chlorinated water has also been linked to high blood pressure and anemia. Anemia is caused by the deleterious effect of chlorine on red blood cells.

Methods to Solve Our Water Problems

Sorting Through the Solutions:

Now that we have established the need for something to guarantee our water quality, what are the alternatives? There are so many water systems being sold that it seems confusing. Let us identify the various processes which are available to us and see what each one's strengths and weaknesses are.

Centralized Water Treatment

Building high-tech water treatment plants to remove impurities aren't the solution. Only 2% of water supplies to our homes is used for human consumption.

A large percentage of our population has small rural or private well supplies for water. These people would not be benefited by large municipal treatment centers.

It isn't logical to build costly plants to treat the water we use for our lawns, to flush our toilets, and to fight fires.

It's evident that it isn't practical to upgrade our treatment plants to treat *all* the water they process. Even if the plants were upgraded, the water has to be piped to our homes. It has the opportunity to pick up materials from the pipes before coming out of the tap.

Boiling Water

Boiling reduces the threat of living organisms. It serves as a method for killing bacteria during emergencies, but it is not recommended for long-term use.

Very little is removed by boiling. You may kill germs, but you still have, sediment, dissolved solids, bad taste, and odor remaining. There may also be many chemical contaminants.

Bottled Water

Is the solution for safe drinking water provided by paying up $2.00 per gallon to drink water prepared and bottled by someone else? This price reflects the costs of bottling, storage, trucking, fuel expenses, wages, insurance, etc.

If you have a point-of-use water system, you eliminate all of these middleman costs, and enjoy purified water for pennies per gallon.

Point-of-Use Water Treatment :

The most efficient and cost-effective approach to the problem of water purity is to treat *just* the water you will consume for drinking and cooking *where* you will consume it. Devices for point-of-use water treatment are available in wide variety of sizes and designs, and have varied claims as to their ability to remove impurities.

Mechanical Filtration

One of the most widely used water quality improvement methods is mechanical filtration which acts much like a fine strainer. Particles of suspended dirt, sand, rust and scale (i.e. turbidity) are trapped and retained, greatly improving the clarity and appeal of water.

When enough particulate matter has accumulated on or within the filter element, it is usually discarded. This type of filter is usually considered a pre-filter.

Activated Carbon Adsorption

Carbon adsorption is probably the most widely sold method for home water treatment. This is because of its ability to improve water by removing many disagreeable tastes and odors including objectionable chlorine.

Activated carbon (AC) is processed carbon. In this form it will remove far more contamination from water than will nonactivated carbon.

AC is made from a variety of materials such as coal, petroleum, nutshells, and fruit pits. These are heated to high temperatures with steam in the absence of oxygen (the activation process) leaving millions of microscopic

---

One pound of activated carbon provides from 60 to 150 acres of surface area.

---

pores and great surface area. One pound of activated carbon provides from 60 to 150 acres of surface area. The pores trap microscopic particles and large organic molecules while the activated surface areas cling on to or adsorb the smaller organic molecules.

While AC theoretically has the ability to remove or reduce numerous organic chemicals like pesticides, THM's, TCE, PCB, etc., its actual effectiveness is highly dependent on the following factors:
1.   The type of carbon and the amount used.
2.   The design of the filter and how slowly water flows through it (contact time).
3.   How long the carbon has been in service and how many gallons it has treated.
4.   The kinds of impurities it has removed.
5.   The water conditions - turbidity, temperature, etc.

One major problem with carbon filters is the growth of bacteria. At first, when the carbon is fresh, practically all organic impurities and even some bacteria are removed.

Once organic impurities accumulate, they can become food for the growth of more bacteria. These can then multiply within the filter to great numbers. While these bacteria may not be disease causing, their high concentration is considered by some to present a health hazard. It is often advised that after periods of non-use (such as overnight) a substantial quantity of water be flushed through the carbon filter to minimize the accumulation of bacteria.

Oligodynamic, Silver Impregnated or Bacteriostatic Carbon

A manufacturer who adds (impregnates) silver compounds to the surface of the carbon granules is trying to inhibit bacteria growth within the carbon bed. However, EPA- sponsored testing of such filters has shown that they are "neither effective nor dependable in meeting these claims" (EPA Report #EPA/600-D-86/232, October 1986).

Some manufacturers have also made misleading claims that their silver-impregnated filters will eliminate bacterial contamination from virtually any water source. The low concentration of silver found in these filters is not capable of destroying influent waterborne bacteria or providing protection from contaminated water under normal flow conditions.

Pyrogens are dead bacteria. All of the bacteria destroyed by the silver-impregnated carbon will still end up in your drinking water in the form of pyrogens.

Because silver is also toxic to humans, such filters are regulated by the EPA under the Insecticide, Fungicide and Rodenticide Act and must be registered and issued a registration number. This registration doesn't imply any EPA approval of the unit or of its effectiveness. It does certify that the carbon will not release more than 50 parts per billion of silver - the maximum safe level.

Another problem with carbon filters is chemical recontamination which can occur when the carbon surface has become saturated with the sum total of impurities it has adsorbed - a point that is impossible to predict. If the use of the carbon is continued, the trapped organics can leave the surface and recontaminate the water with more impurities than those contained in the raw tap water.

To get the most out of carbon, it should be kept scrupulously clean of sediment and heavy organic impurities such as the byproducts of decayed vegetable matter and microorganisms. These impurities prematurely use up the carbon's capacity, preventing it from doing what it does best - adsorbing lightweight toxic organic impurities like THM's and TCE, and undesirable gases such as chlorine.

Solid Block Carbon

This is obtained when very fine pulverized carbon is compressed and fused together with a binding medium (such as a polyethylene plastic) into a solid block. The

Lake Atitlan, Guatemala. Photo by Sean Sprague

intricate maze developed within the block ensures contact with organic impurities and therefore more effective removal. The problem of channeling (open paths developing because of the buildup of impurities, and rapid water movement under pressure) in a loose bed of granulated carbon granules is eliminated by solid block filters.

Block filters can also be fabricated to have such a fine porous structure that they are capable of mechanically filtering out coliform and other associated disease bacteria. This is not to say that all bacteria or any viruses are filtered out. Solid block filters with this feature will require replacement more regularly.

Among the disadvantages of compressed carbon filters is the reduced capacity due to the inert binding agent and their tendency to plug up quickly with particulate matter. They are also substantially more expensive than conventional carbon filters.

Limitations of Carbon Filters

A properly designed carbon filter has shown itself capable of removing many toxic organic contaminants, but they fall far short of being an overall water treatment system for providing protection from the wide spectrum of impurities found in water.

1.   They are not capable of removing any of the undesirable excess total dissolved solids.

2.   They have only a limited and unreliable effect on toxic metals like lead, cadmium, mercury, and arsenic. Large suspended materials will be removed by some filters. Dissolved materials will not be removed at all by carbon filtration. The vast majority of solids are dissolved in water, not suspended.

3.   They have no effect on harmful nitrates, or high sodium and fluoride levels.

4.   For any carbon filter to be effective (even for organic removal), water must pass through the carbon (whether it be granular or compressed) slowly enough for complete contact to be made between the carbon and the impurities. This all-important factor is referred to in the industry as contact time. At useful flow rates of 0.5 - 1 gallon per minute, the flow rate is determined by the amount of carbon, and few manufacturers use an acceptable amount of carbon.

One must read carefully the claims which are made by carbon filter companies to really see what they can and can't do. After claiming that their filter will remove arsenic, cadmium, lead, mercury, and more, there is a note beneath the test results of one manufacturer's carbon filter system: "IMPORTANT - These data are furnished as documented results from specific testing and are generally regarded as indicative of the effectiveness to be expected from the _____ drinking water system but are not specific claims of performance."

Carefully reading this statement tells us that there is no guarantee that the system will perform for us the way the test results say that it will. One has to read between the lines to detect the truth.

Another supplier of carbon block filters restricts their claims of toxic metal removal to the *precipitated* form - that is, particles. One will find, however, that heavy metals occur mostly in *dissolved* forms. A carbon filter

Photo by Sean Sprague

will have no effect on dissolved heavy metals. Before we leave AC filters, there is another area to be addressed

Minerals in Drinking Water

Purveyors of AC filter systems usually bring up the point that "We need minerals in water - these are essential for good health." The only problem with this statement is that there have never been any scientific studies conducted to once and for all prove that minerals *in water* are essential for good health. Frankly, it isn't a priority in the scientific community to spend the vast amounts of money necessary to conduct the investigation needed to arrive at the foundation of this matter.

Therefore, the value of minerals in drinking water remains a moot point - no one really knows for sure. Everybody may have an opinion regarding this matter, but the fact is that nobody knows for sure. One making a dogmatic statement that "minerals in drinking water are bad," or "minerals in drinking water are good" really is showing only ignorance of the issues involved.

The reason filter dealers bring up this point is because their product will not remove dissolved solids. This means that while AC-filtered water retains all of the dissolved calcium, magnesium, etc. commonly called "minerals", it *also* retains all of the dissolved lead, cadmium, arsenic, etc.

Only methods of purification which remove dissolved solids remove dissolved heavy metals.

In summary, AC filters are an important piece of the purification process, although a piece of the puzzle doesn't make a completed puzzle. AC removes chemicals and gases. This makes AC an integral part of legitimate water purification systems. AC won't remove those materials which are dissolved in the water.

Distillation

Distillation is the process of heating water to steam and recondensing it back to water by cooling it. Distillation mimics the hydrologic cycle of nature (the sun causes evaporation over the earth's bodies of water and condensation/precipitation occurs over the land masses).

Distillation will remove impurities such as sediment, dissolved solids, nitrates, sodium, toxic metals, and microorganisms. These are basically left behind as the water turns to steam.

Some toxic organic chemicals will vaporize with the steam and be carried over into the distillate with the water. To solve this problem, an activated carbon filter should be incorporated into the distiller either before or after the boiling chamber.

Sophisticated fractional distillers will remove these organics by heating water in fractions until the boiling point is reached. The organics are vented out at each step of the heating process.

Even with the problem of organics addressed, there are still disadvantages with distillers:

1. Distillers are time-consuming to maintain and clean. The impurities and total dissolved solids are left behind in the boiling chamber. A hard scale builds up on the heating element and in the boiling chamber which must be removed. If this scale is left in the system, the efficiency will be impaired and eventually diminish.

2. The product water should be cooled quickly as its elevated temperature encourages the regrowth of airborne bacteria. This is a problem of convenience.

3. The process of rapid distillation will drive away free oxygen dissolved in the water. Many scientists and doctors refer to distilled water as dead water. The absence of free oxygen will also give the water a flat taste.

4. Research is being performed on the characteristics of drinking water in Europe. Much is being discussed in these circles about the importance of balanced water (balanced in pH, oxidation-reduction, and resistivity). Distillation doesn't provide these balanced characteristics in the product water, according to these health professionals. More information is available on this by request.

5. Distilled water costs a lot to produce because of the energy required to vaporize all of your drinking and cooking water (an exception to this is a solar distiller). Every rate increase from the utility company will make distilled water even more expensive.

Deionization

The process of deionization (DI) is worth discussing even though it isn't a practical water treatment method for household use. It has appeared in several home water treatment devices, however. DI is a chemical process that utilizes minute plastic beads called resins. As untreated water flows over these treated resins, the ions of total dissolved solids are leached from the water.

When the resin beads become saturated, they must be

The value of minerals in drinking water remains a moot point - no one knows for sure.

---

removed, and regenerated with acid or caustic chemicals.

DI removes only *charged* particles (total dissolved solids). DI is not capable of removing dirt, rust, sediment, pesticides, organic toxins, asbestos, bacteria, or viruses at all. It is, therefore, used in conjunction with other water treatment methods.

The resins also will provide an environment that encourages bacteria growth. Some scientists are concerned that the removal of the ions of dissolved solids by deionization affects the basic ionic configuration of the water. They believe that this alters the water in a negative sense for consumption and assimilation by the human body.

Water softeners work by the principle of exchanging one ion for another. The resin beads in a water softener substitute two ions of sodium for an ion of calcium or magnesium. With the removal of the calcium and magnesium ions, the water is no longer hard.

Ultrafiltration/Reverse Osmosis

Osmosis occurs in living organisms, in which there is a piece of tissue or a membrane with fluids on either side of it. Fluids having a lesser concentration will be drawn through the tissue/membrane to mix with fluids having a greater concentration. This is to equalize the concentration of substances in the fluids on both sides of the tissue/membrane.

This can be illustrated if you cut open an avocado and salt the surface of the flesh. In a short time you will notice water has been drawn out of the avocado as osmosis tends to equalize the concentration of salt.

Osmosis occurs when there are two fluids of differing concentration separated by a semipermeable membrane. The fluid will pass through the membrane in the direction of the most concentrated solution.

Osmosis is the process by which oxygen goes from our lungs into the bloodstream, and by which water and nutrients penetrate the root structure of a tree and cause it to grow.

In the natural world surrounding us and inside of us, there is a vast network of biological membranes. These screening barriers govern the selection and passage of chemicals and fluids. In essence, these membranes control the traffic of the life processes themselves.

Membranes help organisms carry out an immense variety of exchanges with their environment. The gills of a fish obtain oxygen from water. Our lungs extract oxygen from the air and place it in our bloodstream. In plants, the cell walls allow photosynthesis to take place by providing the medium for the transfer of carbon dioxide and oxygen.

Our blood is recycled and renewed by many osmotic processes. One integral function is that of the kidneys. As the blood enters the kidney, it flows in small arteries in close contact with tiny excretory units of the kidney known as nephrons. From the blood, water is extracted along with wastes to become an essential component of urine. Water can remain in the nephron to become reabsorbed back into the blood- stream if not enough water is consumed to be excreted freely.

Without our kidneys we would not be able to survive.

The time required to purify one gallon of RO water is 3-4 hours.

The first artificial kidney was built from a cellophane membrane in 1944. In the early 1950's, Drs. Sidney Loeb and S. Sourirajan from UCLA Medical School developed the first synthetic membrane made from cellulose acetates. This had commercial reverse osmosis capabilities.

Reverse osmosis is exactly the opposite of Osmosis. In reverse osmosis (RO), water having a lesser concentration of substances is derived from water having a higher concentration of substances. Tap water with dissolved solids and other materials in it is forced by the water pressure inherent in our water pipes against a membrane. The water is removed from this concentration of materials by penetrating the RO membrane, and leaving the materials behind. This will be up to 99% removal of dissolved solids.

The RO membrane is an ultimate mechanical filter or ultrafilter. It strains out virtually all particulate materials, turbidity, bacteria, viruses, asbestos, even single molecules of the heavier organics.

To appreciate the fineness of this membrane or ultrafilter, its pore size would be two hundred millionths of an inch in diameter. That's smaller than what can be seen by an optical microscope.

By the remarkable phenomenon of RO, particles smaller than water molecules themselves are removed! The molecules diffuse through the membrane in a purified state and collect on the opposite side.

Ultrafiltration/RO membranes remove and reject such a wide spectrum of impurities from water using very minimal energy - just water pressure. RO gives the best water available for the lowest price expended.

Reverse osmosis effectively removed the following:

- Particulate matter, turbidity, sediment, etc.
- Colloidal matter (permanently suspended particles)
- Total dissolved solids (up to 99%)
- Toxic metals
- Radioactive elements
- Microorganisms
- Asbestos
- Pesticides and herbicides   (coupled with AC)

Reverse Osmosis and Activated Carbon Adsorption

Ultrafiltration/RO alone will not remove all of the lighter, low molecular weight volatile organics such as THM's, TCE, vinyl chloride, carbon tetrachloride, etc. They are too small to be removed by the straining action of the RO membrane. Their chemical structure is such that they aren't repelled by the membrane surface. Since these are some of the most toxic of the contaminants found in tap water, it is very important that activated carbon be used in conjunction with the membrane.

In some applications, AC is used before the membrane. In all applications with quality RO systems, there is AC after the membrane. This means that most AC filters don't have to contend with bacteria and all of the other materials which cause fouling and impair performance if AC follows a well-maintained membrane.

Not all RO systems are created equally. That is why you'll see such a variation in price. The engineering and experience behind the RO design is crucial to its overall performance and dependability.

The typical time required to purify one gallon of RO water is three to four hours. RO uses water to purify water. This is what's known as the rate of recovery. Superior RO's use four gallons of brine (waste water) to make one gallon of purified water. Some systems use up to twenty gallons of brine to purify one gallon of product water.

Brine is necessary to remove excess accumulated materials from the RO membrane. These materials have been rejected from the purified water, and if left in the system impair efficiency.

Our bodies also have a waste water elimination system through the kidneys. If we can't purge our systems of these waste materials, we die. Many owners of RO systems direct brine outside and use it in an additional drip line for their gardens, for example.

The cost of water energy for a fine RO system will amount to about $1.33 per month if one pays for water at the rate of $1.00 per 100 cubic feet!

You Can Guarantee the Quality of Your Drinking Water

You can see from the material presented here that there is much to be aware of regarding the purchase of a purification system. All you must do is to decide how comprehensive you want your water treatment system to be. A system which combines many of the technologies will give you better product water than a system which incorporates just one.

Choose the technologies which you can live with for a long time. You might have to purchase another water treatment device if you don't acquire one as sophisticated as you'll eventually need.

- Randy Wimer

Water Filtration Systems

Laboratory Water Test

For Private Water Supplies Only (Private wells, springs, catchment, etc.)

We've found through lots of experience that if you're not on city/municipal or pretreated water, we can't conscientiously sell you a water purification system without first knowing these specific particulars about the water you have: pH, total dissolved solids (TDS), hardness, iron, manganese, copper, and tannin. This test normally costs $40, but we're providing it to our customers for only $17. This analysis tells us enough about your water so that we can make an informed recommendation to you regarding what system will work best with the water you have. Upon receipt of your $17 we will send you a questionnaire and a small plastic bottle. Fill out the questionnaire, return the water sample, and you should have your results back in a few weeks. It's the only way we can both be sure you're getting the right filter for your system! This test does not show contamination problems such as from coliform or chemicals. Refer to our water check with pesticide option for this.

We need to stress that if you're on a municipal or pretreated water system we don't need to test your water to recommend a filtration system for you.

42-000   Water test (Non-city water) $17

Man and boy washing, Jakarta, Indonesia.
Photo by Sean Sprague

National Testing Laboratories (NTL) More Extensive Water Sample Testing

We are working with National Testing Laboratories (NTL) to provide a thorough analysis of your water. Now that the EPA has determined that the ground water in more than 30 states is seriously polluted, we feel it's essential to thoroughly test your water. NTL has two laboratory facilities, one in Michigan and one in Florida, which perform the full range of drinking water analysis, inorganic, organic, and bacteriological. Their laboratories are certified in 13 states to perform analyses of drinking water, using only US EPA approved methods, and a strict quality assurance program. *We've compared National Testing Laboratories with lots of other testing services and find them to be the most comprehensive, accurate, and reasonable available anywhere in this price range.* Their program is simple - you order the test kit from us; we mail it directly to you; you fill the sample bottles and ship the kit to the NTL lab. NTL will analyze the samples according to the test series purchased and will send the full report and explanatory letter back to you within 7 - 10 working days. Your test will show if your water contains any of the listed pollutants in amounts higher than EPA limits. A cover letter interprets your test and clearly explains what action (if any) is recommended to insure that your water is safe to drink. We're offering two of their complete water sample tests for you: the *City Scan* and the *Water Check Plus Pesticides Test.*

City Scan NTL Test

Designed specifically for municipal water systems, the City Scan offers you a way to simply test your own drinking water. The test has 17 different parameters to assist you in determining the quality of your water. Twelve parameters are regulated by the Safe Drinking Water Act and the remaining five are useful in determining the esthetic quality of the water. Four of the tests are done in you own home. You will have instant results for pH, hardness, alkalinity, and free and total Chlorine. The analysis performed in the NTL laboratory check for the following Inorganic materials:

Iron	**Sodium**	**Lead**
Manganese	**Fluoride**	**Chloride**
Sulfate	**Total Dissolved Solids**	

and test for the following Organic materials:

Bromoform	**Bromodichloromethane**
Chloroform	**Dibromochloromethane**

42-002   NTL City Scan Water Test $69

NTL Water Check With Pesticide Option

This is the most comprehensive and accurate analysis of your water available anywhere in this price range. Your water will be analyzed for 73 items, plus 20 pesticides. If you have any questions about your water's integrity, this is the test to give you peace of mind. The kit comes with five water sample bottles; blue gel refrigerant pack (to keep bacterial samples cool for accurate test results), and easy-to-follow sampling instructions. You'll receive back a two-page report showing 93 contaminant levels, together with explasnations of which contaminants, if any, are above recommended values. You'll also receive a follow-up letter with a personalized explanation of your test results, plus knowledgeable, unbiased advice on what action you should take if your drinking water contains contaminants above EPA-recommended levels. The test will check for the following parameters:

METALS:
Arsenic
Barium
Cadmium
Chromium
Copper
Iron
Lead
Manganese
Mercury
Nickel
Selenium
Silver
Sodium
Zinc

INORGANICS
Total alkalinity (as $CaCO_3$)
Chloride
Fluoride
Nitrate (as N)
Nitrite
Sulfate
Hardness (as $CaCO_3$)
pH (standard units)
Total dissolved solids
Turbidity (NTU)

VOLATILE ORGANICS
Bromoform
Bromodichloromethane
Chloroform
Dibromochloromethane
Total trihalomethanes

Benzene
Vinyl chloride
Carbon tetrachloride
1,2-Dichloroethane
Trichloroethylene (TCE)
1,4-Dichlorobenzene
1,1-Dichloroethylene
1,1,1-Trichloroethane

VOCs (cont'd):
Acrolein
Acrylonitrile
Bromobenzene
Bromomethane
Chlorobenzene
Chloroethane
Chloromethane
O-Chlorotoluene
P-Chlorotoluene
Dibromochloropropane (DBCP)
Dibromomethane
1,2-Dichlorobenzene
1,3-Dichlorobenzene
trans-1,2-Dichloroethylene
cis-1,2-Dichloroethylene
Dichloromethane
1,1-Dichloroethane
1,1-Dichloropropene
1,2-Dichloropropane
trans-1,3-Dichloropropane
cis-1,3-Dichloropropane
2,2-Dichloropropane
Ethylenedibromide (EDB)
Ethylbenzene
Styrene
1,1,2-Trichloroethane
1,1,1,2-Tetrachloroethane
1,1,2,2-Tetrachloroethane
Tetrachloroethylene (PCE)
1,2,3-Trichloropropane
Toluene
Xylene
Chloroethylvinyl ether
Dichlorodifluoromethane
cis-1,3-Dichloropropene
Trichlorofluoromethane
Trichlorobenzene(s)

MICROBIOLOGICAL:
Coliform bacteria

and it will check for the following pecticides:

PESTICIDES AND HERBICIDES:
Alachlor
Aldrin
Atrazine
Chlordane
Dichloran
Dieldren
Endrin
Heptachlor
Heptachlor Epoxide
Hexachlorobenzene

Hexachlorapentadiene
Lindane
Methoxychlor
PCBs
Pentachloronitrobenzene
Simazine
Toxaphene
Trifluralin
Silvex 2,4,5,-TP
2,4-D

42-003   NTL Water Check w/Pesticide Option $129

Filter Pure Water Filters

We've been searching for a lower cost water filter line to incorporate with our R.O. and solid carbon block systems. Filter Pure fits the bill. They are all made of high grade ABS Plastic (many lower priced filters are built of lower grade PVC). Filter Pure systems are sealed ultra-sonically rather than with glue that can potentially leak into your water, and they come with a medical grade filtration disk (to alleviate channeling) - the same used in kidney dialysis machines. Filter Pure has not scrimped materials for quality and still offers a competitively priced filter system.

Filter Pure has developed a Redox media called Magnalyte, which appears to be a positive new development within the water filter treatment industry. Independent E.P.A. registered laboratory analysis shows levels of contaminant reduction that builds upon the capacities of carbon, to increase longevity and percentage of reduction. These tests show nearly 100% reduction of chlorine, lead, and volatile organics. Magnalyte appears to be one of the most effective lead reducing filter available today.

CT-M Filter Pure

The CT-M (for Magnalyte) is a counter-top water filter that is the most effective and efficient filtration unit of its kind. It is the best filter for lead reduction, will reduce nearly 100% of the chlorine in your water, and does wonders for taste and odor. It reduces heavy metals, algae, fungi, and organic volatiles. The CT-M is not effective with giardia. The flow rate is 1/2 to 1 gpm - it takes a minimum pressure of 15 psi and a maximum of 65 psi. It has a filtration capacity of up to 20,000 gallons. **For a typical family of four this is up to 20 years.** Nevertheless, we recommend that you re-test your water for chlorine within five years to assure that your filter continues to operate properly. This unit retails for $199 but we're offering it at great price!

42-601   CT-M Filter $129

US-M Filter Pure

The US-M is an under-the-sink filter that is extremely effective at reducing chlorine, lead, bad taste, odor, organic volatiles, algae, and fungi. The US-M will handle more severe water problems than the CT-M (still will not handle giardia) and is also preferred by many people because it provides more counter space. A counter top chrome faucet set is included with every unit. It measures 14.5" x 5", has a flow rate of 0.6 gallons to 1.5 gallons, and a minimum pressure of 15 psi. It has a capacity of 25,000 gallons. **(For a typical family of four this is up to 25 years.)** Nevertheless, we recommend that you re-test your water for chlorine within five years to assure that your filter continues to operate properly.

42-602   US-M Filter $209

TR-1 Portable "Traveler" Filter

The Filter Pure TR-1 will provide filtered water wherever you travel. It attaches easily to any faucet, is lightweight and compact. It reduces chlorine, bad taste, odor and organics, but not giardia. It is fully E.P.A. registered and bacteriostatic. It will handle up to 1,000 gallons. The active ingredient is 1.05% silver carbon. It uses a medical grade filtration disk to filter down to 20 microns.

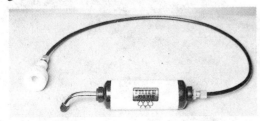

42-605   TR-1 Portable Filter $49

US-2 Filter Pure

The US-2 is an under-the-sink filtration system with a replaceable cartridge. It comes with a sub-micron silver-ceramic cartridge that removes bacteria, asbestos, giardia, and sediment to below one micron in size or is available for $10 extra with an optional carbonized-ceramic cartridge (for city dwellers to remove chlorine). The US-2 is not effective in reducing lead. It comes with a counter top chrome faucet set. The replaceable cartridge is good for 1,000 gallons.

42-603-S	US-2 w/Silver-Ceramic Filter	**$155**
42-603-C	US-2 w/Carbonized-Ceramic	**$195**
42-613	Silver-Ceramic Repl. Filter	**$ 65**
42-623	Carbon-Ceramic Repl. Filter	**$ 89**

US-3 Filter Pure

The US-3 is like the US-2 in that its an under-the-sink model. With three separate filters, it is the closest thing to Reverse Osmosis quality in a standard filter on the market. It removes everything that the US-2 removes as well as reducing chlorine, bad taste, odor, organic volatiles, lead, giardia, and heavy metals. The replaceable cartridges are good for 1,000 gallons of filtration. Of the three filters used in the US-3, the first is a re-cleanable ceramic pre-filter; the second is a Magnalyte replaceable cartridge; and the third is a replaceable carbon cartridge. It comes with the chrome faucet set. *Although the photograph shows gauges, they are not included with the unit.*

42-604	US-3 Water Filter	**$245**
42-614	Replacement Carbon Filter	**$ 65**
42-624	Replacement Magnalyte Filter	**$ 55**

Katadyn Pocket Filter

The Katadyn Pocket Filter is *standard issue* with the
International Red Cross and the armed forces of many
nations and is essential equipment for survival kits.
Manufactured in Switzerland for over half a century, these
filters are of the highest quality imaginable - reminiscent
of Swiss watches! The Katadyn system uses an extremely
fine 0.2 micron ceramic filter that thoroughly blocks
pathological organisms from entering your drinking water.
The Katadyn systems also protect the user from viral
infection as well. The pocket filter is a self-contained and
very easy to use filter the size of a 2-cell flashlight (10" x
2"). It will produce a quart of ultra-pure drinking water in
90 seconds with the simple built-in hand pump. It weighs
only 23 oz. It comes with its own travel case, and a special
brush to clean the ceramic filter. The replaceable ceramic
filter can be cleaned 400 times lasting for many years with
average use. This filter is indispensible for campers,
backpackers, fishermen, mountaineers, river runners,
globetrotters, missionaries, geologists and workers in
disaster areas.

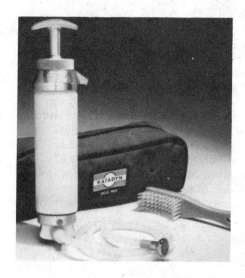

42-608	Katadyn Pocket Filter	$275
42-609	Replacement Filter	$140

Survivor 06 - Hand Water Purifier

The new Survivor 06 is a hand operated water purifier that
will convert sea water, brackish water, or contaminated
water into fresh and pure drinking water. It will produce
one cup every thirteen minutes at 30 strokes per minute,
or six gallons per day. This is the only manually-operated
desalinator in the world. It uses no batteries, generators,
or alternators. It combines reverse osmosis technology
with energy recovery technology. A pre-filter removes
larger particles and debris. The unit measures 5" high x 8"
long x 2-1/2" wide. It will suck to a maximum suction
height of ten feet. It will work under a temperature range
of 33 to 120 degrees farenheit. The unit will perform for
approximately 1,000 hours.

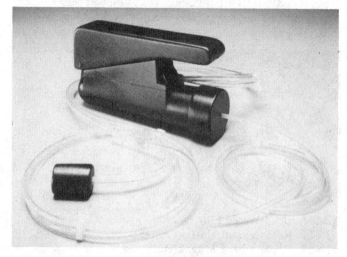

■42-302   Survivor 06 Watermaker $650

PowerSurvivor 12V Watermaker

Fresh water is a necessity at sea and in remote areas. In
many third world countries getting fresh, clean water often
poses a problem. The PowerSurvivor is the most compact,
most efficient desalinator in the world. Drawing only 4
amps at 12 volts it will turn sea, brackish, or contaminated
water into crystal-pure drinking wter at the rate of 1.4
gallons per hour. In an emergency, it can be operated
manually with an easy-to-attach handle. It is highly reliable
with few moving parts and only takes up one cubic-foot of
space. The PowerSurvivor combines reverse osmosis (RO)
and energy recovery technology. The Powersurvivor will
last approximately 2,500 to 3,000 hours when it will need
an overhaul from the company.

■42-301   PowerSurvivor 12V Watermaker $1995

Purwater Reverse Osmosis (RO) Filters

The process of reverse osmosis was developed at great expense by the U.S. Government and is incredibly simple: Polluted tap water is forced against a semipermeable membrane using only water pressure as the power source. No electricity or other energy source is required. When water is applied under pressure, the properties of the membrane allow the pure water molecules, but NOT the pollutants, to pass through.

Reverse Osmosis (RO) works like the human kidney. The RO membrane is the ultimate mechanical filter. RO will remove particulate matter, turbidity, sediment, colloidal matter, total dissolved solids (up to 99%), toxic metals, radioactive elements, microorganisms, asbestos, pesticides, and herbicides.

Purwater 2

The Purwater 2 yields 6-8 gallons daily. It uses a two-gallon holding tank with an easy quick disconnect. It features a built-in flush valve and is of a compact size fitting nicely on a countertop. It measures 12" high by 16-1/8" wide by 8-5/8" thick. The Purwater 2 incorporates an R.O. membrane (either CTA or TFC) and a carbon filter. There is a three year pro-rated warranty on the RO membrane, which is renewed with each new RO membrane installation. The Purwater 2 includes all you need to hook it up. The costs listed below are out of warranty replacement costs.

■42-202	Purwater 2	$375
■42-204	CTA Membrane	$ 75
■42-205	TFC Membrane	$135
■42-206	Carbon Replacement Filter	$ 18

Purwater 3

The Purwater 3 yields 3-5 gallons of purified water daily and comes with a built-in flush valve. It can be permanently installed under your sink easily, so all you can see is the chrome faucet on your sink. It works well with an ice-maker - the tank measures 11" in diameter by 15" high. The unit itself is 15-3/4" high by 7-5/8" wide by 4-1/4" thick. The Purwater 3 incorporates an R.O. membrane (either CTA or TFC) and a carbon filter. There is a three year pro-rated warranty on the RO membrane, which is renewed with each new RO membrane installation. The Purwater 3 includes all you need to hook it up. The costs listed below are out of warranty replacement costs.

■42-203	Purwater 3	$595
■42-207	CTA Membrane	$ 75
■42-208	TFC Membrane	$125
■42-209	Carbon Replacement Filter	$ 24
■42-210	Pre-Sediment Filter	$ 24

Purwater RO units normally require 40 psi to operate - if your water pressure is lower request the "low pressure membrane," which will accommodate pressure as low as 20 psi for the Purwater 2 only.

Solar Distillation

Written by Horace McCracken and Mary McClellan

In the coming years, as serious concern over environmental quality continues to grow, one of our most valuable resources - clean water - will be in the spotlight more and more. Potable water shortages are on the rise. This and the progressive degradation of our water supplies are creating a growing market for

water purification equipment. As early as forty years ago, researchers such as Professor Everett Howe in California and Roger Morse in Australia demonstrated that solar distillation is an effective method of purifying water. In recent years, its increasing cost effectiveness has made solar distillation a more popular water purification process.

Distillation is the process of heating water to produce vapor, then cooling the vapor to condense it back to water. It leaves behind impurities such as sediment and dissolved solids (metals, salts, minerals), and microorganisms. A solar still, properly constructed, installed and operated produces water of extremely high quality. Dissolved solids are routinely reduced to levels of less than one part per million (ppm) total dissolved solids (tds). This is higher quality water than is normally sold in

bottles as distilled water, or "for all distilled water uses". Dozens of metals and minerals are reduced to "non-detectable" levels.

Polluted water can be purified by solar distillation because bacteria and other living organisms do not evaporate along with the individual water molecules. It is important to protect this pure water from contamination by outside substances, whether in the collection trough or the reservoir. Some researchers have suggested catching rainwater off the glass cover of the still to increase the volume of distilled water. However, rainwater carries with it contamination from bird manure and insects, and reservoirs commonly attract rodents and frogs. Some people prefer to use a solar still to purify water from their rain catchment cisterns or reservoirs. A properly constructed and installed distilled water reservoir will remain free of bacteria for years.

In solar stills with only a gentle steaming action (causing almost no breaking bubbles), the amount of organic chemicals - which are occasionally present in the feed water - are greatly reduced in the condensate. Tests to date show reduction in organic chemicals by amounts of 75% to 99.5%, and post treatment of the condensate with charcoal has been found to be unnecessary thus far.

High temperature distillation processes, especially boiling, will drive off oxygen from the system, and there may be no way - as on ships - to replace the oxygen in the distilled water. In a

solar still, the potable water condenses in the presence of air, thus absorbing oxygen. The flat taste usually associated with bottled distilled water is rarely present in solar distilled water.

There has been much talk about the value of minerals in drinking water. Certain minerals in appropriate quantities are desirable and essential for good health, but the scientific community has reached no general agreement that some or any of these minerals are best obtained from drinking water. In light of this, distilled water certainly can be good for your health, all your life, provided you have an adequate diet.

Solar distillers operate with the heat from the sun and continue to work, to some extent, even on slightly overcast days. Most solar stills include a mirror as standard equipment. Attached permanently to the back of the still, it will increase the amount of sunshine actually hitting the still. These are especially valuable in the higher latitudes where the sun's rays hit the earth at a low angle during the cooler seasons.

How it works:

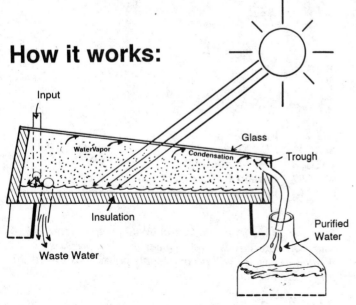

Input

WaterVapor

Glass

Condensation

Trough

Insulation

Purified Water

Waste Water

The cost of solar distilled water is low over the long run. Your initial cost is your only cost and that investment can provide a return for decades, since properly constructed stills can last 20 years or more. Comparing the investment of a solar still to the price of bottled drinking water makes it a great investment. Bottled water costs have risen from 25¢ per gallon in the sixties to an average of $1.25 per gallon today. This is largely due to the escalating cost of energy required for purification and transportation, rising costs of employees, and inflation. In contrast, the sun's energy is "delivered" to your house every day, free. In virtually any location on earth where you must transport drinking water any distance, in any kind of container, solar distillation is advantageous. Considering the high quality of solar distilled water, the simplicity of the still's operation and zero energy cost, solar distillation makes a great deal of sense, both technologically and economically.

Sunwater™ Solar Distiller

For pure drinking water, use a Sunwater Solar Still. Heat from the sun does all the work! The Sunwater Solar Still removes sediment, dissolved solids and microorganisms. Once the still is properly installed, there are absolutely no energy costs - ever - and virtually no expenditures of time or money for filters, parts, and other maintenance costs associated with other purification systems. All you have to do is fill it - manually or automatically.

Horace McCracken, originator of the Sunwater Solar Still, has been a producer of solar stills since 1959. He is a pioneer in the field of distillation and desalination, and was elected to the Solar Hall of Fame in 1984.

The way the Solar Still works is very simple: the feed water goes into the input pipe and inside the still. The sun's rays penetrate the glass, causing the water inside to become hot and evaporate, condensing on the underside of the glass as purified water. From there it flows into the catching reservoir - pure, clean, and ready to drink! Extra water fed into the Still flushes out the concentrated waste. The Sunwater Solar Still comes with a five-year guarantee, and many stills have been in operation for 20 years or more.

Sunwater Solar Stills come in a variety of sizes, for all applications. From automatic residential to industrial and labora-

tory requirements, a Solar Still can be put together to meet any need. Currently installed commercial systems are producing as much as 15 gallons per day. One 15 g.p.d. installation in California produces about $4,000 worth of water a year from brackish well water - enough for 40 people. It will pay for itself in savings (compared to bottled water) in 2 years, is guaranteed for 5 years, and is expected to produce pure, clean water for 20 years.

42-420	26" Complete still	$339
42-420-1	26" Complete still no mirror	$289
42-421	26" Complete kit w/mirror	$269
42-421-1	26" Complete kit no mirror	$229
42-422	26" Partial kit	$189
42-422-1	26" Partial kit no mirror	$159
42-440	50" Complete still	$469
42-440-1	50" Complete still no mirror	$399
42-441	50" Complete kit w/mirror	$379
42-441-1	50" Complete kit no mirror	$319
42-442	50" Partial kit	$229
42-442-1	50" Partial kit no mirror	$199

42-460	78" Complete still	$599
42-460-1	78" Complete still no mirror	$499
42-461	78" Complete kit w/mirror	$479
42-461-1	78" Complete kit no mirror	$409
42-462	78" Partial kit	$279
42-462-1	78" Partial kit no mirror	$239
42-480	98" Complete still	$719
42-480-1	98" Complete still no mirror	$609
42-481	98" Complete kit w/mirror	$579
42-481-1	98" Complete kit no mirror	$489
42-482	98" Partial kit	$319
42-482-1	98" Partial kit no mirror	$269

Note: all partial kits are without lumber, insulation, screws and glass.

42-450	Auto Feed System all sizes (Add to above prices)	$369
42-451	Reservoir w/sight tube(20 gal)	$339
42-452	Reservoir w/sight tube(120 gal)	$489
42-453	Reservoir w/sight tube(300 gal)	$775

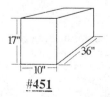

#451

#452

#453

New Rainshower Shower Filter

The Rainshower is a great shower filter that removes chlorine and other contaminants from your water. Since our introduction of the original Rainshower our Winter '89 catalog, customers have given unsolicited raves about the product. Now, there is a new and improved Rainshower that addresses the main problem with the original - its length. The new design in molded ABS plastic is only five inches long including our finest showerhead. The new filter still removes 90% or more of residual free chlorine from shower water and aids in the control of fungus and mildew through its unique use of non-toxic KDF-55 and its "redox" technology.

Chlorine and residual chlorine is very hazardous to hair, skin, eyes, and lungs. We actually can take in more chlorine from one fifteen minute shower (*but don't ever waste that much water!*) than from drinking eight glasses of the same water in one day. Chlorine plays havoc with our skin and hair chemically bonding with the protein in our bodies. It makes hair brittle and dry and can make sensitive skin dry, flaky, and itchy.

Rainshower's use of electrochemical oxidation technology makes it effective for thousands of gallons of water, not the hundreds of gallons of the less effective activated carbon shower systems. The filter converts free

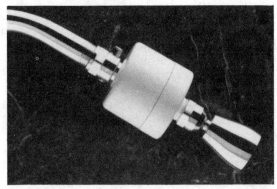

chlorine into a harmless water soluable chloride which washes out of the filter. The copper-zinc "redox" mixture can reduce inorganic compounds in the water such as lead and cadmium. Also, zinc is generally regarded as a nutrient for the skin. The product meets FDA standards for copper and zinc in drinking water.

The "new" Rainshower is easy to install and comes with our lowest flow shower head. Savings in water and energy costs can pay for this product in less than six months for the average family. A 240" roll of teflon tape is included for a leak-free installation. Each unit comes with a back-flushing adapter. A one-year warranty is provided.

42-701   Rainshower Shower Filter $79

Rainshower Garden Filter

Rainshower has gone one step further in de-chlorinating our lifestyles. Now your plants can be chlorine-free even if you're on a chlorinated municipal water system. Using the same technology as the original Rainshower, this device screws on simply to your garden hose. Your plants will love you! As an interesting point of information, we have heard that organic fertilizers are not tested with chlorinated water - only with well water. We have discovered that organic fertilizers are not tested with chlorinated water. We know that chlorine kills micro-organisms and beneficial bacteria in the soil. Therefore the soil building that can take place is supressed by chlorine unless it is removed by the filter. In many cases, this filter will also make a great inexpensive whole house filter!

42-702   Rainshower Garden Filter $69

Inline Sediment Filter:

Sometimes your water isn't dirty enough to mess with fancy and expensive filtration systems and all you need is a simple filter. Our inline sediment filters accept standard 10" filters with one inch center holes. They are designed for cold water lines only and meet National Sanitation Foundation (NSF) standards. Easily installed on any new or existing cold water line (don't forget the shutoff valve), they feature a sump head of fatigue resistant Celcon plastic. This head is equipped with a manually operated pressure release button to relieve internal pressure and simplify cartridge replacement. They're rated for 125 psi maximum and 100 degrees F. They come with a 3/4 inch FNPT inlet and outlet and measure 14" high x 4-9/16" in diameter. It accepts a 10" cartridge and comes with a 5 micron high density fiber cartridge.

41-137   Inline Sediment Filter $45

Our rust and dirt cartridge is made of white cellulose fibers with a graduated density. These filters collect particles as small as 5 microns (two ten-thousands of an inch). These are NSF listed components that take a maximum flow of 6 gpm. Our taste and odor filters are made with granular activated carbon. These filters effectively remove chlorine, sulfur, and iron taste and odor. These are NSF listed components. Maximum flow is 3 gpm. Note: filters should be replaced every six months to prevent bacterial growth or as needed. This is the cartridge to use with the inline sediment filter.

41-138   Rust & Dirt Cartridge (2) $14
41-436   Taste & Odor Filter (2) $34

Ozone Spa & Pool Purification

Ozone was first discovered in 1840. It is produced by nature in the stratosphere when oxygen is exposed to ultraviolet light from the sun, or from an electrical charge, ie. thunderstorms. Ozone's main function is to purify the air we breathe, and screen us from the harmful rays of the sun.

Ozone made its debut in 1906 in the field of drinking water purification, and is still the primary method used in Europe.
Many large municipalities in the US are converting their antiquated water treatment plants and turning to ozone. In 1986, the city of Los Angeles replaced its outdated chlorine system with the worlds largest water purification plant to use ozone (price tag - $106 million).

Some of ozone's other uses are to disinfect pools, spas, and recreational vehicle fresh water holding tanks. Sea World, Monterey Bay Aquarium, and fish hatcheries use ozone for water purification treatment. In short, ozone has proven its worth many times over for safe water treatment.

When used in your pool or spa, the benefits of ozone purification are: Improved bather comfort through reduced skin and eye irritation, improved water clarity and virtual elimination of chloramines. Ozone will reduce your chemical usage by at least 50%, and will make your pool or spa easier to maintain.

Clearwater Tech Ozone Purifiers

Clearwater Tech makes several different ozone generators for different applications. The ozone is manufactured in the generator by intaking air, which is composed of 20% oxygen (O_2) and bombarding it with a specific light frequency. This frequency causes the oxygen molecules to disassociate and reassemble as ozone (O_3). Ozone is the most powerful oxidizing agent available. Ozone has a half-life of 20 minutes, and then it reverts back to pure oxygen. When ozone is drawn into the spa or pool water, it will kill any bacteria, virus, or mold spore that come in contact with it.

S-1200 Ozone Purifier

The S-1200 features a polished stainless steel reaction chamber, thermally protected self-starting ballast, weather tight (outdoor approved), and a 17" specially designed high output ultraviolet lamp. The S-1200 is wired to the pump circuit to be on when your pump is on. The unit has convenient mounting brackets, comes with all necessary fittings, and is easy to install. If the spa is run daily, as recommended by most spa manufactures, four separate one hour cycles in a 24-hour period will generate a sufficient amount of ozone to keep the spa free of biological contamination. The 12-volt S-1200 has the same features, and works the same. It will treat the same capacity of water as the AC model.

42-811	S-1200 Ozone Purifier (110V)	$295
42-812	S-1200 Ozone Purifier (12V)	$360

Each uv lamp is rated at 9,000 hours. Running the S-1200, the PR-1300, or the CS-1400 for four 1 hour intervals per day equates to five years of operation, with a power consumption of approximately 85 cents per month.

PR-1300 Ozone Purifier

This system comes with its own 24-hour timer and compressor (on the AC system ONLY, the 12-volt unit doesn't have the timer). This means that the PR-1300 can run independently of the circulation pump in your spa or pool. It comes with all necessary tubing, check valve and fittings for installation. There is also a diffuser stone which can be attached to the ozone delivery line and submerged into any vessel of water.

You can treat yourself to a lavish chlorine-free bath by using the system in your bathroom, or anyone else's bathroom since the system is totally portable. You can treat your friends' spa before you use it with this portable water treatment system. For our readers in California who are preparing for "The Big One," use the PR-1300 to keep your stored drinking water bacteria free. For those of you who store your water whether you have a spring, or a catchment system, treat your water with ozone instead of chlorine, when you are trying to deal with bacteria, iron, or other problems in your storage tank. The PR-1300 features a GFCI (Ground Fault Circuit Interruption) circuit breaker on 110V systems only. This gives state of the art electrical protection. It has a weather tight cabinet - coated with a baked on enamel finish for years of corrosion free service. The UV lamp is encased in a polished stainless steel reaction chamber, and can be replaced by the homeowner in minutes. The compressor rating at 12V is 2.5 psi. Average lamp life is 9,000 hours. Power consumption is 40 watts. The units are rated up to 1,000 gallons for spas and 2,000 gallons for pools. The size is 20" x 9" x 4". Maximum pressure is 20 psi.

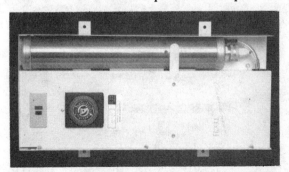

42-821	PR-1300 Ozone Purifier (110V)	$395
42-822	PR-1300 Ozone Purifier (12V)	$435

CS-1400 Ozone Purifier

The CS-1400 is an UV ozone generator designed to be used on swimming pools up to 15,000 gallons or spas up to 2,500 gallons. You can double the output and capacity by adding another CS-1400. The CS-1400 has a polished stainless steel reaction chamber, a thermally protected self-starting ballast, weather tight enclosure (outdoor approved), and a 29" specially designed high output ultraviolet lamp. With no moving parts, the CS-1400 requires virtually no maintenance, and will provide you with years of uninterrupted service. Wire the CS-1400 to the pump circuit, so each will work together to keep your pool or spa perfectly clear. It will also work with a compressor.

The CS-1400 comes with a 1-1/2" venturi injector suitable for use with a single speed pump.

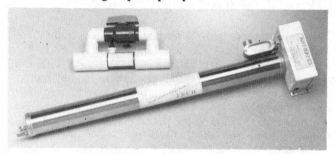

42-831	CS-1400 Ozone Purifier - 110V	$545
42-832	CS-1400 Ozone Purifier - 12V	$595

Floatron Solar Pool Cleaner

The Floatron is a safe and economical alternative to the chemical marinade in which most of us soak in our pools. When floating on the water, sunlight is converted into harmless low power electricity by the solar panel. This energizes the mineral electrode, resulting in ionization. Floatron is solid state with no moving parts, no batteries, portable and cost-effective. The alloyed electrode depletes after 2-3 seasons (depending upon pool size) and can be replaced in one minute for $50. No more chlorine allergies, red eyes, discolored hair, or bleached bathing suits! Floatron will typically reduce chemical expenses by 80%. The Floatron measures 12" in diameter by 6" high. A water test kit is included with each purchase as well as a one-year warranty from the manufacturer. The Floatron is effective for pools up to approximately 40,000 gallons.

• Safe and Non-Toxic

42-801	Floatron	$299

Here's an article by one of our dealers, Kent Williams, who has studied water chemistry and been a commercial swimming pool consultant for 15 years.

Will Ionization Eliminate Chlorine?

Of the dozen or so non-snake-oil alternatives available for pool use, only a very few show up as viable -- and their effectiveness is clearly related to organic load, not size or type of the body of water. For home or light-duty public use, more options are available, such as ionization, bromine, ozone, UV rays, and hydrogen peroxide. Let's look at this concept called "ionization", since some very unique and refined equipment, including the product offered herein, has recently come onto the market.

What th' heck are "ions" anyway? Simplified, they're molecular-sized charged portions of compounds or elements hanging around ready to react or combine with other oppositely charged stuff. Some are biocidal in nature. Free copper and silver free ions, electrolytically released from a source (electrode), have been proven very effective in sanitizing pool water. There is afforded, with this method, the distinct advantage of in-pool disinfecting, not relying on "outboard sterilization" to handle pathogens introduced into the pool itself. (Ozone and ultraviolet systems require waiting for the circulation to cycle all the water through the mechanical system -- a lengthy process requiring up to four hydraulic cycles.) Since we don't want anything alive in the pool but the kids, we can't wait three to ten hours for the demise of bacteria that should have happened in seconds or fractions of a second -- a measure of performance provided by these easily available copper/silver ions. And generated "ions" are exceedingly cheaper than costly hydrogen peroxide or many of the halides (fluorine, chlorine, bromine and iodine).

So where's the catch? Its oxidation, not sanitation, that is the tougher job any chlorine substitute must perform; killing the bugs is the easy part. If your organic introduction (soils, urine, sweat and decomposing organic matter) is quite low, an ionizer will fill the bill beautifully, and your water will remain sanitary and clear. If soil influences are a bit less than ideal (especially outdoors), or "bather preparation" (showers, foot baths, etc.), is weak or absent, you will likely have to supplement your shiny new ionizer with a modicum of bleach (sodium hypochlorite) now and then. This addition of chlorine is in direct relation to the need to keep the water crystal clear through the oxidation properties of hypochlorous acid, the working compound ultimately resulting from any chlorine product.

So if you don't rent out your spa by the hour, or open the pool up to the public five days a week, an ionizer is probably for you. Maintaining water that is sanitary is assured, and operating costs are zilch. And sun-powered? What more can we ask for? "See you in the deep end..." - **Kent Williams**

Sorry, no returns can be accepted on this product because so many have abused the privilege as a way of renting.

Air Purification

Our homes are a source of environmental pollution. Various forms of contaminates intrude into our living space from smoke, sprays, stale air, conditioned and heated air, kitchen and bathroom chemicals, radon, moisture bred bacteria, insulation, paint, and many other sources. We're convinced that *Allotropic Oxygen is by far the best way to purify your air.*

Allotropic Oxygen is a different form of oxygen. Have you ever noticed how pure and clean air smells after a lightning storm? A lightning flash will convert some oxygen in the air to a more active form of oxygen - ozone. The ultraviolet rays from the sun also change some of the oxygen in the upper layers of the atmosphere into ozone. Ozone is a three atom allotrope (O_3) of oxygen (O_2). It is generated by solar energy and exists in the gaseous form at room temperature.

Ozone reacts with most organic materials to form carbon dioxide gas and water vapor. It is an outstanding bacteriacide and virus deactivant. By-products of ozone reactions are ecologically benign. Unreacted ozone decomposes rapidly to simple molecular oxygen.

It is estimated by the E.P.A. that indoor air pollutants may be responsible for tens of thousands of deaths and illnesses each year. According to John Bond, executive director of the National Council for Clean Indoor Air in Washington, DC, indoor air pollution will be the largest emerging environmental health issue in the next five to ten years.

Blue Springs Air Life Systems

Our Blue Springs air purification systems produce ozone by passing oxygen over an ultraviolet lamp that emits light having a wavelength spectrum between 1800 and 2700 Angstroms. These air purification systems help control indoor air pollution and microbiological contamination. They will eliminate:

Cigarette smoke   Cooking odor
Bathroom odors   Pet smells
Perfume odors   Coffee smells
Mold & mildew Factory smells
Airborne & surface bacteria
Airborne & surface molds
Airborne & surface yeast
Airborne and surface virus

They also aid in preventing the spreading of colds, and infectious diseases through a cleaner and healthier environment; and relieves or prevents runny noses and eye irritation. It helps add higher kinetic energy oxygen molecules to your living and working environment. This may improve the productivity and attitudes of the participants in the work place and the home settings. The ozone generator partially removes polluting inorganic metals such as lead, cadmium, mercury, and others from the environment by oxidizing to the metal oxide form. It aids in the elimination of toxic compounds such as aldehydes and ketones that are emitted from walls, insulation, carpets, and furniture.

■53-701	Air Life I	$459
■53-702	Air Life II	$579
■53-703	Air Life III	$899

Here is a specification chart outlining the three air purifiers we're carrying. **Note to our customers who produce their own power: all units will operate easily on the small Trace or Heart Inverter.**

SPECIFICATIONS	AIR LIFE I	AIR LIFE II	AIR LIFE III
Voltage	115V	115V	115V
Watts	345	345	575
Fan	0-100 cfm	0-100 cfm	0-100 cfm
No. of Quartz Lamps	1	2	1
Power Cost/Month	$0.76	$1.52	$1.90
Dimensions (L x W x H)	All Models - 15" x 8" x 7"		
Ozone Output (grams/hr)	0.04	0.06-0.12	1.0
Size of area	one room	1500 sq. ft. house	3000 sq. ft. office

Energy Conservation

Cord Caulk Plugs Damaging Leaks

Cord Caulk is a great draft and weather sealing substance that comes in a 100' roll. The cord itself is 3/16" in diameter. It stops drafts, window rattles, and moisture damage to sills and floors. This is a very simple product: a soft fine fiber yarn that's saturated with synthetic adhesive wax polymers. It can be applied at below freezing temperatures, and can be removed and re-used numerous times without affecting the sealant characteristics required for weatherization.

Cord Caulk adheres well to wood, aluminum, steel, paint, glass, rigid vinyl, rigid plastic sheet, and nylon. It will adhere poorly to plastic foams, poly films, dirty, flaking surfaces, and silicone. The inventor won an award for this energy invention at the Boston Edison Centennial Invention Competition in 1986. Every house utility closet should contain at least one roll of Cord Caulk!

56-501-D Cord Caulk (wood grain brown) $15
56-501-W Cord Caulk (white) $15

Frantz Oil Filters

Consider the following facts relating to oil disposal and the environment:

- *We unnecessarily dispose of 1billion gallons of used crankcase oil annually.*

- *1 gallon of used oil pollutes 2 million gallons of ground water.*

- *50% of our oil is imported.*

- *It takes 42 gallons of crude oil to make 2-1/2 quarts of premium motor oil.*

- *If everyone in America used a Frantz filter, we would save 800 million gallons of motor oil per year!*

The Frantz oil filter eliminates the need to ever again change your vehicle's oil and is a product that can have enormous benefit to our environment. *Our head technician, Jeff Oldham believes that the Frantz oil filter has more potential benefit to our environment than any product that we sell!* Adding the Frantz filters to your cars extends engine life up to 500% by eliminating all dirt (down to .08 microns) and all acid from your oil. You still need to change your car's regular oil filter every 15,000 miles, replace the Frantz's filter element and add a quart of oil every 3,000 miles *but you don't need to change your oil.* The Frantz cleans your oil continuosly, with absolutely no loss of oil pressure.

With a standard system, oil becomes contaminated within two to three thousand miles and loses a substantial amount of its fundamental attributes. Using a process known as edge filtration, the Frantz filter removes dirt and up to six ounces of water. A normal full flow filter only captures larger particles of 20 microns and above. As much as 58% of all engine wear is caused by microscopic dirt which can be effectively filtered between one and five microns. The Frantz takes over where the full flow filter leaves off.

The Frantz filter has been around for 30 years, has 3 million users, and is approved by the FAA (Federal Aviation Administration) for use in piston fired aircraft. It has been fully tested by Southwest Research Laboratories, an EPA approved laboratory. You can easily install the Frantz filter system in your vehicle yourself in one hour.

To order the Frantz filter, we must have the following information:
1. Make and year of your car
2. Engine size.
3. Mfg. & part # of your standard oil filter **(Optional)**

55-104 Frantz Oil Filter Pkg $159
55-105 Replacement Filter $ 8

Send SASE for a thorough report.

Reflectix Insulation

Reflectix is a wonderful new insulating material with dozens of uses. It is lightweight, clean, and requires no gloves, respirators, or protective clothing for installation. It is a 5/16" thick reflective insulation which comes in rolls and is made up of seven layers. Two outer layers of aluminum foil reflect most of the heat which hits them. Each layer of foil is bonded to a layer of tough polyethylene for strength. Two inner layers of bubblepack resist heat flow, while a center layer of polyethylene gives Reflectix additional strength.

Reflectix can be used wherever standard fiberglass insulation is used, without the necessity of wearing goggles or a face mask for installation. It has an R-value comparable to standard insulation (see chart). Reflectix inhibits or eliminates moisture condensation while providing no nesting qualities for birds, rodents, or insects. Other benefits include reductions in heating and cooling costs that accelerates the payback time of the cost of installation over ordinary insulations. It is Class A Class 1 Fire Rated and non-toxic.

Reflectix BP (bubble pack) is used in retrofit installations. For optimum performance and insulating value a 3/4" airspace needs to be put on both sides of any Reflectix installation.

R-Value Table for Reflectix

R-Value ratings need to be clarified for different applications and more specifically for the **direction of heat flow**. "*Up*" refers to heat escaping through the roof in the winter or heat infiltration up through the floor in the summer. "*Down*" refers to preventing solar heat gain through the roof in the summer. "*Horizontal*" refers to heat transfer through the walls.

Reflectix Type	Up	Down	Horizontal
Reflectix BP	8.3	14.3	9.8

As well as its most common usage as a building insulator, Reflectix has a myriad of other uses: Pipe wrap, hot water heater wrap, duct wrap, window coverings, garage doors, as a camping blanket or beach blanket, cooler liner, windshield cover, stadium heating pad, camper shell insulation, behind refrigerator coils, and a camera bag liner.

Reflectix BP (Bubble Pack) comes in 16", 24" & 48" widths and in lengths of 50' & 125'. (We cannot UPS rolls larger than 125.)

We believe in Reflectix and are introducing it at a very attractive price - .39 cents per square foot!

59-501-BP	16" x 50' (66.66 sq. ft.)	$ 31
56-502-BP	16" x 125' (166.66 sq. ft.)	$ 75
56-511-BP	24" x 50' (100 sq. ft.)	$ 49
56-512-BP	24" x 125' (250 sq. ft.)	$109
56-521-BP	48" x 50' (200 sq. ft)	$ 89
56-522-BP	48" x 125' (500 sq. ft)	$215

Attn: Alaska & Hawaii Customers: *Please add $15 additional freight for first class shipment.*

56-529 Add'l freight to HI & AK $15

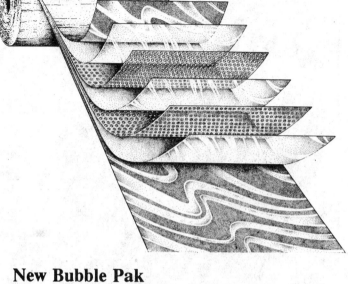

New Bubble Pak Staple-Tab Reflectix

This brand new Reflectix product is the same bubble-pack insulation as the standard Reflectix insulation with the added feature that it's made to be installed between framing members as opposed to on the surface. This product makes installation far easier as it eliminates the need to add the furring strips that are needed to form an air space with standard Reflectix. It comes with easy-to-install staple tabs. **Same pricing as BP insulation above - but be sure to specify Staple-Tab by ending the product number with ST instead of BP.**

59-501-ST	16" x 50' (66.66 sq. ft.)	$ 31
56-502-ST	16" x 125' (166.66 sq. ft.)	$ 75
56-511-ST	24" x 50' (100 sq. ft.)	$ 49
56-512-ST	24" x 125' (250 sq. ft.)	$109
56-521-ST	48" x 50' (200 sq. ft)	$ 89
56-522-ST	48" x 125' (500 sq. ft)	$215

Hawaii & Alaska Customers: *You must add additional freight for first class shipment of $15.*

56-529 Add'l freight to HI & AK $15

Tape for Reflectix

Reflectix makes an aluminum tape that is excellent for bonding two courses of Reflectix together. It works far better than masking tape or duct tape. It also has reflectability and is highly recommended for any Reflectix installation. Two sizes of tape rolls are available: 30' x 2" and 150' x 2".

56-531 30' Roll - 2" Reflectix Tape $3
56-532 150' Roll - 2" Reflectix Tape $9

Chapter 6 - Tools & Appliances

Makita Cordless Power Tools

Makita cordless tools have been a "hit" since their introduction in the Fall/Christmas 1988 Real Goods Catalog. We've had to re-order weekly in large quantities. Makita's 12V battery charger has brought great tool versatility to the remote home! For off-the-grid users, the real advantage of Makita power tools is their new innovation of a 12-Volt DC battery charger that plugs directly into a cigarette lighter type 12V socket.

Makitas are incredibly wellbuilt, energy-efficient, rechargeable power tools. In a very short time they have set the industry standard for cordless tools. Makita has over 70 years of research and development behind their power tools, which excel in quality, durability, and versatility.

For off-line users, the real advantage of Makita power tools is their new innovation of a 12-Volt DC battery charger that plugs directly into a cigarette lighter type 12V socket. We can get all Makita tools (110V as well as cordless) at great prices, but we stock only the cordless line that uses the 9.6V rechargeable nicad battery.

3/8" Cordless Driver Drill (6093D)

The 6093D is the newest and most professional 3/8" cordless drill in the Makita line. It features quick and easy adjustment of six different torque settings and two-speed gear selection with variable speed in either range. Low speed is 0-400 rpm and high speed is 0-1,100 rpm for diverse drilling and driving applications. It will drill 3/8" in steel and 11/16" in wood. It will drill up to 350 5/16" holes or drive up to 400 #9 X 3/4" wood screws in medium hard wood from a single one-hour charge! It comes with a drill chuck and key set and a phillips bit. The battery and charger must be ordered separately. The complete kit consists of the drill, a 9.6V battery, a battery charger (either 12V or 110V), and a steel case.

63-109	Cordless Drill (6093D)	$119
63-110-KT	Drill w/complete kit (12V)	$239
63-111-KT	Drill w/complete kit (110V)	$239

3/8" Cordless Angle Drill

This new cordless angle drill from Makita gets the job done in tight places with its 90 degree head. It comes with a 3/8" keyless chuck and operates on the same 9.6V battery (not included) that most of our Makita tools use. The DA390D has a reversing switch and auto overload protection with a no load speed of 800 rpm. The unit weighs only 2.9 lbs. with battery.

63-101   3/8" Cordless Angle Drill $149

Cordless Circular Saw (5600DW)

We have been urged to carry a heavy duty circular saw to complement our cordless line. The new 5600DW is one heavy duty cordless circular saw. It uses a 10.8V rechargeable battery for stronger operation. It will cut up to 125 pieces of 2" x 4" pine stock from a single one hour charge! We used this saw extensively in setting up all of our Earth Day booths. Standard equipment includes the 10.8V battery, a 110V one-hour fast charger (operable from a PowerStar inverter at 12V), a carbide tipped blade, socket wrench, wrench, and rip fence. Suggested retail is $350.

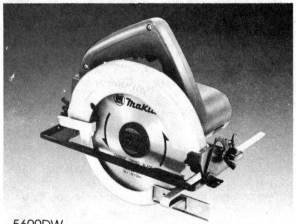

5600DW

63-129	**Circular Saw Kit (5600DW)**	**$295**
63-130	**Extra 10.8V battery**	**$105**

Cordless Reciprocating Saw (4390D)

This tool's usefulness is limited only by its 3" blade, which can be easily replaced. Equipped with a longer blade (up to 8") this new reciprocating saw is ideal for cutting drywall (sheetrock), and for cutting through walls. The saw will cut through 39 feet of 1/2" plywood on a single 1-hour battery charge. It comes with two saw blades; one for wood cutting, and one for mild steel, plastics, formica, etc. The recipro saw is very light weight (3.1 lb) and portable. It operates off of the 9.6V battery which, along with the charger, must be ordered separately.

63-114   Cordless Recripro Saw $109

Cordless Circular Saw (5090D)

This 3-3/8" cordless circular saw is great for cabinetry, shelving, plywood paneling, fencing, sheetrocking, and countertop cutout work. It operates on one 9.6V battery. The 5090D's base will bevel cut up to a 45° angle. It will cut through wood up to a maximum of 3/4" thick. It features a shaft lock for easy blade replacement, and concrete and glass cutting are possible with the optional diamond wheel. The saw will cut up to 55 feet of 3/8" standard plywood from a single one-hour charge. Probably the greatest of all features is its lightweight portability (3.7 lb) and the fact that there is no cord to get tangled up! Battery and charger must be ordered separately.

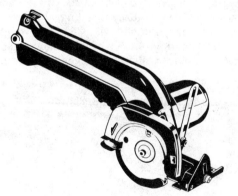

63-116   Cordless Circular Saw(5090D) $119

I just finished installing 300 square feet of oak parquet floor, and the cordless saw was indispensable. Small, powerful, and accurate, it zips through the toughest inside cuts with very little waste. My big Makita power saw would have ripped the wood to shreds.

Rochelle Elkan, Casper, WY

Cordless Jig Saw (4300D)

The 4300D jig saw will perform fast and smooth cutting with 2,700 strokes per minute (like the recipro saw) and will cut up to 39 feet of 1/2" soft wood from a single one-hour charge. Its adjustable, sturdy steel base will perform bevel cutting up to 45° right or left. The jig saw has a lock-off switch that helps prevent accidental starts. Standard equipment includes two saw blades and a hex wrench. It operates off of the 9.6V rechargeable battery, ordered separately.

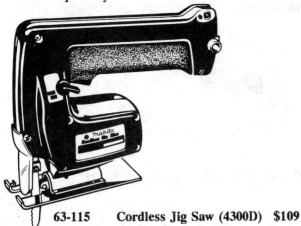

63-115   Cordless Jig Saw (4300D) $109

Rechargeable Flashlight

The Makita rechargeable flashlight (ML900) is one of our most popular items. It's an indispensable add-on if you already have Makita tools, or a good start if you don't. It uses the same 9.6V rechargeable battery as all the other cordless tools and will provide up to 120 minutes of continuous illumination from a single one-hour charge. The flashlight features a focusing knob that allows either a concentrated or diffused light beam, and it comes with a red lens cap for emergency signaling. It also has an integral carrying strap and wire stand and weighs only 1.6 lb for easy handling. Order battery and charger separately. Comes with one replacement bulb inside lens.

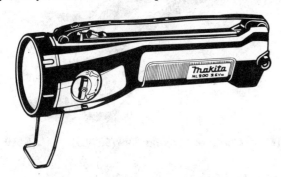

63-302 Rechargeable Flashlight (ML900) $59
63-301 Replacement Bulb $ 3

Cordless Blade Sharpener/ Motorized Tool

Makita's new cordless sharpener and motorized tool (903D) utilizes the standard 9.6VDC rechargeable battery to sharpen chain saw blades quickly and effectively with cordless convenience, operating at 19,000 RPM for a fine finish. It is lightweight (1.5 lbs.) for easy handling and will provide up to 70 minutes of operation from a single one hour charge. As a bonus the tool will also accept the standard tool attachments manufactured by the Dremel Division of Skil Corporation*. Standard equipment includes a sharpening attachment, gauge, and wheel points. Length is 11-5/8".

63-131 Cordless Blade Sharpener $69
* Dremel and Skil are registered trademarks of Skil Corporation

Makita Battery Charger

The 12-volt one-hour fast charger is the reason we got into this line of power tools in the first place! It will fully charge a 9.6-volt battery in one hour's time, and comes with a cigarette lighter plug on the end. Could anything be simpler? You'll find these chargers at most hardware stores for over $60!

63-303   12V Makita Battery Charger $59
63-304   110V Makita Battery Charger $59

9.6V Battery

These 9.6V nickel-cadmium rechargeable batteries - you will need at least two - are the heart of the Makita cordless power tool line. They charge up in one hour using either the 110V or the 12V Fast Charger. Our T+1 rule of thumb for battery buying: one battery for each tool in active use, and one in the charger at all times. You'll find these batteries at most hardware stores for $42 to $55!

63-305   9.6V Makita Battery $45

3/8" Keyless Chuck

Lost chuck things wil become a nightmare of the past with Makita's innovative keyless chuck. This 3/8" chuck easily replaces the existing 3/8" chuck on any Makita drill. Bits are held securly in place by simple hand tightening.

63-108   3/8" Keyless Chuck $25

Driver-Drill Accessories

This seven-piece insert bit kit contains: #1, #2, #3 Phillips bits, 4-6, 6-8, & 8-10 slotted bits, and a 2-1/8" magnetic bit holder. These accessories can be used with all Makita drills and screwdrivers.

63-102   Driver-Drill Accessories $15

Extra Fine Jig Saw Blades

These extra fine blades are designed to cut mild steel, thin plywood, formica, and masonite. They have 24 teeth per inch with an overall length of 3-5/32". The come five per pack.

63-133   XF Jig Saw Blades (5) $9

General Purpose Jig Saw Blades

These general purpose blades have fifteen teeth per inch and are designed for cutting wood. They come five per pack.

63-134   GP Jig Saw Blades (5) $11

General Purpose Recipro Saw Blades

These general purpose wood blades have nine teeth per inch. Overall length is 4-3/4". They come five blades per package.

63-105   GP Recipro Saw Blades (5) $11

Extra Fine Recipro Saw Blades

These extra-fine blades are designed to cut mild steel, thin plywood, masonite, plastics, and formica. There are 24 teeth per inch and the overall length is 3-15/16". Five blades come per package.

63-104   XF Recipro Saw Blades (5) $11

Combination Saw Blade (3-3/8")

This general purpose - rip or cross-cut blade has 50 teeth for all 3-3/8" Makita circular saws. It features a special thin kerf design for more cuts per battery charge.

63-103   Combination Saw Blade $12

Beckman Industrial Test Meters

In order to compliment our new Klein tool line and to help our customers test and troubleshoot their electrical systems, we've added the fine line of Beckman test meters. Beckman is widely known as one of the best meter manufacturers in the world. Their value is unmatched in the industry.

Beckman AM 10 Meter

This inexpensive meter has 4 functions with 16 ranges and is small enough to fit in your shirt pocket or tool box. Accuracy is + or - 4% f.s. Sensitivity is 2,000 OHM/V. Overload protection is 2x full scale for 30 sec. AC/DC volt rages are 10,50,250 and 500V. Resistance ranges are RX 1000, 1M OHM max. DC current ranges are 0.6, 50, 250 mA. "O" adjustment for OHMs and meter movement. Includes "AA" battery and test leads.

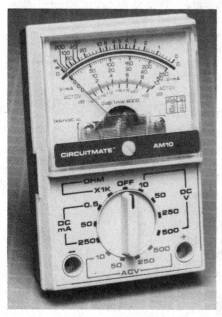

63-155 Beckman AM 10 Meter $14

DUMP a toxin. You don't have to keep those rusty containers in your garage. Pour 'em down the drain. How can that hurt anybody?

Beckman RMS 225 Meter

This top of the line RMS 225 meter does just about everything at an incredible price. Some of the features included are *true RMS* for accurate inverter AC Voltage readings, auto-ranging, and automatic last reading hold (called "Probe Hold"). It also has a 41-segment bar graph, auto minimum-maximum with unlimited recording time that "beeps" with every update, an audible overload alert, and self-resetting fuse. This fine meter comes with a rugged probe holding case and stand. It also comes with a 3-year "No fault" warranty.

Electrical Specifications.
(Ta = 25°± 5°C, 80 % RH max)

DC Voltage	**AC Voltage** (True RMS)
Ranges: 1V, 10V, 100V, 1000V	10V, 100V, 750V
Basic Accuracy*: 0.25% +1	1.5% +3 (45Hz - 1kHz)
Max Resolution: 0.1mV	1mV
Input Impedance: 10MΩ	10MΩ; <110pF
Overload Protection:	
1000VDC or AC Peak	750VAC or 1000V Peak
Response Time: <1 second	<2 second
Conversion Type:	AC coupled True RMS

DC Current	**AC Current** (True RMS)
Ranges: 10mA, 40mA, 10A	10mA, 40mA, 10A
Basic Accuracy*: 0.75 % +1	1.7 % +3 (45Hz - 1kHz)
Max Resolution: 0.001mA	0.001mA

Voltage Burden:
20mV/mA (10mA and 40mA ranges)
0.03V/A (10A range)
Input Protection:
250V RMS Self-Resetting PTC Resistor (40mA input)
Max Input:
40.5mA (40mA input)
20A up to 30 seconds (10A input)

Resistance
Ranges: 1k, 10k, 100k, 1M, 10M, 40M
Basic Accuracy*: 0.5 % +1
Max Resolution: 0.1Ω
Open Circuit Voltage: 3.3V (max)
Input Protection: 500VDC or RMS AC

Diode Test/Continuity
Test Current: 4.5mA (max)
Open Current Voltage: 3.3V (max)
Continuity Threshold: <100mV
Continuity Indicator: 2kHz tone
Input Protection: 500VDC or RMS

63-156   Beckman RMS 225 Meter $175

Klein Tools - Toolkits For the Independent Electrician

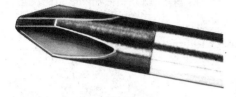

We've been searching for a state-of-the-art tool set to go along with our state-of-the-art remote home power systems. Established in 1857, Klein is the recognized leader in electrician's hand tools and their tools will greatly facilitate systems installations. **All tools carry a lifetime warranty.**

Kit #1 includes: universal side cutter pliers, needle nose pliers, long nose stripper, standard screwdriver 1/4", #2 Phillips screwdriver, standard skinny screwdriver 1/8", small instrument screwdriver 3/32".

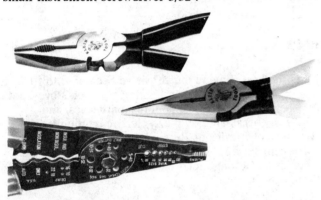

Kit #2 includes: *All of Kit #1 plus the following*: diagonal cutters, water pump pliers - 10", cable cutter - 2/0, heavy standard screwdriver - 3/8", screwdriver cabinet-tp 3/16", and six in one tapping tool 6-32, 8-32, 10-32, 10-24, 12-24, 1/4-20.

Also available is the Tool Pouch.

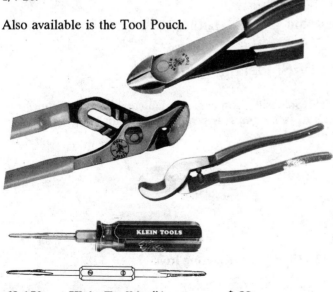

63-150	Klein Toolkit #1	$ 89
63-151	Klein Toolkit #2	$159
63-152	Tool Pouch	$ 39

Leatherman Pocket Tool

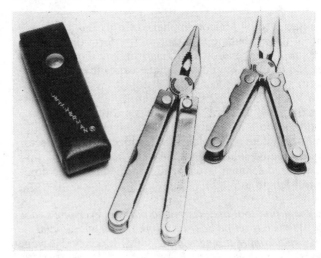

The Leatherman "pliers in the pocket" tool was developed and perfected over 7 years by a machinist who realized his Swiss Army knife was severely limited for many tasks. The pocket survival tool is constructed of 100% stainless steel, the optimum grade/hardness for each tool/blade. Its main attribute is its full-sized, full-strength "fan handle" pliers (regular & needlenose) and wire cutters inside. The tool is designed to military specs and all parts are interconnected. The stainless steel file/saw cuts wood and metal and sharpens fish hooks. It has a full range of screwdrivers (small, medium, large, and Phillips). It also features a knife blade, 8" ruler, can/bottle opener, and an awl/punch. Closed, it's only 4" long and 5 ounces, and comes with its own leather carrying case. Guaranteed for 25 years.

63-201 Leatherman Pocket Pliers Tool $45

Mini-Leatherman

Leatherman has introduced a second tool for those who want the full-size, full-strength pliers, but like a lighter tool in their pocket. It features needlenose pliers, regular pliers, wire cutters, knife blade, ruler, can opener, bottle opener, 1/4" tip screwdriver, and a metal file. It is constructed of 100% stainless steel, guaranteed for 25 years, measures 2-5/8" x 1" x 1/2" and weighs only 4 ounces.

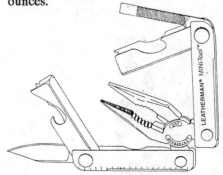

63-202 Mini-Leatherman Pocket Tool $38

Ultra-Torch Butane Heating Tool

Manufactured by Master Appliance Corp., this is the state-of-the-art butane tool that will perform three jobs. It works as a combination soldering iron and flameless heat tool and torch. It employs a unique catalytic combustion system and fingertip controls can adjust the temperature from 394 to 932° F for soldering; up to 1,292° F for heat shrinking; and to 2,372° F for use as a torch. it will operate up to 3 hours at 700° F on a single filling. It comes complete with soldering/heat ejector, torch ejector, tapered needle soldering tip, heat tip, solder sponge, tip cleaner, and spanner wrench, all in a metal storage box. Overall length is 9-1/2".

"This is a fine tool and deserves a place in everyone's tool crib. We use it on pc boards, shrink tubing with it, and even sweat copper tubing with it." J.M. Mooney, Heavener, OK

"I have used a UT-100 for almost two years, now, with no problems. In my line of work, I am soldering all day long, sometimes. From small gauge to 2/0 cable." D. Ritchie, Sausalito, CA

63-309   Ultra-Torch Heating Tool (UT-100) $89

Uncle Bill's Tweezers

In keeping with Real Goods' tradition of offering only the finest quality products available, we're introducing the "Sliver Gripper" made by Uncle Bill's Tweezer Company. These are quite simply the finest tweezers you'll ever use. Made of spring-tempered stainless steel, the precision points are accurately ground and hand-dressed. With these tweezers it's easy to find and grip even the tiniest splinter or stinger. No pocket, purse, first aid kit, or tool box should be without a pair! All tweezers come with a lifetime moneyback guarantee and a convenient holder that fits on your keychain! Our local Lyme disease control is now recommending Uncle Bill's Tweezers for removing ticks.

63-427   Uncle Bill's "Sliver Gripper" Tweezers $4

Ohaus Triple Beam Scales

The Ohaus triple beam has been the state-of-the-art scale for years. It weighs up to 610 grams with a readability down to 0.1 gram. An optional weight set increases the weighing capacity to 2,610 grams (5.75 lb) The 1600 Dial-O-Gram provides for faster and easier weighing.

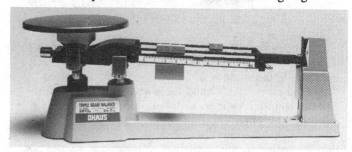

63-430   Ohaus 750-S Triple Beam Scale $100
63-419   Ohaus 1650 Dial-O-Gram $175
63-420   Weight Set for Ohaus Scales $ 30

Insta-Match

This is the best "butane match" we've been able to find. It saves many a burnt finger in lighting gas cook stoves, hot water heaters, woodstoves, lanterns, barbecues, and campfires. It has an easy piezo-electric ignition so it never needs batteries or flints. It refills with a standard butane canister and is very easy to refill and adjust the flame. It has a long metal barrel and is priced right!

63-310   Insta-Match $12

12V Soldering Iron

This soldering iron comes complete with a 12V plug. Draw is 1.9 amps. A good basic tool for your 12V toolbox!

63-118   12V Soldering iron $18

Bicycle Powered Generator

This 100-watt bicycle powered generator/battery charger appears to be the only high quality and dependable small generator manufactured that can be run by a bicycle. It comes with a stand built to accept any three or ten-speed bicycle. Also included is the generator, an ammeter, and all the wires needed to hook up to your battery system. The bicycle generator will produce 8 amps at 12VDC when pedaling at full strength. A human adult can only provide about 100 watts (1/7 of a horsepower) for any period of time. This isn't much, but combined with the 12V batteries, it is plenty for dependable clean lights, pumps, fans, and entertainment centers.

We have several customers who put their kids (or themselves) on a strict discipline - No TV unless you generate the power yourselves on the bicycle generator - It Works!

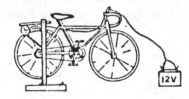

63-308 Bicycle Generator $295

Pruning Saw and Pouch

This 10" saw has an advanced blade design that makes cutting wood easier and safer than using a conventional saw. It cuts through 3" oak branches like butter! The cutting tooth works like the cutting chain on a chain-saw. The teeth control the depth of the cut and position the wood for a precise, smooth cut. This saw is ideal for pruning trees and shrubs because it won't rip or tear wood. The cut is cleaner and there's less danger of disease and scarring setting in the plant. This saw is easily resharpened by the user, another strong advantage over other pruners. It is a dream to use - it really flies through wood. It has a folding, locking hardwood handle (a safety plus). The wood used in the handle comes from small woodlots in southwest Brazil - not from the Rainforest. As an option we recommend the bright red cordura nylon pouch. This tough fabric will outlast leather. The sheath has a rust proof snap and a belt loop.

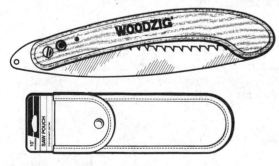

63-512 Pruning Saw $9
63-513 Pouch $7

Max 7-in-1 Tool Kit

The Max incorporates seven basic hand tools into one compact and versatile unit: A Hudson Bay style axe head permanently attached to a 36" American hickory handle, with six quick-attach tools: shovel, mattock, pick, proad pick, fire rake and hoe. Each component slips into a specially designed socket on the axe head and is secured by either a hitch pin or thumbscrew tightener. All components are drop forged from the highest quality tool steel. All attachments fit into a compact sturdy olive drab canvas case that can be carried on a belt or strapped to a pack. The entire kit weighs only 12-1/2 lbs. Ideal for ranchers, homesteaders, foresters, or just as a backyard all purpose tool.

63-511 Max 7-in-1 Tool $145

12V Beverage Heater

Our 12V beverage heater will heat coffee, tea, soup or water very quickly. It comes with a 12V plug and has a 32 long cord. Amp draw is approximately 8 amps at 12V.

63-365 12V Beverage Heater $8

12-Volt Waring Blender

This is by far the strongest, most durable, and most attractive 12V blender ever made. It was developed by Waring's commercial division. It makes the two other 12V blenders on the market (that we used to carry) seem like jokes. It draws up to 11.5 amps to blend a very hefty load. The base is made of chrome. The 45 oz shatter-resistant plastic carafe has a snug-fitting vinyl lid with a removable center insert (so you can add ingredients while blending) which doubles as a 2 oz measure. Stainless steel blades are removable for easy cleaning. A 15-foot cord with attached cigarette lighter plug are included.

63-314   12V Waring blender $139

Corona Stone Mill (4CKC)

The Corona 4CKC is the deluxe corn and grain mill. It's a stone mill with interchangeable metal plates that's easy to convert back and forth. With the stones you can grind fine flour in one operation with no heating problems. You can easily adjust from coarse to fine grind. It will mill any dry grain. The stones are manufactured especially for the Corona hand mill and are bound with a special bonding that will allow no flaking off. The stones are made of a vitrified carbon material that will last a lifetime. When you need to crack cereal or grind moist items simply attach the metal plates. Comes with high hopper and two augers for easier changing of plates.

63-410   Corona Stone Mill (4CKC) $85

12V Fillet Knife

Our 12V fillet knife is great for filleting fish either from a 12V car battery while fishing, or from a 12V home. It uses only 1.7 amps at 12VDC. It features thin stainless steel blades and a pointed tip for fast cutting.

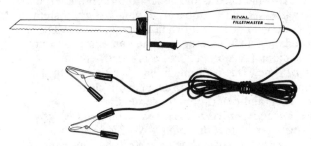

"This knife is great! Easily filets trout, snapper, & even catfish. The removable blade is easy to clean and sharp enough for the job." - **Bruce Fargo, Ukiah, CA.**

63-340   12V Fillet Knife $39

Champion Juicers

Champion juicers are the finest, most reliable, and most versatile juicers on the market. They work great on carrots, all vegetables, apples, and they make great nut butters too. They are 110VAC only and run great off of Trace inverters (612 and larger). Available in almond, white, yellow, brown, avocado. All units come with a 5-year warranty.

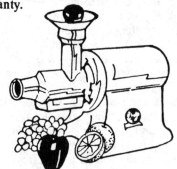

63-401   Champion juicer $239
63-402   Grain mill attachment $ 79

Lifestream Juicer

We feature the Lifestream because we believe in the nutritive properties of wheat grass. This cast iron wheat grass and vegetable/fruit hand juicer has stainless steel auger, nylon bushings, complete with plunger, non-scuff pads, and easy turn handle.

63-405   Lifestream juicer $109

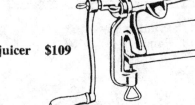

12V Baby Bottle Warmer

Manufactured by Gerber, the 12V baby bottle warmer comes with a cigarette lighter plug on the end, a zipper top and a strap for hanging. It also can serve to keep liquid cool until feeding time.

"We have discovered that us dads can give moms a break with the chow chores if Mom just expresses or pumps some milk ahead of time. Works great if Mom needs to or wants to be away. As for heating milk with this product, it works fine. I have no figures to back me up, but I'm certain it warms a bottle with less energy than a microwave (which may nutritionally alter breast milk.)" - **C.S. Tonkin, Raymond, OH**

63-334   12V Bottle Warmer $15

Remington 12V Shaver

Several years ago we carried a Hong Kong made 12V shaver that we discontinued due to poor quality and we've been searching for a replacement ever since. The Remington has two ultra thin micro screens for an extremely close shave and it comes with a 12V lighter-type plug. It comes with a travel pouch, protective headcover, and a cleaning brush.

63-338   12V Remington Shaver $39

12V Hairdryer/Defroster

Our 12V hairdryer is probably most commonly used as a 12V defroster. It beats using your credit card for scraping icy windshields, and is found in many of our customers' glove boxes. It comes with a 54" cord and measures 4.5" X 4.5". It draws 12 amps from your 12V battery.

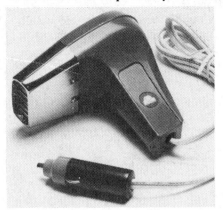

63-323   12V Hairdryer/defroster $17

12 Volt Heaters

These 12V fan forced heaters are perfect for our customers who have surplus power (like hydro-electric systems) and who would like to put that power to work. They're also ideal for use in cars, vans, and RVs without functional heaters. With two sizes to choose from - 180 watts - 12.5A and 300 watts - 25A, they will work with most systems. If you're an energy tycoon, you may parallel as many as you can power to divert a power surplus (load shedding, they call it). Both heaters come equipped with a safety limit switch and the cabinet stays cool to the touch.

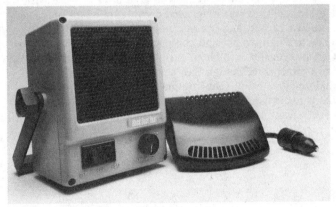

63-351   12V Heater - 180 watts $ 39
63-352   12V Heater - 300 watts $125

James Washers

The James hand-washing machine is made of high-grade stainless steel with a galvanized lid. It uses a pendulum agitator that sweeps in an arc around the bottom of the tub and prevents clothes from lodging in the corner or floating on the surface. This ensures that hot suds are thoroughly mixed with the clothes.

The James is sturdily built. The corners are electrically spot-welded. All moving parts slide on nylon surfaces, reducing wear. The faucet at the bottom permits easy drainage.

63-411   James Washer $195

(wringer attachment pictured is available at additional charge)

Hoky 23T Carpet Sweeper

Here is the perfect **low-cost** answer to the low-voltage vacuuming blues. Even folks with 110V current have put their vacuum cleaners to deep storage after a brush with the Hoky. They are tough and durable, efficient on hard floors or carpet, and operate **noiselessly** and effortlessly. They are lightweight and compact, so they store neatly, transport easily, and clean hard-to-reach places. All parts are replaceable, though they seldom if ever require maintenance other than routine wiping clean or combing the rotor brush. The Hoky is the state-of-the-art carpet sweeper and comes with the Real Goods stamp of approval!

| 63-413 | Hoky 23T Carpet Sweeper | $39 |
| 63-449 | Replacement Rotor Brush | $10 |

Hand Wringer

The hand wringer will remove 90% of the water, while automatic washers remove only 45%. It has a rustproof, all-steel frame and a very strong handle. Hard maple bearings never need oil. Pressure is balanced over the entire length of the roller, by a single adjustable screw. We've sold these wringers without a hitch for over thirteen years.

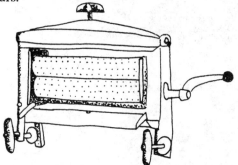

63-412   Hand Wringer $119

"I thought you might be interested that we have had a James washer for almost a year now. I do about two loads a week and I find it's not very difficult work at all. I do all my son's diapers on it and they get very clean and soft, without using any bleach or softener. It's nice to be able to do laundry at home, and even nicer to do it outside in nice weather."

Sue Calhan, Waldoboro, ME

Hoky Commercial Carpet Sweeper

By popular demand we've added the heavy duty Hoky to our product line. Much heavier-duty than the 23T, this Hoky NT features new patented Rotorblades with spiral-wound rubberized blades to pick up everything fast along its 12-1/2" wide path. Lint, feathers, pet hair, nails, broken glass, string, sand, gravel and dirt can be picked up even on the plushest shag carpet or the barest wood or vinyl. Horsehair corner brushes get right up to the wall and push dirt into the path of thr Rotorblades. Soft rubber wheels glide over concrete for outdoor use. The resilient bumper guard protects furniture. There are no bags to empty. The 43" long center-pivoting aluminum handle makes it easy to maneuver. It folds up to 3" flat.

63-451   Hoky Commercial Carpet Sweeper $69

Dustbuster II - Vacuum by Black & Decker

The Dustbuster II is an upright power brush vacuum cleaner that is rechargeable either by 110VAC or by 12VDC. It has two powerful motors to clean both carpets and bare floors. It's cordless so it can be used anywhere without being restricted by a power cord. The long bristles are gentle to bare floors, and its short bristles groom high-pile carpets. The handle is retractable for compact storage.

We supply the Dustbuster converted to 12VDC charging by adding a resistor into the charging mechanism. The 12V charger consumes 110 mA (1/10 of an amp). We have found that the vacuum will operate for a full 15 minutes on a single charge, which doesn't sound like a long time, but it is a long time for vacuuming. It will recharge overnight.

63-322 Dustbuster II $125

12V Vacuum Cleaner

Our canister-type vacuum cleaner has replaceable cloth filters and will suck up wet as well as dry material. Its suction is strong enough to pick up marbles! The optional filter pack has three new replacement elements.

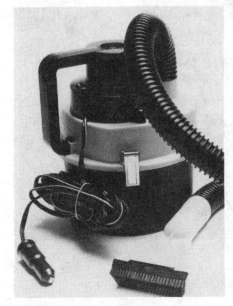

63-321 12V Vacuum Cleaner $39

Eureka Freedom 12V Vacuum

The Eureka Freedom Upright 12V rechargeable vacuum cleaner is as powerful as a standard 110V vacuum. It makes the Black & Decker Dustbuster seem like a toy. It will operate a full 30 minutes on a single charge with heavy vacuuming or 45 minutes under normal use. It will recharge from totally dead in 16 hours or in six hours from a normal discharge. The charger uses a peak of 6.5 Amps @ 12-volt. *The vacuum comes factory equipped with a transformer "cube." For direct 12V operation and charging you must "amputate" the cube and splice on a 12V cordset, such as our fused cordset for $2.25. (order #26-109).* Of course it will also recharge on 110V standard house current. Five sequential LED display lights show, at a glance, the charge remaining. The 3-position handle adjusts to vertical for storage, intermediate for normal operation, and low to reach under furniture. The large-capacity, clog resistant disposable dust bag has 3.5 quarts usable capacity. Built of sturdy ABS housing and polycarbonate resin base. The motor is 12-volt DC and the color is red. While the Eureka is quite a bit more expensive than the "Dustbuster" it is definitely worth the money.

63-325 Eureka Freedom 12V Vacuum $249

FLUSH a toilet. Whether it needs it or not, make a habit of turning the handle whenever you go by. Keeps the pipes clean.

12-Volt Mini-Wash

The 12V Mini-Wash is the ideal appliance to use if you don't have an inverter and are tired of hand-washing down at the river. It has a very efficient rotating drum that washes clothes beautifully clean, is gentle enough for delicate fabrics, and uses only 2.3 amps without a load and an absolute maximum of 5.5 amps at full load. The Miniwash needs no special plumbing, fits on the drain board of your sink, and connects to the tap via the filling/emptying flexible hose provided. It comes with a safety lid that shuts down the machine when it's opened. A 2-10 minute spring-wound automatic timer is included to time all wash loads. The Miniwash will wash up to a maximum of 4.5 lb of dry laundry at one time using a maximum of 3 gallons of water. The casing drum and lid are made of tough polypropylene so there is no possibility of rusting, staining, or dents. A standard 12V cigarette lighter plug is included. The entire unit is very compact, measuring just 20" high by 18" deep by 17" wide and weighing only 17 lb.

Dear Real Goods,
    I am very glad that you are now carrying the 12VDC washing machine in your catalog. This machine has been keeping our clothes clean for over 5 years. It is not only efficient powerwise (5.5 amps), it is also economical, especially compared to laundromat costs. It's compact size is deceptive. I have washed area rugs, king size sheets, and a week's worth of bath towels in it. Another bonus is that I can use the discharge water for my trees and garden (bio-degradable soap used) which is so important during this drought. My husband is very happy with the (genuine!) fresh air smell all our linen has too. This machine is the one that allowed us to decide to go entirely 12VDC in our home. - **Katcha Sanderson, Paicines, CA**

63-319   12V Miniwash $175

Pyromid Portable Barbecue

The Pyromid is a portable barbecue, stove, oven, smoker, and a roaster that's great for picnics, beaches, camping, or backyard use. It's easy to carry (folds to 1" thick by 12" square and comes with a tote bag), lightweight (6 lb) and sets up without tools ready to use in 15 or 20 seconds.

    The Pyromid's surfaces are all stainless steel, making it rust-free and easy to clean. These surfaces cool down quickly, unlike hibachis or kettle grills, which take up to a half hour to cool. Standard accessories include a hood which lets you bake, regulate heat, smoke, or just shield from the wind.

    All units come with a lifetime warranty. The Pyromid is a very ingenious and simple piece of technology.

Our Grill-Stove slips neatly into a saddle bag for trips to the wilderness. Up to now we have never carried a stove.
                                        Carl & Erla Kemp, Denver, CO

63-418   Pyromid portable barbecue $75

The Sun Oven

The Sun Oven is a new solar cooker that has just completed a successful product evaluation. It is very portable - one piece - and weighs only 21 lbs. It is ruggedly built with a strong fiberglass case and tempered glass door. It is completely adjustable with a unique levelling device that keeps food level at all times. The interior oven dimensions are 14"W x 14"D x 9"H. It ranges in temperature from 360 degrees to 400 degrees F. This is a very easy oven to use and it will cook anything! After pre-heating, the Sun Oven will cook one cup of rice in 35-45 minutes.

63-421 Sun Oven $195

Here's an article about the testing of the Sun Oven that we've excerpted from *Home Power #19*, written by Karen Perez:

Cooking with the sun works! There's an inner satisfaction eating a meal grown and cooked by the sun. It just plain tastes better, too. The Sun Oven is long lasting and has several features that make it very easy to use.

The documents include care, setup, utensil, and a few recipes and solar cooking hints. The Sun Oven is well constructed, a first class unit. The body of the cooker is fiberglass and is insulated by dead air. The inside is painted with black non-toxic paint. The door is thermal pane, tempered glass that's very easy to open and close and has a whale of a good seal. The reflectors are very shiny (spectral-quality) aluminum, hinged for easy transport and keeping food warm. The oven also has an adjustable leg for easy sun focusing. A really nifty gamboled shelf helps keep pots level. There's even a carrying handle and a strap to secure the reflectors. Suave!

In the last few weeks I've cooked everything from stews

Solar cooker in Burkina Faso. Photo by Sean Sprague.

to bread. So far, everything has cooked wonderfully. I love baked potatoes, seven medium ones were cooked to perfection in one hour and fifteen minutes. WOW! A few days ago I wanted to cook rice, chicken and bread. A cast iron dutch oven held the chicken and rice. I inverted the lid and used it as a platform to bake the bread. It worked great.

I love it!! In the last three weeks of use the Sun Oven has noticeably reduced our propane consumption. A standard snickered question is "What do you do when there's no sun?" My answer is, "Use SOME of the propane I saved."

The Sun Oven really fits our time, too. In the morning, when I'm energetic, I put dinner in the oven, face it south and forget it. In the evening, when I'm beat from working on Home Power all day, dinner is hot and ready.

Sun Oven is well-made and should last a long time. I think it is worth its price because it works, is portable and durable."

ShowerStar

This brand new lighted shower head is actually a mini-Hydro electric power plant. It uses no batteries. The power of the water through the water wheel develops 2.5 volts at .31 amps - enough to light the PR-6 bulb. It puts on a great show by actually filling the streams of water with light. Installation is simple. Each ShowerStar comes with a "Mood Pak" including red, blue, green, amber, and two clear bulbs. It's available in either chrome or brass.

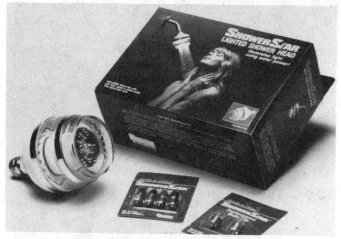

| 63-312 | Chrome ShowerStar | $45 |
| 63-313 | Brass ShowerStar | $65 |

Vegetarians are warned against reading the next item, which may contain offensive material

12-Volt Porta-Pluck

The Hatfield Porta Pluck is a 12V portable duck plucker that works just like the big ones, plucking chickens, ducks, turkeys, and other fowl (vegetarians: you were warned!) fast and easy while pulling 15-30 amps. The Porta Pluck uses patented rubber duck plucking fingers that are flexible and will not hurt your hand or bruise the meat. Different speeds and torque are used to pluck large and small fowl. The average duck can be plucked clean in 90 seconds. The machine partly disassembles, making it easy to store and transport. Two-year conditional guarantee.

63-316 12V Porta Pluck $195

Bedwarmers

When it gets *really cold*, there is no other bed warming appliance that can put so much heat in your bed so fast as our bedwarmers. These units are far more efficient than electric blankets because they put all the heat IN the bed rather than on top of the bed where it dissipates rapidly. The radiant heat keeps the bedding dry, and warms the mattress and spaces beside you. It gives you a soothing feeling of relaxation like sitting in warm bath water. There are many sizes to choose from for different widths of beds. All units are 5 feet long, except where noted. The M76 King size has dual controls for either side of the bed. The rated amp draws are only valid while the heater is cycling on. Otherwise they draw nothing!

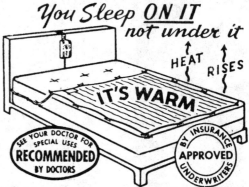

12V Bedwarmers				
Item #	Stock no.	Width	Amps	Price
63-328	B24 (Bunk)	24"	5.0	$49
63-329	B48 (Double)	46"	6.7	$59
63-330	B60 (Queen)	60"	6.7	$85
110V Bedwarmers				
63-331	CD (Twin/Dbl)	40"x54"		$45
63-332	CD60 (Dbl/Qn)	54"x60"		$59
63-333	M76 (King)	76"x60"		$99

Important Note: If you are concerned about electromagnetic radiation (EMF) you should be aware that the 12V blankets are completely radiation free. The 110V models, however, should be regarded the same way that you regard an electric blanket. Ideally, use it as a preheater and shut it down when you go to bed (if EMF concerns you). For more thorough informatin on EMF see our book section on *Cross Currents* and *The Body Electric*.

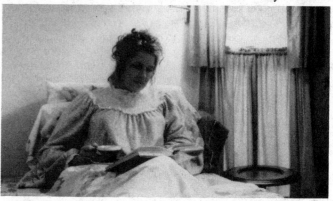

"These bed warmers are everything they're cracked up to be! It makes getting into bed on a frosty night naked just like getting into a hot bath!" JS, Ukiah, CA.

Whale Humidifier

The whale humidifier is a real work of art in beautiful polished brass. Fill it up with water, set it on your woodstove, and watch the steam come out of the whale's blowhole. It's a great gift item and moistens the dry air created by a woodstove too!

63-414   Polished brass whale $150

Indoor/Outdoor Thermometer/Clock

This great thermometer features a huge LCD display (3/4" numbers) that can be read across the room, and a 10' cable with a weatherproof remote temperature probe. This can be used to monitor a fishtank, freezer, doghouse, or outside temperature. A simple switch toggles from inside to outside temperature. It reads from -58 to 122 degrees F. and comes with a one year warranty. Also included is an accurate quartz clock. Battery included.

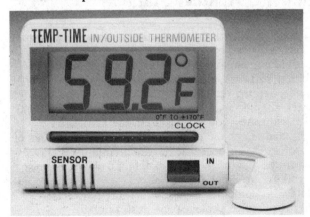

63-501   Indoor/Outdoor Thermometer $25

Dragon Humidifier

Fill this cast iron medieval sculpture with water and set it on top of your woodstove. As the stove heats up, the dragon breathes steam out its nostrils creating not only a dazzling sensation but humidifying the air as well.

63-415   Cast iron dragon $105
63-416   Antique brass dragon $155
63-417   Polished brass dragon $165

Kettle Humidifiers

These traditional humidifiers look great on any woodstove and provide a moistening steam to mitigate the dry wood heat. They are available in cast iron or brass.

63-424   2-1/2 qt. cast iron kettle $ 39
63-425   2 quart solid brass kettle $135
63-426   3-1/2 qt. solid brass kettle $145

Stick-Up Thermometer

This is an ideal simple thermometer for inside refrigerators, RVs, houses, or cars. It comes with a magnet and adhesive disc to stick anywhere. It measures accurately from -20 to 120 degrees F. One year warranty. 2" round dial.

63-502   Stick-Up Thermometer $4

Digitar Weather Master

The "world's smallest weather station" is a feature packed device. The unit is small enough to fit in your shirt pocket, and keeps you informed of the barometric pressure and trends, wind speed and direction, highest gust, wind chill, temperature at the device and at a remote location, the high and low temperature at the remote, the time, elapsed time, and a variety of alarms. If you are traveling, it can tell you your altitude. A built-in lamp lets you read the display in the dark. Powered three ways - built in nicad batteries which self-charge when outside energy is applied, and either 12VDC or 110VAC adapters allow you to use the device in any setting.

Now if you try to run this whole thing from your shirt pocket, you'll have some problems. The elegant little base unit only takes the temperature and runs the barometer, altimeter, clock, elapsed timer, and alarm by itself. To measure windspeed and direction it needs an anemometer and wind vane. (Perhaps these could go on your hat. For our next issue, we will try to perfect this and get you an illustration.) The remote temperature sensor also plugs into the base unit, as does the external power and the optional rain gauge. When fully loaded, five wires plug into the unit's left side, making that side look a little like a porcupine. Even placed on a wall, the unit bristles with wires.

Operation is simple and elegant. In the scan mode it shows you your choice of indicators, or any single condition at a key-touch. Resetting the highs and lows is straightforward. Even initial calibration, usually an hour-long hassle for complex electro-meteorological instruments, is a piece of cake. In side-by-side testing with professional quality instruments, the Digitar Weather Master held its own, excepting the rain collector - resolution to a tenth of an inch is simply not adequate.

If you need windspeed or barometric graphing, this is not the unit for you - see the unit below. If you are interested in the weather, and want to measure it all with one elegant device, check this one out.

63-345 ALT-6 Weather Master $285
63-341 Rain collector option $ 49

Send an SASE for more information.

Digitar Weather Pro

The Weather Pro has the ability to precisely record and display wind and temperature data all in a unit that fits in the palm of your hand. With the optional Rain Collector option, it even provides daily and accumulated rainfall readings. Included are the following functions:

- Wind Speed to 120 miles per hour
- Wind Direction
- Temperature from -60 to 230 degrees F.
- Windchill to -150 degrees F.
- High and Low temperatures
- Daily and accumulated rainfall (using optional rain collector)
- 12 or 24 hour clock

The Weather Pro comes complete with anemometer and 40 feet of shielded cable, external temperature sensor with ten feet of cable, and a detailed instruction booklet.

63-357 Digitar Weather Pro $179
63-341 Rain Collector Option $ 49

Digitar PC Weather Station

If you need to know more about the weather, or want to study its sweep and cycle, and you own an IBM compatible personal computer, this is the ultimate tool. Combining software, a board that fits inside your computer, and the requisite outboard sensors, this unit lets you track climatic indicators over time. With the optional advanced software, you can program the hardware yourself, and hijack the data for analysis with programs of your choice. We suspect that the main circuitry is closely related to the standalone ALT-6 detailed above, because the functions are much the same.

While your computer is off, the battery-powered weather card stores peaks - peak gust, rainfall, temperature extremes - which are retrieved and saved when you turn the computer back on. To gain the fullest benefit of the Weather Card's abilities, automatic posting of weather conditions every half hour, you leave your computer running all the time - which the hardware folks tell us is not a bad idea.

63-326 PCW Computer Weather System $299
63-327 Expanded Software $ 99

Send an SASE for more information.

Farm, Garden & Irrigation

At first glance, drip irrigation systems might not seem to be a logical choice for an alternative catalog. However, water pumping can be one of the major energy demands placed on a home power system. The more that you are able to reduce your demand for water pumping, the more you will cut your energy costs.

Drip irrigation is basically the most efficient method of watering that is available. Drip systems operate at lower water pressures than conventional spray type systems, so they can be operated with a simpler type of pumping system. They are also extremely well suited for the more isolated and mountainous regions that are often the site of the modern-day homesteader.

The wide-scale usage of drip irrigation systems is a relatively recent phenomenon. Its history goes back to the time of the Babylonians, who used a simple system of aqueducts to carry water and small clay pots with holes in them to release the water at individual plants. The purpose of modern drip systems is still basically the same, the slow and even application of water directly to the root zone of the plant.

Drip irrigation systems are perhaps one of the most important technological innovations of modern times. With the development of plastics during and after World War II, it became possible to create flexible and affordable tubing. Much of the early innovation of modern drip systems was by Israeli engineers who were able to extract record harvests from land that had previously been desert.

Drip irrigation systems permit the development of mountain and arid land that is otherwise idle. As more and more arable land in the world becomes desert through poor farming practices, overgrazing and logging, or is taken over by urban sprawl, the importance of drip systems increases.

With proper management, commercial users have achieved water savings of up to 60%, relative to traditional flood-and-soak watering methods. Irrigation labor costs are reduced to almost zero and the electrical bills on pumping can be reduced. With a drip system, virtually all the water that is applied is available to the plants, avoiding excessive weed growth or loss to wind and evaporation. In areas of limited water supply, a successful drip system can mean the difference between agricultural success or failure.

Using drip irrigation systems, California growers have increased yields from 20 to 40 percent. They have also been able to bring steep slopes into production without erosion problems. In addition, fertilizer costs can be reduced by applying them through the drip system. Fertilizers reach only the desired plants, also reducing weed growth and the risk of nitrogen burning. By maintaining a relatively constant moisture level, nutrients remain in the root zone, permitting the plant to obtain the most value from each fertilizer application. A constant moisture level also reduces salt build-up in the soil and keeps salt deposits beyond the root zone of the plant.

Healthy Plants

Plants grow best when they don't lose their momentum. If they can start growing and keep going at their own pace, the maximum harvest results. If they dry out too much, it slows them down and harms their health. When the soil is too wet, the small pockets of air naturally in the ground are pushed out and the plants can't breathe.

In an ideal system, the level of moisture in the soil around a plant will remain constant, so that the amount of water applied is equal to the moisture loss due to transpiration of the leaves and evaporation from the soil. When the soil moisture is maintained at the proper

Man watering onions in Tengkodogo, Burkina Faso. Photo by Sean Sprague

level, air is always available to the root system of that plant. With other watering methods, there are extreme fluctuations in air and water content of the soil, ranging from total saturation after watering, where little or no air is available, to the end of the cycle, where there is little or no water.

This cycle of too-wet-too-dry also subjects plants to extreme temperature variations. Too much water also tends to erode away valuable topsoil as well as causing nutrients to percolate below the root zone.

Drip irrigation maintains a constant moisture zone around the roots of your plants, creating an ideal growing situation.

Criteria for Choosing an Emitter

The four main questions to ask in evaluating different drip systems emitters are:
(1) How resistant are they to clogging?
(2) Will they accurately discharge their rated amount of water?
(3) Do they compensate for variations in pressure?
(4) How does the relative cost correspond to particular advantages and disadvantages?
Generally, for a small backyard garden on flat ground with short rows, simple orifice emitters will work fine. For larger systems that are still on flat ground laminar flow emitters would be a better choice. In situations where long lengths of pipe are involved or if you are growing on a hillside, vortex emitters work much better than the simple orifice or laminar types. For systems that are located on steep mountains and in situations where it is necessary to run long lengths of pipe, diaphragm emitters are the best ones to use.

Types of Emitters

Simple orifice emitters limit the rate of discharge by the size of their opening. The main advantage of this type of emitter is that they are usually inexpensive. The disadvantage is that they tend to clog more easily and don't compensate for pressure variations.

Laminar flow emitters, instead of using a small orifice, limit flow rates by sending water through a long spiral maze. This method permits a large orifice which makes them less likely to clog. However, they don't compensate for pressure variations, which limits their usefulness on long laterals and hillsides

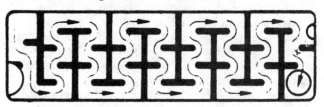

Vortex emitters limit flow by bringing water at an angle into a circular chamber. This diffuses pressure to the outside walls creating lower pressure in the center of the vortex. This permits a larger, less likely to clog orifice and partially compensates for pressure variations.

Diaphragm emitters have a silicone or rubber diaphragm inside a small chamber. The diaphragm remains open when there is a little pressures and gradually closes as pressure increases. In this way, a relative constant discharge rate is maintained, even when pressure vary significantly. Because the diaphragm opens when the system is turned off, the diaphragm emitters are self-cleaning and the most resistant to clogging problems.

Full Flush   Partial Flush   Drip Mode

Because of complications with stocking, and our desire to concentrate on alternative energy supplies, we no longer stock drip irrigation equipment. We recommend that you contact either **Sparetime Supply** in Willits, CA: (707) 459-6791 or **Harmony Farm Supply** in Sebastopol, CA: (707) 823-9125. Both of these companies are reputable and have catalogs.

Mighty Mule Solar Gate Opener

We've finally come up with a heavy-duty, durable automatic gate opener. The Mighty Mule will open gates up to 16' wide or up to 250 lb. It is very easy to install and doesn't require an electrician. It comes with an adjustable timer for automatically closing the gate, and will automatically latch the gate shut. A 6.5 amphour battery is included that can be easily charged with the optional 5-watt solar module. The unit can also be powered by 110V with a step-down transformer if you don't need the solar panel option. All units come with a wireless remote transmitter so that you can open the gate without getting out of your vehicle. Additional transmitters are often purchased so that you can keep one in each car. It takes 18 seconds for the gate to open and 18 seconds to close.

Mighty Mule's electronics are state-of-the-art. The gears are all metal, not nylon or plastic. The highly efficient motor provides enough power to operate ornamental iron and commercial chain link gates up to 250 lbs. Even in high cycle applications the Mighty Mule will not overheat.

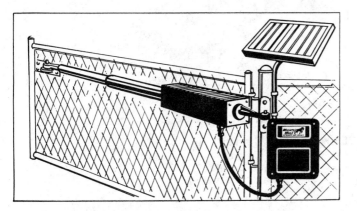

63-126	Mighty Mule	$675
63-127	5-watt solar panel option	$ 89
63-142	Horizontal Gate Latch	$ 69
63-144	Extra Transmitter	$ 29
63-146	Digital Keypad	$ 49

Send SASE for free brochure.

Parmak Solar Fence Charger

The 6 Volt Parmak will operate for 21 days in total darkness and will charge up to 25 miles of fence. It comes complete with a solar panel and a 6-volt sealed, leakproof, low internal resistance gel battery. It's made of 100% solid-state construction with no moving parts. Fully weatherproof, it has a full two-year warranty.

| 63-128 | Parmak solar fence charger | $195 |
| 63-138 | Replacement 6V battery for Parmak | $ 55 |

Minibrute 12V Chainsaw

The Minibrute is made with an American Bosch motor that's warranted for 10 years. It has a 14" bar with 20 feet of battery leads and heavy-duty clips on the end. It draws 82 amps maximum at 12 volts and is extremely quiet. It operates on direct drive with 6,250 rpm. It's ideal for limbing trees by the road from your vehicle. While the Minibrute is a great product, the manufacturer is often flaky with production. Please check with us for availability before ordering.

63-125　Minibrute 12V Chainsaw $179

BUY a poison. You never know when you'll need to kill something. It's good stuff to have on hand, just in case.

Compost Container

Our new compost container has been designed to meet the needs of the urban gardener with a medium amount of waste as well as the large scale, serious gardener who needs a continuous supply of compost. It's flexible, expandable, and constructed of a 14 gauge galvanized wire mesh with green PVC vinyl coating. This material was originally designed to be immersed in salt water for extended periods of time; consequently it is extremely durable. The basic model is a five sided container which can be used alone, or by adding additional panels, be made into a three container system. This simple container system will get you back into composting even in the city!

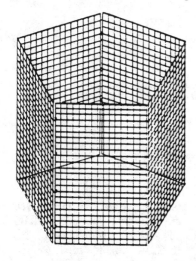

54-301   Compost Container $45

Manual Lawn Mower

After numerous requests from customers, we decided to research the manual lawn mower market. We've found a great manual lawn mower made by the oldest lawn mower manufacturer in the U.S. This mower is safe, lightweight, and very east to push. It's perfect for small lawns and hard-to-cut landscaping. The reel mower provides a better cut than power mowers - keeping lawns healthier and green and it doesn't create harmful fumes or noise pollution. The short grass clippings from the mower can be left on the lawn as natural fertilizer or you can purchase the optional grass catcher and add the grass to your compost pile.

63-505   Lawn Mower $95
63-506   Grass Catcher $19

Rainmatic 2000

The Rainmatic is great for summer vacations and peace of mind as it will automatically water your garden - up to four times per day and from one to seven days per week. On and off watering times can be easily programmed to last from one minute to 24 hours. It can be switched to manual with one touch, and it uses your existing hose or faucet. It's ideal for automating drip systems and is simple to install. It uses 4 "C" alkaline batteries (not included) that will last one full season.

43-601   Rainmatic 2000 $55

Rainmatic 3000

The following letter from Pam Wilkinson says it all:
"Dear Real Goods, I just ordered a Rainmatic 2000 water timer from you, & would like to suggest that you carry the Rainmatic 3000 instead of or in addition to the 2000. The 3000 model is far more versatile, offering 8 waterings/day (instead of 4) and an additional misting option, which can be programmed to water in any interval from 1 minute to 24 hours, for time periods from seconds to hours. The 3000 is wonderful, and is the only battery-powered water timer with a misting feature on the market!"

Pam Wilkinson, Santa Barbara, CA.

43-602   Rainmatic 3000 $69

Nomadics

Pacific Yurts

Because of the incredible success of our Shelter Systems Portable Shelters, we've decided to take one step further in introducing Pacific Yurts. These portable, yet semi-permanent structures meet all of our criteria for quality.

The yurt is an architectural wonder ...a legendary dwelling that has been in continuous use for centuries. It is ideal both as a recreation retreat or for year-round use. It may be kept simple, or include modern amenities such as plumbing, electricity, and multi-level deck systems. It can be insulated and equipped with a woodstove. It heats efficiently and comfortably even in extreme cold. They are easily ventilated and are ideal in warmer tropical climates as well as on the northern Alaskan slope.

The low cost per square foot makes the Pacific Yurt an outstanding value and an economical alternative to higher priced standard frame structures. The Pacific Yurt can be easily transported in a station wagon or small pickup and set up quickly virtually anywhere for a comfortable stay. You can take it home, or leave it up year-round.

Other uses for the Pacific yurt include a hot tub/spa enclosure, a workshop/studio, a resort/conference center, a ski hut, a remote base camp, or as temporary housing for an owner/builder. Naturally strong, the yurt can be reinforced to withstand high winds and heavy snow.

The yurts are easy to install. The 30' yurt takes two people less than a day to erect, and the 12' model only a few hours. Materials are of the finest quality available, including a center ring of cross-laminated kiln-dried fir, select fir rafters, galvanized steel tension cable, electronically bonded vinyl-laminated polyester top cover, and large clear vinyl windows. All top covers come with a five-year pro-rata warranty. Many custom options are available like extra windows, solar skylight arc, insulation and fabric colors.

Size (diameter)ft.	Sq. ft.	Center height	weight	Price
▪12'	115	8'0"	350#	$2,095
▪14'	155	8'9"	450#	$2,745
▪16'	200	9'3"	550#	$3,145
▪20'	314	10'0"	700#	$4,095
▪24'	452	11'6"	900#	$4,645
▪30'	706	13'	1,200#	$6,195

Shipping and packing are extra. Allow 4-6 weeks for delivery. All yurts are shipped freight collect FOB Oregon. *Do not send money to us; we will refer you directly to the manufacturer.*

Send SASE for color brochure and prices on options.

"The yurt manufactured by Pacific Yurts is one of the most beautiful structures I have ever encountered."
                                        -Joan Halifax, PhD, Author.

Shelter Systems Portable Shelters

One of our longstanding customers has a business called "Shelter Systems" that makes a wonderful portable shelter that is as easy to set up as a tent but far more durable. Portable shelters have a wide variety of uses. They are used often as housing for families and will provide shelter from the wind, rain, and cold. They can be used to make a shop or studio, or to give children a room of their own. They make great inexpensive room additions, and greenhouse models are available. They are currently being tested by the U.S. Government as emergency relief shelters.

The Crystal Cave

The Crystal Cave is tunnel shaped and has a variety of outdoor uses including carports, workshops, greenhouses, or gardening tool sheds. The sheathing is a very strong woven ripstop film that has been treated with ultraviolet inhibitors to ensure extra long life when used in the constant sun. The Crystal Cave's doors have hook closures, designed to bypass the need for zippers which are notorious for breaking after a short working life. Included are stakes, hooks for hanging potted plants, and instruction manual. Two-year guarantee.

Crystal Cave 9

Size	9' X 9'
Height	7'
Weight	40#

▪92-204   Crystal Cave 9' $375

Alaska & Hawaii Customers: *Due to high freight costs, please add the following additional charges and we'll send first class.*

▪92-209   **Add'l freight to HI & AK $60**

CRYSTAL CAVE 11

Size	11' X 11'
Height	7'
Weight	51#

▪92-205   **Crystal Cave 11'** $495

Alaska & Hawaii Customers: *Due to high freight costs, please add the following additional charges and we'll send first class.*

▪92-209   **Add'l freight to HI & AK $60**

"...quite attractive design. Very roomy inside. It is in its third year of use. Cover has survived quite well. Fairly easy to set up and compacts nicely for storage."
                                        W. Brooks, Kotzebue, AK.

The Lighthouse

The Lighthouse goes up in 30 minutes and comes down in 5 minutes. Walls are of a 5.5 oz polyester canvas resistant to sun degradation, watertight, flame retardant, breathable, and will not rot or mildew. The translucent skylight above the sidewalls is constructed of a woven, ripstop UV resistant film. It creates a pleasing interior light and opens to provide a great fresh air vent that is fully rain proof. The Lighthouse 18 has four tipi-tyle doors spaced evenly around the dome (the Lighthouse 12 has three). Clear vinyl windows above the doors let you see out in all directions.

The LightHouse comes complete with stakes, guylines, vent tubes, spare parts, and an instruction booklet that details floors, site selection, cooling, winterizing, and stove installation. Two-year guarantee.

Lighthouse 18

Diameter	18'
Height	9'
Weight	60#
Packed size	5' x 18" x 18"

■92-201	Lighthouse 18	$650

Alaska & Hawaii Customers: *Due to high freight costs, please add the following additional charges and we'll send first class.*

■92-209	Add'l freight to HI & AK	$60

OPTIONS for the Lighthouse 18:

■92-256	Ripstop 18' Floor	$109
■92-252	Mosquito Netting for Doors (ea.)	$ 9
■92-251	Porch	$ 59
■92-259	Net wall for tropics (18')	$ 75
■92-253	Full liner for winterizing (18')	$575

Lighthouse 12

Diameter	12'
Height	6.5'
Weight	38#
Packed size	4' x 12" x 12"

■92-202	Lighthouse 12	$450

Alaska & Hawaii Customers: *Due to high freight costs, please add the following additional charges and we'll send first class.*

■92-209	Add'l freight to HI & AK	$60

OPTIONS for the Lighthouse 12:

■92-255	Ripstop 12' Floor	$ 45
■92-252	Mosquito Netting for Doors (ea.)	$ 9
■92-251	Porch	$ 59
■92-258	Net wall for tropics (12')	$ 48
■92-210	Full liners for winterizing	$365

Lighthouse 10

Diameter	10'
Height	6.5'
Weight	18#

■92-203	Lighthouse 10	$265
■92-254	Ripstop 10' Floor	$ 39
■92-257	Net Wall for Lighthouse 10	$ 49

Alaska & Hawaii Customers: *Due to high freight costs, please add the following additional charges and we'll send first class.*

■92-209	Add'l freight to HI & AK	$60

OPTIONS for the Lighthouse 10:

■92-252	Mosquito netting for doors (ea.)	$ 9
■92-251	Porch	$ 59

"I have been completely satisfied with the quality and appearance of our dome. So far it has handled 20" of rain and wind storms very well. The dome is my year round home. I have a bright, airy but warm home inexpensively. The tipi style doors are 100% improvement over zippered doors. Living in the dome, one becomes intimate with the sun, clouds, and waning and waxing moon. I love it."
Paul Guree, CA

Chapter 7

Education & Consumer Products

Books

We find ourselves constantly recommending reference materials to callers beginning and maintaining independent energy systems. The books listed below provide a good foundation for would-be alternative energy users and a great reference library for those already unconnected from the grid. Books do change rapidly and go in and out of print, with new revisions occurring frequently. Check with our periodic *Real Goods Catalog* updates for up-to-the-minute information on our reference materials.

Alternative Energy Books

The Solar Electric House, by Steven Strong. The author has designed more than 75 PV systems. This fine book covers all aspects of PV from the history and economics of solar power to the nuts and bolts of panels, balance of systems equipment, system sizing, installation, utility-intertie, stand-alone PV systems, and wiring instruction. A good starter book for the beginner. 276 pages.

80-800 $20

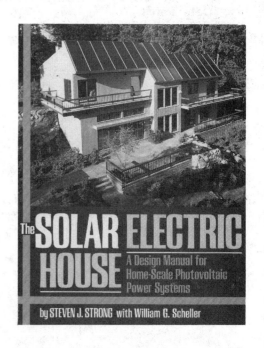

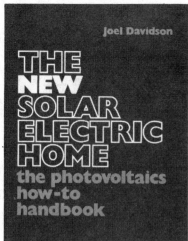

The New Solar Electric Home, by Joel Davidson. Gives you all the information you need to set up a first time PV system, whether it be remote site, grid connect, marine, or mobile, stand-alone, or auxiliary. Good photos, charts, graphics and tables. Written by one of the pioneers. Perhaps the best all-around book for getting started with alternative energy. 408 pages. **80-101 $19**

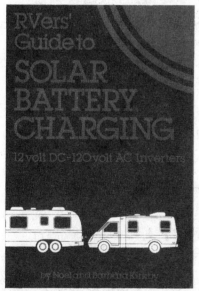

RV'ers Guide To Solar Battery Charging, by Noel & Barbara Kirkby. The authors have been RVing for over 20 years and have applied photovoltaics to their independent lifestyle. This book includes numerous example systems, illustrations, and easy to understand instructions. The *Whole Earth Catalog* calls it "A finely detailed guide to installing PV systems in your motorhome, trailer, boat, or cabin...this book has what you need to know." 176 pages.

80-105 $13

Wiring 12 Volts For Ample Power, by David Smead & Ruth Ishihara. The most comprehensive book written on DC wiring to date written by the authors of the popular book *Living on 12 Volts with Ample Power*. This brand new book presents system schematics, wiring details & troubleshooting information not found in other publications. Chapters cover the history of electricity from 600 BC to the modern age, DC electricity, AC electricity, electric loads, electric sources, wiring practices, system components, tools, and troubleshooting. 240 pages.

80-111 $18.50

Remote Home Kit Owner's Manual, by Real Goods Staff. We wondered in our weaker moments whether our Remote Home Kit Owner's Manual would really be worth all the hours we spent on it. We're happy to report it was! Our 48 page installation manual gives step-by-step instructions for the complete novice to electrify his homestead with one of our Remote Home kits. There are lots valuable charts, diagrams, and reference material contained making it indispensable for any AE library. We encourage all of our new customers to check one out before purchasing one of our turn-key packages - we're getting lots of compliments on this one!
48 pages. Fully revised in Fall 1990.

80-400 $5

Living on 12 Volts With Ample Power, by David Smead & Ruth Ishihara. This book has become the DEFINITIVE work on batteries and DC refrigeration systems. It thoroughly covers all aspects of power systems from the workings of solar panels to the optimization of a balanced energy system. This book will show you why the battery charger you may be using is killing your batteries! The best book available for marine applications. Highly recommended for serious AE users. 368 pages.

80-103 $25

the
Real Goods Trading Corporation

REMOTE HOME POWER KITS INSTALLATION MANUAL

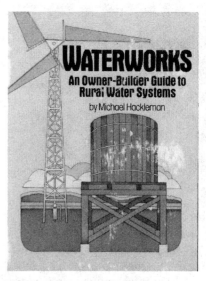

More Other Homes and Garbage, by Leckie, Masters, Whitehouse & Young.

Other Homes & Garbage, first published in 1975 revolutionized the the literature of appropriate technology and sold over 100,000 copies. This update revises & updates the original with the latest information on technological improvements. It includes alternative architecture: house site selection & orientation, solar heating, lighting insulation, ventilation, & fireplace design; photovoltaics, solar heating, waste-handling systems, water supply, agriculture, aquaculture, and much more. Written by Stanford University professors, the is the appropriate tech bible! No other book has so successfully translated sophisticated engineering concepts into comprehensible language for the layman.

80-115 $15

At Home With Alternative Energy, by Michael Hackleman.

The title is accurate. Emphasis on putting together your own "renewable" energy system from solar (no PV) to domestic hot water heating and cooling. Lots on Wind Energy - Fun to read, exciting language. 145 pages, *Out of print until late 1991.*

80-106 $11.95

Planning For An Individual Water System, by American Association of Vocational Instructional Materials (AAVIM).

This is the definitive book on water systems. It was a surprise and complete delight when one of our customers pointed it out to us. Incredible graphics and charts. Discusses water purity, hardness, chemicals; presents a thorough discourse on all forms of water pumps, discusses water pressure, pipe sizing, windmills, freeze protection, and fire protection. If you have a new piece of land or are thinking of developing a water source, you need this book! 160 pages.   **80-201 $19**

The Solar Electric Independent Home Book, by Paul Jeffrey Fowler.

A good, very basic, primer for getting started with PV, written for the layman. Lots of good charts, a good glossary and appendix. It has just been freshly updated for 1991 including 25 additional CAD diagrams (making 75 total) that are more detailed and much improved. The text has been updated to reflect recent changes in PV technology.and is far better than Fowler's original book! Great charts & graphics. This is one of the best all-around books on wiring your PV system. 200+ pages.

80-102 $19

Water Works, by Michael Hackleman.

This is Hackleman's most recent book (1983) and is required reading for folks seeking knowledge in water systems. He draws on the fine book mentioned above, *Planning for an Individual Water System,* extensively, but provides a more down home approach to all aspects of water from source to pump to tank to delivery. A fine book! 172 pages. *Out of print until late 1991.* Confirm availability before ordering.

80-202 $14.95

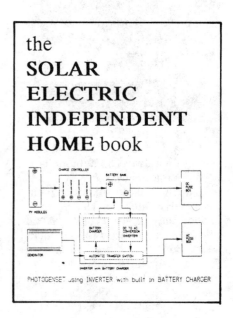

UNDERGROUND BUILDINGS

Underground Buildings, by Malcolm Wells. An architects sketchbook of 26 years of work. This is a delightful, irreverent, and very personal account of the struggle against the current of the architectural mainstream. Mr. Wells includes hundreds of sketches to illustrate the potential, the successes

and the failures of solar energy efficient, underground buildings. Handwritten, easy-to-read, 200 pages.　　　　**80-146　$15**

Basic Electronics Course, by Norman Crowhurst. This book thoroughly explains necessary fundamentals with highly readable language on electron flow, magnetic fields, resistance, voltage, and current. Gives a firm foundation in electronics. 400 pages.

80-305　$18.00

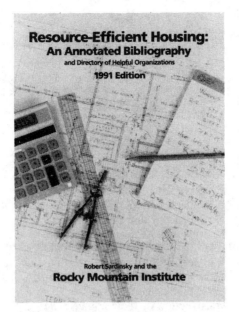

Resource-Efficient Housing: An Annotated Bibliography and Directory of Helpful Organizations, by Robert Sardinsky and the Rocky Mountain Institute. "Most of the entries are excerpted deftly enough to be considered information sources themselves. You'd have to subscribe to a truckload of periodicals to keep up with what's presented here in one book." - J. Baldwin, *Whole Earth Review*, Summer 1988. This directory lists and critiques - complete with addresses, prices, and contact names - the periodicals, books, schools, organizations, and agencies that deal with all aspects of resource-efficient house design, construction, retrofit, and much more. This is the fully updated and revised 1991 edition. 160 pages.　　　　**80-108　$17**

Practical Photovoltaics, by Richard Komp. This book presents the theory and practice of photovoltaics in a non-technical manner. It runs the gamut from explaining the physics of PV to offering hands-on instruction for the assembly of your own PV modules from salvaged cells. It's a unique combination of technical discussion and practical advice.　215 pages.

80-107　$17

Shelter, This is a classic book put out by Shelter Publications originally in 1973, when it sold 185,000 copies. Out of print since 1978, the book has just come back. It features 176 pages on architecture - from bailiwicks to zomes. You won't find any palaces, pyramids, temples, cathedrals, skyscapers, or Pentagons. This is instead a brilliant book on homes and habitations for human beings in all their infinite variety. It is lavishly illustrated, with over 1,000 photographs, numerous drawings, and 250,000 words of text. It is a real piece of environmental drama and a great tool for the person just starting to fantasize a dream living space.　　　　**80-145　$17**

PRACTICAL PHOTOVOLTAICS
Electricity from Solar Cells

Richard J. Komp, PhD
Skyheat Associates
English, Indiana

Second Edition

aaTec

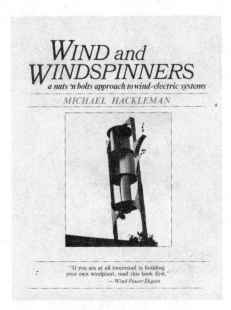

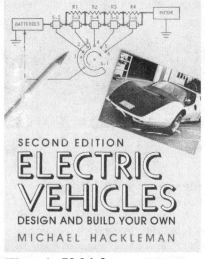

Wind and Windspinners, by
Michael Hackleman. A do-it-yourself manual on building working wind-electric units, simply and inexpensively. Includes an extensive chapter on S-Rotor construction, one of the most versatile back-yard aeroturbines. 140 pages.

80-121 $12

Passive Solar Homes, by the U.S. Dept. of Housing and Urban Development and the Dept. of Energy. This is an indispensable reference tool for home builders who seek to understand what passive solar means and take advantage of passive solar concepts in the construction of their homes. The book gives an easy-to-understand explanation of passive solar energy and the various functions that passive solar systems can perform. It is heavily illustrated with detailed plans and drawings of the 91 winning solar home designs. It describes how to design a home suitable to the climatic conditions of your area. It outlines how to choose and prepare the building site, how to plan, orient, and insulate your home for maximum energy conservation, and gives detailed specs for building the five major components of any passive solar system - collector, absorber, storage, distribution, and control. An excellent glossary and bibliography are provided. We got a great price on a quantity of these books. *Cover price is $12.95.*

80-147 $8

Electric Vehicles, by Michael Hackleman.
A no-nonsense look at how to design, build, power, and license your own non-polluting electric vehicle. Shows how to optimize torque, power, capacity, range, weight, and top speed, what to look for in batteries, vehicle codes applicable to electric cars...and more. 145 pages. *The third edition will not be available until atleast July, 1991.* Confirm availability before ordering. **80-401 $11.95**

Better Use of...Your Electric Lights, Home Appliances, Shop Tools - Everything That Uses Electricity by Michael
Hackleman. This is Michael Hackleman's best book on AE. It's interspersed with real life experiences of off grid people. There are chapters on low voltage alternatives, refrigeration, lighting, music systems, motors, relays, and timers, converting your flashlight to rechargeable. MUST READING FOR SERIOUS OFFLINE USERS! 166 pages, 1#. *Out of print until late 1991.* **80-303 $11.95**

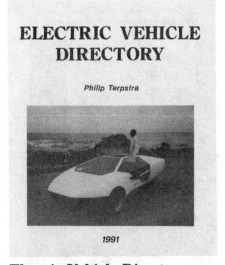

1991

Electric Vehicle Directory, by Philip Terpstra. A brief introduction to electric vehicles. This book shows like no other we know of five different vehicles on the market today complete with specifications and suppliers. It further lists suppliers of other electric vehicles as well as associations and newsletters available to the E.V. enthusiast. 14 pages. **80-403 $6**

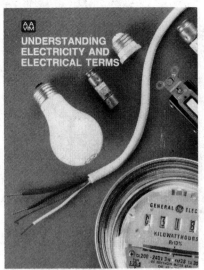

Understanding Electricity and Electrical Terms, by AAVIM. Explains electricity with a descriptive text and good illustrations. A good basic book on electricity. 40 pages.

80-301 $14.00

Electrical Wiring, by the AAVIM staff. Thoroughly covers standard electrical wiring principles and procedures. This book has become an industry standard in training students, teachers, and professionals. It uses over 350 step-by-step color illustrations, covering circuits, receptacles and switches, installing service entrance equipment and more. Revised in 1991 to include changes made in the 1990 National Electrical Code. 188 pages.

80-302 $19

Heaven's Flame Solar Cookers, by Joe Radabaugh. A solar cooker "primer" detailing types of oven designs, specific directions for the author's favorite, the peculiarities of the typical solar cooker, and general cooking tips. The author has over 15 years experience in designing, building, and using solar ovens. A great starter book for solar cookery! 41 pages.

80-701 $9

50 Simple Things You Can Do To Save The Earth , by The Earthworks Group. This is a wonderful brand new book that should be part of everyone's Earth Day library. We have quoted liberally from this book throughout this *Real Goods News* (sometimes annotated, othertimes not.) It's chock full of amazing facts regarding the greenhouse effect, air pollution, ozone depletion, hazardous waste, acid rain, vanishing wildlife, groundwater pollution, and garbage. It offers numerous concrete steps that can be taken to make each one of us a contributor to solutions to the energy problem. **Highly Recommended!**

80-801 $5

50 Simple Things Kids Can Do To Save The Earth, by The EarthWorks Group. Earthworks has written a great sequel to 50 Simple Things... This is a book to help kids learn about our relationship with other living things. This book is truely unique in that it explains the ways kids can help all of us clean up. It is full of facts, experiments, puzzles and activities designed to help childern be aware of and create lifelong habits about the environment. This book also demonstates how common actions and the products we use every day (from pop cans to light bulbs), affect the world we live in. An excellent gift for every child. 96 pages.

80-814 $7

Whole Earth Ecolog, edited by J. Baldwin, et al. Introduction by Stewart Brand. Big and floppy, like the Whole Earth Catalogs used to be, the Ecolog focuses solely on the environment. Whole Earth looked at the plethora of Earth-Saving books and made sure that the Ecolog was not just another version of *99 Recyclable Bottles of Beer on the Wall!* The Ecolog celebrates all groups doing something to change the world. Among the features: *Village Homes* - a parklike solar neighborhood now 15 years old where energy use is 50% less than the U.S. average; *The best sources of safe, environmentally righteous goods*, tools and hardware; *CoHousing*, attractive and affordable especially for single parents and seniors; *Exceptional books and videos* for getting the kids off to a good start. 126 oversized pages.

80-803 $16

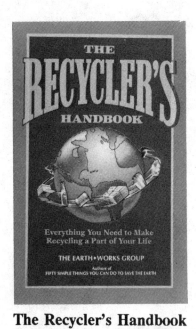

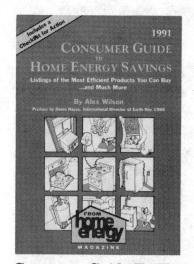

The Recycler's Handbook
From the folks at EarthWorks (50 Simple Things...), comes another excellent *how to* book - on recycling. This is a wonderful guide to all facets of recycling - from finding recyclers (for everyday products & difficult materials such as scrap metals) to organizing recycling programs for businesses and communities. It's written in a simple, effective style. This book is printed on recycled paper (of course). **80-804 $5**

Shopping For a Better World
A perfect guide for the environmentally and socially responsible consumer. This book lists all major companies and a vast array of common household products. The companies are rated using criteria based upon social consciousness, employment policies, and environmental positions. Printed on recycled paper (of course), this book is portable and a valuable tool for informed shopping.   **80-802 $5**

Consumer Guide To Home Energy Savings, by Alex Wilson.
A very comprehensive guide to energy efficient products. From light bulbs to furnaces, air conditioners to washing machines, water heaters to refrigerators: it's all covered in this guide. It shows you how to calculate energy savings and discusses which products pay for themselves. It discusses which products work and which are hype. According to *50 Simple Things...* Home Energy is "the best magazine in America on home energy." 251 pages.

80-816 $7

Nontoxic, Natural, & Earthwise, by Debra Lynn Dadd.
This book helps you identify products that are safe for your home and shows you how to protect the environment as you shop. It contains the most comprehensive listing of healthful products available and u ses a rating system that indicates both safety and environmental impact. It evaluates air & water filters, organic foods, biodegradable cleaners, cosmetics, pest controls, energy-saving appliances, clothing, gardening supplies, baby products and more. *Recommended by The Green Consumer, 50 Simple Things..., & the Whole Earth Ecolog.* 360 pages.

80-817 $13

We Can Do It!, presents images
and concepts that are carefully chosen to be informative, positive, and non-threatening. This book promotes in children the belief that they are not powerless, and that by their actions and words they can and will effect a change in our world. Encourages children to strive for world peace and peaceful solutions to their personal problems. 36 pages.
80-808 $5

Good Planets Are Hard To Find!

Good Planets Are Hard To Find!, By Roma Dehr and Ronald M. Bazar. This is a great environmental information guide, dictionary, and action book for kids (and adults). It gives children information about the problems facing our ecosystem, & offers a large range of activities geared to solving them. It explains complex environmental concepts from CFC's and clearcutting to pesticides and polystyrene in very simple understandable language. 44 pages.

80-806 $7

Ecology Action Workbook And Dictionary, The same authors as "Good Planets..." listed above have created this workbook and dictionary with positive activities for children to help save the environment. Chapters include "How to Write an Effective Letter," "How to Set Up an Ecology Action Club," and "How My Actions Can Make a Difference." This is an excellent workbook for schools. 21 pages.

80-807 $4

Complete Trash, by Norman Crampton. This is the A to Z guide to disposing of all varieties of common household trash, from broken appliances to spent smoke detectors. With humor and an equal eye to environmental and practical concerns, Crampton points out the best alternative among the four major disposal methods: burning, burying, recycling, or composting. It also discusses market value of various refuse and much more. 136 pages.

80-811 $9

The Green Consumer, by Elkington, Hailes & Makower. This is a brand-specific handbook of attractive, cost-competitive, and easily available products, that make great environmental sense. It includes baby products (diapers too), long-lasting light bulbs, biodegradable detergents, soaps, garbage bags, wood products that won't destroy the rainforests, groceries that aren't overpackaged and won't overburden our waste disposal sites, and much more. Printed on recycled paper. Our only complaint is that they don't list *Real Goods*. Write to them! 342 pages.

80-809 $10

Going Green, By Elkington, Hailes & Makower. This comprehensive guide introduces young readers to the major concepts of ecology and provides them with ways in which they can make a contribution to saving the planet. It explores such issues as the greenhouse effect, holes in the ozone layer, endangered species, rainforests, waste disposal, farming & food shortages, and energy use and misuse. It's ideal for generating discussions at home and at school, and suggests lots of projects for children to pursue. 96 pages.

80-810 $9

The Lorax, by Dr. Seuss. Who would have thought Dr. Seuss could be controversial? This book was actually banned by one of our local school districts as being too hard on loggers! This great story about the deforestation of the Truffula trees is helping to awaken an entire generation of young environmentalists. 61 pages.

80-815 $12

Homesick Syndrome, by RCI Environmental Institute. This is a 30 page booklet designed to help home owners recognize common health hazards and carcinogens in soil, water, and building materials found around their homes. It also covers various types and sources of indoor air pollution. Toxic items are identified, and the maximum recommended exposure is listed along with a brief description of detrimental health effects which can occur when that limit is exceeded.

80-821 $5

Magnetic Radiation Books

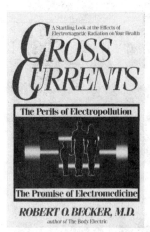

Cross Currents, by Dr. Robert Becker. This is the definitive work on Energy Medicine (electro-medicine) the rapidly emerging new science that promises to unlock the true secrets of healing. Beginning with our historic link to the earth's natural pulsating magnetic fields, we are led through experiments and discoveries which have produced our present new knowledge of electromagnetic medicine and its effect on the healing process. In the last 30 years electro-pollution and cancer have (not coincidentally) both risen at the same alarming rate. The book does an excellent job of getting vitally important information out of the scientific priesthood and into the hands of the general public - who should be making the important environmental decisions! Dr. Becker is *the authority* on the biological effects of magnetism, a pioneering researcher in biological electricity and regeneration, and a professor at two medical universities. 336 pages.

80-612 $22

Business & Reference Books

Pocket Ref, by Thomas J. Glover. This amazing book measuring just 3.2" x 5.4" x 0.6" is like a set of encyclopedias in your shirt pocket! Here is a very small sampling of the hundreds of tables, maps, and charts within: battery charging, lumber sizes & grades, floor joist span limits, insulation R values, periodic table, computer ASCII codes, IBM PC error codes, printer control codes, electric wire size vs. load, resistor color codes, US holidays, morse code, telephone area codes, time zones, sun & planet data, earthquake scales, nail sizes, geometry formulas, currency exchange rates, saw blades, water friction losses, and a detailed index! Have we said enough? **Indispensable!** 480 pages, 0.5#.

80-506 $10

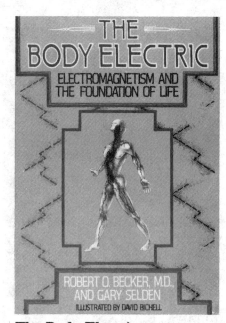

The Body Electric, by Dr. Robert Becker & Gary Seldon. Electro Magnetic Radiation (EMR) can affect the human immune system, the rate of cancer growth, even thoughts & emotions. Dr. Becker discusses the hazards of EMR produced by everything from ordinary household appliances, to utility power lines, radar, & microwave transmitters. This invormation has been suppressed for years. Our present AC electrical distribution network is putting the entire planet at risk. *"Changes would have been made years ago but for the opposition of power companies concerned with their short-term profits, and a government unwilling to challenge them...the entire power supply could be decentralized by using wind, flowing water, sunlight...greatly reducing the voltages and amperages required to transmit power over long distances."* This book is fascinating reading and will give you background information and motivation to clean up the magnetic pollution in you local environment, live longer, and avoid cancer! 352 pages.

80-610 $13

The Partnership Book, by
Denis Clifford & Ralph Warner.
Lots of people dream of going into
business with a friend.
Unfortunately, more often than not
the dream turns into a nightmare!
The best way to avoid the nightmare
is to have a solid partnership
agreement. This book shows how to
write an agreement that covers
evaluation of partner assets,
disputes, buy-outs, and the death of
a partner.   **80-505 $25**

Nolo's Simple Will Book, by
Denis Clifford. Seventy percent of
adult Americans don't have wills. If
people don't make their own
arrangements, the state will give
their property to a few close
relatives. It's easy to write a legally
valid will using this book and even
easier to update or replace it as your
circumstances change.

80-504 $18

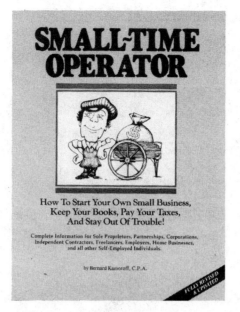

Small-Time Operator, by
Bernard Kamaroff, C.P.A. This is
probably the best book ever written
on starting up a small business. In
fact it's the book we at Real Goods
used to start our first store in 1978.
It's the book that nurtures your first
ideals of going into business yet sets
your reality firmly in concrete.
Written in a personal, non-technical
style, it gives you everything you
need to start up and stay in business
for the first year; providing a giant
boost to avoid the pitfalls of the
90% of new businesses that belly-up
in year one! Over 330,000 in print!
192 pages.   **80-502 $13**

**Step-By-Step Computer
Assembly Video,** by Roy Maul.
This video won the "Course of the
Year" national award from the
Learning Resource Network. It
offers viewers everything you need to
know to build a 100% IBM XT, AT,
or 386 compatible computer. There
are less than a dozen major
components under a PC's hood, and
with just a screwdriver you can put it
together in two to four hours. This
video greatly helps to demystify
computers and can easily save you
from $200 to $500.   **81-101 $45**

Ecopreneuring, By Steven J.
Bennett. This is the first book
written on how to start an
environmental business. It is a start-
up guide offering business
opportunites in: ecologically sound
packaging, child & baby care
products, household cleaning
products, foods & beverages,
education & training and more. It
goes through many case histories of
successful eco-businesses (including
Real Goods!) 320 pages.

80-507 $18

BASED ON THE NATIONAL AWARD WINNING
1987 "COURSE OF THE YEAR!"

Entertainment & Electronics

Electronics items are difficult to sell in an annual catalog like our AE Sourcebook because models are changing fast as the technology continues to improve. We urge you to check with us on availability before ordering electronics products. Further, warranty problems are very difficult on electronic products. There is currently a trend in the industry to have manufacturers and their service centers deal exclusively with product warranty problems; eliminating the dealer from this loop. We have been forced to pay for certain warranty repairs and have had to reconsider our policy. The specific warranties on many products that you buy from us will be honored only by the manufacturer or their designated repair stations. For this reason **DO NOT SEND ANY ELECTRONIC ITEM DIRECTLY BACK TO US!** *We will simply return it to you and have you deal directly with the manufacturer wasting your postage and ours!* We are sorry for having to impose this policy. We carry electronics products as a convenience to our customers who are unable to procure them locally. We hope you understand. *We have recently taken on five new products that can be serviced and warranted locally by the Sound Company at 187 Orchard Plaza Center, Ukiah, CA 95482. These products reference the Sound Company in their product descriptions. Do not send these products back to us!*

AC/DC Telephone & Answering Machine With All the Features

The Code-A-Phone 3450 combines the best of both worlds: Digital and microcassette technology. A 10-number auto dialer dials frequently called numbers automatically. The telephone also features a speakerphone, last number redial, and programmable pause. The answering machine features a digital message counter (tells you how many calls you've received at a glance, digital recording with your voice on a microchip for better reliability, 60 minute message capacity, one touch message playback and a 13-function beeperless remote that can be accessed from any pay phone. The unit will operate on 110V AC or 12VDC using a scant 0.2 amps.

68-107 Phone/Machine $119

Code-A-Phone 2710
12V/110V Answering Machine

Our AC/DC Code-A-Phone 2710 features a digital announcement with microchip technology, plus a separate tape for 30 minutes of messages. It has a personal memo record feature for personal reminders or messages for your family. It has a 60-minute message tape, for recorded messages (30 minutes on each side). One-Touch message playback allows you to playback your messages with a single touch, resetting the machine to answer your calls. Unlike some answering machines, our Code-A-Phone features voice activated recording, recording the caller's message as long as he speaks. You can choose whether messages are limited to a fixed amount of time or unlimited in length. You can choose the number of times the phone rings before the answering machine takes the call. The blinking message counter indicates the number of recorded messages. A new feature is **"Announcement Breakthrough"** allowing you to bypass your announcement to save time and long distance charges. You can also tell your callers how to break through, too.

Remote features include a 10-function tone remote access from any touch-tone phone using your 3-digit security code. Toll saver allows you to save money on long distance charges by telling you if there are messages before the system answers. You can update your announcement when you're away from home. You can retrieve all your messages even when the tape is full. The unit operates on 12VDC or 110VAC. It uses a microscopic 100 milliamps at 12V. While the machine says "9V" on it, we have had detailed discussions with the manufacturer, who assures us the machine will operate reliable from 9 to 14V.

68-104 Code-A-Phone (2710) $109

Cordless AC/DC Freedom Phone

Our 12V/110V rechargeable Southwestern Bell Freedom Phone is the finest cordless telephone we've ever used! The leading consumer magazine (who wishes to remain anonomous) rates this as the best cordless telephone on the market, bar none. The new FF-1725 has more features and replaces the old FF-1700. It has two separate dialing pads, one on the remote handset and the other on the base unit itself. It has a hands-free speakerphone, volume control, and an 18-number memory. It comes with a built-in intercom and digital security coding. The cordless range is 1,000 feet. It's the only cordless we've ever used that doesn't have lots of static. A 12-volt power cord along with a 110V cord is supplied with the unit for recharging the handset on either 12V or 110V. Retail price is $195 in 110V only. *We're offering our AC/DC model for an attractive price.*

We have one of these units in our warehouse and it's the only cordless we've ever owned with no static, and never a customer complaint about it sounding funny. All around a great telephone! JS

68-102   Cordless Freedom Phone (12V/110V) $185

Emerson VCT-120 AC/DC Color TV with VCP

This combination TV and video cassette player has incredible picture quality and is very simple to use. It makes hookup simple by having a TV and a VCP all in the same unit and it's also wonderfully portable. It has a 10" screen, auto color control, slide volume control, still frame, and 3-speed playback. It also features automatic fine tuning, and automatic eject when the videotape has finished rewinding. Draw is 5.5 amps with a tape playing. We love having the unit in our showroom to demonstrate videos to our walk-in customers. This unit has allowed our off-line power network a chance to jump into the twenty-first century. Size 14"H x 11.5"W x 13.9"D.

68-302 Emerson AC/DC TV/VCP $545

FOUL a stream. Push a shopping cart into it if nothing else is handy. Or find an environmentalist and push her in.

12V Video Cassette Player

Here is our most popular video cassette player that, depending on your viewpoint, either revolutionized entertainment for independent power producing folks, or destroyed family life and created a generation of 12-Volt couch potatoes. The VCP3000 manufactured by MGN (Magnin) features a front-loading VHS tape system, 7 key / 9 function wireless remote, auto repeat and picture search. It comes standard with both 110-volt and 12-volt DC power cords. It draws a scant 0.8 amps and provides crystal-clear video viewing.

68-301   12V VCP $275

Universal Camcorder Battery Charger & Power Supply

This unique universal camcorder charger can quickly recharge virtually any camcorder battery - 6V, 7.2V, 9.6V, and 12V. It features an automatic power shutoff circuit that operates when a full charge is attained. It will operate from standard 110VAC or from 12VDC. Its electronic circuitry can reverse the "memory effect" that is common to nicad batteries. You can also operate your camcorder directly from this unit using either 110V or 12V. Power output is 400 mA on the auto cycle and 800mA on the turbo-charge/auto cycle. Size: 7.5"L x 3.5"W x 2"H. Weight: 11 oz.

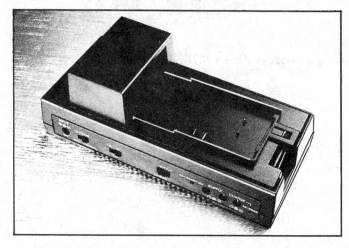

50-217 Camcorder Charger $95

Starlite 12" AC/DC B&W TV

This is our economy 12V TV. It features 82 channels, AC/DC operation and 100 solid state chassis. It has a carrying handle, an ear phone jack and rotary tuning. It comes with a DC 12V cigarette lighter type plug. It draws 1 amp at 12V. TV brands change rapidly so we may substitute a comparable unit if this model is discontinued.

68-304   12" B&W TV $109

RCA 13" AC/DC Color TV with Wireless Remote

This is the top-of-the-line RCA 13" ColorTrak TV with 24-button digital remote control. It comes standard AC or DC with a 12-volt power cord included. Gray Texture finish. Size: 13-7/8"H x 14-1/2"W x 14-7/8"D. Amp draw at 12VDC is 5.5 amps. One year warranty covered by the Sound Company (see above).

68-307   RCA 13" Color TV $449

Starlite -10" AC/DC Color TV

This is our economy model 12V color TV. It features a quick-start picture and automatic fine tuning. It draws about 5 amps @ 12V. It has a built-in antenna and comes with a DC 12V plug. TV brands change rapidly so we may substitute a comparable unit if this model is discontinued.

68-305   Starlite 10" Color TV $255

12V TV Signal Amplifier

This solid state TV signal amplifier increases signals to pick up stations at greater distances. 300 ohm input and output, with an on/off switch and indicator light. Mounts flush in wall, requiring a 12V DC hook-up.

68-321   12V TV Signal Amplifier $39

Fisher 3000 AC/DC Cassette Deck

Here is an AM/FM 12-volt or 110V portable stereo with double cassette deck and surround sound. It comes with detachable 3-way speakers, high speed dubbing, and built in 4-band equalizer. It has inputs for a CD player, microphone, and a headphone jack. It comes standard with a 12V lighter plug for DC use. Very low amp draw. Will use typically 1/10 the power of an AC stereo running off of an inverter and there's no inverter noise! One year warranty provided by the Sound Company (see above).

68-351   Fisher 3000 AC/DC Cassette Deck $189

Fisher 5000 AC/DC Cassette Deck

One step up from the 3000, the Fisher 5000 is a 12-volt double portable cassette stereo with auto reverse and 4-band digital radio - including shortwave. It has continuous or sequential tape play with synchronized tape dubbing. The built-in clock/timer allows play or record at a preset time. It has inputs for a CD and microphone and headphone jack. It comes standard with a 12V lighter plug for DC use. Very low amp draw. Will use typically 1/10 the power of an AC stereo running off of an inverter and there's no inverter noise! One year warranty provided by the Sound Company (see above).

68-353   Fisher 5000 AC/DC Cassette Deck $249

Fisher 9000 AC/DC Cassette Deck w/CD Player

This is our ultimate 12-volt stereo system. It includes a compact disc player with 4-band radio and double cassette deck with one side auto-reverse. It comes with a 16-function wireless remote control for running all functions including the CD track select. It has full Dolby noise reduction on tape play and 3-way/2-way switchable surround speakers for super sound. One further great feature that this player has is *Line Out Jacks* so that you may use a high power car amp and drive full sized home speakers from your 12-volt system! These line out jacks also work for audio and headphones. The unit comes standard with a 12V plug for DC use. Gray texture finish. Very low amp draw. Will use typically 1/10 the power of an AC stereo running off of an inverter and there's no inverter noise! One year warranty provided by the Sound Company (see above).

68-355   Fisher 9000 w/CD player $489

Sony AC/DC Discman

Our Sony D180K portable Discman can be used in your car or your 12V or 110V home. It is a compact disc player with digital filtering, 3-way repeat, shuffle play, and music search. It comes standard with a 12V plug, and a cassette adaptor so that you can play CDs through your car stereo. It also comes with an audio system connection cable and AC adaptor. It will also run on 4 "AA" batteries.

68-361   Sony AC/DC Discman $379

Phantom Loads

Richard Perez

Just because the switch says "OFF" doesn't mean a device is off. Many modern appliances are never really OFF. They contain clocks, memories, remote controls, microprocessors, and instant ON features that consume electricity when plugged in. That's 24 hours a day, 7 days a week... While these Phantom Loads are often small, they add up if several are constantly on line. Some Phantom Loads are easy to spot- things like clocks and and timers have displays. Other Phantom Loads are truly hidden- the device seems OFF when switched OFF, but it really isn't.

Obvious Phantom Loads. Consider a clock. Many appliances contain a clock or timer. The electronic clock/timer and its display consume very little (≈0.5 Watts). However, there is a power supply in the appliance that converts 120 vac into low voltage DC for the clock/timer. This power supply is very inefficient at low power, consuming many times the power actually needed. This consumption is about 40 to 75 Watt-hours daily- enough to run a lightbulb for TWO hours. Most of this consumption goes to do a job better accomplished by low voltage DC directly from the batteries. At night, when hardly any power is used, the inverter may stay on just to supply small Phantom Loads. This operation is very inefficient. One, these jobs are better done via DC. And two, it's forcing the inverter to operate for extended periods in its least efficient mode.

Sneaky Phantom Loads. Some Phantom Loads appear to be truly OFF when switched off. There are no lights or indicators showing power consumption, but the device is still using electricity. Offenders in this category include, stereos, VCRs, computers, calculators, computer printers, satellite TV systems, and any device powered by a "wall cube". Wall cubes are power supplies in plastic boxes that plug into 120 vac outlets. Let's visit a few of these Phantoms where they lurk.

Filters and Line Conditioners. Many 120 vac business appliances like computers, printers, typewriters, FAX machines, and copy machines use filters on their power input. These filters serve a useful purpose- protecting the device from overvoltage, surges, noise and other electric trash that may wander onto the grid supplied electrical lines. Unfortunately, most of these filters are wired in ahead of the power switch, and are on line all the time. They consume power from the inverter- about 8 to 40 Watt-hours daily.

The Primary is Alive!. Many 120 vac appliances contain power supplies. These convert 120 vac, either inverter or grid produced, into low voltage DC for the appliance's electronics. On some appliances the ON & OFF switch is placed on the secondary (low voltage side) of the supply's transformer. The primary is not switched and is always connected to the 120 vac source. See the diagram below.

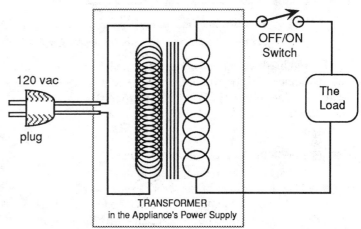

TRANSFORMER
in the Appliance's Power Supply

The inverter or commercial power grid sees the primary of the transformer as a constant load. Power consumption on these devices may run between 50 to over 200 Watt-hours daily.

Wall Cubes. These small black boxes are really Phantom Loads. Wall cubes are actually small power supplies. Consider the case of a telephone answering machine powered by a wall cube. The wall cube is plugged into an electrical outlet and feeds the answering machine via a low voltage power cord. The ON/OFF switch is located on the answering machine itself. Even if the answering machine is turned OFF, the wall cube still consumes electricity. This is electrically the same as having a power switch on the transformer's secondary- the primary is alive all the time. A wall cube uses 20 to 50% of its rated power even when its device is switched off.

How to detect Phantom Loads. We find Phantom Loads by watching for signs of current flow where there should be none. If you are a technician with a meter, then break into the appliance's power circuit and measure consumption when switched OFF. This involves working with live 120 vac wiring and is dangerous if you don't know what you are doing. 120 vac will shock you whether it's produced by the grid or your inverter.

What follows is a very simple circuit for detecting phantom loads. It can be assembled from hardware and Radio Shack parts for under $6. It can be wired before it is connected to 120 vac power, and is as safe to use as a wall socket. The schematic is below:

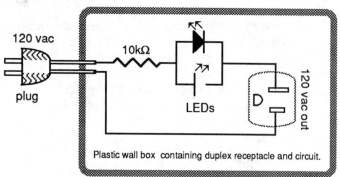

120 vac plug   10kΩ   LEDs   120 vac out

Plastic wall box containing duplex receptacle and circuit.

Everything is assembled into a plastic wall box used in 120 vac homes. The two Light Emitting Diodes (LEDs) can be in any color. Use very small gauge lamp cord (≈18 gauge) wired to a male plug.

Use the Phantom Load Detector as follows:

- Turn the appliance to be tested OFF.
- Unplug the appliance from the 120 vac power source.
- Plug the appliance into the Phantom Load Detector.
- Plug the Phantom Load Detector into the 120 vac power source. If you are using an inverter as a 120 vac source, then make sure that the inverter is "booted" and operating, not in standby mode.

If the appliance is a phantom load and consuming power, then the LEDs will light. The more power the appliance is consuming, then the brighter the LEDs. This device is for testing power consumption in appliances when they are switched OFF.

While the Phantom Load Detector will survive switching the appliance ON, the appliance will probably not operate. The entire circuit is current limited by the 10kΩ resistor. With the 10kΩ resistor, we were able to detect Phantom Loads as small as 0.3 watts. If the 10kΩ resistor is replaced with a 3kΩ resistor, then the detector becomes more sensitive (0.1 watts), but the LEDs will not survive operation into an appliance that is turned ON during testing.

The Phantom Load Detector is very easy to build with only three electronic components. One detector should cover a neighborhood. Either a 120 vac appliance is a phantom load or it isn't. Once we have determined this we don't need to test it anymore, and the detector can circulate for others to use.

Dealing with Phantom Loads. Unplug the appliance! This works for sure because it is disconnected from its power source. However, constantly plugging and unplugging is a

pain and wears out the hardware quickly. Just about every hardware or discount store sells extension cords with multiple female plugs that are SWITCHED on the plug strip. They sell for $5 to $10. When the plug strip is switched OFF, all the appliances plugged into the strip are disconnected from the 120 vac power source.

We use these plug strips for all phantom loads. Here on Agate Flat, we have three SL Waber (Model EP7S, costing $7.99 at the local discount house), seven outlet, plug strips with neon indicator lights. The neon indicator glows when the plug strip is turned ON and supplying power to all the phantom loads connected to it. We have two Mac computers, an ImageWriter printer, an HP DeskWriter printer, a wall cube powering a modem, and a hard drive plugged into these strips. I don't mind feeding these appliances when they are actually operating, but I don't want them flattening our batteries when they are supposed to be OFF.

Selecting Appliances that are NOT Phantom Loads. Any appliance with a built-in clock or timer is a constant and obvious phantom load. If you want a clock, then buy a clock, not a microwave or VCR. Avoid appliances with electronic memories **unless** these memories are kept alive by small batteries within the device. As a last resort, take a Phantom Load Detector to the store when you buy your appliance and check it out.

In many cases all appliances of a particular type are phantom loads. VCRs, for example, all contain clocks and timers that are alive even if their displays are not lit. All appliances using wall cubes are phantom loads. Every piece of electronic office equipment is a micro phantom load because of its filtration. Here the switched plug strip comes to our rescue.

The Bounty on Phantom Loads

If you live on the commercial grid, you're paying an average of 7.75¢ per kilowatt-hour for electricity. A small phantom load of 4 watts costs you about $2.70 yearly.

If you make your own electricity, then the savings situation is even better. Site produced power costs much more, about 90.¢ per kilowatt-hour. The 4 watt phantom load costs home power producers about $31. per year. The plug strip pays for itself in less than 3 months. And we get to use our power elsewhere.

The bottom line here tells only part of the story. Sure we can save some money by disconnecting inoperative appliances that still consume power. We can also save resources for use elsewhere. Regardless of the electrical power source, phantom loads waste energy because they don't do anything in return for their consumption. While in the individual sense, these phantom loads are small; in the collective sense, we're wasting enormous amounts of electricity.

Solar Toys, Maps, & Gifts

Gift Certificates

For those of you enamored by our products but unable to decide what to give, we offer **Real Goods Gift Certificates**. Many of our customers have used these for wedding presents for would-be independently powered newlywed homesteads. We'll be happy to send a gift certificate to the person of your choice along with our latest edition of the *Real Goods News* in the amount of your choice.

REAL GOODS TRADING COMPANY
3041 Guidiville Road
UKIAH, CALIFORNIA 95482
(707) 468-9214 FAX (707) 468-0301

№ 1101

Gift Certificate

05-000   Gift Certificate   Specify $ Amount

Rainforest Crunch

Rainforest Crunch is an incredibly delicious, all-natural, highly addictive cashew and Brazil nut buttercrunch that best-of-all helps preserve the rainforest. The creation of Ben Cohen (Ben & Jerry's Ice Cream), this is part of a larger project to show that the rainforest can be more profitable as a living rainforest than by cutting, burning, and transforming it into plantations and ranches. 40% of the profit goes to rainforest-based preservation organizations and international environmental projects. This very tasty candy provides an ideal vehicle for putting your money where your mouth is! The Crunch comes both in an attractive and colorful one lb. reusable tin, or a colorful box.

"I caught my girlfriend having some for breakfast." -**Ben Cohen, Ben & Jerry's.**

03-801   Rainforest Crunch (1 lb. can) $14
03-802   Rainforest Crunch (8 oz. box) $ 5

Bumper Stickers

Stop Solar Energy

Guaranteed to crank heads and provide lots of double-takes. With the words: "Stop Solar Energy; Mutants For Nuclear Power, Village Idiots For a Toxic Environment," this bumper sticker will go great on the bumper of your electric vehicle!

> ## STOP SOLAR ENERGY
> ## MUTANTS FOR NUCLEAR POWER ☢
> VILLAGE IDIOTS FOR A TOXIC ENVIRONMENT

90-741   Stop Solar... Bumper Sticker $2

I Get My Electricity From the Sun

Black letters on a yellow background. Flaunt your energy independence!

> ## I GET MY ELECTRICITY
> ## FROM THE SUN
> REAL GOODS TRADING CO. ☼ 966 MAZZONI ST. UKIAH, CA. 95482   (707) 468-9214

90-742   Electricity From...Bumper Sticker $1

Lunar Phase Poster

This has been one of our best sellers for years. Every phase of the moon for the entire year is displayed in beautiful silver on a deep blue background. The prize winning graphic is astronomically correct and printed with incomparable sharpness. Each silver crescent is keyed to the date and day of the week - even children can easily read it. 16" x 37"; printed on 100# stock and it's shipped in a sturdy plastic chart tube.

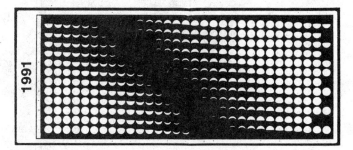

90-717   Lunar Phase Poster $8.50

Solar Clock

The Solar Clock has an amorphous solar panel for its face that generates enough energy to recharge a nicad battery with regular office or natural light. The clock has an attractive solid hardwood frame and solid brass hands and markers. It will provide years of trouble-free service without battery replacement. We put one in our office nine months ago and it has kept perfect time since, as well as being a novel conversation piece! It measures 5" by 5" square and is 2" deep.

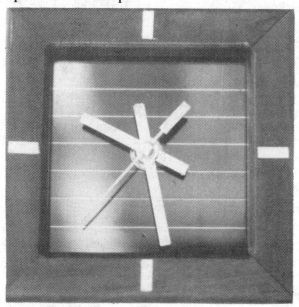

90-201   Solar Clock $49

Solar Watch

Casio's solar watch is powered by an amorphous solar cell, eliminating the need and the expense of a battery. It comes with a daily alarm, can be used as a stop watch, can be set for a 12 or 24 hour format, and has an automatic calendar that automatically determines and sets the number of odd or even days in the month. The watch is also water resistant.

68-405   Solar Watch (Casio) $39

Solar Tea Jar

The small solar panel on the top of this 1 gallon plastic unbreakable tea jar will stir your tea with the paddle and keep your tea warm in the sun. A built-in spigot is provided at the bottom for easy pouring. This brewing method reduces the acidity and bitterness of the tea. The unit measures 10-1/4" high by 6" diameter.

90-202   Solar Tea Jar $18

Solar Wooden Model Kits

Our most popular toy. Our wooden model kits have been incorporated into many other popular product catalogs. These easy to assemble (about 1 hour) models are a perfect demonstration to children (and adults) of the wonder of solar power. The parts snap together easily; glue is included. They make great science projects. We've sold over 4,000 of these over the last two years.

90-402	Airplane Kit	$18
90-403	Helicopter Kit	$18
90-404	Windmill Kit	$18

Solar Construction Kit

This popular 4-in-1 construction kit is entertaining and very educational for kids from five years old. Children can construct a helicopter, windmill, airplane, or water wheel - each with a solar-powered moving part. The kit is complete with an electric motor, a small solar panel and over 100 plastic pieces.

90-400   Solar Construction Kit $22

150 Solar Experiments

With this kit, you can build a solar furnace, a stroboscope, moire patterns, an electronic thermometer, electroplating, photosynthesis, solar energy mobiles and lots of other items that runs on solar electricity. Components include a DC electric motor, a 600 milliamp solar cell, a 50 micro amp volt meter, a solar water heater, a parabolic reflector, a strobe disc, a magnifying glass, a thermometer, diodes, LEDs, and a test tube. It's the most comprehensive solar experimentation kit that we've been able to find.

90-401   150 Solar Experiments $39

High Power Solar Project Kit

This educational kit comes complete with a 5" x 5" amorphous solar panel with an eight-volt 100 mA nominal output. The small solar panel works great for nicad battery charging, running small fans and motors, and countless other solar experiments and hobbies - limited only by your imagination! Instruction sheet, motor, propeller and wiring clips are included. Some soldering required.

90-422   High Power Solar Project Kit $14

Solar Educational Kit

Here is a simple and inexpensive educational kit that will get any child (or adult) started with the graphic basics of solar power. A small solar (PV) panel is included along with a motor, a small fan, and a pinwheel for making a variety of objects that are really functional.

90-405   Solar Educational Kit $9

Solar Cool Caps

These caps work as miniature evaporative coolers. The cool breeze blowing on your moist forehead quickly acts to cool down your entire body. A switch allows you to select solar or battery and a battery compartment is included for 2 AA cells.

These hats are definitely more than a gadget and a conversation piece - they really work! We recently equipped an entire roofing company with them for the hot Ukiah summers.

Specify your first and second choice of colors: Red (90-410-R), Blue (90-410-B), Green (90-410-G), or White (90-410-W).

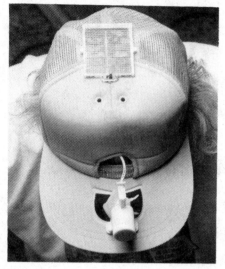

90-410   Solar Cool Cap $24

Solar Safari Hat

Our Safari style solar cooling hat is similar to the Solar Cool Cap but more attractive and a better conversation piece. Switch from solar power to battery power for those humid, steamy nights in the African savannah. A real eye grabber and they really do work wonderfully!

90-411   Solar Safari Hat $42

Solar Mosquito Guard

Our new solar mosquito guard really works to repel mosquitoes. It puts out a high-frequency wave that will repel mosquitoes in a 12' radius. The battery will recharge in three hours of sun and it comes with an on/off switch. Be forewarned that the Solar Mosquito Guard's high pitched sound is definitely audible. With no other noise it can be annoying, but with a little background music or commotion it works great, and is certainly less annoying than the music of a mosquito buzzing in your ears! We must be honest that of the first 775 units we've sold since our Spring '90 catalog was released, we have had five complaints that they do not work - we like to think that these five cases are due to deaf mosquitoes!

90-419   Solar Mosquito Guard $8

TAKE a bath. Haven't bathed since morning? shame on you! Fill the tub to the brim with scalding water, then wait a half hour till it's cool enough to use. Get all your friends to bathe more often.

Solar Speed Boat

Whenever the sun shines on the boat the tiny motor propels it across the water or around in circles depending on where you set the rudder. It's the closest thing ever invented to a perpetual motion machine! A great educational toy.

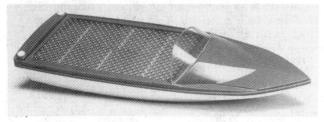

90-406   Solar Speed Boat $24

Solar Musical Keychain

The Solar Musical Keychain is a functional keychain with a small LED light for finding your keyhole in the dark. When the PV panel is directed toward the sun or a bright light it plays a song. A great PV educator.

90-407   Solar Musical Keychain $5

Solar Micro AM Radio

This small AM solar radio really works. It comes complete with clip and sponge earphone for listening while jogging. The solar panel charges the small nicad battery so that the radio will operates in times of darkness.

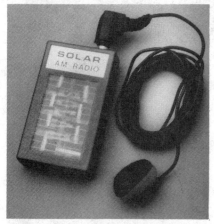

90-408   Solar Micro AM Radio $11

AM/FM Solar Radio

Our solar powered AM/FM radio folds out to recharge the nicad batteries in the sun. Good quality sound out of the built-in high fidelity cobalt mylar speaker and it will accept 3.5mm headphones as well. The unit measures 4.5 cm wide x 11 cm tall. It works directly from the sun.

90-409   AM/FM Solar Radio $19

Solar Visor
AM/FM Speaker Radio

This solar visor AM/FM radio is great for jogging or working in the garden. The radio can be used at night or indoors with the included battery pack. One hour of sun allows for approximately six hours of indoor use.

90-421   Solar AM/FM Visor Radio $19

Dynamo & Solar
AM/FM Receiver

The dynamo radio needs no electricity or disposable batteries other than two "AA" nicads. It has a small solar panel for solar charging and a hand-cranking dynamo for hand charging. The dynamo charges at a 10:1 ratio. One minute of cranking gives you ten minutes of radio operation.

90-418   Dynamo Receiver $39

The Mistic Arch

We just discovered this wonderful cooling product at the Solar Energy Expo in Willits. It was the single most popular item there on a 100 degree day. We bought it for our hardworking overheated warehouse staff and they use it for cooling down on hot days. It puts out a very fine mist that is extremely soothing yet not wet enough to really soak your clothes. It hooks up simply to a garden hose. It consists of 1/2" PVC pipe with a brass ball valve shutoff. It has eight 3-gallons-per-hour misters equally spaced and will operate on any water pressure over 20 psi. The easy-to-read blow up diagram allows you to assemble the arch in about ten minutes! A can of PVC cement is included. This is a major improvement over "running through the sprinklers!"

46-150   The Mistic Arch $55

Solar Shower

This incredible low-tech invention uses solar energy to heat water for all your washing needs. The large 5 gallon capacity provides ample hot water for at least four hot showers. On a 70° day the Solar Shower will heat 60° water to 108° in only 3 hours for a very hot shower. The shower nozzle is an improvement over the leaky old "Sun Showers" that we used to sell. This unit is built of 4-ply construction for greatest durability and efficiency. These have been one of our catalog's fastest sellers. We've sold over 5,000 in the last two years alone!

90-416   5 gallon Solar Shower $12

Stellarscope

This handy "kaleidoscope" type star chart is the easiest to use we've ever found. It provides an hour by hour adjustable system for recognizing and locating the 35 main stars and 60 constellations visible in both the Southern and Northern hemispheres. (4 latitude adaptors are included.) The Stellarscope is the new and improved version of the old *Cosmic Star Chart.*

90-349   Stellarscope $33

Predator - Food Chain Game

The Predator game as well as *AC/DC* and other environmental games were all designed by Marie Lowell, an environmental teacher in the Berkeley school system. She designed the games for her students and the demand from all the other teachers to borrow the game became so great that she coerced her husband who was a printer to go into the business! **Predator** is a whole new idea in card decks based upon the natural food relationships in a temperate zone forest. Each of the 38 beautifully illustrated cards represent a plant or animal commonly found in the forest. An animal takes what it eats and is taken by what eats it. An ideal way to teach children (8 and up) and adults the order of the forest.

90-440 Predator Game $9

Dymaxion World Puzzle

This puzzle is adapted from Buckminster Fuller's world map. It gives a unified and unique view of the world and provides an eye opening education of continental relationships for all ages. The puzzle allows you to explore the relationships of the continents and oceans. It consists of 20 triangles which can be put together in dozens of different ways. You can center the map on any one of the twenty pieces and then see views of countless different worlds.

90-414   Dymaxion puzzle $22

AC/DC Game

This games takes the mystery out of electrical circuitry. Players use both 6 volt and 110 volt components to complete workable electrical circuits without getting "shocked" or "shorted out" by opponents. Circuits can be simple series circuits (power source, switch, appliance and fuse) or elaborate parallel series circuits. No knowledge of electricity is needed to play. Included with the deck of 84 cards is a graphic introduction to electric circuits. 2 to 6 players; ages 9 to adult.

90-443 AC/DC Game $9

Oh Wilderness

A card game for people who enjoy nature and the outdoors, **Oh Wilderness** is about four areas of backcountry lore: wilderness skills, wildlife, plants, land & sky. Small enough to tuck into a back pocket, this game will provide fun for all ages during camping trips or car trips. For two players, up to large groups and easy enough for novices to enjoy as much as experienced folks.

90-441 Oh Wilderness Game $9

Good Heavens! Game

No background in astronomy is needed for this game, and a 24-page booklet is included for a great introduction into astronomy, that offers tips on how to observe and appreciate the wonders of the sky. 54 playing cards each include a question, answer, and informative paragraph as well as a list of possible answers to questions on other cards.

90-442 Good Heavens! Game $9

Tyvek World Map

This is probably the nicest map of the world ever made. The colors are brilliant and the countries and mountains seem to jump out at you. It appears 3-D even though it isn't. It's printed on virtually untearable Tyvek. The size is 30" X 53", the scale is 1=28,000,000, and the projection is Van der Grinten.

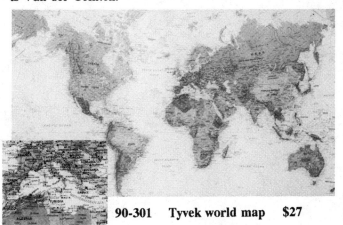

90-301   Tyvek world map $27

Tensegritoy

Tensegritoy is an ingenious new construction puzzle that provides fun and intellectual challenge for children over 10. Based on R. Buckminster Fuller's ideas of tensegrity, (tension and integrity) over 100 intriguing shapes can be built. The structures can bounce, roll, or seemingly float in the air. With the kits' colorful components you can construct a basic four-sided figure, a helix or a geodesic dome, or explore architecture and the arrangement of DNA! The 32-page illustrated instruction booklet provides lots of how-to ideas. This is truly an affordable learning experience.

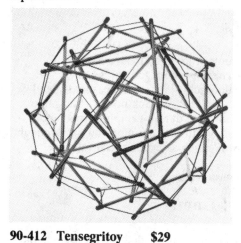

90-412   Tensegritoy $29

Stik-Trix

Stik-Trix is the junior, introductory version of the original Tensegritoy (above). It's made with half-size sticks that can be made into 20 different shapes. It is recommended for children 8 years old and up and is a great introduction into the king of construction toys - Tensegritoy.

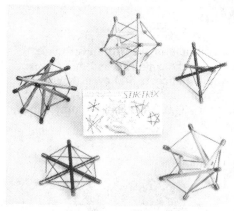

90-413   Stik-Trix $8

EarthSpace Enterprises Products

One of our good customers from Oregon makes the dynamic postcards, posters, and t-shirts of the earth and galaxy that are so frequently copied. The *galaxy design* shows an arrow pointing to our star, the Sun, which is only one of 200 billion other stars in the Milky Way galaxy, pictured here. This galaxy design is accompanied by the words: "You are Here." The *earth design* is the wonderful view of earth from space with the words: "Good Planets Are Hard to Find." (Incidentally both designs are copyrighted by Harrell Graham.)

Both designs are available in either T-shirt (100% cotton - specify size: S,M,L,XL), poster, or post card. For T-shirt sizes: order 90-601-S for small "You are Here," 90-601-M for medium, 90-601-L for large, and 90-601-XL for extra large. Use the same code suffix for the Good Planets T-Shirt

YOU ARE HERE

Good Planets Are Hard To Find

90-601	"You Are Here" T-Shirt	$14
90-602	"Good Planets" T-Shirt	$14
90-701	"You Are Here" Poster	$ 5
90-702	"Good Planets" Poster	$ 5
90-703	"You Are Here" Post Card	$ .60
90-704	"Good Planets" Post Card	$ .60

The Earth Flag

The Earth Flag's popularity gained a lot of momentum from Earth Day 1990, and can now be seen flying everywhere. It continues to be an inspiring symbol to peace, environmental justice, and global vision.

The Earth Flag is now available in three different sizes. Each flag consists of a four-color, photo-like image of the Earth based directly on the now-famous NASA photograph, printed on a dark blue background. The **Large Earth Flag** is 3' x 5' and is printed on nylon, with a canvas header and brass grommets. The **Medium Earth Flag** is 2' x 3' and is printed on cotton, with a canvas header and brass grommets. It is recommended for parades, indoor display, and classroom use. The **Small Earth Flag** is 6" x 9" and is printed on cotton and mounted on a an unfinished stick made from white birch.

90-731	**Earth Flag - large (3' x 5')**	**$39**
90-732	**Earth Flag - medium (2' x 3')**	**$19**
90-733	**Earth Flag - small (6" x 9")**	**$ 4**

All One People Buttons

These buttons are designed to stimulate global awareness. Combining the Apollo 17 Earth photo with the simple message "ALL ONE PEOPLE," this powerful image will hopefully help to unify the people of the planet.

90-706 All One People Button $2

Whole Earth Globe

This extraordinary 16" globe depicts Earth as seen from space, without borders or boundaries. It can be used by educators, peacemakers, as a beach ball, or as an all-purpose toy for kids of all ages everywhere. It comes complete with the Global Handbook and makes a wonderful gift. We've just found a second version of the globe - this one (the "Wildlife Version") has native animals depicted beautifully in their natural habitats.

90-415   Earth Ball (Nasa Photo) $8

90-430   Earth Ball (Wildlife Version) $8

SHOOT a bird. They get too complacent with us tiptoeing around, feeding them, and taking their pictures. They'll survive in the wild only if they know you can't be trusted.

The Earth Puzzle

This 93 piece puzzle of the NASA view of the earth makes great family entertainment without TV! It's designed by a family just north of us and is made from 100% recycled and recyclable materials. Their goal was to demonstrate the touching of our home planet with the idea of putting it back together piece by piece. Inside the puzzle are 10 simple steps that can be taken to save the planet. On the back are gems of inspiration written about our planet by visionaries throughout history. A great gift!

90-435 The Earth Puzzle $6

Celestial Jig Saw Puzzle Glows In Dark

Here's a jig-saw puzzle that's both educational and challenging. This 1,000 piece **Celestial Puzzle** proved to be a match for our resident jig-saw puzzle expert *(she's gotta be smart - she recently made it onto Jeopardy!)*. When complete it depicts the solar system, the Milky Way, and all the major constellations present in the northern and southern skies. *Plus* it even glows in the dark. This puzzle is ideal for older children and adults - educational and fun.

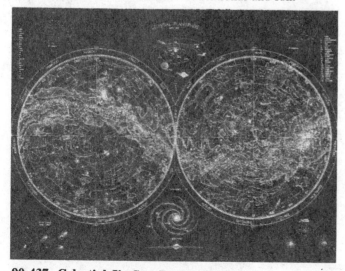

90-437 Celestial Jig Saw Puzzle $16

Night Sky Star Stencil

The Night Sky Star Stencil is one of the most fun and exciting new products we've found. It's like camping out in your own home. The stars are not visible in normal light, but when your room is darkened, the ceiling becomes a spectacular star display: more than 350 stars become instantly visible with completely accurate position and brightness.

Application is simple (1 or 2 hours): put up the reusable stencil with the provided adhesive, paint the stencil holes, take down the stencil, turn off the lights, and the stars come out. Stars will glow 30 minutes to 2 hours depending on light exposure prior to room darkening, and the phosphorescent glow paint lasts for years.

It's available in either summer sky or winter sky stellar configurations and in two sizes: 8' x 7.5' and 12' x 12.5'. Having tried all the various paste-on star kits, we find the Night Sky Star Stencil to be the queen of home planetaria, and have one displayed in our showroom. It's ideal for helping kids conquer their fear of the dark! Glow paint is a water-based non-toxic, non-radioactive material.

90-341	Winter sky (8')	$25
90-342	Winter sky (12')	$30
90-343	Summer sky (8')	$25
90-344	Summer sky (12')	$30

Raven Computer Maps

Raven's computer-generated terrain plots give you strikingly unusual views of landforms, ranging in scale from the very local to the global. With a selection of images that take you all the way from looking at Earth from outer space to probing the ocean floor depths off Hawaii, these beautiful productions bridge the gap between scientific precision and graphic art.

The computer plots define the land surfaces shown, with both the precise accuracy of topographic maps and the graphic clarity of a bird's-eye view.

The Rockies, the High Plains, and the Intermountain West

Looking north and west from above the Texas Panhandle. Available in small (24" x 41.5") and large (28.5" x 63.75") - both views include the small detail. Printed in four inks. Key map is included.

90-351	Computer Rocky Mountains (small)	$24
90-381	Laminated	$49
90-352	Computer Rocky Mountains (large)	$39
90-382	Laminated	$69

Hawaiian Islands

The Islands are tips of a great volcanic arc that extends 1,500 miles across the Pacific. Two oblique views show all the exquisite detail of the ocean floor. Printed in eight inks. 28" x 43.5".

90-356	Computer Hawaiian Islands	$24
90-386	Laminated	$49

Blue Whale's Hawaii

The sea-level view of Hawaii, looking north and east above and below sea-level. The below sea-level colors deepen at 1000-m intervals. Printed in seven inks. 32.5" x 41.75".

90-357	Computer Blue Whale's Hawaii	$24
90-387	Laminated	$49

One World Map

New from Raven, this map showing three views of the world is truly stunning! It's the answer for all those map lovers who have wondered whether Greenland is really as big as South America. It isn't, and these three views make the point clearly without the distortions of conventional flat maps. Physical features and important cities are named. Elevations and ocean depths are shown by color. Political boundaries are not included. Bright clear colors give these globes a dramatically three-dimensional appearance. The map measures 35" x 63" with a scale of 1:23 million.

90-388	One World Map	$39
90-388-L	One World Map (Laminated)	$64

Raven State Maps

These are the best state maps we've ever seen anywhere of each of the 11 Western states, Hawaii and now New England (AZ, AK, CA, CO, ID, MT, NV, NM, OR, UT, WA, WY, HI, ME, VT-NH, & MA-CT-RI.). Raven maps give you the startling illusion of three-dimensional imaging right on a flat surface. They offer remarkably clear views of physical features, and a wealth of detail on towns, roads, and railroads. Maps vary slightly in size but are approximately 42" x 55".

All Raven maps retain the accuracy of the U.S. Geological Survey materials from which they are built. Intended for wall display, all maps are printed in fade-resistant inks on sturdy 70# paper.

Because of the startling realism of these maps, people are highly encouraged to touch them. For this reason, Raven also makes a lamination available of 1.5 mil vinyl coating on both sides. It is lightweight enough so that the map is easily rolled for shipping in a sturdy tube. *Rather than charge an additional map tube shipping charge, we have simply increased the prices on our maps $4.*

ATTN: Hawaiian customers: Raven makes a new exquisite map of all the Hawaiian Islands in multi-colored blue format. It is one of their most impressive maps! (Order the Hawaii State map.)

90-302	**State maps**	**$24**
90-303	**Laminated state maps**	**$49**

Specify state (90-302-AZ, AK, CA, CO, ID, HI, MT, NV, NM, OR, UT, WA, WY, ME, VT-NH, MA-CT-RI.)

Nickel-Cadmium Battery Chargers

Anyone who uses cameras, flashlights, a walkman, or battery-operated toys can testify to the horrific sums of money spent on disposable batteries every year (not to mention the ensuing ecological disaster). Even alkalines don't seem to last as long as we'd like. Nickel-cadmium (nicad) batteries can cut down battery costs from 9 cents per hour to one-tenth of a cent per hour.

One needn't fret because one lives off the power grid. There are now easy ways to charge those money-saving nicad batteries both with 12-volt house current and with mini solar panels directly from the sun.

The ideal way to buy nicad batteries is to buy two sets. That way you have one set in the solar charger at all times charged up and one set in the appliance. When the appliance gets weak, just swap batteries with the fresh ones and you'll never run out of batteries (for a thousand switches anyway!). Some people get frustrated because nicads only hold about one-third the charge of alkaline batteries; this is all the more reason to keep two sets of nicad batteries for every appliance. Here is a chart to help you better understand a nicad's capacity and recharging time. Many people have a basic misunderstanding of nicad batteries, and expect them to last as long as alkaline batteries. *You need to understand up front that nicads will only last approximately 1/3 as long as alkalines per charge, so don't be surprised when you need to recharge often!*

Nicad battery	Storage voltage	Charger capacity	Hours output	Required to recharge in sun
AA	1.25 V	0.5 A	70 mA at 3 V	5-7 hr
C	1.25 V	1.2 A	70 mA at 3 V	10-14 hr
D	1.25 V	1.5 A	70 mA at 3 V	up to 18 hr

4AA Solar Nicad Charger

This is our most popular solar battery charger. We've sold over 8,000 of these in the last two years. We've even seen them selling in other catalogs for up to $20! It couldn't be more simple to operate - just leave the unit in the sun for 10-15 hours and your dead AA nicads will become fully charged! The most common use is for "Walkman" portable tape players. With a set of AA cells in the Walkman and a set of batteries always in the charger, you'll never be without sounds.

50-201   4 AA Charger $14

Metered Solar Nicad Charger

This great solar battery charger goes one up on the four AA charger: At one time you can charge four AA's, and two C's or two D's. It will also partially rejuvenate alkaline batteries for a short period of time. It's the only solar charger that comes standard with a testing meter that shows battery strength (for AA's only) at a glance before, during, or after charging. The MSNC comes highly recommended.

50-203   MSNC Nicad Charger $26

12-Volt Charger

The 12V charger will recharge AA, C, and D nickel-cadmium batteries in 6-10 hours from your 12-volt power source. It will charge two AA or two C or two D batteries at one time. It comes with a 12-volt male cigarette lighter-type plug to go into any 12-volt socket.

50-214   12V charger $24

Solar Button Battery Charger

Our solar button-battery charger is designed especially for charging Mercury button-type batteries. Button-batteries typically power hearing aids, cameras, alarms, calculators, watches, hand held electronic games, and many other appliances. Of the 20 million hearing impaired people in the USA, many use as many as two batteries per day! Charging time is 2-6 hours depending on light intensity, size, and condition of battery. The unit is supplied with a suction cup for window handling, plastic carrying bag, and a complete instruction booklet. It measures 2-1/4"L x 1-3/8"W x 3/8"D.

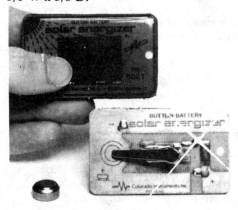

50-205   Solar Button Battery Charger $19

Nicad Batteries

We carry various brands of nicad rechargeable batteries from time to time. Most commonly we carry the Panasonic brand and have had very good luck with them. Read the note above regarding nicad charging before purchasing. The "AA" batteries have a capacity of 500 mA, the "C" and "D" nicads have a capacity of 1200 mA (1.2 amphours).

50-101   AA nicads (ea.) $2.75
50-102   C nicads (ea.) $4.75
50-103   D nicads (ea.) $4.75

Important Note on Nicad Charging: *Nicads have a memory.* Typically one charge will last about 1/3 as long as comparable alkalines (but you get 1,000 charges!). When you receive them they have been "asleep" on the shelf. They need to be awakened by getting fully charged and discharged for 3 to 4 cycles before their memory is "stretched" enough to hold a full charge. They should then always be fully discharged before each recharging. Customers new to nicads should try this simple method before calling us and saying your batteries are defective!

Environmentally Friendly Rechargeable "AA" Batteries

Our brand new *Nickel Metal Hydride* batteries are often referred to as *green* rechargeable batteries. They don't contain any toxic materials such as cadmium (like standard nicads), lead, mercury or lithium. They're considered a significant breakthrough in battery technology. At 1,000 mAh (1.0 amphours), they will store more than twice the capacity of our standard rechargeables which are rated at 500 mAh. They will also last twice as long. These batteries are the answer for those that get frustrated at the usual 1:3 ratio of standard rechargeables to alkaline batteries. The nickel metal hydride batteries will last roughly 2/3 as long per charge as an alkaline but have an infinitely longer life. Further, these new batteries don't have the undesirable memory effect which exists with conventional rechargeables.

50-105   Rechargable "AA" Batteries - NMH (each) $8

Real Goods T-Shirts

Marathon Man T-Shirt

This T-Shirt was the hit of Earth Day booths everywhere. It features cover artist Winston Smith's rendition of a marathon man leaping over the earth carrying a solar panel, with the words: Real Goods, We Can Change the World. Our local T-Shirt printer Bob Perkowski said it was the nicest T-shirt he's done in 10 years! Very vivid colors. Specify S,M,L,XL.

90-605 Spring '90 Real Goods T-Shirt $12

Sunrider Expedition T-Shirt

Our **SUNRIDER EXPEDITION** design is the official T-shirt for this incredible expedition beginning in Spring, 1991 partially sponsored by Real Goods. The 52,000 mile voyage around the world will be in a 24' boat powered entirely by renewable energy. The expedition hopes to raise global environmental awareness by offering educational programs in more than 200 cities in 80 countries over a 2-1/2 year period.

90-671 Sunrider Expedition $14

Cavemen in Tool Heaven T-Shirt

Featuring another cover gem by Winston Smith, our ever-popular Spring '89 *Real Goods News* cover has the enlightened Neanderthals in tool heaven with the words: "Real Goods Brings You Out of the Stone Age Into the Solar Age." It's printed on 100% cotton and comes in S, M, L, and XL.

90-600 Real Goods Caveman T-Shirt $12
Be sure to specify size

Earth - Love it or Lose It

Our new **EARTH - LOVE IT OR LOSE IT** T-Shirt is a striking image of the earth as a delicate balloon with a child fragilely perched on top wondering why the air is coming out. An innovative artist in Colorado makes this beautiful air brushed design for us. 100% Cotton. Specify S, M, L, or XL.

90-672 Earth - Love it or Lose It $14

Pocket Falcon

A falcon is perched on your pocket! A great illusion, this shirt is lots of fun to wear. Also known as a sparrow hawk, the kestrel is America's smallest and most colorful falcon. Our best all time seller in T-shirts! 100% cotton - Specify size (S,M,L,XL)

90-631   Pocket Falcon $14

Pocket Parrot

A blue and gold macaw has perched on your pocket. *Ara araruana* is known for its dazzling plumage and size - the male can reach a length of 34". A fruit and nut eater, this parrot is found in the rain forests of Central and South America. 100% cotton - Specify size.

90-632   Pocket Parrot $14

Pocket Iguana

A green iguana has scaled your T-shirt! Wear this design only if you don't mind drawing attention to yourself. Our model, Stymie, was one of our most interesting subjects. Found in Mexico, Central America and tropical south America, *Iguana iguana's* favorite perch is on a tree limb overhanging water. 100% cotton - Specify size.

90-633   Pocket Iguana $14

Pocket Kitten

Now you can take a warm, soft kitten with you wherever you go. The domesticated *Felis catus* is the result of 3500 years of man's cross-breeding of European and African wildcats, different races of *Felis sylvestris*. 100% cotton - Specify size (S,M,L,XL)

90-634   Pocket Kitten $14

Red-Tailed Hawk

Backlit by the sun, the flash of pale rust through the tail of this soaring hawk is unforgettable. One of America's most common hawks, *Buteo jamaicensis* preys on rodents and has variable coloration in different regions of the country. 100% cotton - Specify size.

90-652 Red-Tailed Hawk $14

Wolf

The WOLF, larger than the coyote and fox, *Canis lupus* is also called the Gray Wolf or Timber Wolf. Due to shrinking habitats and the persecution by man, significant populations of wolves have been reduced...the entire Northern hemisphere was once their range.

90-659 Wolf $14

Crude Addiction

Here is a brand new T-shirt from Randy Johnson that highlights America's obsession with foreign oil, and hesitation to break loose with our energy-saving products. In bold letters on the back printed in red is: "No Known Rehab for the Addiction to Petroleum." Here are excerpts from the six paragraphs on the back written by John McKenzie:

   To be addicted is to "give oneself up to some strong habit." I've never seen any stronger addiction than humanity's complete giving up to the quick fix of petroleum. It's only been since Henry Ford that we became addicted. We are vampires on the earth, sucking dry our limited, finite resource so we can move so much faster, feed our habit, get our fix, maintain our high. We are hooked on a primitive fuel, a dirty fuel; and the earth shows tracks like a junky's arm; and the earth struggles to live in the squalor and degradation and petrol-grime and contaminated atmosphere. As we all well know the petroleum age will end. 100% cotton - be sure to specify S, M, L, or XL.

90-669 Crude Addiction $14

Green and blue ink on white
90-608 Give Peas a Chance $14

Multicolor on light blue
90-609 Love Your Mother $14

multicolor on green
90-606 Harvest the Sun $14

100% Cotton T-Shirts
Be sure to specify S, M, L, or XL

White background - Two-sided
90-661 Animal Chain $14

White background
90-662 Birds of the Tropics $14

White background - two-sided
90-663 Marine Life $14

White background
90-660 Great Cats $14

Mint background, two sided.
90-664 Charging Elephant $14

90-665 Earth Day, Everyday $14

Blue Leopard on White
90-673 Leopard $14

Black on white
90-611 Mental floss $14

Chapter 8
Electric Vehicles

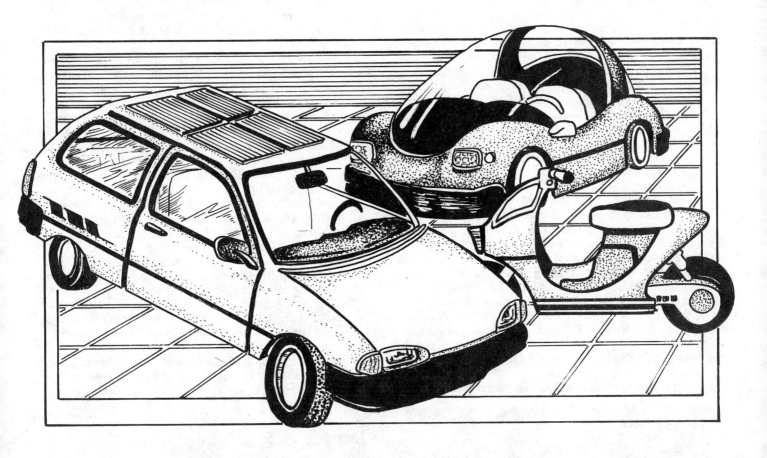

Introduction

We at Real Goods have a strong history in the support of electric vehicles and in the encouragement for the phasing out of the internal combustion engine and its resultant environmental devastation. We sponsored the Stanford solar car, the SUnSUrfer, at the 1990 Solar Energy Expo and Rally in Willits, California. To our delight, our car was victorious in the first ever Tour de Mendo race! We have greatly expanded this section of the Sourcebook even though the selection of products that we're able to offer you is extremely limited!

As electric vehicle techology continues to come of age, we intend to further expand this section and the products we're able to offer you. As far as products go, at present all we can offer is the Sinclair C5 and the Electric Motorcycle. Several manufacturers we're working with have EV's in the prototype stage which we expect to introduce in the next few months in the Real Goods Catalog. We are also working with several vehicle conversion concerns that hope to have electric vehicle conversions available very shortly. We intend to remain on

the cutting edge of electric vehicle development in the future. By focusing here on education, we hope to do our part to encourage manufacturers to develop electric vehicles for the coming market. We get close to twenty inquiries per week from folks ready to buy - it's time to get out of R&D and into production!

We want to convey our sincere thanks to the folks who helped put this section together. First, thanks to Michael Hackleman, a pioneer in bringing electric vehicle technology to the forefront. Michael has generously provided most of the material for this section (much of it is copyrighted ©), and continues to spend most of his time working to promote EV consciousness. His new publication <u>Alternative Transportation News</u>, should serve as the forum for electric vehicle networking. Thanks also to Joe Stevenson who has helped in the editing of this section and has provided articles and insight through his publication <u>Solar Mind</u>. Thanks to Creighton Hart Jr. for his fine introductory piece which leads off this chapter.
JS

Electric Vehicles, Past & Present

The following article was submitted by Creighton Hart Jr. formerly of Hawaii. It presents a good overview of Electric Vehicles past and present.

Can the Earth survive our addiction to oil and the automobile? Will we survive it? Since 1950 the worldwide population of autos has risen from 50 million to a staggering 500 million, more than a tenfold increase. Even with better gas mileage, consumption of crude oil continues to rise, peaking at over 225 billion gallons a year in 1990 - an eightfold increase in consumption since 1950. Cars are the largest contributor of greenhouse and ozone gases. Auto population is expected to double as Eastern European, Soviet and Asian markets open in the next 10 to 20 years.

So ingrained is our habitual addiction to the automobile that the current "crisis" and military excursion in the Persian Gulf can be seen as a direct result of our inability to shake our chemical dependency on oil. Before the latest escalation of troops in September 1990, the United States Department of Defense was subsidizing the protection of the U.S. reflagging of Kuwaiti oil tankers at a cost of $50 billion a year to maintain relatively low prices on Persian Gulf oil for U.S. and allied markets (Japan, Germany, France.) How much will the Persian Gulf crisis and troop build up cost? Nobody knows. This war-like stance could be softened if we could kick our unhealthy addiction to oil. But it doesn't have to end our love affair with the automobile.

Oil and automobile statistics are storm clouds hanging over our future - either we accept the reality or we may die in the storm. Society has reached a moment of transition where it must let go of its oil addiction and seek a new dawn and a new way. Alternative solutions do exist, and they are as bright as daybreak after a stormy night.

Electric cars are not new. The first one was built in 1836 by Thomas Davenport, before the early 1900's they outnumbered gasoline powered vehicles. What's new is the advent of photovoltaic (PV) cells which convert sunlight into electricity rather than drawing solely from the batteries. In theory, a non-polluting solar car can be driven continuously without having to refuel, providing the sun is shining - that would be quite a nice feature as the cost of petroleum continues to escalate and places further environmental strains on the planet.

The solar automobile is different from the autos we are accustomed to in many ways. They are the most aerodynamic and lightest vehicles on the road. Their design includes the features of most autos; a chassis - frame, wheels and suspension system. Instead of gas tank, filler tube (and filling stations along the road) substitute an array of photovoltaic solar cells, amp and volt meters plus a battery. Instead of a huge, gas-hungry internal combustion engine, try an electric motor rated from 1.5 to 6 HP. In comparison to the engines of the 80's and 90's, designed with over 300 parts working at heightened temperatures (with all the associated costly repairs and their ever-more complicated recirculating exhaust and control systems) PV engines are simple, consisting of

A very early electric vehicle.

three or four moving parts, and easily replaced in ten minutes with just a wrench.

Initially, the majority of electric vehicle development, design and research has been done by a collection of grassroots tinkers, inventors and visionaries, placing their emphasis on solar race cars, working on the same problems that challenged Karl Benz and Henry Ford a hundred years ago. Australian Hans Tholstrup has been one of the most active and enthusiastic cheerleaders for solar auto technology. Inspired initially by the 1981 flight of the Solar Challenger over the English channel, Tholstrup and team tackled a solar transcontinental land journey. With a $45,000 grant from British Petroleum of Australia, the project developed a 276 pound auto dubbed the Quiet Achiever. The goal was to traverse Australia from Perth to Sidney in less time than the original motor vehicle trip of 28 days by Francis Birtles in 1912. And that they did, proving in 20 days the viability of solar power and its uses within the automobile sector.

...design and research has been done by a collection of grassroots tinkers, inventors and visionaries.

After that success, a plethora of designs, advancements and competitions appeared, and industry (car manufacturers General Motors and Honda, Siemens Electric of West Germany and Arco Petroleum, both PV manufacturers) stepped forward with sponsorships and research results. Tholstrup organized the first Pentax World Solar Challenge, held in Australia in November of 1987, which put solar vehicles on the map. The race, from Adelaide to Darwin, followed a South to North oriented course to maximize the sun's rays. It was won by the Sunraycer designed by General Motors. On the second day of competition Sunraycer pulled ahead of the rest of the field, and (fortuitously) ahead of a thick cloud bank and dust storm which shaded and hampered the performance of the remaining 21 cars. Sunraycer miraculously managed to stay just ahead of the cloud bank. According to Jon

Tennyson, the designer of the 1987 Mana La solar car sponsored by the late shampoo magnate Paul Mitchell, "We're a better car than G.M. but an act of God foiled us, but we'll be back." Some of the cars in this race were still on the road 2 months after the winner arrived in sun-scorched Darwin.

Tennyson got his return match at the 2nd Pentax Solar Challenge, held November 11 - 22, 1989, when a 34 car field selected by invitation or from solar races held worldwide in the last year. This event was dominated by a Swiss team, The Spirit of Biel. Traversing the 3007 mile Australian desert course in 47 hours and 10 minutes, averaging 40.7 mph, the Swiss entry regularly clocked speeds of 65 mph plus. This university-built auto out-performed Honda's Honda Dream, the prerace favorite, by a winning margin of 8 hours. The remaining 18 finishers had, at the race's completion, average speeds ranging from 34 mph to 20 mph. Twenty-seven of the initial 34 autos crossed the finish line on solar power, although 9 finished during the week after the race officially ended.

The number of individuals involved and the quality of technology has improved dramatically over the last three years. The Spirit of Biel credited its victory to the new laser grooved solar cell developed by professor Martin Green of the University of New South Wales. These new laser grooved cells have an efficiency of 18.5%, almost equal to the 20% output of the gallium arsenide (GAs) cells used by the Sunraycer in the first Solar Challenge, and have several advantages over the toxic and costly GAs cells, making them more practical for widespread public use.

The most recent big-time race in this country started July 9, 1990, at Epcot Center in Buena Vista, Florida, and followed a 1650 mile south to north course through the heartland of America, to end 11 days later at the G.M. Technical Center in Detroit, Michigan. Among the participants was Sunraycer, co-sponsored by General Motors, the U.S. Department of Energy and the Society of Automobile Engineers. Other competitors came from thirty-two North American university teams, representing two and four year schools from 19 different states, Canada and Puerto Rico. The prizes went to the top three positions, which bore with them invitation to the World Challenge. The University of Michigan's Sunrunner won, averaging 22.5 MPH, nipping Western Washington University's Viking XX by minutes. Both of these cars went on to place in the top five at the world solar challenge.

The innovative designs and race strategies employed were as varied as the school's geographic representation. Team captain Paul Jeran of Rochester Institute of Technology calculated that a pound in auto weight equaled an additional five minutes in race time. With this in mind, teams relied on lightweight materials like Styrofoam, Kevlar, graphite fibers, and NOMEX (a featherweight honeycombed materials by DuPont). "The real technology is in design, it is hoped the new design approaches employed so successfully by these solar cars can help make electric cars more practical and automotive technology in general more efficient," says G.M.'s Jerry Williams. "These kids are on the cutting edge."

While the future of solar automobiles is as bright as the sun, according to the optimistic visionaries, automotive industry analysts paint a cloudier picture. Despite successful solar races and advancements, General Motors spokesmen predict solar cars won't be available for several decades, citing lightweightedness and lack of muscle power as two prime unmarketable reasons. Independent builders and solar proponents are out to prove them wrong, and it seems to this observer they are accomplishing this amazing feat. Besides solar race car designs, several independent manufacturers offer solar vehicles for purchase on a limited availability. In Switzerland and elsewhere, solar vehicles geared for urban-innercity commuter market are appearing. Tennyson has converted two conventional pickups to solar using abandoned jet fighter starter engines, one of which has hauled more than half a ton up the steep hill that leads to his farm on the island of Hawaii. The other has a low gear range ratio of 37:1; Tennyson says "This'll climb a tree." He is also working on a completely new vehicle, the Elexes, planned as a production vehicle. Tennyson admits that speeds in solar vehicles are not adequate for today's highways, but foresees this changing rapidly. In the next two to five years, advancements in engines and batteries will be the prime factors. Dr. Fred Sidler, head of the Biel University Engineering School, believes that a commercial solar vehicle could be developed for widespread use within five years.

One design package disclosed by a solar thinktank stretches beyond just solar autos to include changes in our infrastructure: while we convert to solar cars we convert our homes and offices to solar as well. Most urban drivers use their cars most during morning and evening commutes, periods of low radiant solar output and collection. We solve this problem by recharging our car at home and at work. At work, parking lots would be oriented to make best use of the sun. Collection panels could expand to shade autos while increasing the collection surface.

Our perception of automobile design will be refined dramatically as solar autos take to the street. Instead of the large 1 or 2 ton metal behemoths designed to accomodate 4 to 6 passengers but carrying only one 70 percent of the time, we will see 1 or 2 passenger vehicles. Aerodynamic shapes and seating designs, reduced sizes (by as much as forty percent over current autos) will ease the burden on our crowded urban freeways and parking facilities. A 1990 Green Peace study stated half the land in the average American city and two thirds of the land in Los Angeles is devoted to automobiles. Newer designs may mitigate the need for constructing larger freeways and parking facilities; resulting savings can be used for solar research or humanitarian needs.

These new cars will be similarly along the line of GM's prototype Lean Machine, a highly acclaimed 3

wheeled 2 passenger vehicle which leans into corners and is powered by a 250 MPG internal combustion engine. Showcased first in 1984 at the Epcot Center's science exhibit, GMC claimed that the public wasn't ready for this car, and let the concept die. Perhaps a war over oil, gas price escalation, deteriorating atmospheric conditions, consumer outcry and education will change the public's mind. Letters of public support of such vehicles may help the designers and finaciers understand that we want them. Let's help them get their priorities straight.

GMC does deserve credit for its work on efficient electric vehicles. In 1991, 500 of their 'G-Vans,' an electric van that can travel 60 miles between charges will hit the streets. By the mid Nineties, plans include a battery powered model, the Impact, with greater charge capabilities and range.

The sun has risen on the solar autos and is setting on the petroleum gas guzzlers of an earlier era. Problems with solar cars will be overcome. Compared to the problems internal combustion autos have generated -

The sun has risen on the solar autos and is setting on the petroleum gas guzzlers

ozone deterioration, massive consumer debt, resource inefficiency, and global belligerence - the inability of solar cars to reach highway speeds becomes less important. I believe that the automotive industry will once again be driven by futurist individuals with the foresight to look for solutions to our current situations. Remember, the first internal combustion autos designed in 1885 by Karl Benz of Mercedes Benz had speeds of less than 15 MPH and were widely ridiculed. It took another 20 years before Henry Ford found ways to mass produce autos cheaply, and still 40 more years before Americans could afford to put a car in every garage. - **Creighton Hart Jr.**

Slowing Down: Freedom from the Car

The following article by Joe Stevenson, editor of *Solar Mind*, is an excellent piece on breaking our addiction to gasoline-powered travel.

A concern of mine is that we, who publicly demonstrate our care for the environment, not be found to be hypocrites. We are often seen cursing the insensitivity of "Big Oil" while thriving on its goods and services.

There is no need for this contradiction to remain if we are willing to 'change lanes.' I believe that we may have confused our freedom to go, with the freedom to be. The renewal and vitality we receive from contact with the pure and awesome ocean is highly cherished, but strangely, we also discover similar uplifting qualities while traveling in the 'gasmobile.' While we wish to preserve the purity of our environment, we also enjoy the independence and exhilaration gleaned from the gasoline car. It seems to me that something is strange when both cars and nature have equal meaning in our lives. In simplest terms, we are actively creating the pollution we struggle to avoid.

When we thrive on an activity that is inherently self-destructive, we must realize that such a contradiction is only possible through subtle self-deception and through overriding certain of our bodily sensitivities. Just as we have become collectively conscious of the cigarette disease, so we must now become awakened to our car and gasoline addictions. We have the capacity to rationalize the smoke coming from a cigarette, and so also the poison from the tailpipe. The next ride and the next puff go on unconsciously. The question seems to be whether we can collectively awaken to our blindness and self-destructive tendencies before we destroy our very life support systems on Earth.

Exactly what is it that we find addictive about gasoline-powered travel? It seems unlikely that many of us are in love with the chemical composition of gasoline, or its volatile fumes. It is more likely that our love stems from what petroleum makes possible. As a society, we have become addicted to 'going.' Driving means freedom, the exhilaration of acceleration, privacy, detachedness, power, control, and automatic single-mindedness. It is clear that the car is highly prized by us; not many objects in life offer so much to the psyche.

The Worldwatch Institute and the Environmental Protection Agency both tell us that at least one-third of our automotive use is for pleasure; driving just for the fun of it. This very conservative estimate doesn't include the many areas of our selfishness: single passenger commuting, over-powered engines, and excessive relative speeds. (Note - from Mother Earth News, January, 1988: Tests in Denmark showed that nitrous oxide emissions - source of acid rain - doubled between the speeds of 30mph and 50mph. There was a 20% difference between emissions at 38mph and at 50mph.)

The drawing of parallels between car use and cigarette smoking may be a useful way to explore our collective addictions, for there are deep psychological attachments in both habits. Just as with the nicotine and tar of the cigarette,

we speak of getting a similar 'lift' from riding at high speeds. It is clearly this lift that is a large part of car addiction. The cigarette, we now know, causes a change in our body/mind chemistry, stimulates the adrenal gland and suppresses the immune system. What is it, then, that gives us this sense of a lift, or well-being, from driving a car?

It seems logical that something chemical may also be happening while driving. Have you ever noticed the way dogs struggle to get their noses outside the auto as it speeds along? How about the baby's predisposition to deep sleep in a moving auto? Some of us are even made ill by car travel. Here are some facts from scientific research and analysis....

There are 'G' forces (gravitational) in rapid automotive acceleration, less but similar to rockets and airplanes. The body's response is psycho/physical in that the imbalance created by sudden propulsion calls our 'fight/flight' chemistry into action. That adrenal response makes some of us 'high,' and others fearful. In this response condition, back and neck muscles tense, blood pressure rises, and we are prepared for action.

The biggest problem is that we have become unconscious of our physical responses to travel, and just as the initial reaction to the first cigarette is usually overridden by our desire to be cool, so it is with the car. We now live our lives in this chemical fight/flight response while wondering why everything keeps moving faster and faster in and around us. This stimulated state has become so familiar that we must now crave it to feel in balance. There is no doubt that this process

---

At least one-third of our automotive use is for pleasure - driving just for the fun of it.

---

is slowly killing us. The energy gained from adrenaline is not ours in a state of relaxation. We have become hooked on the speed rush and the reactive freedom it offers; but we must change if we are to have healthy planet. This change must be rapid, as we are running out of time.

I am not saying all of us must suddenly stop driving, thus bringing our society to a complete halt, but that we must now begin the process of weaning ourselves from the gasoline car and the fast life. How? You ask. Remember the gas crunch a few years back? When gas was scarce we heard about alternatives like alcohol, propane and hydrogen fuels; and there were even people building and selling electric cars. So what happened? Why is nothing happening with these alternatives now? It seems that with the ending of the 'oil shortage' we generally found it easier and more comfortable to return in full force to the gas pumps. There was also the subsequent improvement in gas mileage and exhaust emissions, which gave many of us the sense that the worst was over. As a result, research and marketing of alternative forms of transportation were no longer supported financially and most could not afford to continue. Twenty electric car manufacturers were in existence in 1981 and in 1990 there are none.

The recurring justification for non-investment in renewable transportation sources and electric cars is the catch phrase "cost effective," and in "Hydrogen as a fuel isn't cost effective yet." What this means is that until the cost of gasoline grows to equal the cost of hydrogen production, let's forget about it. It is odd that the cost of gasoline is the only measure for our choice of energy. When will we begin to see that the true cost of oil use is not limited to the pump price? If we consider our health and environmental costs along with the military and other governmental costs, I believe we will find that gasoline is at least ten times as expensive and hydrogen. Even at today's gas prices, hydrogen can be produced for a little more than thirty cents per gallon more than gasoline. And hydrogen doesn't pollute or create more CO_2. It doesn't burn up engines and carburetors and other engine components. It is clear there are other ways to get around on this planet if we want to. The first step to be taken it to find the clarity within ourselves that we are ready to change. The next step is to see that if we demand vehicles that are clean and quiet, there will be a supply.

Electric cars have existed since the turn of the century and offer the most immediate option (see electric vehicle section). Sad to say, there are only ten thousand or so in the entire USA. The technology now exists (batteries, motors) to build an electric car that could travel 500-1,000 miles on a single overnight charge, and also at freeway speeds (see 750 mile electric car article). Because there has been no demand for such materials, their current cost is high. Also on the horizon are solar-powered and -charged vehicles which represent the most exciting promise (see solar-powered car section). Just think of a highway full of cars propelled by wind, sun and hydrogen; all renewable and clean.

Alcohol fuels, which offered so much promise in the 1970's, have failed the test in Brazil. They create new toxic substances (aldehydes), and do little to help the greenhouse effect. Propane is also an alternative. It isn't perfect, but it is a vast improvement over gasoline. It cuts all emissions by large percentages compared to gasoline, and is largely derived from natural gas. Conversion of existing autos is simple and in many cases costs under $1600, and much of this expense is recovered in the longevity of the engine and related parts.

Hydrogen is the purest of all the combustible fuels. It can be prepared from water at a cost similar to gasoline and can be made in decentralized locations (your home). When combusted, it gives up only water vapor. In electrolysis (separation of water) the by-products are oxygen and fresh water. The fear of explosion associated with hydrogen had been relieved with the discovery of what are known as hydrides. These are tanks which contain materials that attract and bond with hydrogen at normal temperatures, releasing at exhaust temperatures. Under rigorous testing, all fires created in and around these tanks self-extinguished (see "fuel cell" article).

It is clear that at present we have a foundation upon which to build, and that it is our mindset about the use of the automobile which stands in our way. We must learn to view the car as separate from the ego, chemical and independence functions we have come to associate with it. We must recognize that it is essentially the function of the car to help man survive. We can no longer proceed innocently in our blindness.

The freedom and wonder found in our oceans and on our earth are a reflection of the perfect relationship and interaction of natural processes. Our lives also must grow to mirror this freedom in all its aspects. To continue to use the gasoline car as we now do can only bring us to certain, collective disaster. Our cars can enrich our lives without making us dependent extensions of them. It is no longer justifiable to wait for the wizards of Detroit to save the day. The wizards are waiting for us.   **Joe Stevenson**

Alternative Engine Fuels

- Biogas is a fuel derived from the decomposition of manure and biomass in the absence of oxygen. Large amounts are required to produce significant amounts of the useable methane gas, the H2S in the biogas is corrosive, and the 2,200 psi liquification pressure makes it difficult to store and use for mobile applications.

- Propane is a relatively clean fuel and engines that burn it stay clean and emit less pollutants. There's a $1,500 (or greater) cost for converting an auto engine to propane, refueling requires visiting facilities designed to handle RV customers, and its cost is closely linked to the oil industry as a whole.

- Methanol is derived from petroleum-based processes (oil and shale). Highly-toxic formaldehyde is added to the exhaust emissions.

- Ethanol is alcohol derived from petroleum processes or distilled from biomass (crops). Acres of crops would be needed to fuel even modest driving habits.

- Hydrogen is produced by breaking down water into hydrogen and oxygen. When burned, water is the main byproduct! Solar and high-temp processes (i.e., electricity, laser) may be used. Storage is an issue, as is the complexity and cost of the production process, but this is rapidly becoming viable and affordable.

The next series of articles are written by Michael Hackleman (MH), mentioned in the introduction.

The Transportation Dilemma

The gasoline-powered ICE (internal-combustion engine) car reigns supreme. Transportation now accounts for more than 70% of oil consumption in the USA annually. The private automobile, at 41%, is the biggest single offender.

As 1991 begins, we are poised at the brink of war. The presence of the USA in the gulf is a perfect reflection of our addiction to oil. Are we willing to wage war for the continued privilege of driving one-person-one-car on crowded freeways? Tens of thousands of people die every year, victims of the wheel. Must sons and daughters die in a war just to support this habit?

More energy than man has used through his entire tenure on this planet strikes the earth every day in the form of solar energy. Solar cells right now can convert 10% of the incoming solar rays into useful form. Solar-thermal devices use the sun's energy directly, efficiently and inexpensively providing water and space heating.

Oil and gasoline are subsidized so heavily that we pay only 1/10 of its real cost to us at the pumps. Alternative technologies and energy-efficient consumption get no subsidies, and so appear impractical. Who bears the cost of cleaning up the mess of a century's use of oil, abating the pollution we experience daily, and providing health care to people sick and dying from this contamination?

Our own sun shows us the way. It works by fusion, not fission like our current nuclear technology. And showing great care for all living things, the sun is positioned 93 million miles away.

Transportation today is a matter of convenience. We want what energy does for us -- heat, light, power, water, food, and fast wheels. We haven't paid much attention to how it was done, and at what price. Our children have awakened to the inevitable horror that we long ago sold their future for a fleeting taste of this convenience.

It's time to give the alternatives a chance. Fast, efficient rapid transit. Solar power plants. Telecommuting. Needing less. Living closer to work. Wouldn't it be nice to wake up one morning to a world that did it this way?

Fortunately, there is a greater awareness of environmental issues today. We are confronted with problems on every front: pollution, resource depletion, garbage, political stagnation, recession, hunger, and homelessness. It should be clear that changes in the way we do transportation will probably yield the biggest bang for the buck. This next section will build a picture of one way that our future can look. - **MH**

Alternatives in Transportation

If you pursue an alternative fuel or technology, choose one that makes sense. No one source of energy can do it all, all of the time. Of course, the energy sources should be natural, renewable, sustainable. Also, a mode of transportation that tries to fit every situation doesn't do any one of them very well. Adopt an arsenal of transportation methods. For example:

● **Rail technology**, particularly light rail and electric rail, is improving overall and is ideal for commuting. If not available, support projects that will bring it in.

● **Carpooling** most of the time saves fuel, fatigue, money, and hassles with parking. This is particularly beneficial with work commuting. The hassles of carpooling are well known; good communication makes it work.

● **Telecommuting**, or working at home through communication lines, works for a large number of jobs and minimizes unnecessary commuting. Assess the possibility of using this technology for your job, encourage and support co-workers who can use it.

● **Bicycling and walking** can account for a larger percentage of our transportation needs.

● **Mopeds and scooters** are good fair-weather options as gasoline prices climb, and are less expensive to maintain and insure than automobiles.

● **Telephones and mail service** deliver direct to the front door of most homes. Both offer the most cost-effective way of getting what you want -- services, goods, etc. -- without driving a car.

● **Electric Propulsion.** A very different approach to transportation is to use electric propulsion. Bicycles, mopeds, scooters, motorcycles, cars, vans, and trucks can be converted to run on electricity.

There is a price tag for convenience and speed. Planning ahead eliminates the thoughtless abuse of the automobile, and a reprieve most yearn for, from a fast-paced and wasteful lifestyle. - **MH**

The Electric Vehicle

What is an electric vehicle, or EV? An electric vehicle is one that uses an electric motor instead of an internal-combustion engine, and batteries instead of a fuel tank and gasoline. The motor is a larger version of the one that powers your hair-dryer, or the refrigerator in your kitchen, or the tape player in your car. The batteries are similar in size and shape to the one used to start your car's engine, only there are many more of them.

The energy of the battery pack is routed to the motor through an electronic controller. Housed in a small black box, this works like a light-dimmer switch (or the speed control on the electric drill in your garage) — smoothly delivering power to the motor and controlling its speed.

Driving a car that has been converted to electric propulsion is virtually identical to driving one that has a gas-powered engine. The same operator controls are used -- accelerator and brake pedals. So, your foot pushes on the accelerator to control the speed of the vehicle, and pushes on the brake pedal to stop it. You will notice differences, though. You don't have to start an electric car like you must start an engine. The accelerator starts up the electric motor, and you are on your way. You don't have to feel guilty about not warming up the motor, either. Unlike engines, electric motors work best when they are cold.

Stopped at a signal light or stuck in traffic, there's another difference, too. If your foot is not on the accelerator pedal, the electric motor is not running. That will seem strange at first but you will get used to it quickly. Also, since the electric motor is not running, it is not consuming or wasting power.

There is a long list of other advantages to be found in the electric vehicle. For example, since no fuel is consumed, there is no exhaust pipe. That's right — there is zero pollution when an electric vehicle is operated.

Parts that move wear out. Engines have hundreds of these. The electric motor has ONE moving part. And this motor does not use gasoline, lubricating oil, or coolant. Electric cars don't need to be tuned, timed, or adjusted. They do not have points, plugs, carburetors, oil pumps, fuel pumps, oil filters, air cleaner filters, fuel filters, fuel injectors, water pumps, generators, voltage regulators, starter motors, starter solenoids, and fan belts. It doesn't matter if you don't know what all of these things do. Since they aren't needed in an electric vehicle, they can't break down and bring your car to a halt at an inopportune time. Or take a bite out of your pocketbook.

In an electric car, the requirement for a smog certificate is a thing of the past. If it hasn't already occurred to you, there's no more waiting in lines at gas stations, either. Or wondering what you'll have to pay for gasoline today if there's a crisis halfway across the globe.

If electric vehicles are so good, you might ask, why aren't they available to the general public? The answer to this question is complex. The automotive industry says that electric vehicles are low-performing and limited in range. They claim that, since the battery pack must be replaced every 2-3 years, electric vehicles are too expensive to be competitive with gasoline-powered cars. They claim that technological breakthroughs in batteries and motors are required before the electric vehicle will meet performance standards that the driving public has come to expect from automobiles.

Although there are limits to the range of a practical electric vehicle, current technology permits travel at legal speeds as far as 50 to 60 miles with a fully-charged battery pack. This is well within the average trip length for cars daily. Think about it! The daytime range of the electric car is easily doubled, too. An onboard battery charger permits worksite re-charging from a standard wall socket. So, the overnight charge gets you to work, and the recharge during work hours gets you home.

A 100 to 120 mile range will meet the needs of more than 90% of the driving public, and 80% of the needs of the remaining 10% of the population.

Electric vehicles are also quite energy-efficient. For each barrel of oil, 3-5 electric vehicles can be powered over the same distance as one gasoline-powered car when the oil is burned in a utility-size power plant to produce electricity.

It is also much easier to monitor and control the pollution of a single smokestack than tens of thousands of tailpipes. Also, utility scale power plants operate 2-3 times more efficiently than car engines.

Finally, the recent demonstration of utility-scale solar-powered plants is a genuine alternative to oil, coal, and nuclear power in the generation of electricity. Periodic battery replacement, even when included in the operating costs of EVs, still competes with the gas-powered car. Why? The electric vehicle has minimal maintenance requirements. Gasoline-fueled vehicles accumulate repair and maintenance bills with more frequent visits to the shop over the same number of years. Finally, gasoline itself is heavily subsidized (by a factor of 10 over what we pay at the pumps) and does not include the cost of the pollution or health care that results from its use.

Industrial leaders claim that the lead-acid batteries used in most electric cars at this point are too primitive and low-performing. They rest their hopes on more exotic types now under development. This casual rejection ignores that a whole industry exists right now to recycle this material at a fraction of the cost and toxicity of other battery technologies. Oil-based technology continues to impact the environment while we wait for breakthroughs and deplete our precious resources. The electric vehicle confronts and addresses most of these issues NOW. That the automotive industry does not share this vision suggests that they have different priorities. - MH

Commercial Electrics

The oil crisis in the '70's brought about many electric vehicle prototypes. Advanced primarily by the automotive manufacturers through DOE funding, their promises of production died when the price of gasoline dropped.

The steady rise of gasoline has again prompted a hard look at electric propulsion. Currently, Japanese auto manufacturers appear to be the most convinced that there is a market in electric vehicles. This is also a good market for entrepreneurs and industry that is already manufacturing components for the automotive trade. Two of the most visible electric vehicles are the Impact and the Horlacher Electric.

The Impact

It was a good sign for the 90's that General Motors opened in 1990 by unveiling its bold new prototype, IMPACT. This is a car that most environmentalists would enjoy - it's pollution-free, quiet, uses 1/5th the energy from fossil fuels as regular cars, and has a recyclable battery pack.

The IMPACT boasts a respectable 120 miles range — whether you're cruising along at freeway speed or in the stop-and-go "urban driving cycle". Here, the aerodynamists and stylists have finally gotten their act together. The IMPACT has a competitive look, yet boasts the lowest drag coefficient in the history of commuter vehicles. (Low drag means the vehicle is slippery moving through the air, consuming less energy to go the same speed.) The battery pack is basically off-the-shelf technology — lead-acid batteries that pack a lot of energy, are low in weight, and recharge quickly (2 hours for 60% of capacity, 4 hours for 95%). Since the batteries are sealed (recombinant), no watering maintenance is required

The IMPACT prototype is a concept advanced and built primarily by Paul MacCready and the gang at AeroVironment. This is the crew responsible for the gossamer-series of human-powered aircraft!

The Impact is a high-performance electric vehicle. The footage shown during its introduction proves that. From a standstill, the IMPACT *blew away* a Miata and a Mazda 300ZX in the quarter mile. Obviously, the design was intended to blow holes through the perception that electric vehicles are golf carts with pretty bodies over them.

The big question on GM's mind seems to be: Is there a market for electric-powered cars? That GM is unsure about this is revealed in the way they announced it, immediately underselling both the vehicle and technology. GM claims that operating expenses for the Impact are double that of a standard ICE (internal-combustion engine) car, that the Impact is handicapped with periodic

battery replacement, and that it is not a get-in-and-drive-anywhere car. The Impact also needs an infrastructure to recharge and service this new type of car. GM appeared genuinely surprised at the excitement and warm reception given by industry, government, and the general public. They probably thought that everyone was going to laugh. Nobody did!

Still, GM faces an interesting dilemma: Its marketing effort must tiptoe through any comparisons that could hurt its greater investment in ICE-based technology. Still, this is potentially the decade of the environment, and no company is better poised to commit to the effort. Will GM hold out for market share and delay aggressive tactics until other car companies have tested the waters?

Horlacher Electric

With gasoline in Europe at $3-5 a gallon, Boris Horlacher's electric is a welcome sight at the annual Tour de Sol events, sprinting up alpine grades with the best of them. While Boris refuses to go into mass-production until he's satisfied with the design, folks who have driven one of the early prototypes say that the Horlacher meets ALL of their expectations. Propulsive power comes from a 12hp Brusa drive — the motor-controller used in all the winning Tour de Sol electrics — and it pushes this peppy 3-wheel, twin-seater at speeds up to 45 mph for an hour or more. There's even a 4-wheel version of the Horlacher now. DOT (Department of Transportation) regulations and Federal Safety standards will play havoc with introducing these small, lightweight vehicles into the USA. How do we get 2-3 tons monster vehicles off the road to make it safe for environmentally-sound cars like the Horlacher? - MH

Solar-Powered Cars

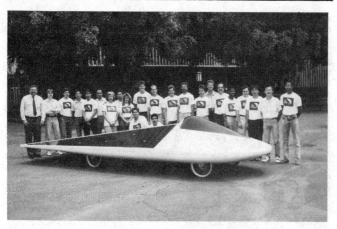

It excites the imagination. Wheels spinning from light. A vehicle achieving speeds of 35 mph on the sun's energy alone. Design so good that, with careful attention to aerodynamics, weight, and efficiency, one horsepower goes a long, long way. Combined with its zero pollution nature, no other system of propulsion offers this unique characteristic.

The past six years has seen a flurry of races and rallies to demonstrate the potential of both electric and solar-powered vehicles. [See Solar Cars Are Here!]

Anyone who has worked on a solar-powered car knows that they are not very practical. The combination of electric propulsion and solar technology is an unlikely marriage, expensive and fragile when the panel is actually mated to the vehicle. However, a solar-powered car does turn people's heads. Anyone who walks up to a solar car for the first time will find that their inner child is alive and well!

Solar cars, then, demonstrate the practicality and simplicity of electric vehicles in three ways. Spectators are frequently amazed by the acceleration and speed these vehicles achieve working with less than 2 horsepower. Also, the solar component of the car drives home the idea that solar is a power source that works anywhere under the sun.

Finally, a solar car project is a fun way to design solutions that consider a scope beyond mere machinery. Coupled with lightweight, aerodynamic bodies, electric propulsion is an elegant solution to the dilemma of smog-ridden basins throughout the U.S. continent, declining oil reserves, and other issues that affect the quality of life.

At the university level, a solar car project sensitizes students to environmental issues by exposing them to technologies based on a non-fossil fuel. In this way, they can bring classroom engineering to bear in designing solutions to today's problems.

At this point, racers are the only electric vehicles that use solar energy as a primary source of propulsive power. EVs designed to operate on the road may have solar cells attached to them, but it would be more accurate to say that they are solar-assisted. This definition will become clearer as the reader delves into the world of solar-electric cars. - **MH**

How does a Solar-Powered Car Work?

In principle, solar-electric cars are fairly simple. Sunlight striking solar-electric modules on the vehicle is converted directly into electricity and routed to the electric motor. Typically, a small set of batteries is carried onboard to help with high loads - acceleration, hill climbing and higher rates of speed. The batteries also assist with propulsion when the panel can't receive much sunlight, i.e., during the early morning and late afternoon hours, and through overcast and shaded conditions. In a racing format, of course, the energy that comes out of the battery must be replenished with electricity generated from the sun's energy.

There is a maximum of 1000 watts of energy in a square meter of area on the earth's surface when exposed to sunlight on a clear day at high noon. Commercial grade cells yield efficiencies of 12-14%, yielding approximately 100 watts of electricity per meter after losses. Racing electric vehicles get a 25% overall increase in power from their panels by using more expensive solar cells. With a 17% conversion efficiency, space-quality cells produce 120 watts or more per square meter.

Drivetrain efficiency - including solar panel, power conditioners, motor and controller, and transmission - is essential when the power source is so restrictive. Aerodynamics, or slicing through the wind without expending much energy to do it, is also vital. Winning solar-powered cars are lightweight. Accelerating the vehicle to some speed, particularly from a standstill, consumes much more energy than it takes to maintain a steady speed. Climbing a hill, even at a constant speed consumes power in big gulps.

Successful solar-powered cars utilize a peak power tracker (PPT), like the Maximizer and LCB. High temperatures, varying light intensity, and changes in the panel's orientation all affect panel output. The PPT's job is to maintain the best match between solar input and the vehicle's battery or motor bus voltage. - **MH**

Australian Electrathon champion Clark Beasley is promoting the first International Electrathon "The 1990 Electrathon Cup" which was the first of many endurance races for "Formula E" electric vehicles held in California in 1990. Pictured is Clark Beasley in his "Australian Challenger."

Solar Cars Are Here!

World awareness of the potential of electric vehicles, particularly one's that could be powered from the sun, started in the mid-80's. Here's a brief summary of this history to date:

1985 - Tour de Sol. Switzerland starts the popular Tour de Sol, a race event that challenges the vehicles over the course of 5 days, including a stretch over the alps! The event is held annually, and attracts entrants worldwide. The first race drew a crowd of 500 spectators, the 2nd saw a gathering of 3,000, and the third drew a crowd estimated at 20,000. Today's races draw many tens of thousands of people.

1987 - World Solar Challenge. Australia expands the solar racing circuit in a grand scale with the World Solar Challenge, challenging its entrants to a transcontinental race, north to south, of over 2000 miles across the starkness and heat of the outback.

1988 - Solar Cup USA. California hosted the Solar Cup in Visalia, launching the first solar car race in the USA.

1989 - Race veteran James Worden piloted his Solectria 5 on a 22-day transcontinental (west-to-east) USA run.

1989 - Tour de Sol USA held its first USA event, and provides organizational support to similar interests all over the country each year.

1990 - Sunrayce, July. The GM-sponsored Sunrayce pits vehicles from 32 colleges and universities against each other and the elements in an 11-day, 1800-mile transcontinental run through 7 states from Florida to Michigan.

1990 - SEER (Solar Energy Expo and Rally), August. A gathering of electric vehicles and alternative energy technology and products from the Western USA. The first annual Tour de Mendo race was won by the *Real Goods*' sponsored SUnSUrfer from Stanford University.

1990 - World Solar Challenge, Nov, 2nd transcontinental Australian race. Over 40 entrants participate, with shorter completion times recorded, and record crowds.

Are Solar Cars Practical?

At this time, there is nothing inexpensive or practical about solar-powered transportation. Add up the cost of a racing solar-powered car and you may wonder why I fume when I hear people claim that a few solar modules will let you zing down the road at 55 mph to work. Putting solar modules on an electric vehicle doesn't make the car a solar-powered vehicle. Solar-assisted, yes. Solar-powered, unlikely. *Currently, a 48-watt module costing $300 and charging in perfect weather for 7 days will only store enough electricity to help a lightweight EV about 5 miles down the road at 45 mph.* The addition of solar panels on electric vehicles makes them fragile, bulky, and prone to theft.

To be solar-powered, a car does not need the panel mounted on the car. There ARE alternatives.

- Plug into the grid, particularly if your regional power comes from hydro-generated electricity.
- Install a grid interface of a home-mounted solar array. This gives you credit for daylight solar charging at home, cashed during the overnight charging of your EV.
- Recharge your EV from a homepower system (pack-to-pack charging); this is easily accomplished, particularly when the home pack is a higher voltage than the EV's pack.
- Utilize battery exchange, switching a daytime charged pack for a depleted one when you get home. - MH

Hybrid Electric Vehicles

The battery pack in an electric vehicle stores a finite amount of power. However large this amount is or how efficiently it is converted into propulsion, the time will come when the vehicle will roll to a stop and require a recharge.

Facilities (infrastructure) that will fast-charge or exchange a depleted battery pack with another are needed to make the "pure" EV as workable as cars with gas engines. However, until this infrastructure is set in place, there is an alternative: the hybrid EV.

An electric vehicle that uses two or more power sources (one of which is a battery pack) is referred to as a "hybrid" EV. The power produced by the extra energy source will vary considerably with the type and design, but it is designed to augment the power from the battery pack, adding to overall range, performance, and other operational characteristics.

Within this broad definition of hybrid EVs, many alternative energy or power sources qualify as an additional energy source. This includes solar panels, a wind arch, and an engine-generator unit. Even a regenerative braking circuit might permit a hybrid classification, since it produces electricity from the vehicle's momentum, and extends range.

Let's look at the hybrid EV that uses a genset, or engine-driven generator. The term genset is an abbreviation for Generator Set, or an engine-generator set. Folks in the construction trade know gensets as standby generators. The 110-volt, 60-cycle AC generator is a typical genset. If none of these terms is familiar to you, imagine hooking up the small engine (i.e., that runs your lawnmower) to the alternator or generator in your car through a V-belt. This would be a low-power unit, but when placed onboard an EV, it becomes an onboard charging system. The engine itself can be fueled with gasoline, propane, alcohol, hydrogen, or diesel, depending on the type and your own preference of fuel.

In the hybrid EV, the ACU (Auxiliary Charger Unit) is a small engine (i.e., 5-8 HP) coupled directly to an alternator. The alternator's output is connected to the batteries. A power cable connects both the ACU's output and the batteries to the motor controller. The motor itself is connected to one of more wheels through a standard transmission and differential. Note that the engine is NOT coupled to the drivetrain mechanically.

A hybrid EV combines the best features of electric motors with the best features of engines. The electric motor contributes its flat torque/RPM and variable-load characteristics, short-term high-power endurance, and its light weight. The engine contributes its high-power density and fuel availability. In the process, each offsets the disadvantages inherent in the other in a mobile propulsive environment.

Hybrid EVs address two fears held by the driving public concerning electric vehicles in general: low performance and getting stuck somewhere with a dead battery pack. Then, too, all energy sources have inherent advantages and disadvantages. Utilizing two or more sources frequently adds the good features of each source and offsets the shortcomings inherent in any one source. City driving favors use of the battery pack; freeway driving favors ACU operation. Short distance suggests battery-only operation, while long-distance driving requires the ACU. Presumably, you know where you're going, how far it is, and how fast you'll drive. This lets you select battery only, or ACU operation. The hybrid EV makes it possible to use the existing infrastructure for fossil-fueled vehicles without actual dependence on it.

Use of an ACU increases overall system complexity and initial costs. It may increase overall vehicle weight, too. Hybrid EVs are really a transition technology, until we are fully able to outlaw any combustion technology, since this is at the root of increased carbon in the atmosphere.

How does the ACU work? When you're stopped at a light or stop sign, all of the ACU's power is going into the batteries. When you're traveling down the road at 15 MPH, some of the ACU's power goes to the electric motor, and the remainder goes into the battery pack. At some speed, say 35 MPH, all of the ACU's output goes into the motors. At 50 MPH, the batteries supply the additional power (above the ACU's output) needed to reach and hold that speed. Note that the engine is relieved of the task of producing PROPULSION and assigned the task of producing POWER toward the propulsive effort, battery storage, or both. Thus, when the ACU is operational, the power it produces is never wasted. It's used or stored. Compare that to an IC-engined car stuck in a traffic jam or waiting for a signal light! -MH

How To Take Action

How can what you do matter? Politicians and successful corporations long ago learned that the REAL vote each individual holds is the dollar. Spend a dollar on gasoline and you vote for it, and whatever policies the people who sell it to you choose. Don't feel guilty about it. Instead, move toward a mode that decreases or eliminates your consumption of oil. Increase your awareness of the issues and technology in transportation. Buy, construct, use, or support electric vehicles and other alternatives!

Publications

Alternative Transportation News

Alternative Transportation News is a 36-page magazine that has the ambitious goal of changing the transportation habits of the world in this decade! It is about transportation issues, alternatives to our current systems, and the means to achieve them. It unites a historical perspective with state-of-the-art technologies to suggest ways we can achieve transportation that is less abusive to life on this planet.

ATN describes how these products and technologies represent a fertile but virtually untapped market to individuals, entrepreneurs, and industry in general. It is written for the layperson, translating otherwise technical information into terms anyone can understand. It is a networking tool, helping individual organizations and corporations worldwide to become aware of each other's efforts in fuel and vehicle technologies and to work together to integrate these efforts. It represents a vision of the future without making promises on how easy, fast, or inexpensive it will be to extract ourselves from our dilemmas.

ATN has departments on human-powered vehicles, air and water transport, fuels and cells, designworks, basic and advanced projects, and history. It reviews books and products, notes calendar events, and provides for reader input in many ways.

An ATN subscription is $12 annually (6 issues). First issue is March 1, 1991. Send subscriptions for ATN to: Earthmind, P.O. Box 743-RG, Mariposa, CA 95338. (Checks payable to Earthmind).

In the First Issue of ATN

• **The Impact:** A prototype designed exclusively to use electric propulsion, the Impact shatters the myth of low performance and range. But -- how will GM market a vehicle that makes its gas-powered vehicles obsolete?
• **Racing with the Sun.** Students from 32 Universities designed and built solar-powered race cars for the GM Sunrayce. Are the lasting impressions ones of hope and promise, or do these futuristic vehicles make EVs appear too expensive and impractical?
• **ATN interviews Paul MacCready.** Paul's team has brought us successful human-powered flight, the winning entry of the first Australian transcontinental race of solar vehicles, and the design of GM's electric commuter. How does he do it?

- **Thunder & Lightning.** Electric vehicle conversions? Low-performers, slow and limited in range? John Sprinkle's 1949 Porsche replica (cover photo) demonstrates how conversions can break that perception through forethought and patience.
- **The Horlacher Electric.** With gasoline at $3-4 per gallon, there's plenty of incentive to explore alternatives in Europe. How can America catch up and take advantage of this technology?
- **Bicycling in the City.** Drivers stuck in traffic jams fantasize about two-wheel transport but worry about the hazards of mixing-it-up with cars, trucks, and buses. Ben Swets made the switch and he's happy about it!
- **Airships.** Lighter-than-air ships (blimps and dirigibles) are coming back. Advances in the technology make this form of aerial transport appealing in the light of increasing cost of fuels, maintenance, and pollution. But -- how can airships compete in a fast world and overcome the stigma of the Hindenberg disaster?
- **The SBLA battery.** The Sealed Bi-polar Lead-Acid battery is just one of several new battery types. ATN evaluates the critical issues about batteries -- what many experts consider to be the only roadblock to successful commercialization of electric vehicles.
- **EVs and Power Plants:** Some environmentalists fear that the commercialization of electric vehicles will lend support to the resurrection of nuclear power plants. What energy sources can we expect to power electric cars?
- **Flying with the Sun:** Eric Raymond's Sunseeker is an airplane designed to use one energy source -- the sun. Close attention to aerodynamic efficiency makes this electric-powered, solar-charged, thermal-riding craft possible. What does it tell us about the integration of down-to-earth energy systems?

Solar Mind

As you can tell from his articles in our Electric Vehicle section, Joe Stevenson is an excellent writer. We highly recommend you subscribe to his publication, *Solar Mind* which is full of vision, insight, and practical solutions to the internal combustion engine. *Solar Mind* is a unique merging of spriit, mind, technology and environmental concerns. It is a place to share and discuss ideas, insights and solutions to earthly problems. To subscribe, send $3 for a sample issue, or $25 for six issues ($12 for students, non-profit groups and retirees), to:

Joe Stevenson
Solar Mind
759-RG South State Street #81
Ukiah, CA 95482

Electric Vehicle Clubs

More and more electric vehicle clubs are sprouting up across the nation. These clubs believe that electric cars are practical and affordable now. Most club members gather monthly and many members drive their electric vehicles to these monthly meetings. This is a great way to see a variety of vehicles, to get a ride in one, or just to experience an electric vehicle first hand. Clubs encourage novices through demonstration, instruction, and assistance. Here is a list of chapters and contact names of the Electric Auto Association around the country and Canada. We encourage you to join!

Lee Clouse
PO Box 11371
Phoenix, AZ 85061
602/943-7950

Mil Stults
2270 Minnie Street
Hayward, CA 94541
415/582-9713

George Schaeffer
211 Ballan Blvd.
San Rafael, CA 94901
415/456-9653

Jean Bardon
540 Moana Way
Pacifica, CA 94044
415/355-3060

Lee Hemstreet
787 Florales Drive
Palo Alto, CA 94306
415/493-5892

Don Gillis
5820 Herma
San Jose, CA 95123
408/225-5446

I.L. Weiss
2034 N. Brighton "C"
Burbank, CA 91504
818/841-5994

Ken Koch
12531 Breezy Way
Orange, CA 92669
714/639-9799

Jim Cullen
Desert Research Institute
2505 Chandler Ave. Ste. 1
Las Vegas, NV 89120

Ken Bancroft
4301 Kingfisher
Houston, TX 77035
713/729-8668

Ray Nadreau
19547 23rd N.W.
Seattle, WA 98177
206/542-5612

Dave Pares
3251 S. Illinois
Milwaukee, WI 53207
414/481-9655

Kasimir Wysocki
293 Hudson St.
Hackensack, NJ 07601
201/342-3684

Bob Batson
1 Fletcher Street
Maynard, MA 01754
508/897-8288

Steve McCrea
101 S.E. 15th Avenue #5
Fort Lauderdale, FL 33301
305/463-0158

VEVA
543 Powell Street
Vancouver, BC, Canada
V6A-1G8
604/987-6188

American Solar Car Assn.
Robert Cotter
PO Box 158
Waldoboro, ME 04572

EVCO
Box 4044 Station "E"
Ottawa, Ontario, Canada
K1S-5B1

DEVC
George Gless
Denver, CO
303/442-6566

Fox Valley EVA
John Stockberger
2S 543 Nelson Lake Road
Batavia, IL 60510
312/879-0207

The 750 Mile Electric Car

We are led to believe by General Motors and others that unlimited range electric cars are futuristic fantasies. We are told that there are no near term batteries or safe fuels that can give us clean air vehicles in the immediate future.

We aren't told that people drive electric cars in many areas such as Holland, Denmark and Switzerland on a daily basis. They are aided with free charging stations and tax credits as well. We aren't told that many of the technical options dismissed by government and corporate powers as "not cost effective" are viable alternatives in most of the world. As we now know, most of the world pays over $3 per gallon of gas. Electrics are real alternatives everywhere but the USA.

Are we really so petroleum-enslaved that choices of convenience define our senses of reality?

Does an energy option become unreal and forgotten simply because of the current cost of gasoline? One seemingly forgotten fuel source for electric cars is the aluminum-air battery. Dismissed because of its comparative cost to gasoline, it now represents a power. I experimented with aluminum-air - on my own - 12 years ago and found it amazing and simple. So does Gordon Stone, President of Voltec Corp. of Belleville, Ill. In 1986, Gordon and Voltec unveiled the A-2 aluminum-air powered car. The intent was to show the amazing potential and simplicity of this technology for everything requiring batteries. Until recently, the interest in this technology has been minor and limited.

What do the car and battery really do?

The A-2 can travel approximately 200 miles between 2-minute water changes and 750 miles before refueling (more aluminum), which takes under 5 minutes. All this while travelling at freeway speeds. The car weighs under 1,200 lbs., seats two (in tandem), an d gains assistance from the wind. Its drag coefficient is .15. The motors are brushless D.C. It is technically a hybrid carrying a small lead acid pack for acceleration.

So why aren't there more of these cars available?

Money! Aluminum power cost is the equivalent of $3 for

gasoline, that is, for a very pure grade aluminum. The other reason, also involving money, is the cost of developing a recharging and distribution infrastructure. To my mind, these arguments against aluminum-air are like comparing Model T Fords to Honda Accords. In 1920, there were so few gasoline stations that farmers sold gasoline to motorists from drums, and the cost of gas went down as demand for it rose.

If we note that many countries in the world already pay well over $3 per gallon for gas, and if we add in the environmental and health costs resulting from gasoline use, we may be close to parity. There are also some efficiency improvements possible with aluminum-air cells as well as the possibility of using lower-grade materials.

We can begin building the infrastructure to make aluminum-air and other clean options viable. I would bet that many ecology -minded businesses and persons would jump at the opportunity to set up service stations, as was the case in the 1920's. Gas was a " Mom & Pop" product, next to the pretzels and bubble gum - the store came first.

Currently, the A-2 and Gordon Stone are proving to a few interested corporations that aluminum-air has a real place in our energy future, and several companies are speaking seriously of moving forward. Stay tuned!

Joe Stevenson

Silence and Technology

As we move towards an orderly and sane world, we also move towards silence. It is not coincidental that the technologies of the future also reflect this movement to order and silence

Our minds are captivated by the seeming effortlessness of photo-voltaic cells, electric cars, and many other promising creations. This obvious elegance in technology is, in part, expressed in its silence.

The history of human technology is deeply grounded in the laws of "brute force." "Clanging and banging" are not assumed of a 3-speed transmission can almost bring a "teen-age mind" to orgasm. We have come to identify with brute force much like a hostage grows to love its abductor. We now feel at home in constant noise.

We read with the TV or CD on. Shop by Musak. Live in

neighborhoods where every 10 minutes the sound of a lawnmower, edger, blower, or chain saw bring us comfort(?). There is clearly a price for this noise and we all know it but can't comfortably admit it.

The nervous system of the human body needs silence. People live tense lives in tense environments and find themselves needing drugs for relaxation. We seek a peace or personal ease even while immersed in our noisy lives.

Safe-Silence is what we seek. A silence which does not remind us of emptiness and death. We seek an institutional silence, the silence of the church, the bathroom, or the drug. Unfortunately we must always resume our lives in the "grating and grinding" world.

The new silent technologies offer us not just a "safe silence," they allow us a choice in our union with noise. The photo-voltaic panel not only lowers the background noise in our home but can also power the VCR and CD player, and the electric car can take us silently to the rock concert. We can witness, however, that overall our world will become more quiet and less scary. It is a process of breaking an addiction to agitation and repetition.

We often create the technologies or make selection of tools based on our openness to ourselves and inner silence. With noisy minds we have built jackhammers and internal combustion engines. We are gradually reflecting an increased collective sensitivity as we awaken from our dinosaur-like practices.

Perhaps as we grow less afraid of silence/ourselves, we can open to the possibilities now manifesting using sound and light waves, magnetics, etc. There are already many examples being used snow. There are lumber mills cutting lumber with light beams, silent refrigeration, and stoves which heat through induction.

We are on the verge of an intelligent technological age. As Bucky Fuller said it, "do more with less." Yes, it is a more complex world, but so is life, and complexity is only a perspective not a static fact. This time around we all need to understand our world and ourselves. Technology must flow through us like song or speech—we are technology.

Whatever it is that is happening to/through us, it shows the promise for a clean and quiet future. A world conscious of its noise and respectful of its quiet space will be a reverent one. It is a world we unconsciously now grope towards anxiously. Let's just allow it to become fully conscious.

Joe Stevenson

Appropriate Transportation: A Proposal for Energy Education

The following article is a draft proposal written by Steve Stollman of the Center for Appropriate Transportation (49 E. Houston St. New York, NY 10012). This proposal is an example of a fresh approach to qualitative change in our environment. We welcome your input.

Participants in the September 1989 National Energy Program Managers meeting in Indianapolis never heard of the Ka'ahele La Project. Few, however, will ever forget the sight of the Naalehu Intermediate School's solar/electric powered vehicle as it spun silently around the track at the Indianapolis Motor Speedway, it's 13 year old chief designer at the wheel. Certainly nobody was surprised when it captured the BEEP (Best Energy Education Promotion) award for this year. Federal DOE officials at the site were impressed enough to hurriedly arrange for the car to be shipped to Washington DC to be exhibited at their offices during October's Energy Month.

There are already plans in a number of States to duplicate Hawaii's innovative program. (They have doubled their efforts this year, are producing a newsletter, and can be reached at DBED - Energy Division, 335 Merchant Street, Room 110, Honolulu, HI 96813 (808) 548-4195 Att: David Rezachek). The following is a suggested, somewhat broader, framework through which this activity may be rapidly expanded. Since this is only a preliminary draft, your comments are most welcome. As an interdisciplinary project, we would be grateful should you be able to unofficially circulate this discussion document to any Education, Energy Transportation, or Economic Development officials whom you feel might care to comment at this phase. In a month or so a more formal presentation will be made.

Progress in the design and construction of solar/electric vehicles has been accelerated in recent years largely thanks to the efforts of a group of idealistic electrical engineers in Switzerland. Six years ago members of "Futurebike", mostly human-powered vehicle builders, decided to extend their efforts into the construction of even more experimental devices. Three-quarters of the nearly one hundred resultant machines were "hybrid" in character, with power being

supplied through a combination of human and solar power. This year the Swiss "Tour de Sol" was joined by nearly forty solar vehicle competitions in Europe alone.

This article outlines a nationwide program to encourage the design and construction of human, solar, and electric-powered vehicles, primarily by groups of students in public and private schools and universities. It varies from conventional approaches to this activity by its insistence that a speed competition not be the sole format. Instead a variety of criteria would be used, such as:

- Aesthetics.
- Functionality (Especially as it applies to a specific purpose).
- Economics (Least expensive/to build as prototype/to produce in quantity).
- Most innovative and creative.
- Most compact (Foldable/Inflatable etc.).
- Most suitable for use by elderly or disabled.
- Most flexible (All terrain etc.).
- Most protective, safest.
- Adaptability to mass transit interface.
- Best use of recycled materials.

Prizes could be awarded by various means: the votes of well-qualified judges, polling of fellow contestants, and through popular balloting. There would, of course, have to be separate prizes awarded according to age, educational level, etc. Cash prizes, as well as donated goods; free travel to awards ceremonies, and more could be offered, in addition to less material rewards, such as recognition, school credits, improved skills, fun etc...

Although schools could be the primary agency for the pursuit of these goals, help and participation by bike stores, service stations, local companies, or motivated individuals could be invited as well. They could work in conjunction with local schools or on their own. Special locally-based contests could attain a status usually reserved for athletic rivalries.

To maximize potential benefits, some designs could be done in order to fill certain specific, actual needs. Their successful adoption would be the most tangible evidence possible of the validity of these exercises. Some of the most rewarding work might be in the area of wheelchairs, and other designs intended for the truly needy. Even "unsuccessful" efforts can still contribute meaningful results and partial solutions can be built on and improved over time.

The primary rationale for engaging in this project is process rather than product. This is an interdisciplinary approach stressing the relationship of science, technology, mechanics, craft, design, communication, construction skills, co-operative work, problem analysis, business needs, math, art, and much else. It is a powerful motivator, something that combines familiar objects like bikes and cars, with exotic concepts like futurism and sun-power. Preferably it would be voluntary in nature, but could become part of classroom work as well.

Involving local businesses, local media, utilities etc. as sponsors certainly is possible, especially as concrete examples of results can be demonstrated. The cooperation of State and County fairs in staging annual exhibits seems likely. Institutions of higher learning can assist private and public schools

while pursuing their own projects (Fifty are already building solar cars). Local associations, the Rotary, Chambers of Commerce, and other booster groups may find these mobile, colorful examples of their involvement in community-based activities beneficial. In some cases, it is even possible that results could be so outstanding that new local businesses might be created.

There are already some teaching materials being prepared, mostly in the form of instructional videotapes. A few different resource directories also exist. There are videotapes of previous events, full of interesting examples of completed vehicles. Prize-winning efforts could be filmed and a running record kept of worthwhile efforts.

Each year, an occasion could be provided for local winners to get together, have a "National Championship", and exchange ideas. The visual nature of this material and the likely wide variety of results should surely attract the interest of network television and perhaps a corporate sponsor. Included as part of this project should be schemes and ideas for improvements in mass transit and the re-allocation of street space, crucial factors in any significant upgrading of (especially urban) mobility. Participation, by Art and Architecture students and professionals, can help provide a fuller context for vehicle design work, as well as colorful, appropriate, and dramatic visual backgrounds.

This year there will be twice as many entries in Hawaii's contest as there were last year. In Europe this summer 36 solar vehicle events took place, more than three times as many as in 1989. It is time for this country and it's educational system to make a commitment to be relevant in these areas into the next century. Technical help is readily available through such organizations as the American Society of Mechanical Engineers, Argonne National Laboratories, American Solar Energy Society, and the International Human Powered Vehicle Association. The potential benefits here, educational, environmental, and economic certainly justify the small resources necessary to implement such a program.

If much of this effort can be undertaken through the redeployment of abandoned bicycles and discarded electric motors, use of cardboard, wood, plastic, and other easily accessible and inexpensive materials, the cost need not be an impediment to success. When needed materials are not easily available, schemes with drawings and explanations, small scale and non-working full sized models would also be permitted. In the lower grades especially, the emphasis could be on scale-model work, giving a foundation in basic principals which can be built on in the higher grades.

The seriousness of these efforts need not close off the opportunity for those so inclined to seize the chance to use it to have a good time too. Even nonsensical entrants (i.e. the "Bionic Taco", an amphibious and perennial winner of Northern California's annual Kinetic Sculpture race) have value and could also qualify. After all, the ultimate goal, as wildly ambitious as it may seem from this early perspective, of the program, would be to transform one of this society's most basic rights of passage - You don't buy your first car, you make it. Instead of perfect consumers, we will have to settle for imperfect creators, crafts people, and producers. Is this such a bad bargain?

-Steve Stollman

Sinclair C5

The Sinclair C5 has been called the most exciting new vehicle since the Model T. Everyone that owns one seems to agree that it's a lot of fun! You can go literally 1,000 miles for about the price of a gallon of gasoline. The Sinclair is 3-wheeled and battery-powered and can also be powered by pedaling. It comes with a 12-volt, 36 amphour battery, requires minimal assembly and weighs only 125 lbs. It has an electric range of 18-22 miles on one charge using the 36 amphour battery. At full speed the Sinclair uses 250 watts or approximately 20 amps. By doubling the battery capacity, you can double the range on one charge. The C5 comes equipped with lights and turn signals.

The Sinclair was designed by Cleve Sinclair (of Sinclair Computers) as a moped for the streets of London. There are several options available: the Solar option will enable you to charge the batteries from a 25W PV panel (it takes about 2 days in the sun to charge back one hour's use.) The 24-Volt option will enable you to increase the maximum speed to 28-30 mph from the standard 15-18 mph for the 12V model. The All Weather Option is a large bubble that fits around the entire drivers seat allowing you to operated the vehicle in the rain and snow. Our supply is limited. Call or write for a complete data sheet.

■91-801	Sinclair C5	$1695
■91-802	24V Option (kit)	$ 195
■91-803	24V Option (assmbld)	$ 250
■91-804	Solar Option (kit)	$ 395
■91-805	Solar Option (assmbld)	$ 465
■91-806	All Weather (kit)	$ 375
■91-807	All Weather (assmbld)	$ 425

Shipped freight collect from either Boston or Ukiah

Electric Motorcycle

The same person who supplies us with the Sinclair C-5 electric cars (we have a few left) has just come across a great 24V electric motorcycle called the Elektro-Mofa from Germany. It is powered by a permanent magnet DC motor at 24V and 560 watts, producing 1200 RPM. It comes with an automatic battery charger, a speedometer, and a battery condition meter. It has a range of 30-35 miles per charge and a maximum speed of 21 miles per hour. Recharging time on AC is six hours. Transmission is a single chain drive from motor to rear wheel. The motorcycle weighs 407 lbs. with batteries. (Batteries not included in price.) Each motorcycle is shipped from Frankfort, Germany - shipping costs run under $200 air freight to any major airport and delivery can be made in one week.

Environmental Umweltfreundlich

Pollution Free Abgasfrei

Low Noise Geräuscharm

Economical Wirtschaftlich

| 91-831 | Electric Motorcycle | $1795 |

Add $200 for air freight from Germany

Convert It

This book by Mike Brown is a step-by-step manual for converting a gas car to an electric powered car. It's a very readable and practical manual for the do-it-yourselfer wanting the fun and educational experience of converting a conventional automobile into an electric vehicle that will be both practical and economical to operate. Included are a generous number of illustrations of an actual conversion, along with instructions for testing and operating the completed vehicle. Mike Brown shares a wealth of his own personal experience in making conversions and includes many practical tips for both safety and ease of construction that will be useful for the novice or the experienced home mechanic alike. Brown does a great job of explaining the cost of the conversion components. This book has been strongly recommended by the *Electric Auto Association, Electric Vehicle Progress, and Alternate Energy Transportation*. The book is 56 pages long. We realize the price seems a bit steep - we have pleaded with the author to re-publish in quantity to reduce his cost. However for the time being this is the *only* book available on the subject and it seems worth the price for that reason.

80-404 Convert It $35

CONVERT IT

BY MICHAEL P BROWN
WITH SHARI PRANGE

A STEP-BY-STEP
MANUAL FOR
CONVERTING
A GAS CAR TO
ELECTRIC POWER

PO #1113 FELTON CA 95018
(408) 429-1989

If I Were To Buy an RV Today...

by Phred Tinseth

Phred Tinseth is a full-time RVer and a scientist. We always look forward to his rig pulling up outside our shop to see his latest discoveries. Phred writes a regular column for for the Escapee's (SKP) newsletter, and periodically tests out products for us. The following is good advice for the would-be RV purchaser.

Wheels and Drums

If you're fortunate, you'll find a trailer that uses standard automotive wheels, not those little 4-bolt things with a funny inside shape that can't be put on a tire changing machine without bending them. Even better, go to a Wheel and Rim shop and for about $55 each get drums that will match the basic hub and spindle on your axle and also match the drums on your tow vehicle (usually they come with bearings and seals). It's surprisingly easy for a good shop to match up the numbers. Apparently, there are relatively few variations. At any rate, it's nice when tow and trailer are all alike.

No matter which way you go, don't let someone impress you with fancy chrome wheel covers. Be it trailer or motorhome, one of the first things you really need to do is to remove any form of wheel cover. Wheel covers do only two things: 1) Keep you from properly taking care of your tires. 2) Keep you from checking your wheel bearings. (If appearance is all that important to you, paint the wheels and put on chrome nuts and grease caps that are sold for just a few dollars in any auto store.)

On the Inside

Assuming you know how to check the basics (e.g., can you stand up in the shower, are walls made of cardboard, etc.), I suggest only that you force yourself not to be impressed by flashy extras. If you're going to buy an RV for many thousands of dollars, <u>don't</u> base your final selection on a "free" TV that you can get anywhere for $300, a VCR, trash masher or similar items. Most standard RV equipment can be found better and cheaper elsewhere.

It would be nice if you could find a "bare bones" model and outfit it yourself, but that's hard to do and you might not want to spend months building an RV. About all most people can do is ensure that everything works in the RV they buy, assume that it will all need to be replaced, and then start researching and shopping for replacements well in advance.

Look for intelligent design, easy access to appliances so you can fix them, storage and counter space, etc. If a TV is in the ceiling and you need to crane your neck to see it, if the water pump is riveted in a box behind the stove, or if water and gas lines come through the middle of a cabinet, think about how you'll fix them.

"Basement" models, at first glance, appear nice; but look closely at some of the deep compartments. Once you pack them full, how do you find the item in the back? The same compartment style often has sewer connects; some with drainage pans that look good but that drain into the underbelly rather than outside. What happens when you make a mess?

If you're buying an RV for full-time use, if you intend to keep it for a long time, and if you want to live comfortably rather than just "make do," you'll surely want to make at least some modifications and additions in the future. Consider the ease of doing so when selecting an RV.

For example: Solar panels and catalytic heaters can be installed in any RV, but how about more and larger batteries? Where will you put them? NO! Not on the back bumper! If you add anything, other than a couple of bicycles (not motorcycles) to the bumper, you're making a serious mistake. (Remember your high school physics?) Can you permanently mount a macerator? No, not just temporarily attach it to the sewer outlet. If you properly mount it, permanently, you'll use it even when hooked up to a sewer, because you'll be able to dump quickly without spilling a drop.

Photovoltaic (PV) Modules

Often called Solar Panels, PV Modules are step number one on the road to independence and self sufficiency. I have quite a few. All are Arco. Not because they're best, there are other good brands; they just had the right price, warranty, etc. when I was shopping. (Arco recently raised their prices. Their chief competitor, Kyocera, of equally high quality, didn't. Probably a moot point since Arco is selling off the PV division.) I have several models all strung together. Unlike most products, you don't have to "trade up", just keep stringing them together.

How many modules do you need? Any good catalog will provide sizing charts. As a rule of thumb: Experienced full-time RVers, with conservative but not stingy use, find that one panel (about 50 watt) and one battery (about 100 AH) per person is adequate. To occasionally use inverter/microwave, most RVers find that 200 watts (4 modules) and about 400+AH of battery capacity is adequate for two people.

My System(s)

I use three, completely separate "banks". Two are on the trailer: Eight modules feed six golf cart (Trojan T-105) batteries (in series-parallel) through an SBC-30 controller. This system powers a Flowlight pump, lights, Sunfrost reefer/freezer, Recair evaporative cooler, a Jabsco macerator, and appliances. Usually I have enough left over power to charge various portable batteries also.

A second system, with its own independent ground, uses two modules, a Battery Guard/Power Guard, and two 12 volt StowAway/Watchman batteries for TVs, VCR, radios and wordprocessor. (I no longer have radio interference.)

The third system, on the van, uses four modules, another Battery Guard, and three batteries for another macerator and Recair, Flojet waterpump, miscellaneous tools/appliances, and Nova Kool reefer/freezer. Any of the systems can be plugged into the other should I run short. I use two trickle-charge panels: One to keep the starting battery charged, one for venting.

Buying a System

PV modules are not cheap, but unlike a pick-up truck, you don't have to buy the whole thing at one time.

Even if you're buying just one panel, read a book. Read a few catalogs. Determine what a full-sized module sells for. If something adequate, e.g., 50 watt, sells for $300+, then don't think you're going to get a bargain by buying something advertised as "All the electricity you need for $200." All you'll get is a little trickle charger. (TANSTAAFL! There Ain't No Such Thing As A Free Lunch.)

On the other hand, all you may want is a trickle charger. Used to be that a vehicle just had an electric clock. If you turned the key off, and had no "leaks", you could let the thing sit for months. No longer. Memory chips in radios, gas detectors in RVs, burglar alarms and a host of other devices can run a starting battery down in less than two weeks.

So-called Self-Regulating Panels

There is lots of controversy here among those who sell panels to RVers. Primarily, because most of those who sell panels are not RVers and don't know what they're talking about-at least when it comes to RVs.

When Arco started selling the 30-cell, reduced-voltage, reduced-wattage, reduced-cost panels, they did not recommend them for RV use, but for direct applications. As time went on people did use them for battery charging and they do work - up to a point. The average RV weekender, or frequent traveler, only needs one or two panels to maintain a charge, the assumption being that frequent driving will "top" the batteries off. Many full-timers, though, use a lot of electricity and may not drive for months. Neither do they want to offend neighbors or violate the atmosphere with a generator.

A self-regulating module does not have sufficient voltage to consistently, fully charge batteries. Obviously, and by definition, it will not then require a regulator (unless connected to batteries when no electricity is being used). Further, as temperatures increase, panel voltage is reduced. Surface temperature of a module on an RV roof can easily exceed 120 ° F.

Look at it this way: A "normal" 33-cell module puts out about 0.47V/cell for 16.3V. At 117° F, by Arco's own figures (and that's surface, not ambient temperature),

voltage will be reduced by about 10%. That comes to 14.4V and, since a typical battery is fully charged at from 14.3 to 14.8V, all is well. However, 30 cells at 0.47V = 14.1V (marginal); and if that is reduced by 10%, it equals only 12.7V, and that is not a full charge. If you disagree with my numbers and want to argue, have at it. I've got plenty more.

All this has been covered in detail by Joel Davidson and Paul Wilkins in *Photovoltaic Network News* and by Kirkby in his newsletter and "RVers Guide". Further, read Arco's own Installation Guide (page 8, Self Regulation) for even more conflicting treats. I am not condemning self-regulating modules. They work for some people. I just recommend that you understand your needs before buying.

Under-rated or reject panels are money savers. Some of my best are rejects. Only reason I can see in one is a hair under the glass. Big deal.

The new Sovonics flexible panels look promising. (And easier to mount on the aerodynamic surface of some RVs.) Verdict is still waiting on their longevity. Will they degrade over time? These are amorphous cells. There have been degradation problems with these in the past. Note that while Kyocera and Arco are warranted for 10 years or more, Sovonics is only 5.

Busted-cell goodies is an all-inclusive term for things made from either small, single cells; partial pieces of cells (i.e., busted), or similar. Often used in silly toys, there are some very useful items being made as well. I have a 4 "AA" Solar NiCad Charger that amounts to a small box with solar cell lid. Load it with 4 NiCads and lay it on the dashboard. Four more NiCads power a "Walkman" type radio/cassette. Switch battery sets once in a while and soon the batteries and charger have paid for themselves. Similarly, a Metered Solar NiCad Charger does the same, but will also charge 2 "C" or "D" cells and has a test meter as well. The Solar Rechargeable Flashlight is another toss-on-the-dashboard item. Not cheap (about $35), it's no toy either. Sealed lead-acid battery (2.5 AH), not "D" cells, includes 12V adapter for charging in no sun and is very bright.

A really splendid item I've come up on is the Sunvent Solar Powered Ventilator. A single cell, is "weather-proof" mounted in a round, also weather-proof cap containing a fan. It can be easily mounted in your reefer roof-hood and will draw air across the rear coils. It can also be mounted on top of holding tank vent pipes. Obvious benefits to this item.

Batteries

Big, 2-cell, industrial chlorides are worth the price but too heavy for an RV. Trojan or equal L-16 are the next best - but too high for many RVs. Trojan T-105 or equal (Exide is good) will fit in most places if you're imaginative.

RV/Marine

12V, usually Group 27 (nowadays usually Group 24 as manufacturers cut corners). By definition: Any battery

Photograph courtesy Arco Solar

with a rope tied to it. Generally not an economically sound choice. Short-term dollar savings at the expense of life and cycles. Next worst thing to automotive batteries. (Actually, that's what they are - a few ounces of lead does not a deep cycle make.)

Auto <u>and</u> RV/Marine

These are not a total disaster. IF you have a well-balanced system and IF you never deep discharge, you can get years out of these things. (My K-Mart, 30-month, teeny battery, purchased in 1974, served as an auxiliary tool battery until 1988!) Maintenance, no overcharge, no undercharge, no deep cycling. Incidentally, a "D-8" battery is not a deep-cycle battery - ask the manufacturer.

Maintenance Free

This is a misnomer as most are maintenance prohibitive. Not all that bad, IF...(as above). Early models, e.g., Delco, were disasters. Most of the current crop; e.g., StowAway/Throwaway, DieHard/DieEZ will be equally unsatisfactory if you deep-discharge or undercharge them.

The Incredicell/Incredibly Silly: It's one of those that are maintenance "friendly" as opposed to prohibitive; i.e., you can service it. However, due to poor design, any crud

that collects on top - including baking soda - will be funneled into the cells when you lift the cap.

Some people report good service from the Delco Voyager - I recommend you not deep cycle it much. For reasonably good service (and price) some people are having good luck with the Interstate brand.

New Stuff

Investigate the West German "Sonnenschein" (translates to Sunshine). Several sizes. One, a 12V, approximately Group 27, estimated 10 year life with 1,000 deep cycles. Costs about $345! TANSTAAFL!

Also look at Nife Sunica. Scandinavian 12V, large amp hour NiCad, of all things. They're using them on remote beacons that they can't even get to as much as 6+ months (PV charged).

So why do I use several StowAways? I have a well-balanced PV system. I don't overcharge and I don't undercharge. I seldom use more than 20% of available power. I almost never use more than 40%. I've deep-discharged these only three or four times since I bought them five years ago. They are as clean as the day I bought them. I treat my Trojan T-105's the same way and expect to get at least 10 years out of them.

GLOSSARY

A

Activated Stand Life • The period of time, at a specified temperature, that a battery can be stored in the charged condition before its capacity falls.

Alternating Current (ac) • An electric current that reverses its direction at a constant rate.

Ambient Temperature • The temperature of the surroundings.

Amorphous Silicon • A type of PV silicon cell having no crystalline structure. See also Single-Crystal Silicon, Polycrystalline Silicon.

Ampere (Amp) (A) • Unit of electric current measuring the flow of electrons. The rate of flow of charge in a conductor of one coulomb per second.

Ampere-Hour (Ah) • The quantity of electricity equal to the flow of a current of one ampere for one hour.

Angle of Incidence • The angle that a light ray striking a surface makes with a line perpendicular to the reflecting surface.

Anode • The positive electrode in an electrochemical cell (battery) toward which current flows. Also, the earth ground in a cathodic protection system.

Array • A collection of photovoltaic (PV) modules, electrically wired together and mechanically installed in their working environment.

Array Current • Current produced by the array when exposed to sunlight. See Rated Module Current.

Array Operating Voltage • The voltage provided by the photovoltaic array under load. See Battery Voltage.

Availability • The quality or condition of being available. PV system availability is the amount of time a PV system is 100 percent operational.

Azimuth • Horizontal angle measured clockwise from true north; 180° is due south.

Base Load • The average amount of electric power that a utility must supply in any period.

Battery • A device that converts the chemical energy contained in its active materials directly into electrical energy by means of an electrochemical oxidation-reduction (redox) reaction. This type of reaction involves the transfer of electrons from one material to another through an electrical circuit.

Battery Capacity • The total number of ampere-hours that can be withdrawn from a fully charged battery. See Ampere-Hour, Rated Battery Capacity.

Battery Cell • The smallest unit or section of a battery that can store an electrical charge and is capable of furnishing a current.

Battery Cycle Life • The number of cycles that a battery can undergo before failing.

Battery Self-Discharge • Loss of chemical energy by a battery that is not under load.

Battery State of Charge • Percentage of full charge.

Battery Terminology

  Captive Electrolyte Battery Type • A battery type having an immobilized (gelled or absorbed in the separator) electrolyte.

  Lead-Acid Battery Type • A general category that includes batteries with pure lead, lead-antimony, or lead-calcium plates and an acid electrolyte.

  Liquid Electrolyte Battery Type • A battery type containing free liquid as an electrolyte.

  Nickel Cadmium Battery • A battery type containing nickel and cadmium plates and an alkaline electrolyte.

  Sealed Battery Type • A battery type with a captive electrolyte and a resealing vent cap, also called a valve-regulated, sealed battery.

  Vented Battery Type • Free liquid electrolyte type battery with a vent cap for free escape of gasses during charging.

Blocking Diode • A diode used to prevent current flow within a PV array or from the battery to the array during periods of darkness or of low current production.

British Thermal Unit (Btu) • A unit of heat. The quantity of heat required to raise the temperature of one pound of water one degree Fahrenheit.

Bypass Diode • A diode connected in parallel with a block of parallel modules to provide an alternate current path in case of module shading or failure.

C

Capacity (C) • The total number of ampere-hours that can be withdrawn from a fully charged battery. See Battery Capacity.

Cathode • The negative electrode in an electrochemical cell.

Charge Controller • A device that controls the charging rate and state of charge for batteries. See Charge Rate.

Charge Rate • The rate at which a battery is recharged. Expressed as a ratio of battery capacity to charge current flow, for instance, C/5.

Cloud Enhancement • The increase in solar insolation due to direct beam insolation plus reflected insolation from partial cloud cover.

Concentrator • A photovoltaic module that uses optical elements to increase the amount of sunlight incident on a PV cell.

Controller Terminology

Adjustable Set Point • A feature allowing adjustment of voltage disconnect levels.

High Voltage Disconnect • The battery voltage at which the charge controller will disconnect the batteries from the array to prevent overcharging.

Low Voltage Disconnect • The battery voltage at which the charge controller will disconnect the batteries from the load to prevent over discharging.

Low Voltage Warning • A warning buzzer or light that indicates low battery voltage.

Maximum Power Tracking • A circuit that keeps the array operating at the peak power point of the I-V curve where maximum power is obtained.

Multistage Controller • Unit that allows multilevel control of battery charging or load.

Reverse Current Protection • A method of preventing current flow from the battery to the array. See Blocking Diode.

Single-Stage Controller • A unit with one activation level for battery charging or load control.

Temperature Compensation • A circuit that adjusts the battery high or low voltage disconnect points as ambient battery cell temperature changes. See Temperature Correction.

Conversion Efficiency • The ratio of the electrical energy produced by a photovoltaic cell to the solar energy received by the cell.

Converter • A unit that changes and conditions dc voltage levels.

Crystalline Silicon • A type of PV cell made from a single crystal or polycrystalline slice of silicon.

Current • The flow of electric charge in a conductor between two points having a difference in potential (voltage), generally expressed in amperes.

Cutoff Voltage • The voltage at which the charge controller disconnects the array from the battery. See Charge Controller.

Cycle • The discharge and subsequent charge of a battery.

D

Days of Storage • The number of consecutive days the stand-alone system will meet a defined load. This term is related to system availability.

Deep Cycle • Battery type that can be discharged to a large fraction of capacity. See Depth of Discharge.

Design Month • The month having the combination of insolation and load that requires the maximum power out of the array.

Depth of Discharge (DOD) • The percent of the rated battery capacity that has been withdrawn.

Diffuse Radiation • Radiation received from the sun after reflection and scattering by the atmosphere.

Diode • Electronic component that allows current flow in one direction only. See Blocking Diode, Bypass Diode.

Direct Current (dc) • Electric current flowing in only one direction.

Discharge • The withdrawal of electrical energy from a battery.

Discharge Rate or C Rate • The rate at which current is withdrawn from a battery. Expressed as a ratio of battery capacity to discharge current rate. See Charge Rate.

Disconnect • Switch gear used to enable or disable components in a PV system.

Dry Cell • A cell with a captive electrolyte. A primary battery cell.

Duty Cycle • The ratio of active time to total time. Used to describe the operating regime of appliances or loads in PV systems.

Duty Rating • The amount of time an appliance can produce at full rated power.

E

Efficiency • The ratio of output power (or energy) to input power (or energy). Expressed in percent.

Electrolyte • The medium that provides the ion transport mechanism between the positive and negative electrodes of a battery.

Energy Density • The ratio of the energy available from a battery to its volume (Wh/L) or weight (Wh/kg).

Equalization • The process of restoring all cells in a battery to an equal state of charge.

F

Fill Factor • For an I-V curve: the ratio of the maximum power to the product of the open-circuit voltage and the short-circuit current. Fill factor is a measure of the "squareness" of the I-V curve shape.

Fixed Tilt Array • A PV array set in a fixed position.

Flat-Plate Array • A PV array that consists of nonconcentrating PV modules.

Float Charge • The charge to a battery having a current equal to or slightly greater than the self discharge rate.

Frequency • The number of reptitions per unit time of a complete waveform, as of an electric current, usually expresssed in Hertz.

Gassing • Gas by-products produced when charging a battery. Also, termed out-gassing. See Vented Battery Type.

Grid • Term used to describe an electrical utility distribution network.

I

Insolation • The solar radiation incident on an area over time. Usually expressed in kilowatt-hours per square meter. See also Solar Resource.

Inverter • In a PV system, an inverter converts dc power from the PV array to ac power compatible with the utility and house loads. An inverter is also called a power conditioner or power-conditioning subsystem.

> **Square Wave** • A waveform that contains a large number of harmonics. A waveform that can be generated by opening and closing a switch.

> **Modified Sine Wave** • A distorted sine wave consisting basically of one single periodic oscillation.

> **Sine Wave** • A waveform corresponding to a single-frequency, periodic oscillation, which can be shown as a function of amplitude against angle and in which the value of the curve at any point is a function of the sine of that angle.

Irradiance • The instantaneous solar radiation incident on a surface. Usually expressed in kilowatts per square meter.

I-V Curve • The plot of the current versus voltage characteristics of a photovoltaic cell, module, or array.

K

Kilowatt (kW) • One thousand watts.

Kilowatt Hour (kWh) • One thousand watt hours.

L

Life • The period during which a system is capable of operating above a specified performance level.

Life-Cycle Cost • The estimated cost of owning and operating a system for the period of its useful life.

Load • The amount of electric power used by any electrical unit or appliance at any given moment.

Load Circuit • The current path that supplies the load. See also Load.

Load Current (amps) • The current required by the electrical unit during operation. See also Ampere.

Load Resistance • See Resistance.

Langley • Unit of solar irradiance. One gram calorie per square centimeter.

Low Voltage Cutoff (LVC) • Battery voltage level at which a controller will disconnect the load.

M Maintenance-Free Battery • A battery to which water cannot be added to maintain electrolyte volume. All batteries require inspection and maintenance.

Maximum Power Point • A mode of operation for a power conditioner, whereby it continuously controls the PV source voltage in order to operate the PV source at its maximum power point.

Module • The smallest replaceable unit in a PV array. An integral encapsulated unit containing a number of PV cells.

Modularity • The concept of using identical complete subunits to produce a large system.

Module Derate Factor • A factor that lowers the module current to account for normal operating conditions.

N NEC • National Electrical Code, which contains safety guidelines for all types of electrical installations. The 1984 and later editions of the NEC contain Article 690, "Solar Photovoltaic Systems."

Normal Operating Cell Temperature (NOCT) • The temperature of a PV module when operating under 800 W/m² irradiance, 20°C ambient temperature and wind speed of 1 meter per second. NOCT is used to estimate the nominal operating temperature of a module in its working environment.

Nominal Voltage • The terminal voltage of a cell or battery discharging at a specified rate and at a specified temperature.

N-Type Silicon • Silicon having a crystalline structure that contains negatively charged impurities.

O Ohm • The unit of electrical resistance equal to the resistance of a circuit in which an electromotive force of 1 volt maintains a current of 1 ampere.

Open Circuit Voltage • The maximum voltage produced by a photovoltaic cell, module, or array without a load applied.

Operating Point • The current and voltage that a module or array produces under load. See I-V Curve.

Orientation • Placement with respect to the cardinal directions, N, S, E, W; azimuth is the measure of orientation.

Outgas • See Gassing.

Overcharge • Forcing current into a fully charged battery.

P

Panel • A designation for a number of PV modules assembled in a single mechanical frame.

Parallel Connection • Term used to describe the interconnecting of PV modules or batteries in which like terminals are connected together.

Peak Load • The maximum load demand on a system.

Peak Power Current • Amperage produced by a module operating at the "knee" of the I-V (current-voltage) curve. See I-V Curve.

Peak Sun Hours • The equivalent number of hours per day when solar irradiance averages 1,000 W/m^2. Six peak sun hours means that the energy received during total daylight hours equals the energy that would have been received had the sun shone for six hours at 1,000 W/m^2.

Peak Watt • A manufacturer's unit indicating the amount of power a photovoltaic module will produce at standard test conditions (normally 1,000 W/m^2 and 25° cell temperature).

Photovoltaic Cell • The treated semiconductor material that converts solar irradiance to electricity. See Cell.

Photovoltaic System (PV System) • An installation of PV modules and other components designed to produce power from sunlight.

Plates • A thin piece of metal or other material used to collect electrical energy in a battery.

Pocket Plate • A plate for a battery in which active materials are held in a perforated metal pocket on a support strip.

Polycrystalline Silicon • Material used to make PV cells which consist of many crystal structures.

Power (Watts) • A basic unit of electrical energy, measured in watts. See also Watts.

Power Conditioning System (PCS) • See Inverter or Converter.

Power Density • The ratio of the rated power available from a battery to its volume (W/liter) or weight (W/kg).

Power Factor • The cosine of the phase angle between the voltage and the current waveforms in an ac circuit. Used as a designator for inverter performance.

Power Loss • Power reduction due to wire resistance.

Primary Battery • A battery whose initial capacity cannot be restored by charging.

Pyranometer • An instrument used for measuring solar irradiance received from a whole hemisphere.

R

Rated Battery Capacity • Term used by battery manufacturers to indicate the maximum amount of energy that can be withdrawn from a battery at a specified rate. See Battery Capacity.

Rated Module Current • Module current measured at standard test conditions.

Remote Site • Site not serviced by an electrical utility grid.

Resistance (R) • The property of a conductor by which it opposes the flow of an electric current resulting in the generation of heat in the conducting material. The measure of the resistance of a given conductor is the electromotive force needed for a unit current usually expressed in ohms.

S

Seasonal Depth of Discharge • An adjustment factor providing for long-term seasonal battery discharge. This factor results in a smaller array size by planning to use battery capacity to fully meet long-term load requirements during the low insolation season.

Secondary Battery • A battery which after discharge can be recharged to a fully charged state.

Self-Discharge • The loss of useful capacity of a battery due to internal chemical action.

Semiconductor • A material that has a limited capacity for conducting electricity.

Series Connection • The interconnecting of PV modules or batteries so that the voltage is additive.

Shallow Cycle Battery • Battery type that should not be discharged greater than 25 percent. See Depth of Discharge.

Shelf Life • The period of time that a device can be stored and still retain a specified performance.

Short Circuit Current (Isc) • Current produced by a PV cell, module, or array when its output terminals are shorted.

Silicon • A semiconductor material commonly used to make photovoltaic cells.

Single-Crystal Silicon • A type of PV cell formed from a single silicon ingot.

Solar Cell • See Photovoltaic Cell.

Solar Insolation • See Insolation.

Solar Irradiance • See Irradiance.

Solar Resource • The amount of solar insolation a site receives, usually measured in kWh/m^2/day. See Insolation.

Specific Gravity • The ratio of the weight of the solution to the weight of an equal volume of water at a specified temperature. Used as an indicator of battery state of charge.

Stand-Alone • A photovoltaic system that operates independent of the utility grid.

Starved Electrolyte Cell • A battery containing little or no free fluid electrolyte.

State of Charge (SOC) • The capacity of a battery expressed at a percentage of rated capacity.

Stratification • A condition that occurs in a deep-cycled liquid-electrolyte lead-acid battery when the acid concentration varies from top to bottom in the battery. Periodic controlled overcharging tends to destratify or equalize the battery.

String • A number of modules or panels interconnected electrically to obtain the operating voltage of the array.

Subsystem • Any one of several components in a PV system (i.e., array, controller, batteries, inverter, load).

Sulfating • The formation of lead-sulfate crystals on the plates of a lead-acid battery. Can cause permanent damage to the battery.

Surge Capacity • The requirement of an inverter to tolerate a momentary current surge imposed by starting ac motors or transformers.

System Availability • The probability or percentage of time a PV system will fully meet the load demand.

System Operating Voltage • The PV system voltage.

System Storage • See Battery Capacity.

T

Temperature Compensation • Allowance made for changing battery temperatures in charge controllers.

Temperature Correction • A correction factor that adjusts the nameplate battery capacity when operating a battery at lower than 20°C. See Temperature Compensation.

Thin Film PV Module • See Amorphous Silicon.

Tilt Angle • Angle of inclination of collector as measured from the horizontal.

Total ac Load Demand • The sum of the ac loads, used when selecting an inverter.

Tracking Array • A PV array that follows the daily path of the sun. This can mean one axis or two axis tracking.

Trickle Charge • A low charge current intended to maintain a battery in a fully charged condition.

U

Uninterrupted Power Supply (UPS) • Designation of a power supply providing continuous uninterrupted service.

V

Varistor • A voltage-dependent variable resistor. Normally used to protect sensitive equipment from sharp power spikes (i.e., lightning strikes) by shunting the energy to ground.

Vented Cell • A battery cell designed with a vent mechanism to expel gases generated during charging. See Vented Battery Type.

Volt (V) • The practical unit of electromotive force or difference in potential between two points in an electric field.

Voltage Drop • Voltage reduction due to wire resistance.

Watt (W) • The unit of electrical power. The power developed in a circuit by a current of one ampere flowing through a potential difference of one volt; 1/746 of a horsepower.

Watt Hour (Wh) • A unit of energy measurement; 1 watt for one hour.

Waveform • Characteristic shape of the ac output from an inverter.

Wet Shelf Life • The period of time that a battery can remain unused in the charged condition before dropping below a specified level of performance when filled with electrolyte.

Water Pumping Terminology

　Centrifugal Pump • A class of water pump using a rotary or screw to move water. The faster the rotation, the greater the flow. This type of pump is used in shallow well applications.

　Dynamic Head • Vertical distance from center of pump to the point of free discharge including pipe friction.

　Friction Head • The energy that must be overcome by the pump to offset the friction losses of the water moving through a pipe.

　Positive Displacement • See Volumetric Pump.

　Static Head • Vertical distance from the top of the static water level to the point of free discharge.

　Storage • This term has dual meaning for water pumping systems. Storage can be achieved by pumping water to a storage tank, or storing energy in a battery subsystem. See Battery Capacity.

　Suction Head • Vertical distance from surface of free water source to center of pump (when pump is located above water level).

　Volumetric Pump • A class of water pump utilizing a piston to volumetrically displace the water. Volumetric pumps are typically used for deep well applications.

Glossary courtesy of the National Technical Information Service; U.S. Department of Commerce

Lightning Protection

Many customers have inquired about lightning protection for their alternative power systems. The following article was written by Windy Dankoff, who also manufactures the Flowlight Slowpump and Booster Pump.

Lightning and related static discharge is the number one cause of sudden, unexpected failures in PV systems. Lightning does not have to strike directly to cause damage to sensitive electronic equipment, such as inverters and controls. It can be miles away or not even visible, and still induce high-volltage surges in wiring, especially long lines. Fortunately, almost all causes of lightning damage can be prevented by proper system grounding. Our own customers have reported damage to inverters (primarily), refrigerator compressor controllers, fluorescent ballasts, TV sets, and motors. These damages have cost thousands of dollars, and *all* reports were from systems *not grounded.*

Grounding means connecting part of your system structure or wiring electrically to the earth. During lightning storms, the clouds build up a strong static electric charge, which may be positive or negative. This attracts an accumulation of the opposite charge in objects on the ground. The charged clouds attract the opposite charge toward them. Objects that are *insulated* from the earth (such as dry trees) tend to *accumulate* the charge more strongly than the surrounding earth. If the potential difference (voltage) between sky and the object, or between the object and earth is great enough, a spark will jump from sky to the object (lightning) or from the object to the earth. If the "object" happens to be your ungrounded PV array and wiring, its path to earth may be through parts of your system that cannot tolerate such high voltage - electronic devices in particular.

Grounding your system does four things: (1) It drains off accumulated charges so that *lightning is not highly attracted* to your system. (2) If lightning does strike, or if high charge does build up, your ground connection provides a safe path for discharge directly to the earth. (3) It reduces shock hazard from the higher voltage (AC) parts of your system, and (4) reduces electrical hum and radio caused by inverters, motors, fluorescent lights, and other devices. To achieve effective grounding, follow these guidelines:

Install A Proper Ground System

Standard practice on any electrical system is to drive a copper-plated ground rod, usually 8 ft long, into the earth. This is a minimum procedure in an area where the earth is moist (conductive). Where ground is dry, especially sandy, or where the array is relatively large and high up, more rods should be installed, at least 10 feet apart. Connect all ground rods together via min. #8 bare copper wire, buried. Use only the proper clamps to connect wire to rods. Do not solder ground wire connections. If your array is some distance from the house, drive ground rod(s) near it, and bury bare wire in the trench with the power lines.

Metal water pipes that are buried in the ground are also good to ground to. Purchase connectors made for the purpose, and connect *only* to cold water pipes, *never* to hot water or gas pipes. Beware of plastic couplings -

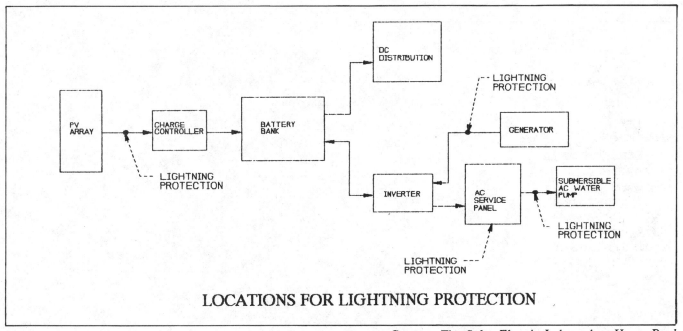

LOCATIONS FOR LIGHTNING PROTECTION

Courtesy The Solar Electric Independent Home Book

bypass them with copper wire. Iron well casings are super ground rods, but you may need to drill and tap a hole to get a good bolted connection. If you connect to more than one grounded object (the more the better) it is essential to electrically "bond" them all together using min. #6 copper wire. Connections made in or near the ground are subject to corrosion, so use proper bronze or copper connectors.

If your site is rocky and you cannot drive ground rods deeply, bury (as much as feasible) at least 150 feet of bare copper wire. Several pieces radiating outward is best. Try to bury them in areas that tend to be moist. If you are in a lightning-prone area, bury several hundred feet if you can, over the broadest area feasible. If you need to run any power wiring over any distance of 30 feet or more, and are in a high lightning, dry area, run the wires in metal conduit and ground the conduit.

What to Connect to Your Ground System

Ground the Metallic Framework of your PV array. Be sure to bolt your ground wire solidly to the metal so it will not come loose, and inspect it periodically. Also ground antenna masts and wind generator towers.

Ground the Negative Line from Your Battery Bank, but *first* make the following test for leakage to ground: Obtain a common multi-tester. Set it on the highest milliamp scale. Place the negative probe on battery neg. and the positive probe on your ground system. No

reading? Good. Now switch it down to the lowest milli- or microamp scale and try again. If you get only a few microamps, or zero, *then ground your battery negative*. If you DID read leakage to ground, check your system for something on the positive side that may be contacting earth somehow. (If you read a few microamps to ground, it is probably your meter detecting radio station signals.)

Ground Your AC Generator and Inverter Frames, and AC neutral wires and conduits in the manner conventional for all AC systems. This protects from shock hazard as well as lightning damage. Follow directions for your generator or inverter or consult an electrician.

Array Wiring should be done with minimum lengths of wire, tucked into the metal framework. Positive and negative wires should be run together wherever possible, rather than being some distance apart. Avoid overhead wire runs; burial gives added protection.

Further Protection may be called for if you have thousands of dollars invested in electric equipment, and are in a lightning-prone area where the grounding potential is poor. We sell high- voltage surge protectors for both your AC and DC side. They have very high current capacity, and a rather high price, but are relatively cheap insurance against expensive damage.

Safety First!! If you are clumsy with wiring, or uncertain how to wire properly, **Hire an Electrician!!**

LIGHT a light. You can help balance the demand down at the power plant and brighten your life at the flick of a switch. Let your house glow with radiance from every window.

National Electrical Code®

ARTICLE 690 — SOLAR PHOTOVOLTAIC SYSTEMS

A. General

690-1. Scope. The provisions of this article apply to solar photovoltaic electrical energy systems including the array circuit(s), power conditioning unit(s) and controller(s) for such systems. Solar photovoltaic systems covered by this article may be interactive with other electric power production sources or stand alone, with or without electrical energy storage such as batteries. These systems may have alternating- or direct-current output for utilization.

690-2. Definitions.

Array: A mechanically integrated assembly of modules or panels with a support structure and foundation, tracking, thermal control, and other components, as required, to form a direct-current power-producing unit.

Blocking Diode: A diode used to block reverse flow of current into a photovoltaic source circuit.

Interactive System: A solar photovoltaic system that operates in parallel with and may be designed to deliver power to another electric power production source connected to the same load. For the purpose of this definition, an energy storage subsystem of a solar photovoltaic system, such as a battery, is not another electric power production source.

Module: The smallest complete, environmentally protected assembly of solar cells, optics and other components, exclusive of tracking, designed to generate direct-current power under sunlight.

Panel: A collection of modules mechanically fastened together, wired, and designed to provide a field-installable unit.

Photovoltaic Output Circuit: Circuit conductors between the photovoltaic source circuit(s) and the power conditioning unit or direct-current utilization equipment. See Diagram 690-1.

Photovoltaic Power Source: An array or aggregate of arrays which generates direct-current power at system voltage and current.

Photovoltaic Source Circuit: Conductors between modules and from modules to the common connection point(s) of the direct-current system. See Diagram 690-1.

Power Conditioning Unit: Equipment which is used to change voltage level or waveform or both of electrical energy. Commonly a power conditioning unit is an inverter which changes a direct-current input to an alternating-current output.

Power Conditioning Unit Output Circuit: Conductors between the power conditioning unit and the connection to the service equipment or another electric power production source such as a utility. See Diagram 690-1.

Solar Cell: The basic photovoltaic device which generates electricity when exposed to light.

Solar Photovoltaic System: The total components and subsystems which in combination convert solar energy into electrical energy suitable for connection to a utilization load.

Stand-Alone System: A solar photovoltaic system that supplies power independently but which may receive control power from another electric power production source.

690-3. Other Articles. Wherever the requirements of other articles of this Code and Article 690 differ, the requirements of Article 690 shall apply. Solar photovoltaic systems operating as interconnected power production sources shall be installed in accordance with the provisions of Article 705.

690-4. Installation.

(a) Photovoltaic System. A solar photovoltaic system shall be permitted to supply a building or other structure in addition to any service(s) of another electricity supply system(s).

(b) Conductors of Different Systems. Photovoltaic source circuits and photovoltaic output circuits shall not be contained in the same raceway, cable tray, cable, outlet box, junction box or similar fitting as feeders or branch circuits of other systems.

Exception: Where the conductors of the different systems are separated by a partition or are connected together.

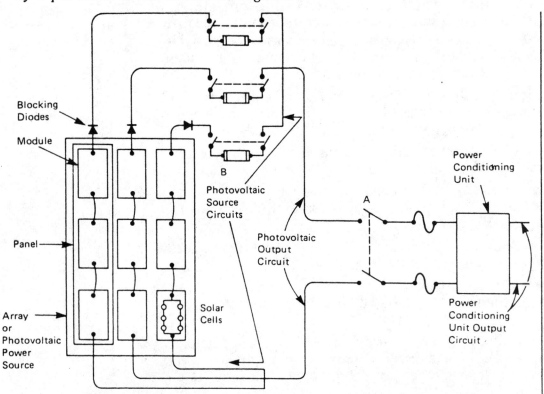

A: Disconnecting means required by Section 690-13.
B: Equipment permitted to be on the photovoltaic power source side of the photovoltaic power source disconnecting means, per Section 690-14, Exception No. 2. See Section 690-16.

Diagram 690-1. Solar Photovoltaic System
(Simplified Circuit)

(c) Module Connection Arrangement. The connections to a module or panel shall be so arranged that removal of a module or panel from a photovoltaic source circuit does not interrupt a grounded conductor to another photovoltaic source circuit.

690-5. Ground Fault Detection and Interruption. Roof-mounted photovoltaic arrays located on dwellings shall be provided with ground-fault protection to reduce fire hazard. The ground-fault protection circuit shall be capable of detecting a ground fault, interrupting the fault path, and disabling the array.

B. Circuit Requirements

690-7. Maximum Voltage.

(a) Voltage Rating. In a photovoltaic power source and its direct-current circuits, the voltage considered shall be the rated open-circuit voltage.

(b) Direct-Current Utilization Circuits. The voltage of direct-current utilization circuits shall conform with Section 210-6.

(c) Photovoltaic Source and Output Circuits. Photovoltaic source circuits and photovoltaic output circuits which do not include lampholders, fixtures or receptacles shall be permitted up to 600 volts.

(d) Circuits Over 150 Volts to Ground. In one- and two-family dwellings, live parts in photovoltaic source circuits and photovoltaic output circuits over 150 volts to ground shall not be accessible while energized to other than qualified persons.

(FPN): See Section 110-17 for guarding of live parts, and Section 210-6 for voltage to ground and between conductors.

690-8. Circuit Sizing and Current.

(a) Ampacity and Overcurrent Devices. The ampacity of the conductors and the rating or setting of overcurrent devices in a circuit of a solar photovoltaic system shall not be less than 125 percent of the current computed in accordance with (b) below. The rating or setting of overcurrent devices shall be permitted in accordance with Section 240-3, Exception No. 4.

Exception: Circuits containing an assembly together with its overcurrent device(s) that is listed for continuous operation at 100 percent of its rating.

(b) Computation of Circuit Current. The current for the individual type of circuit shall be computed as follows:

(1) Photovoltaic Source Circuits. The sum of parallel module short-circuit current ratings.

(2) Photovoltaic Output Circuit. The photovoltaic power source current rating.

(3) Power Conditioning Unit Output Circuit. The power conditioning unit output current rating.

Exception: The current rating of a circuit without an overcurrent device shall be the short-circuit current, and it shall not exceed the ampacity of the circuit conductors.

690-9. Overcurrent Protection.

(a) Circuits and Equipment. Photovoltaic source circuit, photovoltaic output circuit, power conditioning unit output circuit, and storage battery circuit conductors and equipment shall be protected in accordance with the requirements of Article 240. Circuits connected to more than one electrical source shall have overcurrent devices so located as to provide overcurrent protection from all sources.

(FPN): Possible backfeed of current from any source of supply, including a supply through a power conditioning unit into the photovoltaic output circuit and photovoltaic source circuits, must be considered in determining whether adequate overcurrent protection from all sources is provided for conductors and modules.

(b) Power Transformers. Overcurrent protection for a transformer with a source(s) on each side shall be provided in accordance with Section 450-3 by considering first one side of the transformer, then the other side of the transformer as the primary.

Exception: A power transformer with a current rating on the side connected toward the photovoltaic power source not less than the short-circuit output current rating of the power conditioning unit shall be permitted without overcurrent protection from that source.

(c) Photovoltaic Source Circuits. Branch-circuit or supplementary type overcurrent devices shall be permitted to provide overcurrent protection in photovoltaic source circuits. The overcurrent devices shall be accessible, but shall not be required to be readily accessible.

C. Disconnecting Means

690-13. All Conductors. Means shall be provided to disconnect all current-carrying conductors of a photovoltaic power source from all other conductors in a building or other structure.

690-14. Additional Provisions. The provisions of Article 230, Part F shall apply to the photovoltaic power source disconnecting means.

Exception No. 1: The disconnecting means shall not be required to be suitable as service equipment and shall be rated in accordance with Section 690-17.

Exception No. 2: Equipment such as photovoltaic source circuit isolating switches, overcurrent devices, and blocking diodes shall be permitted on the photovoltaic power source side of the photovoltaic power source disconnecting means.

690-15. Disconnection of Photovoltaic Equipment. Means shall be provided to disconnect equipment, such as a power conditioning unit, filter assembly and the like from all ungrounded conductors of all sources. If the equipment is energized (live) from more than one source, the disconnecting means shall be grouped and identified.

690-16. Fuses. Disconnecting means shall be provided to disconnect a fuse from all sources of supply if the fuse is energized from both directions and is accessible to other than qualified persons. Such a fuse in a photovoltaic source circuit shall be capable of being disconnected independently of fuses in other photovoltaic source circuits.

690-17. Switch or Circuit Breaker. The disconnecting means for ungrounded conductors shall consist of a manually operable switch(es) or circuit breaker(s): (1) located where readily accessible, (2) externally operable without exposing the operator to contact with live parts, (3) plainly indicating whether in the open or closed position, and (4) having ratings not less than the load to be carried. Where all terminals of the disconnecting means may be energized in the open position, a warning sign shall be mounted on or adjacent to the disconnecting means. The sign shall be clearly legible and shall read substantially: WARNING - ELECTRIC SHOCK - DO NOT TOUCH - TERMINALS ENERGIZED IN OPEN POSITION.

Exception: A disconnecting means located on the direct-current side shall be permitted to have an interrupting rating less than the current-carrying rating when the system is designed so that the direct-current switch cannot be opened under load.

690-18. Disablement of an Array. Means shall be provided to disable an array or portions of an array.

(FPN): Photovoltaic modules are energized while exposed to light. Installation, replacement, or servicing of array components while a module(s) is irradiated may expose persons to electric shock.

D. Wiring Methods

690-31. Methods Permitted.

(a) **Wiring Systems.** All raceway and cable wiring methods included in this Code and other wiring systems and fittings specifically intended and identified for use on photovoltaic arrays shall be permitted. Where wiring devices with integral enclosures are used, sufficient length of cable shall be provided to facilitate replacement.

(b) **Single Conductor Cable.** Type UF single conductor cable shall be permitted in photovoltaic source circuits where installed in the same manner as a Type UF multiconductor cable in accordance with Article 339. Where exposed to direct rays of the sun, cable identified as sunlight-resistant shall be used.

690-32. Component Interconnections. Fittings and connectors which are intended to be concealed at the time of on-site assembly, when listed for such use, shall be permitted for on-site interconnection of modules or other array components. Such fittings and connectors shall be equal to the wiring method employed in insulation, temperature rise and fault-current withstand, and shall be capable of resisting the effects of the environment in which they are used.

690-33. Connectors. The connectors permitted by Section 690-32 shall comply with (a) through (e) below.

(a) **Configuration.** The connectors shall be polarized and shall have a configuration that is noninterchangeable with receptacles in other electrical systems on the premises.

(b) **Guarding.** The connectors shall be constructed and installed so as to guard against inadvertent contact with live parts by persons.

(c) **Type.** The connectors shall be of the latching or locking type.

(d) Grounding Member. The grounding member shall be the first to make and the last to break contact with the mating connector.

(e) Interruption of Circuit. The connectors shall be capable of interrupting the circuit current without hazard to the operator.

690-34. Access to Boxes. Junction, pull and outlet boxes located behind modules or panels shall be installed so that the wiring contained in them can be rendered accessible directly or by displacement of a module(s) or panel(s) secured by removable fasteners and connected by a flexible wiring system.

E. Grounding

690-41. System Grounding. For a photovoltaic power source, one conductor of a 2-wire system rated over 50 volts and a neutral conductor of a 3-wire system shall be solidly grounded.

Exception: Other methods which accomplish equivalent system protection and which utilize equipment listed and identified for the use shall be permitted.

(FPN): See the first Fine Print Note under Section 250-1.

690-42. Point of System Grounding Connection. The direct-current circuit grounding connection shall be made at any single point on the photovoltaic output circuit.

(FPN): Locating the grounding connection point as close as practicable to the photovoltaic source will better protect the system from voltage surges due to lightning.

690-43. Size of Equipment Grounding Conductor. The equipment grounding conductor shall be no smaller than the required size of the circuit conductors in systems where the available photovoltaic power source | short-circuit current is less than twice the current rating of the overcurrent device. In other systems the equipment grounding conductor shall be sized | in accordance with Section 250-95.

690-44. Common Grounding Electrode. Exposed noncurrent-carrying metal parts of equipment and conductor enclosures of a photovoltaic system shall be grounded to the grounding electrode that is used to ground the direct-current system. Two or more electrodes that are effectively bonded together shall be considered as a single electrode in this sense.

F. Marking

690-51. Modules. Modules shall be marked with identification of terminals or leads as to polarity, maximum overcurrent device rating for module protection and with rated: (1) open-circuit voltage, (2) operating voltage, (3) maximum permissible system voltage, (4) operating current, (5) short-circuit current, and (6) maximum power.

690-52. Photovoltaic Power Source. A marking, specifying the photovoltaic power source rated: (1) operating current, (2) operating voltage, (3) open-circuit voltage, and (4) short-circuit current, shall be provided at an accessible location at the disconnecting means for the photovoltaic power source.

(FPN): Reflecting systems used for irradiance enhancement may result in increased levels of output current and power.

G. Connection to Other Sources

690-61. Loss of System Voltage. The power output from a power conditioning unit in a solar photovoltaic system that is interactive with another electric system(s) shall be automatically disconnected from all ungrounded conductors in such other electric system(s) upon loss of voltage in that electric system(s) and shall not reconnect to that electric system(s) until its voltage is restored.

(FPN): For other interconnected electric power production sources, see Article 705.

A normally interactive solar photovoltaic system shall be permitted to operate as a stand-alone system to supply premises wiring.

690-62. Ampacity of Neutral Conductor. If a single-phase, 2-wire power conditioning unit output is connected to the neutral and one ungrounded conductor (only) of a 3-wire system or of a 3-phase, 4-wire wye-connected system, the maximum load connected between the neutral and any one ungrounded conductor plus the power conditioning unit output rating shall not exceed the ampacity of the neutral conductor.

690-63. Unbalanced Interconnections.

(a) Single-Phase. The output of a single-phase power conditioning unit shall not be connected to a 3-phase, 3- or 4-wire electrical service derived directly from a delta-connected transformer.

(b) Three-Phase. A 3-phase power conditioning unit shall be automatically disconnected from all ungrounded conductors of the interconnected system when one of the phases opens in either source.

Exception for (a) and (b): Where the interconnected system is designed so that significant unbalanced voltages will not result.

690-64. Point of Connection. The output of a power production source shall be connected as specified in (a) or (b) below.

(FPN): For the purposes of this section a power production source is considered to be: (1) the output of a power conditioning unit when connected to an alternating current electric source; and (2) the photovoltaic output circuit when interactive with a direct current electric source.

(a) Supply Side. To the supply side of the service disconnecting means as permitted in Section 230-82, Exception No. 6.

(b) Load Side. To the load side of the service disconnecting means of the other source(s), if all of the following conditions are met:

(1) Each source interconnection shall be made at a dedicated circuit breaker or fusible disconnecting means.

(2) The sum of the ampere ratings of overcurrent devices in circuits supplying power to a busbar or conductor shall not exceed the rating of the busbar or conductor.

Exception: For a dwelling unit the sum of the ampere ratings of the overcurrent devices shall not exceed 120 percent of the rating of the busbar or conductor.

(3) The interconnection point shall be on the line side of all ground-fault protection equipment.

Exception: Connection shall be permitted to be made to the load side of ground-fault protection provided that there is ground-fault protection for equipment from all ground-fault current sources.

(4) Equipment containing overcurrent devices in circuits supplying power to a busbar or conductor shall be marked to indicate the presence of all sources.

Exception: Equipment with power supplied from a single point of connection.

(5) Equipment such as circuit breakers, if back-fed, shall be identified for such operation.

H. Storage Batteries

690-71. Installation.

(a) General. Storage batteries in a solar photovoltaic system shall be installed in accordance with the provisions of Article 480.

Exception: As provided in Section 690-73.

(b) Dwellings.

(1) Storage batteries for dwellings shall have the cells connected so as to operate at less than 50 volts.

Exception: Where live parts are not accessible during routine battery maintenance, a battery system voltage in accordance with Section 690-7 shall be permitted.

(2) Live parts of battery systems for dwellings shall be guarded to prevent accidental contact by persons or objects, regardless of voltage or battery type.

(FPN): Batteries in solar photovoltaic systems are subject to extensive charge-discharge cycles and typically require frequent maintenance, such as checking electrolyte and cleaning connections.

690-72. State of Charge. Equipment shall be provided to control the state of charge of the battery. All adjusting means for control of the state of charge shall be accessible only to qualified persons.

Exception: Where the design of the photovoltaic power source is matched to the voltage rating and charge current requirements for the interconnected battery cells.

690-73. Grounding. The interconnected battery cells shall be considered grounded where the photovoltaic power source is installed in accordance with Section 690-41, Exception.

Calculating Your System Loads

In the following format systematically list all the appliances that you intend to use (refer to chart above), the wattages of each one, and the hours per day you intend to use them. Remember that some appliances like microwaves, trash compacters, and blenders will be used far less than one hour per day, so use the fraction of the hour. List the appliance regardless if it is to run DC direct or AC from the inverter.

Appliance	Watts	Hrs/Day	Watthours/Day

TOTAL WATTHOURS PER DAY _____

Multiply by 1.2 to account for inverter & battery efficiency losses

Total Watthours per Day (with efficiency loss) _____

Cubic Inches of Box Size Required for the Number of Connections for Common Wire Sizes

Additions		2	3	4	5	6	7
	#14	2.0	4.0	6.0	8.0	10.0	14.0
	#12	2.3	4.5	6.8	9.0	11.3	15.8
Conductor Sizes	#10	2.5	5.0	7.5	10.0	12.5	17.5
	#8	3.0	6.0	9.0	12.0	15.0	21.0
	#6	5.0	10.0	15.0	20.0	25.0	35.0

According to the National Electrical Code (NEC), ground wires need only be counted as one conductor. Box fittings like fixture studs and cable clamps are counted as one conductor for each type. We recommend that you add 2 cubic inches for each cigarette type outlet. Calculation of the proper box size for 2 #10/2 wires with ground coming into one box with one outlet proceeds as follows:

1. 2 #10/2 wires (one in, one out) equals 4 #10 wires, plus 1 #10 wire for the grounds totals 5 #10 wires.
2. Add 2 cubic inches for one outlet.
3. 5 #10 wires in the table requires 12.5 cubic inches; add 2 cubic inches yields a total of 14.5 cubic inches as the minimum capacity.

Most boxes are stamped with their cubic inch capacity. For our example, a deep switch box will serve.

The maximum number of conductors that can be in a particular size of conduit is given in the next table.

Maximum Number of Conductors for a Given Conduit Size.

Conduit size		½"	¾"	1"	1¼"	1½"	2"
	#12	4	8	13			
	#10	4	6	11			
	#8	1	3	5	10		
Conductor size	#6	1	2	4	7	10	
	#4	1	1	3	5	7	12
	#2	1	1	2	4	5	9
	#0		1	1	2	3	5

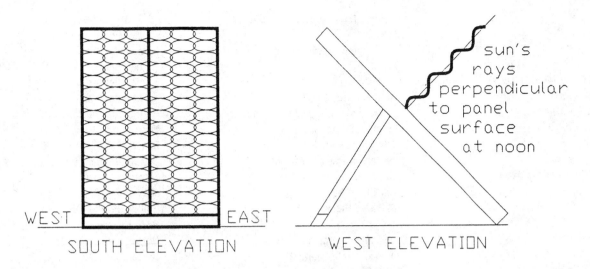

WEST | EAST
SOUTH ELEVATION

WEST ELEVATION

sun's rays perpendicular to panel surface at noon

Figure 1 - Northern Hemisphere Solar Panel Alignment

Note that figure 1 is for Northern Hemisphere sites; for the southern hemisphere, the directions should be reversed.

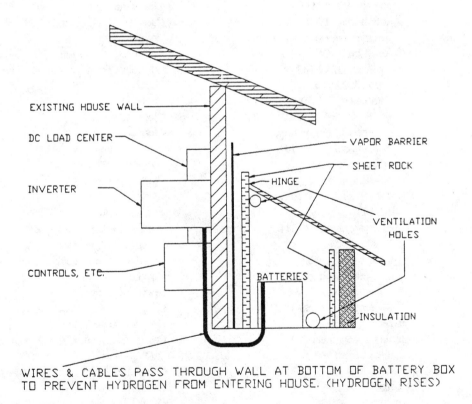

EXISTING HOUSE WALL

DC LOAD CENTER

INVERTER

CONTROLS, ETC.

VAPOR BARRIER

SHEET ROCK

HINGE

VENTILATION HOLES

BATTERIES

INSULATION

WIRES & CABLES PASS THROUGH WALL AT BOTTOM OF BATTERY BOX TO PREVENT HYDROGEN FROM ENTERING HOUSE. (HYDROGEN RISES)

Figure 2 - Recommended Battery Enclosure

Index